Handbook of Chemicals *and* Safety

Handbook of Chemicals *and* Safety

T.S.S. Dikshith

CRC Press
Taylor & Francis Group
Boca Raton London New York

CRC Press is an imprint of the
Taylor & Francis Group, an **informa** business

CRC Press
Taylor & Francis Group
6000 Broken Sound Parkway NW, Suite 300
Boca Raton, FL 33487-2742

First issued in paperback 2017

© 2011 by Taylor and Francis Group, LLC
CRC Press is an imprint of Taylor & Francis Group, an Informa business

No claim to original U.S. Government works

ISBN 13: 978-1-138-11661-0 (pbk)
ISBN 13: 978-1-4398-2060-5 (hbk)

Visit the Taylor & Francis Web site at
http://www.taylorandfrancis.com

and the CRC Press Web site at
http://www.crcpress.com

To my parents

Gowramma and Turuvekere Subrahmanya Dikshith

and

my wife, Saroja Dikshith

A hundred times every day I remind myself, that my inner and outer life depended

on labors of other men, living and dead, and I must exert myself in order to

give in the same measure as I have received and am still receiving.

—Albert Einstein

Contents

Foreword

This excellent work, *Handbook of Chemicals and Safety*, does not need an extraneous Foreword. It deserves appreciative attention and careful study. Professor T. S. S. Dikshith belongs to the rare genre of meticulous, authentic scholars. His treatment and presentation of the subject is meticulous and thorough and speaks of his dedication to the field of toxicology. The author's erudition and thoroughness is matched only by his great humility and urbanity. His earlier works, *Industrial Guide to Chemical and Drug Safety, Safe Use of Chemicals: A Practical Guide*, and others have attracted appreciative attention. He has established a reputation as a thorough, greatly admired, and widely read author. He brings to bear on his works his enormous practical experience and expertise in the field of toxicology.

The next decade of the twenty-first century will be a stunning one and the most decisive bend of human civilization. The exploits of science and technology will change everything—our worldview, our lifestyle and interests, our physical, intellectual, and moral environment. Everything will change.

All human knowledge is just a small sphere in the vast expanse of the unknown. The bigger this sphere becomes, the greater will be its contact with the unknown. This thought compels a becoming humility in understanding the inextricably intertwined and extremely complex inter-relationship between human life on the planet and the ecosystems. The bravado of science that flaunts the approach of a conqueror and that nature is out there to be conquered and subjugated has only led to immense human misery. Man has a better chance of survival if, instead, he chooses to live more in harmony with nature than in perpetual conflict with it. Science is knowledge; technology is ability. The synergy of the influence of the two unleashes enormous power. Power is either for good or for evil. If the growth of knowledge is not tempered by a corresponding growth in wisdom, then power becomes one for evil.

Industrial chemicals and pharmaceuticals have improved the quality of human life and enhanced the standards of living. Modern medicine has brought great blessings to humanity. Breakthroughs in the basic sciences, immunology, antibiotics, and nutrition have changed the health-care scenario for the better. It has overcome the great and crippling potential of epidemics. It has overcome the scourge of smallpox, bubonic plague, and has almost eradicated polio. At the same time, indiscriminate and injudicious use of solvents, pesticides, chemical fertilizers, metals, minerals, antibiotics, and toxic gases has opened up the prospects of a nightmare, compelling man to wonder whether modern civilization, which emphasizes an unending pre-occupation with higher and higher standards of living at the cost of quality of life and of those finer feelings and sentiments to which life owes its savour, has made man's life on Earth only busier and not necessarily better. This is the dilemma of the *Homo economicus* today. A learned author speaking of the hazards of an unwise choice in even small innocent pleasures of life says,

> Consider a box of microwavable, butter-flavored pop corn. The label assures buyers that it has zero grams of trans-fat and zero mg cholesterol. But the ingredients list fails to mention that the savory butter taste and mouth watering aroma comes courtesy of di-acetyl, a flavoring long known by pulmonary specialists to cause *bronchiolitis obliterans,* a disease in which the small air-ways in the lungs become swollen, scarred, and eventually obliterated. Victims can breathe in deeply but have severe difficulties exhaling. More commonly known as 'Pop-corn worker's lung.'

A writer assessing the enormity of the problem of chemical pollution said, "Roughly 400 million tons of chemicals are produced in tens of thousands of varieties every year. Most have never been tested for their effects of health and environment. According to Vyvyan Howard, a senior lecturer at the University of Liverpool, there are currently 100,000 man-made chemicals in use with another 1000 added each year. The majority have been foisted without any test or regulations. The aberrations caused by these chemicals are going unnoticed."[*]

A correct scientific understanding of the chemical entities that pervade our lives is imperative. The effective way of ensuring sanity in this bizarre world of industrial chemicals is to understand analytically the effect and role of each of these chemical entities, both their beneficial and baneful implications. But we cannot throw away the baby with the bathwater. Works of scientists like Professor Dikshith will provide sensitive instruction to the industry, so that a wise decision as to "risk-benefit" ratio is possible. Act of friendship in science, it is said, is honest criticism. Authentic scholars like Professor Dikshith can influence the choices that can avert and alleviate the consequences of misguided industrial enthusiasm and enterprise dedicated only toward the crude prizes of the business world.

The importance of Professor Dikshith's works, in a non-trivial sense, is their influence on the right choices at the right places. They are an invaluable contribution to what a scholar calls, the need for 'collective eye opening'.

M. N. Venkatachaliah

Former Chief Justice of India
Government of India
Bangalore, India

---

[*] Daniel Goleman, *Ecological Intelligence*, London: Allen Lane, 142; Dr. Madhav Mehra, "Building a strategy for sustainability," *Quarterly Times*, Vol. XV, No. 4, 2010.

Preface

Let noble thoughts come to us from every side

Rig Veda (I–89–I)

Asato ma sadgamaya [lead me from untruth to the truth];
Tamaso ma jyotirgamaya [lead me from darkness to light];
Mrtyorma amrtam gamaya [lead me from death to immortality];
Om shanthi shanthi shanthihi [Let there be peace, peace, peace].

—Brhadaranyaka Upanishad—I.iii.28

ANITYAANI SHAREERANI [human life is very temporary];
VIBHAVO NIVA SHASWATAM [luxuries are never permanent];
NITYAM SANNIHITO MRUTYUHU [death approaches fast and closely];
KARTAVYAM DHARMA SNGRAHAHA [purpose of human life/Society, is to be virtuous].

—Science of Ethics

Science without religion is lame; Religion without science is blind.

—Albert Einstein

Chemical substances and their multiple applications have become an essential part of human activities and indicators of societal progress. The benefits of proper and judicious use of chemical substances have been several: providing human society with good food, grains, fruits, and vegetables, and provided protection and relief from hunger, thirst, famines, diseases, and infections. The synthesis and formulations of newer drugs and pharmaceuticals offered benefits of health care to society. In short, proper and judicious application of a variety of chemical substances improved the quality of life. By contrast, improper use, often called misuse, of chemical substances resulted in health disorders and fatalities. Furthermore, improper use caused chemical disasters both at the workplace and in the living environment.

Societal development requires understanding of proper use of chemical substances with pragmatism, proper methods, quality regulations, and good management. The users of chemical substances must be well aware of the implications of improper use of candidate chemical substance either alone or in combination. In fact, all chemical substances are toxic and there is no absolute safety. It is the manner of use of a chemical substance that brings either the good or danger to the user, to the immediate workplace, and to society at large. Improper use and waste disposal of chemical substances endangers human health and causes environmental pollution and chemical disasters. The adverse health effects and human fatalities because of environmental disasters and incidences due to chemical substances are many. These depend on many factors that include, but are not limited to, the toxicity of the candidate chemical, the duration or period of exposure, the individual's age, and the health status besides other factors. To contain the adverse health effects of chemical substances, workers and the chemical management team must be aware of the properties, mechanisms of action, elements of good laboratory practice (GLP), and good manufacturing practice (GMP). The term "toxic industrial chemical" refers to a variety of chemical substances used in industries and in various processes of product development. Any chemical substance can be toxic or harmful in some dose to human health. Toxic industrial chemicals are known to pose risks when they are stored in large quantities in one location. An act of sabotage or an accident can result in a large-scale release of toxic chemicals/degraded products into the air, people living nearby breathe this air and may develop health disorders and the living environment may be polluted. Chemical substances have been involved in several human health and safety and environmental disasters around the world. These include, but are not limited to, the London smog disaster of 1952 that killed approximately 12,000 people, mainly children, elderly people, and people suffering from chronic respiratory or cardiac disease. The number of deaths during the smog disaster was three or four times that on a normal day. They could be attributed to lung disease, tuberculosis and heart failure. Mortality from bronchitis and pneumonia increased more than sevenfold. The Love Canal disaster in 1953, when chlorine gas was released from a large tank into the surrounding air, was one of the worst environmental disasters in American history. The 1976 explosion that occurred in a TCP (2,4,5-trichlorophenol) reactor at the ICMESA chemical company in Meda, Seveso, Italy. The toxic cloud, containing high concentrations of TCDD, a highly toxic form of dioxin, escaped into the atmosphere. The victims of the Seveso chemical disaster suffered from a directly visible symptom known as chloracne and from genetic impairments. The Bhopal tragedy in India in 1984 because of the methyl isocyanate (MIC) gas. In 1989, Exxon Valdez, the American oil tanker spill killed approximately 250,000 sea birds, 2,800 sea otters, 250 bald eagles, possibly 22 killer whales, and many others. In 1998 the dam of the mining residual tank of a pyrite mine in Aznalcollar, Spain, suffered a rupture, releasing sludge and contaminated wastewater. The wastewater entered the Guadiamar river, polluting the river with heavy metals, such as cadmium, lead, zinc, and copper that contaminated an area of 4.634 ha, and polluted 2.703 ha with sludge and with acidic water. The Chernobyl nuclear power plant disaster in Ukraine in 1986. Errors in the reactor design and errors in judgment of the personnel of the power plant caused cooling water to start boiling, eventually causing the reactor stress, and increasing energy production to ten times the normal level. Temperatures reached more than 2000°C, causing fuel rod melting. Extreme pressures in water pipes resulted in cracks and caused steam to escape. In the middle of the night the escaped steam caused an explosion, starting a major fire, and releasing a huge cloud (185–250 million Ci) of radioactive material into the atmosphere.

As stated earlier, societal progress and development depend on the knowledge and proper use of chemical substances with a pragmatic approach, and certainly not by misuse or reckless imposition of bans on chemicals substances. There are no safe chemical substances. Huge amounts of time, money, and human effort have been spent by thousands and thousands of researchers and workers to identify newer molecules for human use. These chemical molecules have been identified, synthesized, and formulated for human use in the form of drugs, pesticides, preservatives, and many other useful products. Any kind of misuse and or negligence by users and occupational workers in management, in storage, during transport and waste disposal of chemical substances does not to achieving human safety and environmental protection. Today, imparting proper education, suitable guidance, and good training to students, workers, and to society at large is very necessary to achieve human safety. The present global requirements are to achieve economic progress in both developing and the developed parts of the world. It needs to be remembered that, "the future progresses depend on the wisdom not to replace one poison with another", and therefore to use chemical substances properly and judiciously.

The purpose of this book is to provide and promote basic and elementary knowledge about chemical substances irrespective of workplace, laboratory, factory, field, or home. Timely availability of scientific knowledge improves the health of worker(s) besides reducing, if not completely eliminating, the possibilities of chemical poisonings, incidents, and disasters. This book offers a comprehensive, integrated, speedy, and easy tool for the management of several chemical substances commonly used, handled, stored, transported, and disposed of as wastes by a large population. The list of chemical substances includes, but is not limited to, industrial solvents, pesticides, metals, air pollutants, toxic gases, chemical substances as drugs and many more other items.

This book also offers guidance to students, basic scientists, toxicologists, occupational workers, professionals, risk assessors, and regulatory agencies. Chemical substances are ubiquitous and their application is universal. Therefore, readers, students, workers, toxicologists, practicing hygienists, safety engineers, and basic scientists often require a standardized, comprehensive, and a ready reference data book.

Information about the use and possible health effects, the immediate and long-term effects and hazards of chemical substances, to users is very important for purposes of better and judicious management of chemical substances. The management includes proper and correct identification of the candidate chemical, judicious use, separation at areas of storage, and waste disposal at the right time and place. Correct knowledge of these factors is essential to protect the health of the user and to contain chemical accidents and disasters both at the workplace and in the living environment.

The author has made every effort to collect and collate information from different published sources as well as his own earlier work about a large number of chemical substances. Essentially, this book provides ready information at times of need to users and occupational workers. The *Handbook of Chemicals and Safety* is a compilation of chemical substances selected and listed in alphabetical order to meet a variety of criteria. The information on each chemical substance is concise and easy to understand. The information includes the name of the candidate chemical substance with the chemical abstract system (CAS) number, IUPAC name, molecular formula, synonyms and trade names, use and exposure, toxicity and health effects, carcinogen or not, the exposure limits, methods of proper storage, precautions, waste disposals, relevant references, and appendices wherever required. In certain chapters of this book, chemical substances are listed in alphabetical order to facilitate speedy and easy access for the reader; the classifications of chemical substances are included separately.

It is important to state here that this compilation is not to discourage the use of chemical substances. As is well known, applications of a variety of chemical substances are very essential and when used properly, chemical substances become useful tools for societal development and improvement of quality of life. This book is to educate students, semi-skilled workers in different occupations, householders, and users about the basic realities of chemical substances, the responsibilities, and the immediate, short- and long-term consequences of improper use and negligence during handling, use, storage, transport, waste disposal, and overall management.

The purpose of the author in compiling *Handbook of Chemicals and Safety* is to provide ready scientific information to students and occupational workers at different workplaces about the possible health hazards of different chemical substances, suggesting the importance of proper management and judicious use of each chemical substance irrespective of the workplace(s), to improve the safety and protection of the user, workplace, and the living environment. It goes without saying that the toxicological profile of chemical substances becomes more complicated because of the presence of a mixture of contaminants, meaning that aggregated chemical substances could lead to aggregated risks. Occupational workers must be well aware of the consequences of the synergist effects of each chemical substance, working environment, and many contaminants in a workplace that are yet to be studied and properly understood. The author is also fully aware of the fact that, despite his efforts to present an up-to-date and comprehensive compilation in one place, many gaps may have occurred. Groups, numbers, and categories of chemical substances are so large and the global information is equally large. The author is therefore well aware that this handbook may touch only a minute tip of the vast knowledge and huge global wealth of chemical substances. The handbook, however, hopes to provide an integrated, yet simple description of several chemical substances commonly used, handled, stored, and transported by workers and householders, and help to contain chemical accidents at workplaces, to improve the safety of users, occupational workers, and the living environment. In brief, the salient features of the handbook include:

- Information on general fundamentals and specific hazards and effects of chemical substances
- Information about the basics of exposures and responses to chemical substances in the work environment
- Evaluation of toxic responses in different body systems
- General perspective on the problem of chemical exposures and the possible health effects

It is hoped that the information in this handbook will educate the general public about the importance of proper management of chemical substances and the implications of negligence in the management of chemical substances to human health and the safety of the living environment.

T. S. S. DIKSHITH

Bangalore
Karnataka State, India

Acknowledgments

This handbook contains information drawn from materials published in the literature. The compilation of this book would not have been possible without the generous permission granted to the author by different agencies, publishers, and international bodies to cite, refer, and use the published scientific literature. The author gratefully acknowledges the copyright permission granted by the US Environmental Protection Agency (US EPA), the Agency for Toxic Substances and Disease Registry (ATSDR), the National Institute for Occupational Safety and Health (NIOSH), the International Registry on Potentially Toxic Chemicals (IRPTC), the International Program on Chemical Safety (IPCS), the World Health Organization (WHO), the Occupational Safety and Health Administration (OSHA), the National Library of Medicine (NLM), the Hazardous Substances Data Bank (HSDB), INCHEM, Centers for Disease Control and Prevention (CDCP), the Central Insecticide Board (CIB), Ministry of Agriculture and Cooperation, Government of India, and many other organizations and agencies.

With great pleasure, the author records his grateful thanks to Honorable Justice M.N. Venkatachaliah, former Chief Justice of India, for encouragement and his scholarly, and scientific and socially relevant foreword to this book. Thanks are also expressed to the administrative authorities of the United States Food and Drug Administration (Joan G. Lytle, US FDA, Jill Smith), to Candice Burns Hoffmann, CDC Division of Media Relations, Centers for Disease Control and Prevention (CDC), the Agency for Toxic Substances and Disease Registry (ATSDR), Atlanta, GA, USA, to the National Center for Environmental Health, Atlanta, GA, USA, to the Organization for Economic Co-operation and Development (OECD), and to Aleksandra Sawicka, Public Affairs and Communications Directorate, Paris, France, for granting copyright permission, to Dr Hugh Cartwright, Department of Chemistry, University of Oxford, Oxford, UK, to Carl J. Foreman, Director, EH&S, One Shields Avenue, Davis, CA, USA, to the International Labor Office, Geneva, Switzerland, to the National Institute for Occupational Safety and Health (NIOSH), Cincinnati, OH, USA, to the California Department of Health Services, to the US Geological Survey, and to the Canadian Centre for Occupational Health and Safety (CCOHS), Hamilton, Ontario, Canada, to Anne Logan, Mallinckrodt Baker, Inc., Phillipsburg, NJ.

The author expresses his deep sense of appreciation to Dr. Srikanta, Medical Director, SAMATVAM, Jnana Sanjeevini Medical Center, Bangalore, India, to Mr. Narasimha Kramadhati, Mrs. Pratibha Narasimha Kramadhati, to Dr. Deepak Murthy, and Mrs. Prerana Murthy for their overall support, encouragement and for sharing recent global thoughts about the subject. Thanks to Cindy Renee Carelli, Senior Acquisitions Editor, to Kari Budyk, Senior Project Coordinator, Editorial Project Development, to Amy Blalock, Project Coordinator, to Glenon C. Butler, Jr., Project Editor, to Brittany Gilbert, Editorial Assistant, CRC Press, Boca Raton, FL, and to Rajesh, Amnet International, for his cooperation and coordination in the publication of the book. The author also expresses sincere thanks to several friends for providing technical support in the completion of this work.

With pleasure, the author records his grateful appreciation to the many scientists, authorities, and individuals who provided support and copyright permissions to cite, refer, quote, and use the already published scientific information of a large number of chemical substances listed in the handbook. Sincere thanks are due to Dr. Steven G. Gilbert of the Institute of Neurotoxicology and Neurological Disorders, Seattle, WA, USA, for granting permission to cite his published work in this book, to Paulette Goldweber, Associate Manager, Permissions, and to Sheik Safdar, Permissions Coordinator, Global Rights, John Wiley & Sons Inc., Hoboken, NJ, USA, for granting copyright permission to use the scientific literature related to different chemical substances from the author's earlier book, *Industrial Guide to Chemicals and Drug Safety*, published by John Wiley & Sons Inc., and to Mindy Rosenkrantz, Permissions Coordinator, CRC Press, Taylor & Francis Group, New York, for granting copyright permission to cite selected literature from the author's earlier book, *Safe Use of Chemicals: A Practical Guide*.

Author

Turuvekere Subrahmanya Shanmukha Dikshith, PhD, was responsible for the establishment of the Pesticide Toxicology Laboratory at Industrial Toxicology Research Centre, Council of Scientific and Industrial Research (CSIR), Government of India, Lucknow, India. Later, he was the director of the Toxicology Research Laboratory, VIMTA Labs Ltd., Hyderabad, India. He was also a consultant for the Pharmacology Division, Indian Institute of Chemical Technology (CSIR), Hyderabad, India. Dr. Dikshith served on committees and expert panels constituted by different ministries, Government of India. He is a member of the World Health Organization Task Group on Environmental Health Criteria Documents and the International Program on Chemical Safety. Dr. Dikshith is the technical specialist for Standards Australia Quality Assurance Services (SAQAS), Australia, and Lloyd Register Quality Assurance Ltd. (LRQA) London, for the quality management of the laboratory and good laboratory practice. He is the recipient of the Chandra Kanta Dandiya prize in pharmacology.

As a fellow of the World Health Organization, Dr. Dikshith worked at the Institute of Comparative and Human Toxicology, Albany Medical College, Albany, New York, and also at the International Center of Environmental Safety, Holloman, New Mexico. He worked in several laboratories in France, Germany, and Canada. Dr. Dikshith edited *Toxicology of Pesticides in Animals* for CRC Press, Boca Raton, FL, authored a book chapter for Plenum Press, New York, and has authored *Safety Evaluation of Environmental Chemicals* for New Age International Publishers, India, *Industrial Guide to Chemical and Drug Safety* for John Wiley, Hoboken, New Jersey, and *Safe Use of Chemicals: A Practical Guide* for CRC Press, Boca Raton, Florida.

1

Introduction

Chemical substances play an important role in daily life. The list of chemical substances is huge, their groups are many, and their contribution to economic growth and progressive societal changes are equally diverse. A cursory glance at elementary classes/groups of chemical substances include, but are not limited to, alcohols, aliphatic hydrocarbons, aliphatic nitriles and cyanates, aliphatic nitrosamines, aromatic nitro compounds, aromatic organophosphorous compounds, azo compounds, chlorine compounds, chemical fumes, dust, toxic gases and vapor of metals, organic solvents, halogenated aromatic compounds, halogenated cresols, metals and metal compounds, phthalates, pesticides, polycyclic aromatic hydrocarbons, and polychlorinated biphenyls. Occupational workers and the public can be exposed to groups of chemicals, such as paint thinners, thin oil-based paints, and cleaning agents. This group of chemical substances includes acetone, mineral spirits, mineral turpentine, wood turpentine, naphtha, toluene, white spirit, xylene, and many more. Unstable/reactive chemicals undergo vigorous polymerization and decomposition; water-reactive chemical substances react with water and release flammable and toxic gas; corrosive chemicals cause visible destruction of living tissue; chemicals as irritants; sensitizers/allergens produce their own kind of adverse health effects on occupational workers. Each group includes individual chemical substances as well. (Readers will find further details on the classification of chemical substances in textbooks of organic chemistry.)

Different laws around the world identify and regulate the hazardous properties of chemical substances. The Globally Harmonized System of Classification and Labeling of Chemicals (GHS) has recently come into effect. Further, the European Commission (EC) proposed the regulation on labeling and packaging of chemical substances and mixtures, and waste disposal. Therefore, the proper use, handling, storage, transport, waste disposal, overall management, safety of chemical substances, elements of toxicology, and precautions should be known. This applies more so to students, semi-skilled workers, occupational workers, and work managers to achieve and safeguard human health and safety, and the protection of the living environment and chemical substances have gained global importance.

Workers are exposed to chemical substances during production. These include paints, pigments, coatings, toxic and compressed gases, cleaning agents, storage, transport, and large-scale disposal of chemical substances, pharmaceuticals production, oil refining and associated onshore oil-related industry, agrochemicals and pesticides, fume from welding or soldering, mist from metal working, dust from quarrying, and gases from silage. These pose health risks and cause adverse effects to human health and the environment. Global actions are in progress to reduce and eliminate the release of these chemicals.

Occupational workers should have safety data sheets at their respective workplaces. It is important to educate occupational workers as well as the general public about chemical substances and provide an updated collection of harmonized data on human poisonings. Chemical safety is inherently linked to other safety issues, including laboratory procedures, personal protective equipment, electrical safety, fire safety, and hazardous waste disposal. It is therefore important to halt the spread of the adverse health effects of chemical substances. Workers and chemical management teams must be aware of the properties, the mechanisms of action, the elements of good laboratory practice (GLP), and good manufacturing practice (GMP). Students, occupational workers, managers, chemical work management teams, and the general public should know that:

Chemical safety = Knowledge + Common sense + Caution

References

Clayton, E.G.D. and Clayton, F.E. (Eds.). 1994. *Patty's Industrial Hygiene and Toxicology*, 4th ed., Wiley, J. New York.

Dikshith, T.S.S. (Ed.). 1991. *Toxicology of Pesticides in Animals*, CRC Press, Boca Raton, FL.

———. 2009. *Safe Use of Chemicals: A Practical Guide*, CRC Press, Boca Raton, FL.

Dikshith, T.S.S. and Diwan, P.V. (Eds.). 2003. *Industrial Guide to Chemical and Drug Safety*, Wiley, J. Hoboken, NJ.

Hayes, W.J. (Ed.). 1982. *Pesticides Studied in Man*, Williams & Wilkins, Baltimore, MD.

Lewis, R.J. (Ed.). 2000. *Sax's Dangerous Properties of Industrial Materials*, 10th ed., Van Nostrand Reinhold, New York.

Sitting, M. (Ed.). 1991. *Handbook of Toxic and Hazardous Chemicals*, 3rd ed., Noyes, Park Ridge, NJ.

Tomlin, C.D.S. (Ed.). 2006. *The Pesticide Manual*, 14th ed., British Crop Protection Council Cambridge.

2

Categorization of Chemical Substances

Introduction to Use and Safety of Chemical Substances

Chemical substances are divided into classes and categories. In brief, chemical substances include, acids (inorganic acids and organic acids), alkali (bases), colors, dyes, and pigments, cyanides, detergents, dopants, fillers, food contaminants, gases (cryogenic gases, industrial gases), metals and metal compounds, oxidizers, pesticides, radioactive chemicals, solvents, and many more. (Readers will find more information in chemistry textbooks.)

Chemical substances used in laboratories and chemical industries have been classified, grouped, and categorized systematically, and the established classification has stood the test of time. This chapter describes several chemical substances that students and semi-skilled workers can use for purposes of easy identification. Many chemicals display more than one type of toxicity. The following are the classes/groups of chemical substances commonly encountered and handled by occupational workers at different workplaces: (1) chemical substances causing irritation effects, (2) chemical substances with corrosive effects, (3) allergens, (4) asphyxiants, (5) carcinogens, (6) endocrine disruptors, (7) reproductive and developmental toxins, (8) neurotoxic chemicals, and (9) chemical substances causing adverse effects to other organs.

Categorization of Chemical Substances

Chemical substances causing irritation effects: Irritants are non-corrosive chemicals that cause reversible inflammatory effects (swelling and redness) on living tissue by chemical action at the site of contact. A wide variety of organic and inorganic chemicals are irritants, consequently, skin and eye contact with all laboratory chemicals should be avoided.

Chemical substances with corrosive effects: Corrosive substances can be solids, liquids, or gases, which cause destruction of living tissue by chemical action at the site of contact. Corrosive effects can occur not only on the skin and eyes, but also in the respiratory tract and, in the case of ingestion, in the gastrointestinal tract as well. Corrosive materials are probably the most common toxic substances encountered in the laboratory. Corrosive liquids are especially dangerous because their effect on tissue generally takes place very rapidly. Bromine, sulfuric acid, aqueous sodium hydroxide solution, and hydrogen peroxide are examples of highly corrosive liquids. Corrosive gases are also frequently encountered. Gases such as chlorine, ammonia, and nitrogen dioxide can damage the lining of the lungs, leading, after a delay of several hours, to the fatal buildup of fluid known as pulmonary edema. Finally, a number of solid chemicals have corrosive effects on living tissue. Examples of common corrosive solids include sodium hydroxide, phosphorus, and phenol. Dust from corrosive solids can be inhaled and cause serious damage to the respiratory tract. Classes of corrosive substances include strong acids, such as nitric, sulfuric, and hydrochloric acid, which can cause serious damage to the skin and eyes. Hydrofluoric acid is particularly dangerous and produces slow-healing, painful burns. Strong bases, such as metal hydroxides and ammonia, make up another class of corrosive chemicals. Strong dehydrating agents, such as phosphorus pentoxide and calcium oxide, have a powerful affinity for water and can cause serious burns on contact with the skin. Finally, strong oxidizing agents, such as concentrated solutions of hydrogen peroxide, can also have serious corrosive effects and should never come in contact with the skin or eyes.

Allergens: A chemical allergy is an adverse reaction by the immune system to a chemical. Such allergic reactions result from previous sensitization to that chemical or a structurally similar chemical. Once sensitization occurs, allergic reactions can result from exposure to extremely low doses of the chemical. Allergic reactions can be immediate, occurring within a few minutes after exposure. Anaphylactic shock is a severe immediate allergic reaction that can result in death if not treated quickly. If this is likely to be a hazard for a planned experiment, advice on emergency response should be obtained. Allergic reactions can also be delayed, taking hours or even days to develop. The skin is usually the site of such delayed reactions, in which case it becomes red, swollen, and itchy.

It is important to recognize that delayed chemical allergy can occur even for a period of time after the chemical has been removed. Contact with poison ivy is a familiar example of an exposure that causes a delayed allergic reaction. Also, just as people vary widely in their susceptibility to sensitization by environmental allergens such as dust and pollen, individuals may also exhibit wide differences in their sensitivity to laboratory chemicals. Examples of substances that may cause allergic reactions in some individuals include diazomethane, dicyclohexylcarbodiimide, formaldehyde, various isocyanates, benzylic and allylic halides, and certain phenol derivatives.

Asphyxiants: Asphyxiants are substances that interfere with the transport of an adequate supply of oxygen to the vital organs of the body. The brain is the organ most easily affected by oxygen starvation, and exposure to asphyxiants can lead to rapid collapse and death. Simple asphyxiants are substances that displace oxygen from inhaled air to such an extent that adverse effects result. Acetylene, carbon dioxide, argon, helium, ethane, nitrogen, and methane are common asphyxiants. It is thus important to recognize that even chemically inert and biologically benign substances can be extremely dangerous under certain circumstances. Certain other chemicals have the ability to combine with hemoglobin, thus reducing the capacity of the blood to transport oxygen. Carbon monoxide, hydrogen cyanide, and certain organic and inorganic cyanides are examples of such substances.

Acids: Acids are corrosive substances widely used in different industries for cleaning, etching, plating, and stripping. They are usually in liquid or powder form. Most are acutely hazardous, especially when concentrated. Acids can penetrate clothing, rapidly causing serious burns and damage to the tissue beneath the skin. Protective gear is essential, especially for the hands, face, eyes, and lungs. All corrosives should be clearly labeled with warning placards.

Organic acids: Acetic acid, adipic acid, citric acid, formic acid, lactic acid, and oxalic acid.

Inorganic acids: Hydrobromic acid, hydrochloric acid, hydrocyanic acid, hydrofluoric acid, nitric acid, phosphoric acid, sulfonic acid, and sulfuric acid.

Alkalis (Bases): Alkaline or base substances are used primarily for cleaning and scouring. Like acids, they are acutely hazardous, especially in concentrated form. Examples of alkaline chemical substances include, but are not limited to, ammonia, ammonia persulfate, ammonium fluoride, ammonium hydroxide, calcium hydroxide, potassium hydroxide, and sodium hydroxide. Most have strong caustic or corrosive action, and as such should be clearly labeled with warning placards. Students, laboratory workers, and occupational workers should always use protective gear, especially for the face, eyes, hands, and lungs, when using alkaline chemical substances.

Carcinogens: A carcinogen is a chemical substance capable of causing cancer. Cancer, in the simplest sense, is the uncontrolled growth of cells, and it can occur in any organ. The mechanism by which cancer develops is not well understood, but the current thinking is that some chemicals interact directly with DNA, the genetic material in all cells, resulting in permanent alterations. Other chemical carcinogens can modify DNA indirectly by changing the way the cells grow. Carcinogens are chronically toxic substances; that is, they cause damage after repeated or long-duration exposure, and their effects may become evident only after a long latency period. Carcinogens are particularly insidious toxins because they may have no immediate apparent harmful effects.

Cyanides: Cyanides are a group of highly irritating and rapidly acting poisons. They are used for cleaning, plating, and metallizing. Notice that most cyanide compounds (salts) contain a metal or mineral molecule. The biggest risk is exposure to cyanide in gas form, although it is often stored in solid or liquid form. Cyanide is quickly absorbed through the skin and lungs. It prevents the body tissues from taking up oxygen, causing sudden death by asphyxiation. Repeated low-level exposure can cause severe dermatitis, thyroid disease, and muscle incoordination. Another highly reactive and poisonous group related to cyanides are isocyanates. Users and occupational workers should be very careful during handling of any cyanide compound and always wear a proper respirator if the process is not completely enclosed. Workers should ensure that containers of cyanide are well labeled with warning placards. Cyanides include calcium cyanide, copper cyanide, hydrocyanic acid, nickel cyanide, potassium cyanide, potassium ferrocyanide, sodium cyanide, and zinc cyanide.

Dopants: Dopants are metal compounds in solid, liquid, or gas form and are used to make chips. Dopants are sometimes called impurities. They are usually injected as a gas or vapor into ovens that are heated to extreme temperatures. When heated, the metal of the dopant is deposited in the semi-conductor wafer, penetrating its surface and giving it the ability to conduct electricity. The metals more commonly used in doping include aluminum, antimony, arsenic, boron, and phosphorous. Dopants are considered potentially the most hazardous group of chemical substances used in electronics. Many dopants are highly toxic chemicals. Any kind of a leak or rupture occurring with a substance like phosphine, arsine, or boranes can result in a chemical hazard in the workplace and fatal injury to the surrounding community. Argon and deuterium are sometimes used as carrier gases.

Gases: Gas is one of the phases of matter. The molecules in gases are more widely spaced than in solids or liquids. Gases undergo occasional collisions with one another. In general, gases have relatively low density and viscosity and undergo relatively great expansion and contraction with changes in pressure and temperature. Gases have the ability to diffuse readily and become uniformly distributed throughout any container. The common gases (at room temperature) include ammonia, argon, carbon dioxide, carbon monoxide, ethane, fluorine, hydrogen, hydrogen chloride, hydrogen cyanide, helium, methane, neon, nitrogen, nitrogen oxide, oxygen, phosphine, and xenon.

Cryogenic gases: Cryogenic means ultra-cold. These gases are usually stored in liquid form under high pressure and are used to heat and cool ovens in the process of semiconductor wafer fabrication. Some are used as "carrier" gases, carrying dopants into the oven chamber. Hydrogen and oxygen are extremely flammable (they ignite and burn very easily). These gases have the potential to explode and thus require special storage and handling precautions. A major leak of liquefied gas can rapidly fill a workroom, displacing oxygen and causing sudden death by asphyxiation. Cryogenic gases include argon, carbon dioxide, carbon monoxide, deuterium, helium, hydrogen, nitrogen, oxygen, and ozone.

Endocrine disruptor chemicals (EDCs): Endocrine disruptors are also called endocrine modulators, environmental hormones, and endocrine active compounds. The endocrine system is one of the body's main communication networks and is responsible for controlling and coordinating numerous body functions. Hormones are first produced by the endocrine tissues, such as the ovaries, testes, pituitary, thyroid, and pancreas, and then secreted in the blood to act as the body's chemical messengers, where they direct communication and coordination among other tissues throughout the body. The EDCs are known to interfere with the body's endocrine system and produce adverse developmental, reproductive, neurological, and immune effects in animals and humans. A wide range of chemical substances, both natural and man-made, have been identified as endocrine disruptors. These include, but are not limited to, pharmaceuticals, dioxin and dioxin-like compounds, polychlorinated biphenyls, DDT and other pesticides, and plasticizers such as bisphenol A. Endocrine disruptors may be found in many everyday products, including plastic bottles, metal food cans, flame retardants, food, toys, cosmetics, and pesticides, causing developmental, reproductive, neurological and immune effects. The National Institute of Environmental Health Sciences (NIEHS) supports studies to determine whether exposure to endocrine disruptors may result in adverse human health effects, including lowered fertility and an increased incidence of endometriosis and some cancers. Research shows that endocrine disruptors may pose the greatest risk during prenatal and early postnatal development when organ and neural systems are forming. Although limited scientific information is available on the potential adverse human health effects, concern arises because endocrine disrupting chemicals, while present in the environment at very low levels, have been shown to have adverse health effects. The difficulty of assessing the public health effects due to EDCs is further increased because people are typically exposed to multiple chemical substances directly or indirectly. The NIEHS and the National Toxicology Program (NTP) support research to understand how these chemical substances work and the effects that they may have in various animal and human populations with the long-term goal of developing prevention and intervention strategies to reduce possible adverse health effects.

Reproductive and developmental toxins: Many chemical substances cause adverse health effects to the reproductive and developmental cycles of animals and humans. They inflict adverse effects on various aspects of reproduction, such as fertility, gestation, lactation, and the growth and development of the embryo or fetus. These effects include lethality (death of the fertilized egg, the embryo), malformations or terata (the chemical substances are also called teratogens), retarded growth, and postnatal functional deficiencies. When a pregnant worker is exposed to a toxic chemical substance, generally the fetus is also exposed because the placenta is an extremely poor barrier. Reproductive toxins can affect both men and women. Toxic chemical substances like dibromochloropropane cause sterility in male workers.

Neurotoxic chemicals: Neurotoxic chemicals can induce an adverse effect on the structure or function of the central and/or peripheral nervous system, which can be permanent or reversible. In some cases, the detection of neurotoxic effects may require specialized laboratory techniques, but often they can be inferred from behavior such as slurred speech and staggered gait. Many neurotoxins are chronically toxic substances whose adverse effects are not immediately apparent. Currently, because of the limited data available in this area, significant uncertainties attend the assessment of risks associated with work with neurotoxic substances.

Greenhouse gases: Greenhouse gases are gases that trap solar radiation and trigger a rise in temperature levels on the planet. There are various greenhouse gases, the most prominent being carbon dioxide, chlorofluorocarbons (CFCs), methane, nitrous oxide, and sulfur hexafluoride.

Inert gases: Inert gases, also called noble gases, include helium (He), neon (Ne), argon (Ar), krypton (Kr), xenon (Xe), and radon (Rn). They are colorless, odorless, and tasteless gases and were once believed to be entirely inert, meaning, forming no chemical compounds. However, some compounds of these elements have produced fluorides of krypton, xenon, and radon.

Fillers: Fillers are powders or tiny fibers added to resins (plastics, epoxies, glues, and paints) to give bulk, strength, and form. They are durable and some resist heat, fire, and electricity. Asbestos and chromates cause cancer, and fiberglass can cause serious lung problems if inhaled over a period of time. These substances can also be highly irritating to the skin and eyes. Fillers are used to make printed circuit boards and plastics. They are easily released as harmful dusts when resin products are shaped, sawn, or drilled. Students and workers should avoid breathing and direct contact of filler materials at workplaces.

Metals and metal compounds: Ordinarily, people do not consider that metals are chemicals. But they are, and many can be very harmful if swallowed or if small unnoticeable amounts are inhaled day after day. Because metals are good conductors of electricity, they are widely used in electronics. Metals are used or occur in many forms—such as bulk solids, powders, and liquid solutions suspended in gas form—and are emitted as a fume when heated and as dust when drilled, sawn, or filed. Exposure to the more dangerous forms of metal (gases, dusts, and fumes) occurs more frequently during doping, soldering, plating, tinning, and other metal work. The known metals include, antimony, arsenic, aluminum, antimony, barium, beryllium, boron, calcium, chromium, chromates, cobalt, copper, gallium, iron, gold, lead, manganese, mercury, molybdenum, nickel, phosphorus, selenium, silver, terrarium, tin, titanium, tungsten, vanadium, and zinc.

Oxidizers: Oxidizers are highly reactive chemicals that can be used to clean or to render a metal surface free from corrosion. During oxidation, oxygen (from the oxidizer or from the air) combines with a metal or semi-conductor surface to form a protective oxide layer. Some oxidizers have strong corrosive action and care must be taken to protect the eyes, skin, and lungs from exposure. Oxidizers are also highly flammable and require special handling and storage arrangements. The list of oxidizers include, but is not limited to, ammonium persulfate, chlorine, chromic acid, hydrogen peroxide, iodine nitrous oxide, oxygen, ozone, potassium iodide, and silver nitrate.

Pesticides: Pesticides come in many different formulations. They are used to control pests of different kinds. Pests may be target insects, vegetation, or fungi. Pesticides are known poisons used specifically for the control of crop pests and rodents. Some are very poisonous, or toxic, and may seriously injure or even kill humans. Others are relatively non-toxic. Pesticides can irritate the skin, eyes, nose, or mouth. The health effects of pesticides depend on the type of pesticide. The important and major groups include, but are not limited to, acaricide, aphicides, fumigants, fungicides, herbicides, insecticides, larvicides, molluscides, miticides, nematocides, repellants, and rodenticides. The organophosphate and carbamate pesticides affect the nervous system. Others cause irritation to the skin, eyes, and mucous membranes. Several pesticides behave as carcinogens. Prolonged periods of exposure to high concentrations of pesticides cause adverse health effects, which include, but are not limited to, (1) reproductive effects, (2) teratogenic effects, (3) carcinogenic effects, (4) oncogenic effects, (5) mutagenic effects, (6) neurotoxicity, and (7) immuno-suppressive agents. There are several classes of pesticides: organophosphate pesticides (OPPs), organochlorine pesticides (OCPs), carbamates, synthetic pyrethroids, biopesticides, and microbial pesticides. The OPPs cause adverse effects and damage the nervous system by disrupting the enzyme that regulates acetylcholine, a neurotransmitter. Many OCPs are known to persist in food, water, terrestrial, aquatic fauna flora, and in the living environment. Global regulatory agencies advocate for the restricted use of OCPs. Carbamate pesticides like OPPs also affect the nervous system and disrupt the enzyme that regulates acetylcholine (however, this is reversible). Synthetic pyrethroids form the synthetic version of the naturally occurring pesticide pyrethrin (found in chrysanthemum flowers). Some synthetic pyrethroids cause adverse effects to the nervous system. (Refer to the literature for more information on pesticides and health effects.) A survey of the literature indicates that a large number of existing pesticides and their formulation products have not been evaluated. More studies are required in this direction.

Resins (epoxies, curing agents, plastics): There are many kinds of resins, e.g., plastics, epoxies, glues, adhesives, paints, waxes, synthetic rubber, and synthetic fibers. With the obvious exception of rosin (colophony) flux used in soldering, most resins used in electronics are man-made organic polymers. Polymers are complex chemical substances. Most contain many poisonous ingredients, such as solvents, dyes, stabilizers, fillers, plasticizers, catalysts, and monomer residue. Some of these ingredients cause allergies, birth defects, and cancer. Polymers are formed from monomers. Epoxides (epoxy resins) are normally cured with a phenol compound, and polyesters are cured with a peroxide compound. Uncured epoxy resins or monomers are very toxic and rapidly penetrate the skin and lungs. After they are reacted, cured, or set, they are much less harmful, though dust created by shaping, cutting, and drilling can be harmful. Resins are widely used in electronics, particularly in making printed circuit boards, molding plastics, bonding, encapsulating, and packaging, and are also used as wire coatings and a variety of other electrical insulation materials. Resins can produce a wide variety of highly toxic vapors and gases when heated or burned. Fires caused by burning plastic are sometimes very difficult to control.

Solvents: In simple terms, water is the most common solvent in everyday life. Most other commonly used solvents include organic chemicals. The vast majority of chemical reactions are performed in solution. The solvent fulfills several functions during a chemical reaction. It solvates the reactants and reagents so that they dissolve. This facilitates collisions between the reactant(s) and reagents that must occur in order to transform the reactant(s) to product(s). The solvent also provides a means of temperature control, either to increase the energy of the colliding particles so that they will react more quickly, or to absorb heat that is generated during an exothermic reaction. The selection of an appropriate solvent is guided by theory and experience. Solvents are used in nearly every phase of electronics manufacturing. They are used primarily for cleaning and degreasing, and for thinning plastics, resins, glues, inks, paints, and waxes. There is a wide range of organic solvents, some very toxic and others only mildly toxic. The subgroups should be considered in order to have a better idea of the specific hazard risks and uses. The aromatic compounds and the chlorinated hydrocarbons are perhaps the most dangerous groups of solvents since many of them are known to cause cancer and other serious diseases. Solvents are chemical substances, usually liquid, that are commonly used to dissolve unwanted substances or material. Solvents are liquids used for the purposes of mixing and to dissolve other substances, such as paints, greases, wax, and oils. Solvents are found in fuels, adhesives, glues, cleaning fluids, epoxy resins, hardeners, lacquers, paints, paint thinners, primers, and even nail polish remover. Prolonged periods of exposure to some solvents, such as acetone, alcohols, benzene, gasoline, mineral spirits, methylene chloride, toluene, turpentine, and xylene, causes acute and chronic health effects. Solvents are among the most frequently used industrial chemicals because of their ability to clean grime and grease. Application of different solvents in industries and homes is very common and has become a global trend. More particularly, polymers, paints and coating industry use solvents in very large quantities around the globe. Human exposure to different solvents in workplace air and the general atmosphere is very common. Uses, the manner of exposures, the health effects, and the environmental impact of different solvents are discussed in related pages of this book. More information is available in the literature. Some of most common and selected solvents include, but are not limited to, acetone, benzene, n-butyl alcohol, butyl glycidyl ether, benzene, carbon disulfide, carbon tetrachloride, chloroform, cyclohexanone, dimethyl formamide, 1,4-dioxane, ethyl acetate, ethyl alcohol, ethyl ether, formic acid, formamide, kerosene, methyl cellosolve, dichlorobenzene, methylene chloride, methyl ethyl ketone, methyl isobutyl ketone, naphtha, pentane, perchloroethylene, petroleum spirits, n-propyl alcohol, propylene glycol, propyl alcohol, styrene, toluene, trichloroethylene, and xylene.

Chemical substances causing adverse effects to other organs: Chemical substances also cause adverse health effects to other organs or non-target organs. The chemical substances include, but are not limited to, aromatic and chlorinated hydrocarbons, metals and metal compounds, carbon monoxide, and cyanides.

Disposal of Chemical Substances

Environmental and waste disposal issues for source reduction, waste minimization, and recycling of materials must be considered in any experiment plan. The chemical composition of all products and waste materials generated by the experiment should be considered, and appropriate handling and disposal procedures for each of these materials should be evaluated in advance. Careful attention to regulatory requirements is essential for waste disposal. Special issues to consider include the frequency and amount of waste generated, the methods to minimize waste, the steps to neutralize waste or render it non-hazardous, procedures for dealing with unstable waste or waste that requires special storage and handling, and the compatibility of materials being accumulated. During the planning stage, particular attention should be given to the minimization of multi-hazardous waste, such as waste that represents both a chemical and a biological hazard.

Chemical Substances and Human Use

Nearly 85% of pharmaceutical products in use require the use of chlorine in their production. As a result of the chemicals used in pharmaceuticals, combination drug therapy reduced AIDS deaths by more than 70% from 1994 to 1997. Chemicals called "phthalates" (there are several kinds of phthalates) are used in polyvinyl chloride (PVC), a vinyl used for medical tubing, blood bags, and numerous other products. While environmentalists have tried to ban these products, vinyl medical devices provided numerous lifesaving benefits. PVC is a safe, durable, sterile product that can withstand heat and pressure, as well as produce tubing that doesn't kink. It is particularly beneficial for vinyl blood bags because it stores blood twice as long as the next best alternative and doesn't break as glass alternatives do. With blood shortages looming, PVC blood bags are an essential tool in maintaining and transporting supplies.

Modern farming with the proper application of chemical substances, such as chemical fertilizers and pesticides (herbicides, fungicides, fumigants, rodenticides, insecticides, and weedicides), provides the benefits of more food production to meet the challenges of the outpaced population growth. Modern farming provides people in both developed and developing countries with more food per capita. Herbicides control weeds, reduce the need for tilling soil, and reduce soil erosion.

Chemical Substances, Global Regulatory Systems and Regulations

Global advances in scientific researches, technology and development, identification and formulations of newer chemical products for societal development, for the improvement of the quality of life along with economic development have created newer avenues and opportunities. This has necessitated a science-based regulatory system for the management of chemical substances and to make decisions to meet the global regulatory challenges. In this direction, international collaborations and partnerships, have enhanced newer scientific innovations and greater mobility of a variety of products around the world. This also has provided increased speed to communicate and share the benefits of technology between governments. The data requirements for the registration and distribution of chemical substances, such as pesticides, drugs and pharmaceuticals, cosmetics, and food products, have been made uniform. There are many regulatory agencies associated with the safe management of chemical substances in both developed and developing countries. The following is a list of some of the well-known global regulatory agencies:

- US Food and Drug Administration (US FDA)
- US Environmental Protection Agency (US EPA)
- Insecticides Act, Ministry of Agriculture and Cooperation, Government of India, India
- Pest Management Regulatory Agency Health, Canada
- Pest Management Regulatory Agency (PMRA), US Environmental Protection Agency (US EPA)
- National Toxicology Program (NTP), United States
- National Administration of Drugs, Food and Medical Technology, Argentina
- The European Agency for the Evaluation of Medicinal Products
- The Federal Institute for Drugs and Medical Devices, Germany
- National Institute of Health Sciences, Japan
- Association of the British Pharmaceutical Industry, United Kingdom
- The Department of Health, United Kingdom
- Medicines and Healthcare Products Regulatory Agency (MHRA), United Kingdom
- The Food and Drug Administration (FDA), United States
- The National Institute of Health (NIH), United States
- Centers for Disease Control and Prevention (CDCP), United States
- Department of Health and Human Services (DHHS), United States
- The National Center for Complementary and Alternative Medicine (NCCAM), United States
- The National Library of Medicine (NLM), United States
- World Health Organization (WHO), Geneva, Switzerland
- International Conference on Harmonization (ICH), Geneva, Switzerland

3

Elements of Toxicology and Guidelines

Introduction

Humans have been using chemical substances of different kinds and classes for a very long time and in practically every aspect of life. Chemical substances have the inherent ability to react with a host of factors and, to multiple physical and biological conditions. The reactive properties of chemical substances differ widely. Chemical substances when not properly used, stored, and/or managed are known to cause harmful health effects to users and occupational workers, are workplace hazards, and damage the living environment. Therefore, it is important to determine the potential hazards of each chemical substance before they come in contact with animals, humans, and the living environment. In other words, chemical substances of different categories (refer to Chapter 1) used and handled by the general public and society should be evaluated.

Toxicology is the branch of science that deals with the gross and intrinsic capabilities of a chemical substance(s) on biological systems, meaning on plants, animals, and humans. Toxicology is a multidisciplinary science and is closely interrelated with many other branches of science. Chemical substances are required for health, progress, and societal development. In the very close linkage with an array of chemical substances and societal development, human health cannot be ignored. Therefore, thinkers of the past and present around the world framed regulations for the manner and methods of using chemical substances. There are no safe chemical substances and all are toxic in one way or another. The safety of a chemical substance depends on the concentration and manner of exposure and use. This is important and should be very well understood by all students, occupational workers, and household users who handle, store, transport, and dispose of different chemical substances. Improper and negligent use and management of chemical substances cause injury, disaster, and death. This chapter focuses on and discusses, in brief, the elements of toxicology vis-a-vis the effects of chemical substances and human use.

Chemical substances as and when they are marketed for human use in the form of drugs, food additives, cosmetics, and many others items, require safety data and detailed quality evaluations. To generate quality data about the candidate chemical substance, different countries and international regulatory agencies have framed elaborate procedures. By understanding the basics of toxicology and correctly adhering to the regulations and observing precautions, the benefits of chemicals could enrich human society, free from hunger and diseases.

Chemical Substances, Exposures, Occupations, and Hazards

There are many occupations associated with different chemical substances, their manufacturing, formulation, storage, transport, waste disposal, and related human exposure(s). With the development and ramifications of industries around the globe, human health and safety and protection of the living environment gains importance. The following are some of the chemicals in brief. Acids, Alkalis (Bases), Cryogenic Gases, Cyanides, Dopants, Insecticides, Pesticides, Metals and their Compounds, Oxidizers, Radioactive Chemicals, Resins, Epoxies, Curing agents, Plastics, Semiconductors, Solvents.

Different Industries

- Automobile industries.
- Battery reclamation.
- Dry cleaning industry.
- Dye industries: Currently, there are approximately some 1200 different commercial dyes. Dyes are produced by a variety of chemical reactions from raw materials and most of these materials are hazardous to humans.
- Electronics and computer industry.
- Elecroplating industries.
- Petrochemical and gas stations.
- Glass manufacturing: Hazardous components of the raw materials include etching agents, typically hydrogen fluoride and fluoride-donating salts. Strong oxidizing corrosives, such as nitric acid. Heavy metals containing arsenic, cobalt, zinc, thorium, and uranium are common pigment materials in addition to specialty organic dyes. Discharges into waterways and sensitive areas present significant threats from the heavy metal pigments, organic dyes, and strong corrosives. Fluoride waste streams are extremely dangerous to human populations.
- Metal fabricators: Various processes are involved in metal fabrications to achieve the desired shapes after casting, cutting operations, shearing operations, drilling, milling, and forging. These processes are known to cause health-related problems among occupational workers. Improper use of cutting oils such as ethylene glycol, degreasing and cleaning solvents like trichlorethane and methyl ethyl ketone, hydrochloric acid and sulfuric acid, alkalis and heavy metals lead to short-term and long-term health disorders among workers.
- Ordnance sites: A wide variety of military munitions and weaponry including rifle rounds, shells, bombs, grenades, mines, and explosives are manufactured and is an area known to cause health hazards. In the manufacturing of explosives, several chemicals are used, i.e., di- and tri-nitro-benzene, strong acids, ethyl alcohol, di- and tri-nitro-phenol, mercury, ethylene glycol, di- and tri-nitro-toluene, reactive metals, phenols, ketones, ethers, formaldehyde, nitroglycerine, ammoniated compounds, and sodium hydroxide. Any kind of improper use or negligence during handling, storage, and disposal of these chemicals results in human fatalities and environmental disasters.

- Paint industry: A wide variety of chemical substances are used in the manufacture of different kinds of paints and coatings. Common chemicals include polymers, organic compounds, heavy metals, epoxies, solvents, mild corrosives, polyurethanes, herbicides, and fungicides. Further, in the manufacture of specialty paints, toxic chemicals such as toluene, xylene, ethyl acetate, lead, acetone, titanium dioxide, cadmium, zinc, and chromium are used. During cleaning, waste products are often washed into a sewer drain, leading to health disorders among workers in different workplaces.

- Pesticide industry: A wide variety of chemicals, almost all of which are known to be health hazards to humans, are used in the manufacture of pesticides of different kinds. Chemicals typically found at a pesticide facility include, but are not restricted to, ammonia, benzene, carbon tetrachloride, hydrogen cyanide, mercury, nitric acid, phosgene, sulfuric acid, and xylene.

- Asbestosis: Asbestosis is a fibrotic disease of the lungs caused by chronic exposure and inhalation of asbestos fibers. Asbestos is a mixture of chemicals that occurs naturally as a fiber substance and is widely used in the building industry for insulation, roofing, and fireproofing. Asbestosis diffusely affects the lungs, predominantly damaging the interstitium (the connective tissue between airspaces). The lungs become fibrotic and stiff. Asbestos exposure can also damage the pleura of the lungs and lead to other diseases such as mesothelioma and lung cancer.

- Silicosis: Workers in the following occupations are at risk for developing silicosis: Dusts highway and bridge construction and repair; building construction, demolition, and repair; abrasive blasting; masonry work; concrete finishing; drywall finishing; rock drilling; mining; sand and gravel screening; rock crushing (for road base).

Toxicology Studies

Toxicological studies are essential to understanding the possible adverse effects that a candidate chemical or combination of chemicals may cause to animals, humans, fauna and flora, and to make relevant, reliable, and reproducible predictions. Thus the generation of toxicological data after conducting short-term and long-term exposures in species of organisms, laboratory animals using different routes of exposures provide substantial and basic guidance to establish safe levels of chemicals follow chemical safety. Depending on the route of exposure, the duration of exposure, and the quantity of the test chemical, the experimental animals develop signs and symptoms of toxicity. The test provides information about

- The nature of toxicity of the test chemical substance
- The dose and concentration(s) of the chemical substance that cause adverse effects in the animal
- The toxicity profile in male and female test animals, oral, dermal, and respiratory routes

- The immediate and long-term health effects
- The effects of two or more chemical substances as additives or synergistic effects

---

History of Toxicology

What is toxicology? What is the history of toxicology? What is the importance of toxicology to modern society? These are some of the questions that need a better and meaningful understanding for the management of chemical substances and to protect human health. Toxicology is a scientific discipline many thousands of years old. Recent reports have traced the history of toxicology from 3000 BC, to the Middle Ages (476–1453), to the period of the Renaissance (1400–1600), and subsequent years. The history of toxicology needs to be traced along with global development. To trace and document the history of toxicology to certain parts of the world alone is both incomplete and incorrect. Therefore, it is necessary to know the origin and global development of the science of toxicology.

The science of toxicology has a very solid and authentic historical base. In fact, elementary knowledge about toxicology dates back to early times of human history and civilization. India is well known as the birthplace of Ayurveda, the ancient Indian system of medicine and human health care and recent documents indicate Ayurveda's origin as ca. 5000 BC according to the Indian scriptures that have stood the test of time dates to much earlier periods of human history. The Ayurvedic system of medicine and health care has valid links to the ancient books of wisdom—the *Vedas*. The word *Veda* in Sanskrit (Samskruta) means knowledge and the language is Samskruta/Sanskrit, or *Devanagari* script, आयुर्वेद. The term Ayurveda is made up of two words, namely, *Ayuh* meaning life and *Veda* meaning the knowledge (the knowledge of longevity/life). Thus, Ayurveda originated in India in the pre-Vedic period—the *Rigveda* and *Atharva-veda* (5000 BC). The texts of Ayurveda, such as Charak Samhita and Sushruta Samhita, were documented about 1000 BC. As has been documented elsewhere, Ayurveda is one of the oldest systems of health care, describing both the preventive and curative aspects of different herbal medicines for the improvement of the quality of life. Ayurveda in a most comprehensive way describes medication for human ailments and bears a close similarity to the principles of health care of the modern era propounded by the World Health Organization.

The ancient seers of India in the *Astanga Hrudaya* of *Vagbhata* and others have paved the way for understanding the concept of human health. Human ailments, including poisoning, are the areas covered by Ayurveda. In brief, Ayurveda discusses the combination of four essential parts of the system, namely, the human body, mind, senses, and the soul, and unravels the effect of toxic chemical substances on the body and the manner of its elimination by different processes. Further, the history of indigenous Indian medical science along with Indus Valley civilization dates back more than 3000 BC. The well-planned cities of Harappa and Mohenjo-Daro not only exemplify the rich cultural heritage of India, but also its advanced systems of hygiene and human health care.

For the people of the Indian subcontinent in particular, the way of life and the association with food and drink was quite different and stringent compared to the human populations in Occidental regions of the world. Elementary knowledge about the use and restricted use of certain substances/food items and drinks were the guiding principles for the maintenance of good health. This is very evident in the dictum of the native language of India, Samskruta. The dictum may be grouped under health and hygiene or Yoga system of philosophy—a path to lead a life of righteousness. The dictum in Samskruta runs as follows: *Ati Sarvatra Varjayet*, meaning avoid excess in eating, drinking, and or other activities, anywhere, any time. Even *nectar* (the *ambrosia*) the drink when consumed in excess can cause adverse effects! There are many well-documented regulations for human health care. The dictum: (a) *Langanam Parmaushadham* (in Samskruta), meaning fasting or moderate food before bed at night is the best medicine to maintain a proper and good health; and (b) *Madyam na Pibeyam*, meaning not to be alcoholic. In fact, Rigveda, the ancient scriptures of India, clearly mention *visha*, a Sanskrit term for poison. Similar references are also made in hymns to poison liquids that produce ecstasy. In the Puranic legend, poison is mentioned during the mythological process of churning the cosmic ocean, before the drink (Amruta) of immortality is won.

Much later (1493–1541), Paracelsus, the father of modern toxicology pronounced a dictum of his own: *Sola dosis facit veneum*, meaning only the dose makes the poison. "All substances are poisons; there is none which is not a poison. The right dose differentiates a poison from a remedy." In other words, no substance is absolutely safe. What a glorified commonness between ancient thinkers from India very much earlier in history and the West in later periods and without knowing each other during the periods of world history. This is the glorious saga of the global history of toxicology.

In the Western world, the ancient Greeks were probably the first to dissociate medicine from magic and religion. Important and valuable contributions of several thinkers of the West improved the quality of human health and our understanding of toxicology. A list of some of the leading thinkers follows.

- Shen Nung (2696 BC) is the father of Chinese medicine, noted for tasting 365 herbs. He wrote the treatise *On Herbal Medical Experiment Poisons* and died of a toxic dose.

- Ebers Papyrus (1500 BC) is the oldest well-preserved medical document from ancient Egyptian records. Dating from approximately 1500 BC, it contains 110 pages on anatomy and physiology, toxicology, spells, and treatment, recorded on papyrus.

- Homer (about 850 BC) wrote of the use of arrows poisoned with venom in the epic tale of *The Odyssey* and *The Iliad*. The Greek *toxikon* means arrow poison.

- Hippocrates (460 BC) was a Greek physician, born on the island of Cos, Greece. He became known as the founder or father of modern medicine and was regarded as the greatest physician of his time. A person of many talents, he named cancer using the Greek word *karkinos* (crab) because of the creeping, clutching, crab-claw appearance of cancerous tissue spreading into other tissue areas. He moved medicine toward science and away from superstition. He is also noted for his founding of an oath of ethics still used today.

- Plato (427–347 BC) reported the death of Socrates (470–399 BC) by hemlock (*Conium maculatum*).

- Death of Socrates in 399 BC by hemlock. Socrates was charged with religious heresy and corrupting the morals of local youth. The active chemical used was the alkaloid coniine that, when ingested, causes paralysis, convulsions, and potentially death.

- Aristotle (384–322 BC) was familiar with the venom of jellyfish and scorpion fishes.

- Theophra Stus (370–286 BC), Hippocrates (400 BC), and Ebers Papyrus (1500 BC). In fact, use of hemlock to by Greeks to execute the great philosopher Socrates (470–399 BC) is an instance of its own. More recently, Ramazzini (1700) documented the possible preventive measures to control industrial hazards among occupational workers. For more information, refer to the literature.

- Mithridates VI (131–63 BC) from a young age period was fearful of being poisoned. He went beyond the art of poisoning to systematically study how to prevent and counteract poisons. He used both himself and prisoners as "guinea pigs" to test his poisons and antidotes. He consumed mixtures of poisons to protect himself, which is the origin of the term "mithridatic." The term mithridatism is well known in pharmacology. It is named after King Mithridates of Pontus (112–63 BC), an enemy of the Roman Empire. To avoid being assassinated, he took small doses of poison to immunize himself against it. He was the first to develop antidotes in his quest for the universal antidote.

- *Sulla 82 BC Lex Cornelia de sicariis et veneficis*—law against poisoning people including prisoners; could not buy, sell, or possess poisons.

- Aulus Cornelius Celsus (25 BCE–50 AD) promoted cleanliness and recommended the washing of wounds with an antiseptic such as vinegar. He published *De Medicina*, which contained information on diet, pharmacy, surgery, and preparation of medical opiods.

- Pedanius Dioscorides (40–90 CE) was a Greek pharmacologist and physician in the time of Nero. He wrote *De Materia Medica*, the basis for the modern pharmacopeia. It was used up to 1600 CE.

- Devonshire Colic 1700s, Devonshire, England. High incidence of lead colic among those who drank contaminated cider. The apple press was constructed in part by lead. Discovered and described in the 1760s by Dr. George Baker.

- Ramazzini (1700) documented the possible preventive measures to control industrial hazards among occupational workers.

- John Jones (1701) extensively researched the medical effects of opium.

- Richard Meade (1673–1754) wrote the first English language book dedicated to poisonous snakes, animals, and plants.

- Percivall Pott (1775) was born in 1714 and apprenticed to Edward Nourse. Pott made some groundbreaking discoveries in the fields of cancer research and surgery techniques. He discovered the link between occupational carcinogens and scrotal cancer in chimney sweeps and wrote multiple scientific articles in his lifetime.

- Friedrich Serturner (1783–1841) was the first scientist to successfully isolate morphine crystals from the poppy plant, in effect creating a much stronger and more effective painkiller.

- Francois Magendie (1783–1855) was born in France and researched the different motor functions of the body in relation to the spine, as well as the nerves within it. In addition, he researched the effects of morphine, quinine, strychnine, and a multitude of alkaloids. He is noted as the father of experimental pharmacology.

- Louis Lewin (1854–1929) was a German scientist who took up the task of classifying drugs and plants in accordance to their psychological effects. The classifications were *inebriantia* (inebriants), *exitantia* (stimulants), *euphorica* (euphoriants), *hypnotica* (tranquilizers), and *phantastica* (hallucinogens).
- Serhard Schrader (1903–1990) was born in Germany. A chemist, Dr. Schrader accidentally developed the toxic nerve agents sarin, tabun, soman, and cyclosarin, while attempting to develop new insecticides. As a result, these highly toxic gases were utilized during World War II by the Nazis. He is sometimes called the "father of the nerve agents."

Branches of Toxicology

Chemicals are used extensively in industries, homes, and crop fields to meet the growing challenges for healthy living. However, it has been reported that a vast majority of chemicals lack basic toxicity data and this is a cause for concern. Generation of quality data on the toxicity and safety of chemical substances, the proper evaluations and meaningful interpretations to human health and environmental safety demands the support of specialized branches of science. In simple terms, the chemical substance(s) under test have to pass through different branches for evaluation. These are (1) analytical toxicology, (2) aquatic toxicology, (3) biochemical toxicology, (4) clinical toxicology, (5) eco-toxicology, (6) environmental toxicology, (7) epidemiological toxicology, (8) genetic toxicology, (9) immunotoxicology, (10) nutritional toxicology, (11) mammalian toxicology, (12) regulatory toxicology, and many other related branches.

Recent advances in toxicology and technology have now taken yet another important turn with the emerging discipline of nanotechnology and nanotoxicology. In fact, nanotechnology is one of the top research priorities of the US government. Nanotechnology involves research and technology development at the atomic, molecular, or macromolecular levels, in the length scale of approximately 1–100 nm. This technology creates and uses structures, devices, and systems that have novel properties and functions because of their small and/or intermediate sizes, and the novel ability to be controlled or manipulated on the atomic scale.

Nanomaterials manufactured in different industries, more particularly drugs and pharmaceuticals, might pose risks to human health and other organisms because of their composition, reactivity, and unique size. Nanotechnology researches and development, particularly in medical research, work at micro- and nanoscale levels to develop new drug delivery methods, therapeutics, and pharmaceuticals. In such areas of research, it is equally important to consider the potential interactions of nanomaterials with the environment and the associated risks. This involves studying the effects of natural nanoparticles in the air and soil, lifecycle aspects of manufactured nanomaterials, and their fate and transport. Risk assessment also includes studies on the toxicity of natural and manufactured nanomaterials, as well as their routes of exposure to humans and other organisms and the potential for bioaccumulation. Also, the nanoscale colloidal particles so produced are involved in the transformation and transport of metals, toxic organic compounds, viruses, and radionuclides in the environment, because nanomaterials have been found to cause toxic responses in test animal systems. In fact, data on the toxicology of nanoparticles and nanotubes (tiny carbon tubes) are very sketchy. Nanoparticles perhaps inflict undesirable effects on the lungs and other body systems. Nanoparticles in food may cross into the gut lymphatic system. Inhalation of nanoparticles has been known to travel from the nasal nerves to the brain, causing health disorders.

The nanomaterials and the structures so formulated with characteristic dimensions (approximately 1–100 nm) contain a variety of unique and tunable chemical and physical properties. In fact, these properties have made nanoparticles central components of the emerging global technologies. The use of nanotechnology is increasing. However, its potential adverse effects on biological systems, with particular reference to human health, are not well understood. In order to accurately conduct hazard assessments, there is a need to know the concepts that apply to pathways of dermal, oral, and respiratory exposure with reference to nanomaterials. This gains added importance in the study of biological systems that include, but are not limited to, membrane transfer, screening methods, and the impact on major body organs and systems.

While there are differences in the methods of data generation from one branch to another, all branches are interrelated to provide complete data about the toxicity and safety of a candidate test chemical substance vis-a-vis human safety. Toxicity of a chemical is the result of several reactions and interactions between the candidate chemical and/or its metabolite(s) and the cellular receptors. These include enzymes, glutathione, nucleic acids, hormone receptors, etc. The degree of toxicity of a chemical can be explained as follows:

Toxicity C Ar (Chemical) (Receptor)

Ar is the specific affinity of the receptor for the toxic chemical **C**. The toxicity of a chemical can also be expressed as **Toxicity = k (C) (R) Ac**, where toxicity is dependent upon **C**, **R**, and **Ac**, where **C** is the concentration of the candidate chemical in the tissue, **R** is the concentration of the endogenous receptor of the tissue, and **Ac** is the affinity of the receptor for the chemical.

The toxicological evaluations related to human safety of chemical substances is a very complex process. It involves the determination of the intrinsic toxicity and hazard of the test chemical(s). Subsequently, this evaluation leads to determining and establishing a "no observed effect level" (NOEL), the highest dose level tested experimentally that does not produce any adverse effects. This dose level is then divided by a "safety factor" to establish an acceptable daily intake (ADI) of the candidate chemical substance. The ADI value is normally based on current research, long-term studies on species of laboratory animals with several doses including high doses. Subsequently the NOEL is scaled by a safety factor based on judgment, experience, and international convention. Typically, the safety factor ranges between 100 and 1000, depending on the biologic relevance and severity of the observed effect, to extrapolate the differences between test animals and humans. This provides a substantially lower level and thus a large margin of safety for humans.

ADI is a measure of a specific chemical substance, pesticide residue, or a food additive, in food, beverages, or drinking water that can be ingested over a lifetime period without an appreciable health risk. ADIs are expressed by body mass, usually in milligrams per kilograms of body mass per day. The higher the value of ADI, the safer the chemical substance is in food, water and for regular ingestion. In fact, the concept of ADI is a measure to indicate the toxicity from long-term exposure to repeated ingestion of chemical substances in foods. This concept was first introduced in 1957 by the Council of Europe and later the Joint Expert Committee on Food Additives (JECFA) of the United Nations Food and Agricultural Organization (FAO) and the World Health Organization. This internationally accepted concept is applied when estimating the safe levels of food additives, pesticides, and veterinary drugs.

Types of Toxicological Studies

All kinds of chemical substances have the intrinsic property of toxicity in one way or another, depending on the quantities involved, the conditions of the system conditions. The purpose of toxicological studies is to define the biological effect(s) of the different chemical substances commonly used by humans. Further, the studies are also required to understand the intrinsic properties of chemical substances on children, animals, and the living environment. The regulatory agencies of different countries require information on the dose(s) of the test chemical substance that produce adverse biological effect in species of test animals as well as the doses that cause no significant toxicological or pharmacological effect (NOEL). The spacing of the doses also provides an assessment of the dose-response relationship.

Acute Toxicity

Acute toxicity tests are conducted on laboratory animals to generate data on the test chemical and its ability to cause systemic damage as a result of a one-time exposure to relatively large amounts through a specified route of exposure. The test substances are administered to animals in specific amounts as either one oral dose or multiple doses within 24 h. Chemical substances that are acutely toxic cause damage in a relatively short time (within minutes or hours). Exposure to a single concentrated dose of a test chemical substance induces irritation, burns, illness, and other signs and symptoms of toxicity including death (Table 1). Commonly used chemicals, such as ammonia and chlorine, cause severe inflammation, shock, collapse, or even sudden death when inhaled in high

TABLE 1

Signs and Symptoms of Toxicity (Predictable from Species of Laboratory Animal Studies)

Clinical side effect		Clinical side effect	
Drowsiness	Yes	Hypertension	Yes
Anorexia	Yes	Nausea	No
Insomnia	Yes	Depression	Yes
Dizziness	No	Fatigue	No
Increased appetite	Yes	Sedation	Yes
Constipation	Yes	Tremor	Yes
Dry mouth	Yes	Tinnitus	No
Perspiration	Yes	Nervousness	Yes
Weight gain	Yes	Dermatitis	Yes
Epigastric distress	No	Hypotension	Yes
Headache	No	Vertigo	No
Vomiting	Yes	Heartburn	No
Palpitation	Yes	Weakness	Yes
Diarrhea	Yes	Blurred vision	Yes
Skin rash	Yes	Lethargy	Yes

Source: From Dikshith T.S.S. (ed.) *Safe Use of Chemicals: A Practical Guide*, CRC Press, Boca Raton, FL, 2009. With permission.

concentrations. Corrosive materials, such as acids and bases, may cause irritation, burns, and serious tissue damage if splashed onto the skin or eyes. Exposures to chemical substances, the development of symptoms of poisoning, methods, standard procedures, and estimation of LD50 values are available in literature.

Chronic Toxicity

Chronic toxicity studies provide information on the long-term health effects of chemical substances. Adverse health effects in exposed animals and subsequent severe damage is known to occur after repeated exposures to low doses over a period of time. The slow accumulation of mercury or lead in the body, or after a long latency period from exposure to chemical carcinogens is an example. Chronic or a prolonged period of exposures to chemical substances also cause adverse effects to animals. The symptoms caused after chronic exposures usually differ from those observed in acute poisoning from the same chemical. In fact, when exposed to low concentrations of chemical substances, as is the case with chronic toxicity studies, the occupational worker and the general public become unaware of such exposures to the chemical substance.

Chronic toxicity also includes exposure to embryotoxins, teratogenic agents, and mutagenic agents. Embryotoxins are substances that cause any adverse effects on the fetus (death, malformations, retarded growth, and functional problems). Teratogenic compounds specifically cause malformation of the fetus. Examples of embryotoxic compounds include mercury and lead compounds. Mutagenic compounds can cause changes in the gene structure of the sex cells, resulting in the occurrence of a mutation in a future generation. Approximately 90% of carcinogenic compounds are also mutagens.

The regulatory agencies of different countries require toxicity profiles of candidate chemical substances. It is mandatory that all such data/information need to be (a) generated through a battery of genetic toxicity tests about the chemical substances, (b) a 90-day feeding study both in a rodent species (usually the rat) and in a non-rodent mammalian species (usually the dog), (c) a two-generation reproduction study with a teratology component in rats, and (d) other specialized testing studies to define adequately the biological effect of the test chemical substance. The specialized studies include testing for (i) neurotoxicity, (ii) immunotoxicity, and (iii) effects following *in utero* exposure. Regulatory agencies also advocate and require data on toxicity tests performed for the safety evaluation of direct food additives and color additives used in food and food products.

The Organization for Economic Co-operation and Development (OECD) Guidelines for the Testing of Chemicals are a collection of the most relevant internationally agreed testing methods used by governments, industries, and independent laboratories to assess the safety of chemical products to man and animals. These guidelines represent a basic set of important tools that are primarily for use in regulatory safety testing and subsequent chemical product notification and chemical registration by different governments around the world.

The details of several other toxicological tests, namely, repeated-dose toxicity, sub-chronic toxicity, chronic toxicity, genotoxicity, mutagenicity, teratogenicity, carcinogenicity, neurotoxicity, and ecotoxicology, the methods, the purposes, and the importance of safety evaluation studies to achieve human health have been discussed in literature. Humans are exposed to chemical substances normally through contamination, food poisoning, accidental ingestion, skin absorption, and/or the respiratory route. To generate toxicity data, species of laboratory animals are exposed to test chemicals through the three major routes. However, more often than not, chemical substances enter through more than one route, meaning through skin absorption, by accidental ingestion, and inhalation into the body of occupational workers who are negligent during work. To generate data on the toxicity profile of the test chemical substance and for further extrapolation of the data to human situations, other routes of exposure have also been used in laboratory animals. These routes include: (a) inhalation (breathing in); (b) absorption (through the skin or eyes); (c) ingestion, oral (eating, swallowing); (d) transfer across the placenta to the unborn baby; (e) intravenous (injection into the vein); (f) intramuscular (injection into the muscle); (g) subcutaneous (injection under the skin); and (h) intraperitoneal (injection inside the membrane that lines the interior wall of the abdomen). These routes are advocated by the regulatory authorities of governments for the generation of quality data about chemical substances and drugs and are subject to specific data requirements. The laboratory animals used for testing should represent the species in which the drug will be used. The most sensitive breed/class of test animal should be selected for testing. The species of test animals should be free of disease and not exposed to environmental conditions and environmental pollutants.

Additional experimental parameters should be included in the animal safety studies when they might reveal suspected adverse properties of the test chemical substance/product. This is to know the species sensitivity to the test product, or related drug product. The test animals should be properly acclimated to the study environment. Subsequent studies should be adequately designed, well controlled, and conducted by qualified investigators to generate meaningful data. Further, the safety evaluation of the test chemical substance(s) should be identical to the product intended to be marketed, meaning (i) same chemical substance, (ii) same particle size, and (iii) same formulation if any. Because the Center for Veterinary Medicine (CVM) regulates the manufacture and distribution of food additives and drugs that are given to animals, a discussion between the sponsor and the CVM prior to use of an alternative drug product is recommended.

The route(s) of administration should be the same as proposed in the protocol as well as by labeling. In order to minimize autolytic decomposition, necropsy should be performed promptly soon after death on all animals that die during the study. The necropsy should be performed by a qualified and experienced person. A complete physical examination should be performed, and baseline data should be collected by a qualified and trained worker. Data should be obtained prior to the start of the trial and at reasonable, predetermined intervals thereafter in accordance with the study protocol.

The clinical observations should be recorded twice daily, seven days a week, during the entire study period, or according to the study protocol. Appropriate clinical pathologic procedures should be conducted on all test groups. This is required on all animals in each group or, when appropriate, on a representative number (usually one half or a previously agreed number) of animals pre-selected at random from each group, at predetermined intervals, and described in the study protocol.

On completion of the studies, tissues should be collected and preserved for histologic examination. Again, all animals or a representative number (usually one half or a previously agreed number) from each group is selected for further studies. All or selected tissue(s) of test animals exposed to the highest dose treatment and from control groups should be examined for possible histological changes. Where microscopic lesions are observed, the corresponding tissues of the test group from the next lower treatment group should be examined until a NOEL is established.

Documentation of all studies should be made, indicating the representative test conditions and the manner of use of the test chemical substance. It is very important to remember that, more often than not, the toxicological effects observed in animals and humans caused by chemical substances involve various modulating factors. Over the years, the potential health risks caused by chemical substances acting in combination have attracted the importance. In fact, the interaction between chemical substances takes many forms. Such interactions between chemical substances have become very relevant to determine the potential health risks vis-a-vis human safety. Some of the known and common forms of interactions include the following four categories:

(a) An additive effect is one in which the combined effect of two chemical substances is equal to the sum of the effects of each, meaning $2 + 2 = 4$.

(b) An antagonistic effect occurs when the toxic effect of the combination of chemical substances is less than what would be predicted from the individual toxicities. The antagonistic effect or antagonism is like adding $1 + 1 = 1$.

(c) A synergistic effect occurs when the combined toxic effect of two chemical substances is much greater, or worse, than the sum of the effects of each by itself. Synergism is similar to adding $2 + 2$ and getting 5 as the result.

(d) The potentiation is the ability of one chemical substance to enhance or increase the simple summation of the two expected activities $(1 + 0 = 1)$.

Toxicological interactions among chemicals substances depend on the chemicals present, their mode of action, and their concentrations. Of the four types of interactions, the additive effect is the most plausible. It requires that the chemicals act through similar mechanisms and affect the same target tissue. For instance, the (combined) action of two or more chemicals causing irritation effects is often an added effect rather than attributable to any one candidate chemical substance.

It is also important to remember that while conducting tissue irritation studies in laboratory animals using different chemical substances, including cosmetic products or injectable drugs, the protocol should include data on the product vehicle and at least two times the use level concentration of the active ingredient. The same volume of both preparations should be administered to all animals of the experimental groups. Observation should be made about tissue inflammation, swelling, necrosis, and other reactions.

Influencing Factors

The toxicological effect(s) of any chemical substance is dependent on a number of factors. In other words, the toxicological tests using species of laboratory animals and generation of data are modulated by different important influencing factors. The data so generated offer valuable guidance for the interpretation and extrapolation of laboratory animal data to human situations at the workplace and elsewhere. In brief, these include, but are not limited to, (1) species and strains of test animals, (2) sex of the test species, (3) the age of test animal, (4) dose of the test chemical substance, (5) the nutritional and health status of the test animal, (6) the route(s) of exposure, (7) the mode of interactions of two or more chemicals to cause (8) synergistic effect and produce toxic effects that are much greater in combination or individual effect, (9) additive effect, and/or (10) antagonistic effect.

Dose-Time Relationship

The most important factor is the dose-time relationship. The amount of a substance that enters or contacts a person is called a dose. An important consideration in evaluating a dose is body weight. Dose is the quantity of a chemical substance that a surface, plant, or animal is exposed to. Time means how often or the duration of exposure of a chemical substance. In simple terms, the dose-time relationship provides information on how much of the test substance is involved and how often the exposure to the test substance occurs. This relationship gives rise to two different types of toxicity of a chemical substance, namely, acute toxicity and chronic toxicity.

Route(s) of Exposure and Toxicity Tests

The major routes through which the toxic chemicals enter the body, under normal workplace conditions, are by inhalation (respiratory route), through skin absorption (dermal route), or through ingestion (oral route). Many chemicals are known to cause the most severe health effects and rapid responses to test chemicals as soon as they enter directly into the blood circulation of animals. Several routes are used to evaluate and determine the toxicity and safety of chemical substances using species of laboratory animals in experimental toxicology studies. These routes of exposure are:

- Inhalation (breathing in)
- Absorption (through the skin or eyes)
- Ingestion, oral (eating, swallowing)
- Transfer across the placenta to the unborn baby
- Intravenous (injection into the vein)
- Intramuscular (injection into the muscle)
- Subcutaneous (injection under the skin)
- Intraperitoneal (injection inside the membrane that lines the interior wall of the abdomen)

Parameters of Toxicity

Occupational workers and the public are regularly exposed to a wide range of chemicals, depending on the nature of the work and workplace. Prolonged periods of exposures to high concentrations of chemicals cause health effects of different types. These can be listed as follows.

- Primary irritants cause local effects such as irritation to eyes, skin, nose, mucous membranes, skin rashes, and dermatitis.
- Lung irritants cause irritation or damage to pulmonary tissue.
- Asphyxiants cause interference or prevent the uptake and transformation of oxygen in the body.
- Narcotics cause mild anesthesia reactions, damage to the central nervous system, loss of consciousness, and death.
- Neurotoxic chemicals interfere with the transfer of signals between nerves of the nervous system.
- Hepatotoxic chemicals cause liver damage, jaundice, and liver enlargement.
- Nephrotoxic chemicals cause kidney damage and renal failure.

- Hematopoietic chemicals interfere with the production of red blood cells symptoms include anemia and leukemia.
- Reproductive toxins cause spontaneous abortions, birth defects, and sterility.
- The design of a toxicity study should meet the objectives intended and minimize the pain, distress, and suffering of the test animals. The study should gather as much information as possible about the substance to be tested.

Parameters and the Safety Evaluation of Chemicals and Drugs

Although the inherent toxicity of any chemical substance cannot be changed, the possibility of poisoning can be minimized by preventing and/or limiting the manner of exposure(s). For this purpose, chemical substances are subjected to different tests. These include, but are not limited to, (a) acute toxicity, (b) cumulative toxicity, (c) absorption from different routes, (d) elimination and accumulation/storage in deep compartments of the body system, (e) penetration of barriers, (f) carcinogenicity, (g) mutagenicity,(h) teratogenicity, (i) sensitization, and (j) local irritation. The risk of harm/danger of a chemical substance is equal to how poisonous it is multiplied by the amount and route of exposure: **Risk = Toxicity × Exposure**.

Good Laboratory Practice and Regulations

Chemical substances of different classes and kinds play an important role in the maintenance and improvement of quality of life. The safety and possible health hazards caused by chemical substances to animals, humans, and the living environment have to be evaluated carefully. Good laboratory practice (GLP) offers valuable avenues for the development and coordination of environmental health and safety activities. In fact, the primary objective of the OECD Principles of GLP is to ensure the generation of high quality and reliable test data related to the safety of industrial chemical substances and preparations in the framework of harmonizing testing procedures for the mutual acceptance of data (MAD).

Non-clinical laboratory studies in target animals should be conducted in accordance with GLP regulations, namely, 21 CFR Part 58. Non-clinical studies that are relevant to animal safety determinations are subject to GLP regulations. Since animal husbandry requirements often differ between laboratory animals and domestic animals, the FDA does not require that domestic animals, including poultry, are maintained under the same conditions as laboratory animals. The regulations include terms such as "when applicable" and "as required" for those situations where differences in acceptable husbandry practices exist. Each non-clinical laboratory study contained in the new animal drug application must be accompanied by a statement declaring whether or not the study was conducted in compliance with the GLP regulations. If the study was not conducted in compliance with such regulations, the statement must describe, in detail, differences between the practices used and those required in the regulations. Although clinical studies contribute data relative to the overall safety assessment of a drug product, GLP regulations do not apply to clinical studies.

Good Laboratory Practice

The Good Laboratory Practice (GLP) is concerned with the organizational processes involving all types of studies in a laboratory or test that should be planned, performed, monitored, recorded, and reported. By adhering to the principle of GLP, a laboratory ensures the proper planning of studies and the provision of adequate means to arrive at meaningful study conclusions. The studies carried out according to GLP ensures the quality and integrity of the data generated and allows their use by government regulatory authorities in hazard and risk assessment of chemicals. This prescribes GLP standards for conducting toxicology studies on agricultural chemicals. Compliance with these standards is intended to ensure the quality and integrity of toxicological data. Primarily, GLP is intended to ensure the quality and integrity of data generated in a laboratory on a product. Any violation would occur if a set protocol is not followed. The US EPA has regulations and guidelines suggesting the studies required and how they are to be performed. In fact, today, GLP standards are recognized throughout the world.

To facilitate proper use of chemical substances, the OECD is developing proposals for classification criteria and labeling of chemical substances in the area of health and environmental hazards and the UN Sub-Committee of Experts on the GHS is playing a significant role. The Task Force on Harmonization of Classification and Labeling has been established to coordinate the technical work carried out by the experts.

To generate quality data of a chemical substance and to comply with GLP, many provisions are set by the OECD. These include, in brief, the following:

- A complete description of how the protocol objectives were accomplished.
- All raw data and an interpretation or analysis of the information collected. This includes procedures used to allocate animals to treatment groups.
- A prior history on all animals used, including source, previous illnesses, and vaccinations (if known).
- Animal management practices (holding facilities, handling techniques, feeding regimen).
- A full description of the complete diet for experimental animals.
- A description of all prophylactic measures and treatments used to prevent or control infectious disease if administered during or just prior to the acclimation period. (If it is anticipated that animals will need to be treated for complicating diseases during the baseline period or during the trial, detailed plans for this treatment should be provided in the protocol.)
- The description of each treatment for each animal should include: (a) identification of the animal, (b) nature and severity of disease, (c) date of first observation and duration of disease, (d) nature of treatment and dates, and (e) outcome of treatments.
- Documentation of protocol changes or any deviation from the protocol. This is a GLP requirement.
- Complete descriptions of equipment, testing, sampling, sample handling, and assay procedures; and statistical evaluation of studies.

Toxicology Test Report

Record keeping is essential and it begins with a protocol that delineates the objectives of the study and outlines the experimental design and methods. To comply with the GLP requirements, the final test report on the toxicological effects of the test chemical substance includes specific descriptions. These may be listed as:

- Identification of the study, the test item, and the reference item.
- A descriptive title
- Identification of the test item by code or name
- Identification of the reference item by name
- Characterization of the test item, including purity, stability, and homogeneity
- Name and address of the sponsor
- Name and address of any test facilities and test sites involved
- Name and address of the study director
- Name and address of the principal investigator(s) and details, if applicable
- Name and address of scientists having contributed reports to the final report
- Experimental starting and completion dates
- A quality assurance statement listing the details (types of inspections, dates, phase(s), and results, reporting date to management and to the study director)
- Description of methods and materials used
- A summary of the results
- All information and data required by the study plan
- A presentation of the results, including calculations and determinations of statistical significance
- An evaluation and discussion of the results and, where appropriate, conclusions
- The location(s) where the study plan, samples of test and reference items, specimens, raw data, and the final report are to be stored
- Signature of principal investigators/scientists involved in the conduct of study and dated
- Signature by the study director and dated indicate acceptance of responsibility for the validity of the data
- The extent of compliance with GLP

Further, it is very important that any kind of corrections and additions to the final test report should be in the form of amendments, clearly indicating the reason for the corrections and/or additions with the signature and date of the study director.

In conclusion, students and workers in different industries and workplaces must have a basic knowledge of the proper use of chemical substances and adhere to regulations and precautions to achieve chemical safety.

References

Anonymous. 1996. *An Introduction to Ayurveda*. Ayurvedic Foundation, India.

Lad, V. 1998. The Ayurvedic Institute, India. Albuquerque, NM.

Dev, S. 1999. Ancient–Modern concordance in ayurvedic plants: Some examples. *Environ Hlth Perspect*. 107: 783.

Dikshith, T.S.S. and Diwan, P.V. 2003. *Industrial Guide to Chemical and Drug Safety*. Wiley, J. Hoboken, NJ.

Dikshith, T.S.S. 1996. *Safety Evaluation of Environmental Chemicals*. New Age International, New Delhi, India.

Organization for Economic Cooperation and Development (OECD). 1999. *Guidance Document on Humane Endpoints for Experimental Animals Used in Safety Evaluation Studies*. OECD, Paris, France.

Steven, G.G. 2004. *A Small Dose of Toxicology: The Health Effects of Common Chemicals*. CRC Press, New York.

Zhao, Y. and Nalwa, H.S. Eds. 2006. *Nanotoxicology: Interactions of Nanomaterials with Biological Systems*. Academic Press, New York.

Organization for Economic Cooperation and Development. 1981–2006. Environment Directorate, Chemicals Testing: OECD Guidelines for the Testing of Chemicals – Sections 1–5. OECD, Paris, France.

National Research Council. 1995. *Prudent Practices in the Laboratory: Handling and Disposal of Chemicals*. National Research Council, Canada.

Occupational Safety and Health Administration (OSHA). 1989. Good Laboratory Standards: Federal Insecticide, Fungicide and Rodenticide Act (FIFRA). Final Rule, Federal Register No. 54: 34052–34074, CFR Title 40 (August 17, 1989). Washington, DC.

Organization for Economic Cooperation and Development. 1992. Testing Guidelines prepared by the Organization for Economic Cooperation and Development of the United Nations, Paris, France.

Organization for Economic Cooperation and Development. 1998. OECD Series on Principles of Good Laboratory Practice and Compliance Monitoring, No. 8. OECD, Paris, France.

Paget, C.E. 1979. *Good Laboratory Practice*. MTP Press, Lancaster, UK.

Paget, C.E. and Thomson, R. 1979. *Standard Operating Procedures in Toxicology*. MTP Press, Lancaster, UK.

US Food and Drug Administration. 2006. Guidance for Industry. Summary Table of Recommended Toxicological Testing for Additives Used in Food. US FDA, Rockville, MD.

US Food and Drug Administration. 2007. Toxicological Principles for the Safety Assessment of Food Ingredients Redbook 2000. CFSAN/Office of Food Additive Safety (July 2000; updated July 2007). US FDA, Rockville, MD.

Kalayanova, F.P. and Dikshith, T.S.S. 1991. Pesticides and regulatory measures. In: *Toxicology of Pesticides in Animals*. (Ed.). Dikshith, T.S.S. Chapter 11, 219–242. CRC Press, Boca Raton, FL.

Dikshith, T.S.S. (Ed.). 2009. *Safe Use of Chemicals: A Practical Guide*. CRC Press, Boca Raton, FL.

4

List of Chemical Substances

A

- Abamectin
- Abate
- Acacia powder
- Acenaphthene
- 1-Acenaphthenol
- Acenaphthylene
- Acephate
- Acetal
- Acetaldehyde
- Acetamide
- Acetamidic acid (Methomyl)
- 2-Acetamidofluorene
- N-hydroxy-2-acetylaminofluorene
- Material Safety Data Sheet4-Acetamidophenol
- Acetanilide
- Acetic acid
- Acetic anhydride
- O-Acetoacetaniside
- Acetochlor
- Acetoin
- 2'Acetonaphthone
- Acetone
- Acetone cyanohydrin
- Acetonitrile
- Acetophenone oxime -Acetotoluidide (4'-Methylacetanilide)
- Acetylacetone
- 2' Acetylaminofluorene
- Acetyl bromide

- Acetyl chloride
- Acetylene
- Acetylene dichloride
- 2-Acetylfluorene
- 2-Acetylfuran
- 3-Acetylindole
- Acetyl iodide
- n-Acetylmorpholine
- Acetyl pyrazine
- 2-Acetylpyridine
- n-Acetylthiourea
- Acrolein
- Acrylamide
- Acrylic acid
- Acrylonitrile
- Adipamide
- Adipic acid
- Adipamide
- Alachlor
- Aldicarb
- Aldicarb products:
- Aldicarb sulfoxide
- Aldicarb sulfone
- Aldoxycarb
- Allene
- Allyl alcohol
- Allylbenzene
- Allyl bromide
- Allyl chloride
- Alpha-naphthylthiourea (α-naphthalene thiourea)
- Allyl glycidyl ether
- Allyl propyl disulfide
- Aluminum compounds
- Aluminum
- Aluminum hydroxide
- Aluminum metallic powder
- Aluminum potassium sulfate
- Aluminum sulfate
- Aluminum potasium sulfate dodecahydrate

- Alpha-Alumina
- Aminocarb
- 4-Aminophenyl ether
- 2-Aminopyridine
- 4-Aminopyridine (Avitrol)
- Amyl alcohol
- Aniline
- Antimony compounds (as Sb)
- Antimony
- Aromatic phenols
- p-Arsanilic acid
- Arsenates
- Arsenic acid
- Sodium arsenate dibasic
- Arsenic (inorganic) compounds (As+3 and As+5)
- Arsenic pentoxide
- Arsenic sulfide
- Arsenic trioxide
- Arsine
- Arsenic and arsenic compounds
    - Arsenic
    - Arsenic acid
    - Arsenous acid
    - Arsenic trioxide
    - Arsenic trihydride (Arsine)
    - Cadmium arsenide
    - Gallium arsenide
    - Lead arsenate
- Sodium arsenate
- Sodium arsenate heptahydrate
- Asbestos
- Asphalt fumes
- Atrazine
- Azinphos-ethyl
- Azinphos methyl

Abamectin (CAS No. 71751-41-2)

Molecular formula: $C_{48}H_{72}O_{14}$ (Avermectin B_{1a}); $C_{47}H_{70}O_{14}$ (Avermectin B_{1b})
Chemical name: Avermectin B1

Synonyms and trade names: Avermectin; Affirm; Avermectin B1; Agri-mek; Agrimek; Vertimec; Zephyr

Use and exposure

Avermectin is a colorless to yellowish crystalline powder. It is soluble in acetone, methanol, toluene, chloroform, and ethanol, but insoluble in water. It is stable, and incompatible with strong oxidizing agents. Abamectin is a mixture of avermectins containing about 80% avermectin B1a and 20% avermectin B1b. These two components, B1a and B1b, have very similar biological and toxicological properties. The avermectins are insecticidal/miticidal compounds derived from the soil bacterium *Streptomyces avermitilis*. Abamectin is used to control insect and mite pests of citrus, pear, and nut tree crops, and is used by homeowners to control fire ants. It acts on the nervous system of insects, causing paralyzing effects. Abamectin is a general use pesticide (GUP). It is grouped as toxicity class IV, meaning practically non-toxic, requiring no precautionary statement on its label.

Toxicity and health effects

Avermectin is an insecticide and miticide. It is very toxic and causes adverse health effects if swallowed and/or inhaled. Emulsifiable concentrate formulations of avermectin cause slight to moderate eye irritation and mild skin irritation. The symptoms of poisoning observed in laboratory animals include pupil dilation, vomiting, convulsions and/or tremors, and coma. Abamectin acts on insects by interfering with the nervous system. At very high doses, laboratory mammals develop symptoms of nervous system depression, incoordination, tremors, lethargy, excitation, and pupil dilation. Very high doses have caused death from respiratory failure in animals. Additionally, avermectin has been reported to cause reproductive effects. Abamectin blocks the nerval conduct system in insects, causing paralysis and death. Laboratory studies have indicated that abamectin may affect the nervous system in experimental animals. A 1-year study with dogs given oral doses of abamectin (0.5 and 1 mg/kg/day) caused adverse health effects, such as pupil dilation, weight loss, lethargy, tremors, and recumbency.

Abamectin and carcinogenicity

Abamectin is not carcinogenic in rats or mice and is not a human carcinogen. Mutagenicity tests in live rats and mice were found to be negative, and abamectin is non-mutagenic in the Ames test.

Exposure limits

The acceptable daily intake (ADI) of avermectin has been set as 0.0001 mg/kg/day. There are no published data on the permissible exposure limit (PEL) and the threshold limit value (TLV).

Precautions

During use of avermectin, occupational workers should use safety glasses, gloves, and protective clothing to prevent prolonged skin contact, and work in good ventilation.

References

EXTOXNET. 1996. Abamectin. Extension Toxicology Network. Pesticide Information Profiles. Oregon State University, USDA/Extension Service, Corvallis, OR.

Material Safety Data Sheeet (MSDS). 2009. The Physical and Theoretical Chemistry Laboratory. Avermetin. Oxford University, U.K.

US Environmental Protection Agency (US EPA). 1990. Avermectin B1. Pesticide Fact Sheet No. 89.2. Office of Pesticides and Toxic Substances, Washington, DC.

Abate (CAS No. 3383-96-8)

Molecular formula: $C_{16}H_{20}O_6P_2S_3$
IUPAC name: O,O'-(thiodi-4,1-phenylene) bis(O,O-dimethyl phosphorothioate)
Synonyms and trade names: Abat; Abate; Abathion; Biothion; Bithion; Ecopro; Difennthos; Lypor; Nimitox; Swebate; Temephos

Use and exposure

Abate is the trade name for Temephos, the known organophosphate larvicide. Abate is used to treat water infested with disease-carrying fleas, to control mosquito, midge, and black fly larvae. Abate quickly controls mosquito and other insect populations because it kills insect larvae before they mature, and the residual activity of abate prevents further insect populations. On decomposition, abate produces oxides of carbon, phosphorous, and sulfur.

Toxicity and health effects

Exposures to abate in workplaces cause adverse health effects (refer: Temephos for details).

Storage

To avoid contamination of water, food, or feed, abate should be stored in a secure, dry, well-ventilated room, building, or covered area. Similarly, the methods of disposal should be in accord with local, state, and federal regulations.

Precautions

During use and handling of abate, occupational workers/applicators should wear long-sleeved shirts, long pants, shoes, and socks, chemical-resistant gloves, and protective eyewear, goggles, or safety glasses. Flaggers must wear chemical-resistant headgear and protective eyewear.

References

Meister, R.T., Berg, G.L., Sine, C., Meister, S. and Poplyk, J. (eds.). 1984. *Farm Chemicals Handbook*, 70th ed. Meister, Willoughby, OH.

Tomlin, C.D.S. (ed.). 2006. *The Pesticide Manual*, 14th ed. British Crop Protection Council, Blackwell, Cambridge, U.K.

Acacia powder (CAS. No. 9000-01-5)

Synonyms and trade names: Gum arabic; Acacia gum; Acacia dealbata gum; Acacia senegal; Acacia syrup; Australian gum; Gum acacia; Gum ovaline; Indian gum; Senegal gum; Wattle gum

Use and exposure

Acacia gum is a white to yellow-white odorless powder. It is soluble in water and incompatible with alcohol, oxidizing agents, and precipitates or forms jellies on addition of solutions of ferric salts, borax, lead subacetate, alcohol, sodium silicate, gelatin, ammoniated tincture of guaiac. It is non-toxic and non-hazardous. A water-soluble gum from several species of the acacia tree, especially *Acacia senegal* and *A. Arabica*, it is used in the manufacture of adhesives and ink, and as a binding medium for marbling colors.

Gum arabic, also known as gum acacia, chaar gund, or char goond, is a natural gum made of hardened sap taken from two species of the acacia tree—*A. senegal* and *A. seyal*. Gum arabic is a natural product of the *A. senegal* tree, occurring as an exudate from the trunks and branches. It is used primarily in the food industry as a stabilizer, but has had more varied uses. It is normally collected by hand when dried, when it resembles a hard, amber-like resin normally referred to as "tears." Gum arabic is widely used in the food industry, as an emulsifier, thickener, and flavor enhancer. It is employed as a soothing agent in inflammatory conditions of the respiratory, digestive, and urinary tract, and is useful in diarrhea and dysentery. It exerts a soothing influence on all the surfaces with which it comes in contact. Gum acacia is an ingredient of all the official Trochisci, and various syrups, pastes, and pastilles or jujubes. During the time of the gum harvest, the Moors of the desert are said to live almost entirely on it, and it has been proved that 6 oz is sufficient to support an adult for 24 h. Gum acacia is a mixture of saccharides and glycoproteins, is highly nutritious, and provides the properties of a glue, and a binder suitable for human consumption. In many cases of disease, it is considered that a solution of gum arabic may, for a time,constitute the exclusive drink and food of the patient. Gum arabic reduces the surface tension of liquids, which leads to increased fizzing in carbonated drinks.

Toxicity and health effects

Exposures to gum arabica dust produce a weak allergen reaction. Prolonged periods of dust inhalation may cause allergic respiratory reaction, headache, coughing, dizziness, dyspnea, respiratory symptoms such as asthma, watery nose and eyes, cough, wheezing, nausea, vomiting, dyspnea, and urticaria. Hives, eczema, and swelling may also occur. Ingestion and inhalation of gum acacia is considered non-toxic, but sensitive individuals may develop symptoms of mild toxicity.

Acacia gum and carcinogenicity

Acacia powder has not been listed by the National Toxicology Program (NTP), the International Agency for Research on Cancer (IARC), the Occupational Safety & Health Administration (OSHA), or others agencies as a human carcinogen.

Precautions

Workers should avoid breathing dust, avoid getting it in the eyes or on the skin, and wash thoroughly after handling the material. Acacia powder should be stored in a dry place away from direct sunlight, heat, and incompatible materials.

References

Chemical Data Sheet. Gum arabic. Chemeo chemicals. Ocean Service, National Oceanic and Atmospheric Administration (NOAA), USA.

Environmental Health & Safety. 2007. Acacia powder. MSDS no. A0012. Mallinckrodt Baker, Phillipsburg, NJ.

Acenaphthene (CAS No. 83-32-9)

Molecular formula: $C_{12}H_{10}$
Synonyms and trade names: 1,2-Dihydroacenaphthylene; 1,8-Ethylenenaphthalene; Peri-Ethylenenaphthalene; Naphthyleneethylene

Use and exposure

Acenaphthene is a tricyclic aromatic hydrocarbon, crystalline solid at ambient temperature. Acenaphthene does not dissolve in water, but is soluble in many organic solvents. Acenaphthene occurs in coal tar produced during high temperature carbonization or coking of coal. It is used as a dye intermediate in the manufacture of some plastics and as an insecticide and fungicide. Acenaphthene is a component of crude oil and a product of combustion that may be produced and released into the environment during natural fires. Emissions from petroleum refining, coal tar distillation, coal combustion, and diesel-fueled engines are major contributors of acenaphthene to the environment. Acenaphthene is an environmental pollutant and has been detected in cigarette smoke, automobile exhausts, and urban air; in effluents from petrochemical, pesticide, and wood preservative industries; and in soils, groundwater, and surface waters at hazardous waste sites. This compound is one of a number of polycyclic aromatic hydrocarbons on the US EPA's priority pollutant list.

Toxicity and health effects

Exposures to acenaphthene cause poisoning and include symptoms such as irritation to the skin, eyes, mucous membranes, and upper respiratory tract. Studies on laboratory animals orally exposed to acenaphthene showed loss of body weight, peripheral blood changes (unspecified), increased aminotransferase levels in blood serum, and mild morphological damage to the liver and kidneys. In chronic exposures, acenaphthene is known to cause damage to the kidneys and liver. Acenaphthene is irritating to the skin and mucous membranes of humans and animals. Oral exposure of rats to acenaphthene for 32 days produced peripheral blood changes, mild liver and kidney damage, and pulmonary effects. However, detailed studies with acenaphthene in humans are limited.

Acenaphthene and carcinogenicity

There is no published information by the NTP, the IARC, the OSHA, or other agencies as to the carcinogenicity of acenaphthene in humans.

References

International Programme on Chemical Safety and the Commission of the European Communities (IPCS-CEC). 2005. Acenaphthene. IPCS Card no. ICSC: 1674. IPCS, Geneva, Switzerland (updated 2006).

US Environmental Protection Agency (US EPA). 1997. Acenaphthene. The Risk Assessment Information System. US EPA, Washington, DC.

1-Acenaphthenol (CAS No. 6306-07-6)

Molecular formula: $C_{12}H_{10}O$
Synonym: 1-Hydroxyacenaphthene

Use and exposure

1-Acenaphthenol is a white to cream solid in appearance. It is almost insoluble in water. 1-Acenaphthenol is a stable and combustible chemical substance. It is incompatible with strong oxidizing agents. It is used in the syntheses of organic compounds.

Toxicity and health effects

Exposures to 1-acenaphthenol cause irritant effects. It is harmful by ingestion, inhalation, or skin absorption. There are no complete data on the toxicology of 1-acenaphthenol.

Precautions

Students and workers should use safety glasses and avoid breathing dust of 1-acenaphthenol.

Reference

The Physical and Theoretical Laboratory. 2009. 1-Acenaphthenol. Oxford University, U.K.

Acenaphthylene (CAS No. 208-96-8)

Molecular formula: $C_{12}H_8$
Synonym: Cyclopenta(de)naphthalene

Use and exposure

Acenaphthylene is a polycyclic aromatic hydrocarbon (PAH) with three aromatic rings. It is used to manufacture plastics. Intermediate for dyes, soaps, pigments, pharmaceuticals, insecticide, fungicide, herbicide, and plant growth hormones. Intermediate for naphthalic acids, naphthalic anhydride (intermediate for pigments), and acenaphthylene (intermediate for resins). The largest emissions of PAH result from incomplete combustion of organic materials during industrial processes and other human activities. These include (a) processing of coal, crude oil, and natural gas, including coal; (b) coking, coal conversion, petroleum refining, and production of carbon blacks, creosote, coal-tar, and bitumen; (c) aluminium, iron, and steel production in plants and foundries; (d) heating in power plants and residences and cooking; (e) combustion of refuse; (f) motor vehicle traffic; and (g) environmental tobacco smoke.

Toxicity and health effects

Acenaphthylene is irritating to the skin and mucous membranes of rabbits. Subchronic oral doses of acenaphthylene caused adverse effects to the kidneys, liver, blood, reproductive system, and lungs of experimental animals. Prolonged period of inhalation at low doses caused pulmonary effects like bronchitis, pneumonia, and desquamation of the bronchial and alveolar epithelium in rats.

Acenaphthylene and carcinogenicity

The US EPA classified acenaphthylene as Group D, meaning not classifiable as a human carcinogen.

References

Faust, R.A. 1994. Toxicity summary for acenaphthylene. Chemical Hazard Evaluation Group. (DE-AC05-84OR21400). Oak Ridge National Laboratory, US Department of Energy, Oak Ridge, TN.

Knobloch, K., Szedzikowski, S. and Slusarcyk-Zablobona, A. 1969. Acute and subacute toxicity of acenaphthene and acenaphthylene. *Med. Pracy.* 20: 210–222 (in Polish).

US Environmental Protection Agency (US EPA). 1987. Health Effects Assessment for Acenaphthylene. Environmental Criteria and Assessment Office, Office of Health, Cincinnati, OH.

Acephate (CAS No. 30560-19-1)

Molecular formula: $C_4H_{10}NO_3PS$
Synonyms and trade names: Acetamidophos; Asataf; Pillarthene; Kitron; Aimthane; Orthene; Ortran

Use and exposure

Acephate is an organophosphate foliar spray insecticide of moderate persistence with residual systemic activity. It is a contact and systemic insecticide and very effective against a large number of crop pests, such as alfalfa looper, aphids, armyworms, bagworms, bean leafbeetle, bean leafroller, blackgrass bugs, bollworm, budworm, and cabbage looper.

Toxicity and health effects

Acephate is a colorless to white, solid organophosphate insecticide. Exposures to acephate cause poisoning to animals and humans. Acephate inhibits acetylcholine esterase (AchE), the essential nervous system enzyme, and causes characteristic organophosphate poisoning. The symptoms of toxicity include, but are not limited to, headache, nervousness, blurred vision, weakness, nausea, fatigue, stomach cramps, diarrhea, difficulty breathing, chest pain, sweating, pin-point pupils, tearing, salivation, clear nasal discharge and sputum, vomiting, muscle twitching, muscle weakness, and in severe poisonings, convulsions, respiratory depression, coma, and death. Acephate causes cholinesterase inhibition leading to overstimulation, respiratory paralysis, and death.

Acephate and carcinogenicity

The US EPA classified acephate as Group C, meaning a possible human carcinogen.

References

EXTOXNET. 1995. Acephate. Pesticide Information Profiles. Extension Toxicology Network. Cornell University, Ithaca, NY (updated 2001).

Routt, R.J. and Roberts, J.R. 1999. Organophosphate insecticides. Recognition and Management of Pesticide Poisonings. US Environmental Protection Agency, National Technical Information Service (US EPA 735-R-98-003; pp. 55–57), Washington, DC.

Tomlin, C.D.S. (ed.). 2008. *The Pesticide Manual*, 15th ed. British Crop Protection Council (BCPC). Blackwell Scientific, Hampshire, U.K.

US Environmental Protection Agency (US EPA). 1987. Acephate. Pesticide Fact Sheet. National Technical Information Service. US EPA, Washington, DC.

Acetal (CAS No. 105-57-7)

Molecular formula: $C_6H_{14}O_2$
Synonyms and trade names: 1,1-Diethoxyethane; Diethylacetal; Ethylidene diethyl ether

Use and exposure

Acetal is a clear, colorless, and extremely flammable liquid with an agreeable odor. The vapor may cause flash fire. Acetal is sensitive to light and on storage may form peroxides. In fact, it has been reported to be susceptible to autoxidation and should, therefore, be classified as peroxidizable. Acetal is incompatible with strong oxidizing agents and acids.

Toxicity and health effects

Exposures to acetal cause irritation to the eyes, skin, gastrointestinal tract, nausea, vomiting, and diarrhea. In high concentrations, acetal produces narcotic effects in workers.

Acetal and carcinogenicity

Acetal has not been classified or listed by the American Conference of Governmental Industrial Hygienists (ACGIH), the IARC, or the NTP as a human carcinogen.

Precautions

Acetal under normal storage conditions form peroxidizable compounds that accumulate and may explode when subjected to heat or shock. This material is most hazardous when peroxide levels are concentrated by distillation or evaporation.

Reference

Material Safety Data Sheet (MSDS). 2008. Acetal. Physical Chemistry at Oxford University, U.K.

Acetaldehyde (CAS No. 75-07-0)

Molecular formula: C_2H_4O
Synonyms and trade names: Acetic aldehyde; Aldehyde; Ethanol; Ethylaldehyde

Use and exposure

Acetaldehyde is a highly flammable, volatile, colorless liquid. It has a characteristic pungent and suffocating odor, and is miscible in water. Acetaldehyde is ubiquitous in the ambient environment. It is an intermediate product of higher plant respiration and formed as a product of incomplete wood combustion in fireplaces and woodstoves, burning of tobacco, vehicle exhaust fumes, coal refining, and waste processing. Exposures to acetaldehyde occur during the production of acetic acid and various other industrial chemical substances. For instance, the manufacture of drugs, dyes, explosives, disinfectants, phenolic and urea resins, rubber accelerators, and varnish.

Toxicity and health effects

Exposures to acetaldehyde liquids and vapors for a prolonged period in work areas cause irritation to the eyes, skin, upper respiratory passages, and bronchi. Continued exposure is known to damage the corneal epithelium, dermatitis, photophobia, a foreign body sensation, and persistent lacrimation or discharge of tears. Acetaldehyde causes bronchitis and a reduction in the number of pulmonary macrophages. The severity of lung damage increases with the build up of fluid in the lungs (pulmonary edema), and respiratory distress in the worker. Occupational workers exposed to high concentrations of acetaldehyde suffer coughing, pulmonary edema, necrosis, photophobia, a foreign body sensation, damage to the nasal mucosa and trachea, and persistent lacrimation.

Acetaldehyde and carcinogenicity

Laboratory animal studies indicated that exposures through inhalation of the vapors of acetaldehyde cause nasal tumors in rats and laryngeal tumors in hamsters. However, there are no adequate data available regarding acetaldehyde as a human carcinogen. The US EPA has classified acetaldehyde as Group 2B, meaning a possible human carcinogen.

References

Budavari, S. (ed.). 1996. *The Merck Index – An Encyclopedia of Chemicals, Drugs, and Biologicals*. Merck, Whitehouse Station, NJ.
Dikshith, T.S.S. 2008. *Safe Use of Chemicals: A Practical Guide*. CRC Press, Boca Raton, FL.
US Environmental Protection Agency (US EPA). 2000. Acetaldehyde Hazard Summary. Technology Transfer Network. Air Toxics web site. US EPA (updated 2007).

Acetamide (CAS No. 60-35-5)

Molecular formula: C_2H_5NO
Synonyms and trade names: Acetic acid amide; Ethanamide; Methane carboxamide; Acetimidic acid; Methanecarboxamide

Use and exposure

Acetamide occurs as hexagonal, colorless deliquescent crystals with a musty odor. It is incompatible with strong acids, strong oxidizing agents, strong bases, and triboluminescent. Acetamide is used primarily as a solvent, as a plasticizer, and a wetting and penetrating agent. Workplace exposures to acetamide are associated with the plastics and chemical industries.

Toxicity and health effects

Acetamide causes mild skin irritation in humans from acute exposure. Laboratory studies indicate that acute oral exposures of rats and mice have low to moderate toxicity. Repeated ingestion may cause liver tumors.

Acetamide and carcinogenicity

After oral exposures to acetamide, animals developed liver tumors. However, no information is available on the carcinogenic effects of acetamide in humans. The US EPA has not classified acetamide for carcinogenicity. The IARC has classified acetamide as a Group 2B, meaning a possible human carcinogen.

Precautions

During handling and/use of acetamide, workers should wear special protective equipment. After leaving work areas, workers should wash hands, face, forearms, and neck, dispose of outer clothing, and change to clean garments at the end of the day.

Storage

Acetamide should be kept stored in a tightly closed container, in a cool, dry, ventilated area. It should be protected against physical damage, away from any source of heat, ignition, or oxidizing materials.

References

Dikshith, T.S.S. 2009. *Safe Use of Chemicals: A Practical Guide*. CRC Press, Boca Raton, FL.
International Agency for Research on Cancer (IARC). 1999. *Acetamide, Summaries and Evaluations*. IARC, Lyon, France.
Material Safety Data Sheet (MSDS). 2006. Acetamide. Environmental Health & Safety. MSDS no. A0120. Mallinckrodt Baker, Phillipsburg, NJ.
US Environmental Protection Agency (US EPA). 2002. Acetamide Hazard Summary. Technology Transfer Network Air Toxics (updated 2007).

Acetamidic acid (methomyl) (CAS No. 16752-77-5)

Molecular formula: $C_6H_{12}N_2O_3S$
Synonyms and trade names: Methomyl; Acinate; Agrinate; DPX-x1179; DuPont 1179; S-methyl N-(methylcarbamoyloxy) thioacetimidate; Flytek; Kipsin; Lannate; Lannate® LV; Lanox; Metomil; Mesomile; Memilene; Methavin; Methomex; Nudrin; NuBait; Pillarmate

Use and exposure

Acetamidic acid (methomyl) is carbamate broad-spectrum insecticide. Methomyl is classified as a restricted use pesticide (RUP). It is a crystalline solid with a slight sulfurous odor, very soluble in methanol, acetone, ethanol, and isopropanol. It decomposes with heat and releases hazardous gases/vapors, such as sulfur oxides, methylisocyanate, and HCN. Acetamidic acid (methomyl) is used both as a contact insecticide and as a systemic insecticide. It is used as an acaricide to control ticks, spiders, as fly bait, for foliar treatment of vegetable, fruit and field crops, cotton, commercial ornamentals, and in and around poultry houses and dairies.

Toxicity and health effects

Acetamidic acid (methomyl) is potentially a highly poisonous material in humans. Exposures to acetamidic acid (methomyl) cause adverse health effects. It is highly toxic and causes inhibition of cholinesterase activity. The symptoms of toxicity include, but are not limited to, weakness, lack of appetite, blurred vision, pupillary constriction, corneal injury, headache, nausea, abdominal cramps, burning sensation, coughing, wheezing, laryngitis, shortness of breath, and vomiting. It may be harmful if absorbed through the skin of the mucous membranes and upper upper respiratory tract and cause chest discomfort, constriction of pupils, sweating, muscle tremors, and decreased pulse. After severe poisoning, occupational workers show symptoms of twitching, giddiness, confusion, muscle incoordination, heart irregularities, loss of reflexes, slurred speech, paralysis of the muscles of the respiratory system, and death. The target organs of methomyl toxicity include nerves, cardiovascular system, liver, and kidneys.

Acetamidic acid (methomyl) and carcinogenicity

Acetamidic acid (methomyl) has not been listed by the IARC, the NTP, the USEPA, the OSHA, or the ACGIH as a human carcinogen.

Exposure limits

There are no published data about the exposure limits of acetamidic acid by the OSHA or the ACGIH. However, countries such as Denmark, Germany, Norway, Switzerland, and the UK have set the occupational exposure level (OEL) as 2.5 mg/m^3 (TWA).

Storage

Acetamidic acid (methomyl) should be kept stored in a sealed container in a cool (below 0°C), dry, area with adequate ventilation. The chemical should be kept away from incompatible chemical substances, strong oxidizing agents, strong bases, other pesticides, and food or feed.

Precautions

Occupational workers should avoid contact of acetamidic acid (methomyl) with the skin, eyes, or clothing, and avoid exposures to vapor or mist. Workers should be careful when storing and/or disposing of acetamidic acid, avoid the dust contaminating water, food, or feed by storage or disposal. Workers should use respiratory protection: Government approved respirator. Hand Protection: Compatible chemical-resistant gloves. Eye Protection: Chemical safety goggles.chemical safety goggles.

References

Hayes, W.J. and Laws, E.R. (eds.). 1990. *Handbook of Pesticide Toxicology*, Vol. 3, *Classes of Pesticides*. Academic Press, NY.

Material Safety Data Sheets (MSDS). 1991. Methomyl. DuPont Agricultural Products. DuPont, Wilmington, DE.

Material Safety Data Sheet (MSDS). 1996. Methomyl. Extension Toxicology Network (EXTOXNET), Pesticide Information Profiles (PIP). Oregon State University, USDA/Extension Service, Corvallis, OR.

Acetone cyanohydrin (CAS No. 75-86-5)

Molecular formula: C_4H_7NO

Synonyms and trade names: 2-Hydroxy-2-methyl propionitrile; 2-Methyl-lactonitrile; 2-Cyano-2-propanol; p-Hydroxyisobutyronitrile

Use and exposure

Acetone cyanohydrin is a colorless, flammable liquid with a faint odor of bitter almond. It is an organic compound used in the production of methyl methacrylate, the monomer of the transparent plastic polymethyl methacrylate (PMMA), also known as acrylic. It is incompatible with sulfuric acid and caustics. Acetone cyanohydrin readily undergoes decomposition by water to form hydrogen cyanide and acetone.

Toxicity and health effects

Acetone cyanohydrin is extremely toxic. Exposures to acetone cyanohydrin cause adverse health effects. The symptoms of toxicity include, but are not limited to, irritation to the eyes, skin, respiratory system, dizziness, lassitude (weakness, exhaustion), headache, confusion, pulmonary edema, asphyxia convulsions, liver, kidney injury, pulmonary edema, and asphyxia. The target organs include the eyes, skin, respiratory system, central nervous system (CNS), cardiovascular system, liver, kidneys, and the gastrointestinal tract.

Reference

International Programme on Chemical Safety and the Commission of the European Communities (IPCS-CEC). 1998. Acetone cyanohydrin. ICSC Card no. 0611. IPCS, Geneva, Switzerland (updated 2005).

Acetonitrile (CAS No. 75-05-8)

Molecular formula: C_2H_3N

Synonyms and trade names: Cyanomethane; Cyanure de methyl; Ethane nitrile; Ethyl nitrile; Methyl cyanide; Methanecarbonitrile; Methylkyanid

Use and exposure

Acetonitrile is a colorless liquid with an ether-like odor and a polar solvent. It is the simplest organic nitrile and is widely used. It is a by-product of the manufacture of acrylonitrile, and acetonitrile has, in fact, replaced acrylonitrile. It is used as a starting material for the production of acetophenone, alpha-naphthalenacetic acid, thiamine, and acetamidine. It has been used as a solvent and in making pesticides, pharmaceuticals, batteries, and rubber products, and formulations for nail polish remover, despite its low but significant toxicity. Acetonitrile has been banned in cosmetic products in the European Economic Area (EEA) since early 2000 and acetone and ethyl are often preferred as safer for domestic use. Acetonitrile has a number of uses, primarily as an extraction solvent for butadiene; as a chemical intermediate in pesticide manufacturing; as a solvent for both inorganic and organic compounds; to remove tars, phenols, and coloring matter from petroleum hydrocarbons not soluble in acetonitrile; in the production of acrylic fibers; in pharmaceuticals, perfumes, nitrile rubber, and acrylonitrile-butadiene-styrene (ABS) resins; in high-performance liquid and gas chromatographic analysis; and in extraction and refining of copper.

Toxicity and health effects

Acetonitrile liquid or vapor is irritating to the skin, eyes, and respiratory tract. Acetonitrile has only a modest toxicity, but it can be metabolized in the body to hydrogen cyanide and thiocyanate. Acetonitrile causes delayed symptoms of poisoning (several hours after the exposure) that include, but are not limited to, salivation, nausea, vomiting, anxiety, confusion, hyperpnea, dyspnea, respiratory distress, disturbed pulse rate, unconsciousness, convulsions, and coma. Cases of acetonitrile poisoning in humans (or, more strictly, of cyanide poisoning after exposure to acetonitrile) are rare but not unknown, by inhalation, ingestion, and (possibly) by skin absorption. Repeated exposure to acetonitrile may cause headache, anorexia, dizziness, weakness, and macular, papular, or vesicular dermatitis.

Acetonitrile and carcinogenicity

There are no reports about the carcinogenicity of acetonitrile in animals or humans. The NTP completed a 2-year carcinogenesis inhalation study on acetonitrile and concluded that there was equivocal evidence of carcinogenicity in male rats, and no evidence in male or female mice or in female rats. The US EPA has classified it as Group D, not classifiable as a human carcinogen.

References

National Institute for Occupational Safety and Health (NIOSH). 2005. Acetonitrile. *NIOSH Pocket Guide to Chemical Hazards.* Centers for Disease Control and Prevention, Atlanta, GA (updated 2008).

National Toxicology Program (NTP). 1996. Toxicology and Carcinogenesis Studies of Acetonitrile (CAS No. 75-05-8) in F344/N Rats and B6C3F1 Mice (Inhalation Studies) (NTP TR-447). Department of Health and Human Services, Public Health Service, National Institutes of Health, Bethesda, MD.

US Environmental Protection Agency (USEPA). 2000. Acetonitrile. Hazard Summary. Technology Transfer Network. Air Toxics web site. US EPA, Research Triangle Park, NC (updated 2007).

Acetophenone oxime (CAS No. 613-91-2)

Molecular formula: C_8H_9NO

Synonyms and trade names: Phenylethan-1-one oxime; 1-Phenyl-1-ethanone oxime; Acetophenone oxime; Acetylbenzene oxime; 4-Acetophenone oxime; Benzoyl methide oxime; Hypnone oxinie; Methyl phenyl ketone oxime

Use and exposure

Acetophenone oxime is a solid but slightly soluble substance. It is incompatible with strong oxidizing agents, moisture, acids, metal, and alkali compounds. On decomposition, acetophenone oxime releases irritating and toxic fumes and gases. There are no published data about the hazardous polymerization of acetophenone oxime.

Toxicity and health effects

Exposures to acetophenone oxime cause irritation to the eyes, skin, and the respiratory and digestive tracts. The toxicological properties of acetophenone oxime are not been fully investigated.

Acetophenone oxime and carcinogenicity

The OSHA has not set any PEL for acetophenone oxime.

Storage

Acetophenone oxime should be kept stored in a cool, dry place and in a closed container when not in use.

Precautions

On exposures to acetophenone oxime, occupational workers should immediately flush the eyes with plenty of water. Workers should use proper personal protective equipment, wear appropriate gloves to prevent skin exposure, and wear appropriate protective eyeglasses or chemical safety goggles as described.

Reference

Material Safety Data Sheet (MSDS). 2007. Acetophenone oxime. MSDS ACC no. 02208. Acros Organics N.V., Fair Lawn, NJ.

p-Acetotoluidide (4'-Methylacetanilide) (CAS No. 103-89-9)

Molecular formula: $C_9H_{11}NO$

Synonyms and trade names: 4'-Methylacetanilide; p-Acetotoluidide; p-Cetamidotoluene; Acetyl-p-toluidine; 1-Acetamido-4-methylbenzene; 4-(Acetylamino)toluene; 4-Acetotoluide; Acet-p-toluide; p-Acetotoluidine; N-(4-Methylphenyl)acetamide; N-Acetyl-p-toluidide; 4-Acetotoluidide; p-Methylacetanilide; p-Acetotoluide; n-Acetyl-p-toluid

Use and exposure

p-Acetotoluidide is an off-white to brown flake, odorless, solid or crystalline powder (pure form), soluble in hot water alcohol, ether, chloroform, acetone, glycerol, and benzene. It is combustible and undergoes self-ignition (at 545°C), but is otherwise stable under most conditions. On decomposition, p-acetanilide releases carbon monoxide, oxides of nitrogen, carbon dioxide, and toxic fumes. p-Acetanilide is used as an inhibitor of peroxides and a stabilizer for cellulose ester varnishes. It is used as an intermediate for the synthesis of rubber accelerators, dyes and dye intermediate, and camphor. It is also used as a precursor in penicillin synthesis and other pharmaceuticals, including painkillers and intermediates. Phenylacetamide structure shows analgesic and antipyretic effects. But acetanilide is not used directly for this application because it causes methemoglobinemia (the presence of excessive methemoglobin that does not function reversibly as an oxygen carrier in the blood).

Toxicity and health effects

Exposures to p-acetotoluidide through ingestion or inhalation cause potential health effects. The symptoms of toxicity include, but are not limited to, irritation to the eyes and redness of the skin, wheezing, cough, shortness of breath, or burning in the mouth, throat, or chest, and respiratory tract irritation.

p-Acetotoluidide and carcinogenicity

The National Institute for Occupational Safety and Health (NIOSH), the NTP, the IARC, the ACGIH, and the OSHA have not listed p-acetotoluidide as a human carcinogen.

Storage

p-Acetanilide should be kept stored in a cool, dry place with the container closed when not in use.

Precautions

During handling and use of p-aetotoluidide, proper personal protective equipment should be worn. Avoid contact with the eyes, skin, and clothing. Avoid ingestion and inhalation. On accidental exposure at the workplace, workers should immediately flush the eyes and exposed part of the body with plenty of water. Workers should avoid using the contaminated clothing and shoes.

Reference

National Oceanic and Atmospheric Administration (NOAA). n-Acetyl-p-toluidine. NOAA, USA.

Acetylacetone (CAS No. 123-54-6)

Molecular formula: $C_5H_8O_2$
Synonyms and trade names: Hacac; Acetoacetone; 2,4-Pentadione; Pentane; 2,4-Pentadione; Acetyl 2-propanone; Diacetylmethane; Pentan-2,4-dione

Use and exposure

Acetylacetone (2,4-pentanedione) is a clear or slightly yellowish liquid with a putrid odor. It is readily soluble in water. It is with other incompatible materials, light, ignition sources, excess heat, oxidizing agents, strong reducing agents, and strong bases. On decomposition, acetylacetone releases hazardous products, such as carbon monoxide, irritating and toxic fumes and gases, and carbon dioxide. Acetylacetone is used in the production of anticorrosion agents and its peroxide compounds for the radical initiator application for polymerization. It is used as a chemical intermediate for drugs (such as sulfamethazine, nicarbazine, vitamin B6, and vitamin K), sulfonylurea herbicides, and pesticides. It is used as a solvent for cellulose acetate, as an additive in gasoline and lubricant, as a dryer of paint and varnish. It is used as an indicator for the complexometric titration of Fe(III), for the modification of guanidino groups and amino groups in proteins, and in the preparation of metal acetylacetonates for catalyst application.

Toxicity and health effects

Exposures to acetyl acetone cause eye irritation, chemical conjunctivitis, corneal damage, and skin irritation (harmful if absorbed through the skin). At low concentrations for long periods, inhalation/dermal absorption of acetyl acetone causes irritation and dermatitis, cyanosis of the extremities, pulmonary edema, and a burning sensation in the chest. Ingestion/accidental ingestion in the workplace can result in gastrointestinal irritation, nausea, vomiting, diarrhea, and CNS depression. Inhalation of high concentrations may cause CNS effects characterized by nausea, headache, dizziness or suffocation, unconsciousness, and coma. The target organ of acetyl acetone poisoning has been identified as the CNS.

Acetylacetone and carcinogenicity

Acetylacetone has not been listed by the ACGIH, the OSHA, the NTP, or the IARC as a human carcinogen.

Storage

Acetylacetone should be stored away from heat, sparks, flame, and from sources of ignition. It should be stored in a tightly sealed container, in a cool, dry, well-ventilated area, away from incompatible substances.

Precautions

Occupational workers should only use/handle acetyl acetone in a well-ventilated area, with spark-proof tools and explosion-proof equipment. Workers should not cut, weld, braze, solder, drill, grind, pressurize, or expose empty containers to heat, sparks, or flames.

Reference

Material Safety Data Sheet. 2003. Acetylacetone. Physical Chemistry at Oxford University, U.K.

2′ Acetylaminofluorene (CAS No. 53-95-2)

Molecular formula: $C_{15}H_{13}NO_2$

Synonyms and trade names: Hydroxy-N-2-fluorenylacetamide; N-Hydroxy-N-acetyl-2-aminofluorene; Fluorenyl-2-acethydroxamic acid; N-Fluoren-2-ylacetohydroxamic acid; N-Hydroxy-2-acetamidofluorene; N-Hydroxy-2-faa; N-Hydroxy-N-2-fluorenylacetamide; NOHFAA; Acetamide, N-fluoren-2-yl-N-hydroxy; N-HO-AAF

Use and exposure

2-Acetylaminofluorene is frequently used in the laboratory by biochemists and technicians as a positive control in the study of liver enzymes and the carcinogenicity and mutagenicity of aromatic amines. Exposure to 2-acetylaminofluorene may occur via inhalation or dermal contact in laboratories where it is being used in the study of carcinogenesis. Occupations at greatest risk of exposure are organic chemists, chemical stockroom workers, and biomedical researchers.

Toxicity and health effects

There is no complete information about the acute or chronic health effects of 2-acetylaminofluorene in laboratory animals or in humans. However, oral exposures of 2-acetylaminofluorene in laboratory mice caused moderate acute toxicity.

2-acetylaminofluorene and carcinogenicity

There is no available information on the carcinogenic effects of 2-acetylaminofluorene in humans. It has been reported that rats exposed through diet to 2-acetylaminofluorene developed tumors of the liver, bladder, renal pelvis, Zymbal gland (in the ear canal), colon, lung, pancreas, and testis. Tumors of the liver, bladder, and kidney have been observed in mice exposed to 2-acetylaminofluorene in their diet. Bladder and liver tumors have also been observed in other laboratory animals. The US EPA has not classified 2-acetylaminofluorene for carcinogenicity. The Department of Health and Human Services (DHHS) has classified 2-acetylaminofluorene as anticipated to be a human carcinogen, based on sufficient evidence of carcinogenicity in experimental animals.

References

Budavari, S. (ed.). 1989. *The Merck Index. An Encyclopedia of Chemicals, Drugs, and Biologicals*, 11th ed. Merck, Rahway, NJ.

National Institute for Occupational Safety and Health (NIOSH). 2005. 2-Acetylaminofluorene. *NIOSH Pocket Guide to Chemical Hazards*. NIOSH. Centers for Disease Control and Prevention, Atlanta, GA.

Sittig, M. (ed.). 1985. *Handbook of Toxic and Hazardous Chemicals and Carcinogens*, 2nd ed. Noyes Publications, Park Ridge, NJ.

US Environmental Protection Agency (US EPA). 2000. 2-Acetylaminofluorene. Hazard summary. Technology Transfer Network, Air Toxics web site (updated 2007).

Acetyl bromide (CAS No. 506-96-7)

Molecular formula: **CH₃COBr**

Synonyms and trade names: Ethanoyl bromide; Acetic Acid, Bromide

Use and exposure

Acetyl bromide is a colorless fuming liquid, with a pungent odor, combustible and turns yellow on exposure to air. It is used as an acetylating agent in the synthesis of fine chemicals, agrochemicals, and pharmaceuticals. It is also used as an intermediate for dyes. Acetylation, a case of acylation, is an organic synthesis process whereby the acetyl group is incorporated into a molecule by substitution for protecting -OH groups.

Toxicity and health effects

Exposures to acetyl bromide cause abdominal pain, sore throat, cough, burning sensation, shortness of breath, and respiratory distress. Contact with the skin causes pain, redness, blisters, dermatitis, and skin burn, severe deep burns, loss of vision, shock or collapse. The occupational worker may show delayed symptoms and lung edema. The vapor is corrosive to the eyes, the skin, and the respiratory tract. The irritation caused by acetyl bromide may lead to chemical pneumonitis and pulmonary edema, and may cause burns to the respiratory tract. The target organs include the eyes, skin, and the mucous membranes.

Acetyl bromide and carcinogenicity

At present, no information is available on the carcinogenicity of acetyl bromide. The ACGIH, the IARC, and the NTP have not listed acetyl bromide as a human carcinogen.

Storage

Acetyl bromide should be stored in a tightly sealed container, in a cool, dry, well-ventilated area, away from water and incompatible substances.

Precautions

Acetyl bromide is combustible and emits irritating or toxic fumes (or gases) in a fire. It should have no contact with naked flames or water. During use, occupational workers should use protective gloves, protective clothing, a face shield or eye protection in combination with breathing protection and should not eat, drink, or smoke. Acetyl bromide decomposes on heating and produces toxic and corrosive fumes, such as hydrogen bromide and carbonyl bromide. It reacts violently with water, methanol, or ethanol to form hydrogen bromide. Acetyl bromide attacks and damages many metals in the presence of water.

References

Anonymous. 2002. Material Safety Data Sheet. Acetyl bromide. Laboratory Reagents.

International Programme on Chemical Safety (IPCS). 1997. Acetyl bromide. ICSC Card no. 0365 (updated 2005).

International Programme on Chemical Safety and the Commission of the European Communities (IPCS CEC) 1994. Acetyl bromide. ICSC Card no. 0365 (updated 1997).

Acetyl chloride (CAS No. 75-36-5)

Molecular formula: CH_3COCl
Synonyms and trade names: Acetic chloride; Ethanoyl chloride; Acetic acid chloride

Use and exposure

Acetic chloride is a colorless to light yellow liquid with a pungent and choking odor. Acetic chloride is highly flammable, reacts violently with dimethyl sulfoxide (DMSO), water, lower alcohols, and amines to generate toxic fumes. Together with air, acetic chloride may form an explosive mixture. It is incompatible with water, alcohols, amines, strong bases, strong oxidizing agents, and most common metals. On decomposition when heated, acetic chloride produces carbon monoxide, carbon dioxide, hydrogen chloride, and phosgene.

Toxicity and health effects

Exposure to acetic chloride causes severe health effects. It is corrosive and causes severe skin burns. On contact with the eyes and skin and accidental ingestion, acetic chloride causes permanent eye damage and serious burns to the mouth and stomach. The spray mist or liquid causes tissue damage (mucous membranes of eyes, mouth, and upper respiratory tract). Inhalation of the spray mist causes severe irritation of the respiratory tract, with symptoms of a burning sensation, coughing, wheezing, laryngitis, shortness of breath, headache, nausea, and vomiting. Prolonged periods of inhalation of acetic chloride may be fatal as a result of spasm, inflammation, and edema of the larynx and bronchi, chemical pneumonitis, and pulmonary edema characterized by coughing, choking, or shortness of breath. However, there is no published information about the carcinogenicity, mutagenicity, teratogenicity, and developmental toxicity of acetic chloride in animals and humans.

Precautions

During handling and use of acetic chloride, care must be taken to keep the container dry and away from heat and sources of ignition. Avoid contact with the skin and eyes, avoid ingestion and breathing the gas/fumes/vapor/spray. During use, the workers should never add water to acetic chloride, and should wear suitable respiratory equipment.

Storage

Acetic chloride should be stored in a segregated and approved area, away from incompatibles such as oxidizing agents, alkalis and moisture. The container of acetic chloride should be kept in a cool, well-ventilated area, tightly sealed until ready for use. Users should avoid all possible sources of ignition, i.e., spark or flames.

Reference

Environmental Health & Safety. 2007. Acetyl chloride. MSDS no. A0806. Mallinckrodt Baker, Phillipsburg, NJ.

Acetylene (CAS No. 74-86-2)

Molecular formula: C_2H_2
Synonyms and trade names: Ethyne; Acetylen; Ethine; Narcylen; Vinylene; Welding gas

Use and exposure

Acetylene (100% purity) is odorless but commercial purity has a distinctive garlic-like odor. It is very soluble in alcohol and almost miscible with ethane. Acetylene is a flammable gas and kept under pressure in gas cylinders. Under certain conditions, acetylene can react with copper, silver, and mercury to form acetylides, compounds that can act as ignition sources. Brasses contain a form acetylides, compounds that can act as ignition sources. Brasses containing less than 65% copper in the alloy and certain nickel alloys are suitable for acetylene. Acetylene is not compatible with strong oxidizers such as chlorine, bromine pentafluoride, oxygen, oxygen difluoride, and nitrogen trifluoride, brass metal, calcium hypochlorite, heavy metals such as copper, silver, mercury, and their salts, bromine, chlorine, iodine, fluorine, sodium hydride, cesium hydride, ozone, perchloric acid, or potassium.

Toxicity and health effects

Prolonged periods of exposure to acetylene cause symptoms including headaches, respiratory difficulty, ringing in ears, shortness of breath, wheezing, dizziness, drowsiness, unconsciousness, nausea, vomiting, and depression of all the senses. The skin of a victim of overexposure may have a blue color. Currently, there are no known adverse health effects associated with chronic exposure to the components of this compressed gas. Lack of sufficient oxygen may cause serious injury or death. The target organs include the kidneys, CNS, liver, respiratory system, and eyes.

Acetylene and carcinogenicity

Acetylene is not a considered or suspected cancer-causing agent, mutagenic agent, or teratogen.

Precautions

Acetylene reacts vigorously and explosively when combined with oxygen and other oxidizers, including all halogens and halogen compounds, and with trifluoromethyl hypofluorite. The presence of moisture, certain acids, or alkaline materials tends to enhance the formation of copper acetylides. Do not attempt to dispose of residual or unused quantities. Return cylinder to supplier. Unserviceable cylinders should be returned to the supplier for safe and proper disposal. To extinguish fire, only water spray, dry chemical, carbon dioxide, or chemical foam should be used.

Storage

Acetylene should be kept stored in a cool, dry place in a tightly sealed container, and should only be used in a well-ventilated area. Cylinders should be separated from oxygen and other oxidizers by a minimum of 20 ft or by a barrier of non-combustible material at least 5 ft high, having a fire resistance rating of at least 30 min. Storage in excess of 2500 cu ft is prohibited in buildings with other occupancies. Cylinders should be stored upright with a valve protection cap in place and firmly secured to prevent falling or being knocked over. The cylinders should be protected from physical damage and avoid dragging, rolling, sliding, or dropping the cylinder. During transport, workers should use a suitable hand truck for cylinder movement. Care should be taken to label "No Smoking" or "Open Flames" signs in the storage or use areas. There should be no sources of ignition. All electrical equipment should be explosion-proof in the storage and use areas.

Acetylene dichloride (CAS No. 540-59-0)

Molecular formula: $C_2H_2Cl_2$
Synonyms and trade names: cis-Acetylene dichloride; trans-Acetylene dichloride; sym-Dichloroethylene; 1,2-Dichloroethylene

Use and exposure

Acetylene dichloride is a colorless liquid (usually a mixture of the cis and trans isomers) with a slightly acrid, chloroform-like odor. It is incompatible with strong oxidizers, strong alkalis, potassium hydroxide, and copper. Acetylene dichloride is highly flammable and in a fire gives off irritating or toxic fumes/gases. (Acetylene dichloride usually contains inhibitors to prevent polymerization.)

Toxicity and health effects

Exposures to acetylene dichloride cause irritation to the eyes, respiratory system, and CNS depression with symptoms of cough, sore throat, dizziness, nausea, drowsiness, weakness, unconsciousness, and vomiting. It involves the eyes, respiratory system, and the CNS as the target organs. Prolonged periods of exposure to acetylene dichloride defats the skin and may have effects on the liver.

Exposure limits

The OSHA has set the PEL as 200 ppm (790 mg/m³) TWA, and the NIOSH has set the recommended exposure limit (REL) as 200 ppm (790 mg/m³) TWA and the immediately dangerous to life and health (IDLH) limit as 1000 ppm.

Precautions

Acetylene dichloride decomposes on heating or under the influence of air. Light moisture producing toxic and corrosive fumes, including hydrogen chloride, reacts with strong oxidants. Acetylene dichloride reacts with copper bases and attacks plastic.

Reference

International Programme on Chemical Safety & the Commission of the European Communities (IPCS-CEC). 2003. 1,2-Dichloroethene. ICSC no. 0436. IPCS CEC, Geneva, Switzerland.

2-Acetylfluorene (CAS No. 781-73-7)

Molecular formula: $C_{15}H_{12}O$
Synonyms and trade names: 2-Fluorenyl methyl ketone; 1-(9H-Fluoren-2-yl)-1-ethanone

Use and exposure

2-Acetylfluorene is a light yellow crystalline powder and is stable under normal temperatures and pressures. 2-Acetylfluorene is incompatible with strong oxidizing agents. On decomposition, it releases carbon monoxide and carbon dioxide.

Toxicity and health effects

Exposures to 2-acetylfluorene cause irritations to the eyes, skin, respiratory tract, and digestive tract. The toxicological properties have not been fully investigated.

2-Acetylfluorene and carcinogenicity

2-Acetylfluorene has not been listed by the ACGIH, the IARC, the NTP, or Chemical Abstract (CA) as a human carcinogen.

Storage

2-Acetylfluorene should be stored in a cool, dry place, in a tightly sealed container.

Precautions

Occupational workers should avoid exposures to 2-acetylfluorene and breathing dust, vapor, mist, or gas. They should avoid contact with the skin and eyes, and avoid ingestion and inhalation. Workers should wear appropriate protective gloves to prevent skin exposure and chemical-splash goggles.

Reference

Material Safety Data Sheet (MSDS). 2009. 2-Acetylfluorene. The Physical and Theoretical Chemistry Laboratory. Oxford University, U.K.

2-Acetylfuran (CAS No. 1192-62-7)

Molecular formula: $C_6H_6O_2$
IUPAC name: 1-Furan-2-ylethanone
Synonyms and trade names: 2-Furyl methyl ketone; Acetyl furan; Methyl 2-furyl ketone; 1-(2-Furanyl) ethanone

Use and exposure

2-Acetylfuran is a gold-colored solid with an odor of sweet balsam almond cocoa, caramel coffee. It is incompatible with strong oxidizing agents, strong reducing agents, and strong bases. It is combustible. On decomposition, 2-acetylfuran releases carbon monoxide and carbon dioxide. It is insoluble in water, but soluble in alcohol, dipropylene glycol, and ethyl ether.

Toxicity and health effects

Exposures to 2-acetylfuran by ingestion, inhalation, and skin absorption cause harmful effects. It causes irritation to the skin, the eyes, coughing, respiratory tract irritation, and respiratory distress. Accidental ingestion causes irritation of the digestive tract. There are no data about chronic exposure and adverse effects.

2-Acetylfuran and carcinogenicity

2-Acetylfuran has not been listed by the ACGIH, the IARC, or the NTP for its carcinogenicity.

Exposure limits

Exposure limits of 2-acetylfuran have been set as 0.1000% in the fragrance concentrate and the usage levels in the flavor concentrate as 100,0000 ppm.

Storage

2-Acetylfuran should be kept away from sources of ignition, heat, sparks, flame, and light. It should be kept stored in a tightly sealed container in a dry area/refrigerated (below 4°C/39°F).

Precautions

During handling and use of 2-acetylfuran, occupational workers should use appropriate and proper personal protective equipment and wear a self-contained breathing apparatus. Workers should avoid generating dusty conditions, minimize generation and accumulation of dust, remove all sources of ignition, and use a spark-proof tool. Workers should use 2-acetylfuran only in a chemical fume hood.

Reference

Material Safety Data Sheets (MSDS). Allylben 2009. Department of physical chemistry, oxford university, Oxford, U.K.

Acrylonitrile (CAS No. 107-13-1)

Molecular formula: $H_2C=CHCN$
Synonyms and trade names: AN; Cyanoethylene; 2-Propenenitrile; Propenenitrile, Vinyl cyanide

Use and exposure

Acrylonitrile is a colorless, flammable liquid. Its vapors may explode when exposed to an open flame. Acrylonitrile does not occur naturally. It is produced in very large amounts by several chemical industries in the United States and its requirement and demand has increased in recent years. The largest users of acrylonitrile are chemical industries that make acrylic and modacrylic fibers, high impact acrylonitrile-butadiene-styrene (ABS) plastics. Acrylonitrile is also used in business machines, luggage, and construction material, in the manufacturing of styrene-acrylonitrile (SAN) plastics for automotive and household goods, and in packaging material. Adiponitrile is used to make nylon, dyes, drugs, and pesticides.

Adipamide (CAS No. 628-94-4)

Molecular formula: $C_6H_{12}N_2O_2$
Synonyms and trade names: Adipic acid amide; Adipic acid diamide; Adipic diamide; 1,4-Butanedicarboxamide; Hexanediamide; NCI-C02095

Use and exposure

Adipamide is white powder and chunks. It is slightly soluble in water and incompatible with strong oxidizing agents. On combustion and decomposition, adipamide produces toxic fumes of carbon monoxide, carbon dioxide, and nitrogen oxides.

Toxicity and health effects

Exposures to adipamide by inhalation, ingestion, or skin absorption cause adverse health effects. The symptoms include irritation of the eyes, skin, mucous membranes, and upper respiratory tract. There is no complete information about the toxicological properties of the chemical.

Precautions

Workers should avoid repeated exposures to adipamide. During use, workers should wear suitable protective clothing, self-contained breathing apparatus, chemical safety goggles, rubber boots, and heavy rubber gloves.

Reference

Material Safety Data Sheet (MSDS). 2007. Adipamide. The Physical and Theoretical Chemistry Laboratory, Oxford University, U.K. (updated 2009).

Adipic acid (CAS No. 124-04-9)

Molecular formula: $C_6H_{10}O_4$
Synonyms and trade names: Hexanedioic acid; 1,4-Butane dicarboxylic acid

Use and exposure

Adipic acid is a white crystalline solid/granule; its odor has not been characterized. It is stable and incompatible with ammonia and strong oxidizing agents. It may form combustible dust concentrations in air. The likely routes of exposure to workers are by skin contact and inhalation at workplaces. It is used in the manufacture of nylon, plasticizers, urethanes, adhesives, and food additives.

Toxicity and health effects

Exposures to adipic acid cause pain, redness of the skin and eyes, tearing or lacrimation. Adipic acid has been reported as a non-toxic chemical. Excessive concentrations of adipic acid dust are known to cause moderate eye irritation, irritation to the skin, and dermatitis.

It may be harmful if swallowed or inhaled. It causes respiratory tract irritation with symptoms of coughing, sneezing, and blood-tinged mucous.

Exposure limits

The exposure limits (PEL) for adipic acid has not been set by the OSHA, while the ACGIH has set the TLV as 5 mg/m^3 (TWA).

Storage

Adipic acid should be kept stored in a tightly closed container, in a cool, dry, ventilated area and protected against physical damage. In storage, adipic acid should be isolated from incompatible substances and dust formation and ignition sources should be avoided.

Precautions

Occupational workers should avoid contact of the adipic acid with the eyes, avoid breathing dust, and keep the container closed. Workers should use adipic acid only with adequate ventilation. Workers should wash thoroughly after handling adipic acid and keep away from heat, sparks, and flame. Also, workers should use rubber gloves and laboratory coats, aprons, or coveralls, and avoid creating a dust cloud when handling, transferring, and cleaning up.

Reference

Material Safety Data Sheet (MSDS). 2008. Adipic acid. MSDS no. A1716. Environmental Health & Safety. Mallinckrodt Baker, Phillipsburg, NJ.

Adipamide (CAS No. 628-94-4)

Molecular formula: $C_6H_{12}N_2O_2$
Synonyms and trade names: Adipic acid amide; Adipic acid diamide; Adipic diamide; 1,4-Butanedicarboxamide; Hexanediamide; NCI-C02095

Use and exposure

Adipamide is powder in appearance and slightly soluble in water. It is incompatible with strong oxidizing agents. On combustion or decomposition, adipamide releases hazardous products, toxic fumes of carbon monoxide, carbon dioxide, and nitrogen oxides.

Toxicity and health effects

Exposures to adipamide cause health disorders such as irritation to the skin, eyes, and respiratory system. No specific adverse health effects have been reported from human exposure to polyhexamethylene adipamide, except for mechanical irritation of the skin and eyes caused by particles. Significant skin permeation and systemic toxicity after contact appears unlikely. The compound is not likely to be hazardous by skin contact, but cleansing the skin after use is advisable. If molten polymer gets on the skin, cool rapidly with cold water. Workers should not attempt to peel polymer from skin, but consult a medical unit for treatment of thermal burn.

Storage

Adipamide should be kept stored in a cool, dry place, with containers tightly closed to prevent moisture absorption and contamination.

Precautions

During use and handling of adipamide, workers should wear self-contained breathing apparatus, rubber boots, heavy rubber gloves, and chemical safety goggles to avoid splashing of material. A full-face mask respirator provides protection from eye irritation. Workers should avoid exposure to high concentrations of dust. No specific intervention is indicated as the compound is not likely to be hazardous by inhalation. Consult a physician if necessary. If exposed to fumes from overheating or combustion, move to fresh air. Consult a physician if symptoms persist.

Alachlor (CAS No. 15972-60-8)

Molecular formula: $C_{14}H_{20}Cl\,NO_2$
Chemical name: 2-Chloro-2′,6′-diethyl-N-(methoxymethyl) acetanilide

Use and exposure

Alachlor is a colorless to yellow crystal chemical substance. It is soluble in most organic solvents, but sparingly in water. Alachlor is an RUP, therefore it should be purchased and used only by certified, trained workers and plant protection applicators. The US EPA categorizes it as toxicity class III, meaning slightly toxic. However, alachlor products bear the signal word DANGER on their labels because of their potential to cause cancer in laboratory animals. Alachlor is an aniline herbicide used to control annual grasses and broadleaf weeds in field corn, soybeans, and peanuts. It is a selective systemic herbicide, absorbed by germinating shoots and roots. It works by interfering with a plant's ability to produce protein and by interfering with root elongation. Alachlor has extensive use as a herbicide in the United States. It is available as granules or emulsifiable concentrate.

Toxicity and health effects

Alachlor is a slightly toxic herbicide. It causes slight to moderate degrees of skin irritation. While a 90-day study on laboratory rats and dogs given diets containing low to moderate amounts of alachlor (1–100 mg/kg/day) showed no adverse effects, a 1-year study indicated that at a dose above 1 mg/kg/day, alachlor causes damage in the liver, spleen, and kidney.

Alachlor and carcinogenicity

Laboratory rats given high doses of alachlor developed stomach, thyroid, and nasal turbinate tumors. An 18-month mouse study with doses from 26 to 260 mg/kg/day showed an increase of lung tumors at the highest dose for females but not males. The oncogenic potential of alachlor is uncertain and alachlor as a human carcinogen requires more confirmatory studies.

Exposure limits

The ADI of alachlor has been set as 0.0025 mg/kg/day.

References

EXTOXNET. Material Safety Data Sheet (MSDS). 1996. Alachlor. Oregon State University, USDA/ Extension Service, Corvallis, OR.
Material Safety Data Sheet (MSDS). 1991. Alachlor Technical (94%). Monsanto, St. Louis, MO.
US Environmental Protection Agency (US EPA). 1987. Alachlor. Pesticide Fact Sheet no. 97.1. Office of Pesticides and Toxic Substances, Washington, DC.

Aldicarb (CAS No. 116-06-3)

Molecular formula: $C_7H_{14}N_2O_2S$
IUPAC name: 2-Methyl-2-(methylthio)propionaldehydeO-methyl-carbamoyloxime
Synonyms and trade names: Aldicarbe; Carbamic acid; Carbanolate; Propanal; Temik; Sanacarb
Toxicity class: USEPA: I; WHO: Ia

Use and exposure

Aldicarb is a white crystalline solid. It is not compatible with alkaline materials and is non-corrosive to metals and plastics. It is sparingly soluble in water, but soluble in acetone, xylene, ethyl ether, toluene, and other organic solvents. Aldicarb has been classified as an RUP and hence should only be used by trained and certified workers. Aldicarb is used to control mites, nematodes, and aphids and is applied directly to the soil. It is used widely on cotton, peanut, and soybean crops. In the mid-1980s, there were highly publicized incidents in which misapplication of aldicarb contaminated cucumbers and watermelons and led to adverse effects in people. Aldicarb is metabolized in the liver of mammals first into aldicarb sulfoxide and later into aldicarb sulfone.

Toxicity and health effects

Aldicarb is the most potent of the commercially available carbamate pesticides and is an unusual source of acute human poisoning. Aldicarb has severe systemic toxicity to animals and humans. Aldicarb is a cholinesterase inhibitor that prevents the breakdown of acetylcholine in the synapse. In the case of severe poisoning, the victim dies of respiratory failure. The primary route of human exposure to aldicarb is consumption of contaminated food and water from contaminated wells. Occupational exposure to high levels of aldicarb is due to product handling, and most cases of aldicarb poisoning occur from loading and application of the pesticide. Aldicarb is extremely toxic both through the oral and dermal route. Absorption from the gut is rapid and almost complete. When administered in oil or other organic solvents, aldicarb is absorbed rapidly through the skin. Its skin toxicity is roughly 1000 times that of other carbamates. In humans, the onset of symptoms is rapid (15 min to 3 h). Symptoms disappear in 4–12 h.

The acute oral LD_{50} of aldicarb in rats, mice, guinea pigs, and rabbits ranges from 0.5 to 1.5 mg/kg when administered in a liquid or oil form. The toxicities of the dry granules are distinctly lower (LD_{50} 7.0 mg/kg), though still highly toxic. Aldicarb is a cholinesterase inhibitor and therefore can result in a variety of symptoms including weakness, blurred vision, headache, nausea, tearing, sweating, and tremors. At very high concentrations, aldicarb causes paralysis, respiratory system damage, eventually leading to the death of the exposed worker.

There is very little evidence of chronic effects from aldicarb exposure. Rats and dogs fed low doses of aldicarb for 2 years showed no significant adverse effects. One epidemiological study suggested a possible link between low-level exposure and immunological abnormalities. The result of this study, however, has been widely disputed. Aldicarb administered to pregnant rats at very low levels (0.001–0.1 mg/kg/day) depressed AChE activity more in the fetuses than in the mother. The aldicarb was also retained in the mother's body for longer periods than in non-pregnant rats. A three-generation study at doses of 0.05 and 0.10 mg/kg/day produced no significant toxic effects, and in another study, a dose of 0.70 mg/kg/day produced no adverse effects. Thus, reproductive effects in humans are unlikely at expected exposure levels.

Aldicarb and carcinogenicity

Reports have indicated that aldicarb is not carcinogenic to animals and humans. Similarly, there are no reports on the teratogenic and mutagenic effects of aldicarb to animals and humans.

Exposure limits

The ADI for aldicarb has been set as 0.003 mg/kg/day. In fact, no Permissible Exposure Limit (PEL) or Threshold Limit Value (TLV) has been established for aldicarb.

Storage

Aldicarb should be kept stored in a dry, cool area, out of reach of children and animals, and away from food, feedstuffs, fertilizers, and seed.

Handling and precautions

Aldicarb is a fast-acting cholinesterase inhibitor, and causes rapid accumulation of acetylcholine at the synaptic cleft. Workers should not get aldicarb on the skin or in the eyes, and should not breathe dusts. Skin contact should be prevented through use of suitable protective clothing, gloves, and footwear. Eye contact should be prevented through use of chemical safety glasses with side shields or splash-proof goggles. An emergency eye wash must be readily accessible to the work area. Face contact should be prevented through the use of a face shield. Workers should note that the dry powders can build static electricity charges when subjected to the friction of conveying, mixing, or sliding. Workers should take adequate precautions in electrical grounding or inert atmospheres when the material is used in the presence of flammable materials, to prevent ignition. Aldicarb should, therefore, be handled by trained and qualified workers in accordance with the appropriate regulatory standards and/or industrial recommendations.

Reference

EXTOXNET. 1996. Aldicarb. Pesticide Information Profiles. ICSC no. 0094. Extension Toxicology Network. Oregon State University International Chemical Safety (updated 2004).

International Programme on Chemical Safety (IPCS). 1991. Aldicarb. Health and Safety Guide No. 64. IPCS, Geneva, Switzerland.

Montgomery, J.H. 1997. *Agrochemicals Desk Reference: Desk Reference.* CRC Press, Boca Raton, FL

World Health Organization (WHO). 1991. Aldicarb, Environmental Health Criteria (EHC) No. 121: International Programme on Chemical Safety (IPCS). WHO, Geneva, Switzerland.

Aldicarb Products

Aldicarb sulfoxide (CAS No. 1646-87-3)

Aldicarb sulfoxide is a breakdown product of aldicarb.

Aldicarb sulfone (CAS No. 1646-88-4)

Synonym: Aldoxycarb.

Aldoxycarb (CAS No. 1646-88-4)

Molecular formula: $C_7H_{14}N_2O_4S$
IUPAC name: 2-Methyl-2-(methylsulfonyl)propanal O-[(methylamino)carbonyl] oxime
Synonyms and trade name: Aldicarb sulfone; Standak; Sulfocarb

Use and exposure

Aldoxycarb is a carbamate pesticide used mainly as an insecticide and acaricide. It is a white crystalline powder with a slightly sulfurous odor. Aldoxycarb is extremely toxic to wildlife. Aldoxycarb (Aldicarb sulfone) is a breakdown product of aldicarb. Aldoxycarb is a degradate/metabolite of aldicarb produced by the oxidation of the thio-moiety.

Toxicity and health effects

Exposures to high levels of aldoxycarb cause adverse health effects and damages the nervous system of humans and laboratory animals. Chemicals that cause adverse health effects in humans and laboratory animals after high levels of exposure may pose a risk of adverse health effects in humans exposed to lower levels over long periods of time. Ingestion and other exposures to the chemical cause various health disorders. The type and severity of symptoms varies depending on the amount of chemical involved and the nature of the exposure. The symptoms include, but are not limited to, eye irritation, skin irritation, dizziness, sweating, malaise, CNS depression, abdominal pain, diarrhea, nausea, vomiting, salivation, headache, ataxia, muscle twitching, muscle weakness, runny nose, tearing eyes, pinpoint pupils or dark vision, wheezing, increased or decreased heart rate, seizures, convulsions, and death. It has not caused any delayed neurotoxicity in animals. The onset of these symptoms is rapid and the severity of symptoms depends on the dose. The immediate cause of death is usually respiratory failure.

Aldoxycarb and carcinogenicity

Aldoxycarb has not been shown to cause oncogenic, mutagenic, teratogenic, delayed neurotoxicity, or reproduction effects.

Allene (CAS No. 463-49-0)

Molecular formula: C_3H_4
Synonyms and trade names: Sym-allylene; Dimethylenemethane; Propadiene; 1,2-Propadiene; Propadiene

Use and exposure

Allene is colorless with a sweet odor. It is a flammable gas and may cause flash fire at room temperature. Allene is the simplest member of the 1,2-diene class of compounds of hydrocarbons of the aliphatic chemical family. It is a benzene and petroleum ether. Allenes are not as stable as conjugated dienes or isolated dienes. Allene is slightly more strained than the isomeric methylactylene. Allene is chiefly used in organic synthesis. It is incompatible with metals and oxidizing materials, therefore it is important to avoid copper and silver and their alloys when used under high pressure or temperature.

Toxicity and health effects

Allene is a simple asphyxiant with slight narcotic properties. Exposures to allene cause irritation to the skin and eyes, nausea, vomiting, headache, symptoms of drunkenness, suffocation, convulsions, and coma.

Exposure limits

No OELs for allene have been established.

Precautions

Allene is a flammable gas and the container must be kept stored away from open flame and smoking areas. Allene should be stored in well-ventilated areas with valve protection caps in place. The gas cylinders should not be dragged or rolled, and the compressed gas cylinders should not be refilled without express written permission. Allene decomposes violently at increased temperature/heat and/or pressure and the containers should be protected from rupture and explosion.

Aluminium (III) acetylacetonate (CAS No. 13963-57-0)

Molecular formula: $C_{15}H_{21}AlO_6$
Synonyms and trade names: Aluminium (III) acetylacetonate; Tris(acetylacetone)aluminium; Tris(2,4-pentadionato) aluminium; Tris(2,4-pentanedione)aluminium; Aluminium complex of acetyl acetone; Aluminium tris(acetylacetonate); Aluminium 2,4-pentanedioate; Aluminum acetylacetonate

Use and exposure

Aluminium (III) acetylacetonate is a white to yellow powder or crystals in appearance. It is stable and incompatible with strong oxidizing agents.

Toxicity and health effects

Exposures to aluminium (III) acetylacetonate cause adverse health effects. Exposures by inhalation, ingestion, and through skin contact are harmful. It causes irritation effects to the eyes, skin, and respiratory tract. Aluminium (III) acetylacetonate may release 2,4-pentanedione, which is a suspected teratogen and causes neurological health effects and harm to the unborn child. However, toxicology is not fully investigated.

Reference

Material Safety Data Sheet (MSDS). 2005. Aluminium (III) acetylacetonate. The Physical and Theoretical Chemistry Laboratory, Oxford University, U.K.

Allyl alcohol (CAS No. 107-18-6)

Molecular formula: C_3H_6O
Synonyms and trade names: Vinyl carbinol; Propenyl alcohol; 2-Propeno-1; Propenol-3

Use and exposure

Allyl alcohol is a colorless liquid with a mustard-like odor. It is used in making drugs, organic chemicals, pesticides, in the manufacture of allyl esters, and as monomers and prepolymers for the manufacture of resins and plastics. It has a large use in the preparation of pharmaceutical products, in organic synthesis, and as a fungicide and herbicide. Occupational workers engaged in industries such as the manufacture of drugs, pesticides, allyl esters, organic chemicals, resins, war gas, and plasticizers, are often exposed to this alcohol.

Toxicity and health effects

Exposure to high concentrations of allyl alcohol vapors causes irritation to the eyes, skin, and upper respiratory tract. Laboratory studies with animals have shown the symptoms of local muscle spasms, pulmonary edema, and tissue damage to the liver and kidney, convulsions, and death.

Allyl alcohol and carcinogenicity

Allyl alcohol has not undergone complete evaluation of carcinogenicity. There are no data on the human carcinogenic potential of allyl alcohol.

Precautions

Occupational workers should be careful during handling and use of allyl alcohol and wear protective clothing.

References

Carpanini, F.M.B., Gaunt, I.F., Hardy, J., Gangalli, S.D., Butterworth, K.R. and Lloyd, H.G. 1978. Short-term toxicity of allyl alcohol in rats. *Toxicology*, 9: 29–45.

US Environmental Protection Agency (US EPA). 1985. Allyl alcohol. Health and Environmental Effects Profile for Allyl Alcohol. Environmental Assessment. Environmental Criteria and Assessment Office, Cincinnati, OH.

Allylbenzene (CAS No. 300-57-2)

Molecular formula: C_9H_{10}
Synonyms and trade names: 2-Propen-1-yl-benzene; 2-Propenylbenzene; 1-Benzylethene; 1-Phenyl-2-propene; 3-Phenyl-1-propene; 3-Phenylpropene

Use and exposure

Allylbenzene is a colorless, flammable chemical, incompatible with strong oxidizing agents. On decomposition, allylbenzene releases carbon monoxide and carbon dioxide.

Toxicity and health effects

Exposures to allylbenzene cause irritation to the skin, eyes, and respiratory system. It causes harmful effects if inhaled or swallowed and causes respiratory tract irritation. The toxicological properties of this substance have not been fully investigated.

Allylbenzene and carcinogenicity

Allylbenzene has not been listed by the ACGIH, the IARC, the National Institute of Occupational Safety and Health (NIOSH), the NTP, or the OSHA as a human carcinogen.

Precaution

Workers while handling allylbenzene should not breathe dust or vapor, or ingest the chemical substance. Allylbenzene should only be used in a chemical fume hood and sources of ignition should be avoided. Workers should wear appropriate protective eyeglasses or chemical safety goggles as described by OSHA protection regulations.

Storage

Allylbenzene should be kept stored in a cool, dry, well-ventilated area with adequate ventilation during use.

Reference

Safety Officer in Physical Chemistry at Oxford University. Material Safety Data Sheet (MSDS). 2008. Allylbenzene. Department of physical chemistry, oxford university, oxford, U.K.

Allyl bromide (CAS No. 106-95-6)

Molecular formula: C_3H_5Br
Synonyms and trade names: 3-Bromopropene; 3-Bromopropylene; 3-Bromo-1-propene; Bromoallylene; 2-Propenyl bromide; UN 1099

Use and exposure

Allyl bromide is a clear to light yellow liquid. As an alkylating agent, allyl bromide is used extensively in the synthesis of polymers, pharmaceuticals, allyls, and other organic compounds. Allyl bromide is a clear liquid with an intense, acrid, persistent smell and is flammable. It is insoluble in water, but soluble in alcohol, aether, acetone, carbon tetrachloride, and chloroform. In fact, allyl bromide is used in the synthesis of other allyl compounds, to synthesize dyestuff, spice, and as a curative in the medicine industry. Allyl bromide has a very high mobility in soil. It is also used as a soil fumigant and as a contact poison. Allyl bromide induces unscheduled DNA synthesis in HeLa cells.

Toxicity and health effects

Exposures to allyl bromide cause severe eye and skin burns, irritation to the eyes, skin, and respiratory system. It is harmful when absorbed through the skin or inhaled in the workplace. Laboratory rats exposed for a prolonged period of time developed symptoms of poisoning, such as excessive salivation in a small number of animals, and severe gastric irritation. Vapors of allyl bromide may cause dizziness or suffocation, headache, coughing, and distressed breathing.

Storage

Allyl bromide should be stored separate from oxidizing materials and alkalis in a cool, dry, well-ventilated location in tightly closed containers.

Precautions

Workers should wear positive pressure self-contained breathing apparatus (SCBA), goggles and a face shield, protective clothing for high concentrations of vapor, chemical protective clothing that is specifically recommended by the manufacturer to avoid poisoning. Workers should be careful as allyl bromide reacts with oxidizing materials and alkalis.

References

Budavari, S. et al. 1989. *The Merck Index*, 11th ed. Merck, Rahway, NJ.
Gosselin, R.E., Hodge, H.C., Smith, R.P. and Gleason, M.N. 1976. *Clinical Toxicology of Commercial Products*, 4th ed. Williams and Wilkins, Baltimore, MD.

Allyl chloride (CAS No. 107-05-1)

Molecular formula: C_3H_5Cl
Synonyms and trade names: 1-Propene, 3-chloro-; AC; Chlorallylene; Chloroallylene; 3-Chloro-1-propene; 3-Chloropropylene

Use and exposure

Allyl chloride is a colorless liquid, insoluble in water but soluble in common organic solvents. Allyl chloride is prepared by the reaction of propylene with chlorine. It is a common alkylating agent relevant to the manufacture of pharmaceuticals and pesticides. It is also a component in some thermo-setting resins. Allyl chloride has been produced commercially since 1945 and is used almost exclusively as a chemical intermediate, principally in the production of epichlorohydrin or as a raw material for epichlorohydrin. It is also used as a chemical intermediate in the preparation of glycerin, glycerol chlorohydrins, glycidyl ethers, allylamines, and allyl ethers of trimethylpropane, sodium allyl sulfonate, a series of allyl amines and quaternary ammonium salts, allyl ethers, and a variety of alcohols, phenols, and polyols. It is also used in pharmaceuticals as a raw material for the production of allyl isothiocyanate (synthetic mustard oil), allyl substituted barbiturates (sedatives), and cyclopropane (anesthetic); in the manufacture of specialty resins for water treatment and to produce babiturate and hypnotic agents such as aprobarbital, butalbital, methohexital sodium, secobarbital, talbutal, and thiamyl sodium.

Toxicity and health effects

Allyl chloride is toxic and flammable. Exposures to allyl chloride cause a cough, sore throat, headache, dizziness, weakness, respiratory distress, abdominal pain, burning sensation, vomiting, and loss of consciousness. After acute inhalation exposures to high levels of allyl chloride, workers developed irritation of the eyes and respiratory passages, loss of consciousness, and fatal injury. Prolonged and intense exposure produced conjunctivitis, reddening of eyelids, and corneal burn, damage to the CNS, causing motor and sensory neurotoxic damage, and the heart and respiratory system, causing the onset of pulmonary edema in humans. Laboratory rabbits exposed to allyl chloride through inhalation developed degenerative changes that included dilation of sinusoids and vacuolar degeneration in the liver, congestion or cloudy swelling and fatty degeneration of the epithelium of the renal convoluted tubules, and thickening of the alveolar septa in the lungs. The exposed cat exhibited only muscle weakness and unsteady gait toward the end of the exposure period.

Allyl chloride and carcinogenicity

Allyl chloride has been tested for its carcinogenicity by intragastric intubation in mice and rats, by skin application in mice, both by repeated application and in a two-stage assay, and by intraperitoneal injection in mice. Following oral administration to mice, a non-significant increase in the incidence of squamous-cell papillomas and carcinomas of the fore-stomach was observed; the experiment in rats was inadequate for evaluation. No case report or epidemiological study of the carcinogenicity of allyl chloride to humans was available to the Working Group. There is inadequate evidence for the carcinogenicity of allyl chloride in experimental animals. In the absence of epidemiological data, no evaluation could be made of the carcinogenicity of allyl chloride to humans. The US EPA has observed allyl chloride as a cancer-causing agent and classified it as Group C, meaning a possible human carcinogen.

References

He, F. and Zhang, S.L. 1985. Effects of allyl chloride on occupationally exposed subjects. *Scand. J. Work Environ. Health* 11 (Suppl 4): 43–45.

Lu, B., Shuwei, D., Airu, Y., YinLin, X., Taibao, G. and Tao, C. 1982. Studies on the toxicity of allyl chloride. *Ecotoxicol. Environ. Saf.* 6: 19–27.

US Environmental Protection Agency (US EPA). 1986. Health and Environmental Effects Profile for Allyl Chloride. Office of Health and Environmental Assessment. ECAO-Cin-P186. Environmental Criteria and Assessment Office, Cincinnati, OH.

Alpha-naphthylthiourea (α-naphthalene thiourea) (CAS No. 86-88-4)

Molecular formula: $C_{11}H_{10}N_2S$
Synonyms and trade names: 1-(1-Naphthyl)-2-thiourea-; ANTU

Use and exposure

Alpha-naphthylthiourea (α-naphthalene thiourea) is a pure white or beige-brown solid/blue-gray powder. It is hard to dissolve in water, acid, and general organic solvents, but dissolves in boiling ethanol and alkaline solution. On decomposition, ANTU releases carbon monoxide, toxic and irritating fumes and gases, and carbon dioxide. It is a rodenticide and a poison bait to lure rodents.

Toxicity and health effects

α-Naphthalene thiourea, a rodenticide, is very toxic and is fatal if swallowed. Exposures to ANTU cause poisoning with symptoms that include, but are not limited to, headache, weakness, dizziness, shortness of breath, cyanosis, blood abnormalities, methemoglobinemia, irritation of the digestive tract, liver and kidney damage, cardiac and CNS disturbances, convulsions, tachycardia, dyspnea, vertigo, tinnitus, weakness, disorientation, lethargy, drowsiness, and finally coma and death. The target organs include the blood, kidneys, CNS, liver, lungs, cardiovascular system, and blood-forming organs.

α-Naphthalene thiourea and carcinogenicity

1-(1-Naphthyl)-2-thiourea has not listed by the ACGIH, the IARC, or the NTP as a human carcinogen. Information about the carcinogenic effect of α-naphthalene thiourea is limited.

Precautions

Workers should use/handle α-naphthalene thiourea with adequate ventilation. During use, dust generation and accumulation should be minimum, and avoid contact with the eyes, skin, and clothing.

Storage

α-Naphthalene thiourea should be kept stored in a tightly closed container in a locked poison room, in a cool, dry, well-ventilated area away from incompatible substances.

Allyl glycidyl ether (CAS No. 106-92-3)

Molecular formula: $C_6H_{10}O_2$
Synonyms and trade names: AGE; Allyl 2,3-epoxypropyl ether; 1-Allyloxy-2,3-epoxypropane; 1,2-Epoxy-3-allyoxypropane; Glycidyl allyl ether; Oxirane, [(2-Propenyloxy) methyl]; Propane, 1-(Allyloxy)-2,3-epoxy; Oxirane

Use and exposure

Allyl glycidyl ether is a stable, colorless, flammable liquid with a pleasant odor. It is incompatible with strong oxidizers agents, acids, and bases. It may form peroxides in storage if in contact with air.

Toxicity and health effects

Occupational workers exposed to allyl glycidyl ether develop severe symptoms of poisoning that include, but are not limited to, irritation of the eyes, redness, pain, blurred vision, deep skin burns, respiratory system; causes damage of the mucous membranes, dermatitis, burning sensation, shortness of breath, headache, drowsiness, dullness, nausea, vomiting, pulmonary edema, narcosis, possible hematopoietic and reproductive effects. Acute exposure may cause CNS depression. The major target organs include the eyes, skin, respiratory system, blood, and the reproductive system.

Allyl glycidyl ether and carcinogenicity

Ingestion or inhalation of allyl glycidyl ether has not been reported as a recognized or suspected carcinogen. The ACGIH (2004) has observed that it is not classifiable as a human carcinogen A4.

Exposure limits

TLV: 1 ppm as TWA; the NIOSH has set the REL as 5 ppm (22 mg/m^3) (TWA) Short Term Exposure Limit (STEL) 10 ppm (44 mg/m^3) (skin); IDLH 50 ppm.

Storage

Allyl glycidyl ether should be kept stored in a cool, dark, fireproof area separated from strong oxidants, strong bases, and strong acids.

Precautions

During use and/or handling of allyl glycidyl ether, occupational workers should not be near open flames, sparks, or smoking areas. For temperatures above 48°C, use a closed-system ventilation and explosion-proof electrical equipment. Workers should use protective gloves, protective clothing, and avoid all contact.

References

International Programme on Chemical Safety and the European Commission (IPCS-EC). 1996. Allyl glycidyl ether. ICSC card no. 0096 (updated 2004).

National Institute for Occupational Safety and Health (NIOSH). 2005. Allyl glycidyl ether. *NIOSH Pocket Guide to Chemical Hazards*. Centers for Disease Control and Prevention, Atlanta, GA.

Allyl propyl disulfide (CAS No. 2179-59-1)

Molecular formula: $C_6H_{12}S_2$
Synonyms and trade names: 2-Propenyl propyl disulfide; Onion oil; 4,5-Dithia-1-octene; 2-Propenyl propyl disulfide; Propyl allyl disulfide

Use and exposure

Allyl propyl disulfide is a pale yellow liquid with a pungent odor and is insoluble in water. It decomposes on burning, producing sulfur oxides, and it also gives off irritating or toxic fumes (or gases) in a fire. It reacts with oxidants. Allyl propyl sulfide is used as a synthetic flavor and food additive. Allyl propyl disulfide is the chief volatile component of onion oil as well as being a volatile component of chives and garlic.

Toxicity and health effects

Exposures to allyl propyl disulfide are absorbed into the body by inhalation and by ingestion. It causes a burning sensation, cough, chest tightness, nausea, vomiting, irritation to the eyes, skin, and respiratory tract. Based on animal and human experimental data, allyl propyl sulfide is regarded as an occupational skin sensitizer.

Allyl propyl disulfide and carcinogenicity

No human or animal information is available about the carcinogenicity of allyl propyl disulfide. The IARC has not evaluated the carcinogenicity of allyl propyl disulfide. Similarly, the ACGIH and the NTP have not listed allyl propyl disulfide as a human carcinogen.

Precautions

During use/handling of allyl propyl disulfide, occupational workers should use safety goggles or eye protection in combination with breathing protection and protective gloves. Many of the chemical reactions may result in a hazardous situation, i.e., generation of flammable or toxic chemicals, fire or detonation.

Exposure limits

The exposure limits for allyl propyl disulfide are as follows: TLV: 2 ppm as TWA 3 ppm as STEL intended change 0.5 ppm (ACGIH 2004).

Storage

Allyl propyl disulfide should be separated from oxidants.

Reference

International Chemical Safety Cards (WHO/IPCS/ILO). 2001. Allyl propyl disulfide. ICSC: 1422. IPCS CEC 2002 (updated 2005). IPCS, WHO, Geneva.

Aluminum Compounds

Aloglutamol, Aloxiprin, Aluminoxane, Aluminum (pyro powders and welding fumes, as Al), Aluminum (soluble salts and alkyls, as Al), Aluminum aceglutamide, Aluminum acetate, Aluminum acetoacetate, Aluminum antimonide, Aluminum arsenide, Aluminum azide, Aluminum borohydride, Aluminum bromide, Aluminum carbide (CAS No. 1299-86-1), Aluminum chloride (CAS No. 7446-70-0), Aluminum chloride anhydrous (CAS No. 7446-70-0), Aluminum chloride hexahydrate (CAS No. 7784-13-6), Aluminum chlorohydrate, Aluminum clofibrate, Aluminum dextran, Aluminum diboride, Aluminum dodecaboride, Aluminum flufenamate, Aluminum fluoride (CAS No. 7784-18-1), Aluminum gallium arsenide, Aluminum gallium indium phosphide, Aluminum gallium nitride, Aluminum glycinate, Aluminum hydride, Aluminum hydroxide, Aluminum indium arsenide, Aluminum iodide, Aluminum isopropoxide, Aluminum molybdate, Aluminum monochloride, Aluminum monofluoride, Aluminum monostearate, Aluminum monoxide, Aluminum nitrate (anhydrous) (CAS No. 13473-90-0), Aluminum nitride, Aluminum oxide (CAS No. 1344-28-1), Aluminum oxide hydroxide, Aluminum oxynitride, Aluminum phosphate, Aluminum phosphide, Aluminum potassium sulfate, Aluminum selenide, Aluminum silicate, Aluminum sulfate, Aluminum sulfide, Ammonium alum, Calcium aluminoferrite, Diisobutylaluminum hydride, Potassium alum, Potassium aluminum borate, Soda alum, Sodium aluminium hydride, Sodium hexa-fluoroaluminate, Sucralfate

Aluminum (CAS No. 7429-90-5)

Molecular formula: **Al**
Synonyms and trade names: Aluminum wire; Aluminum foil; Aluminum shot

Use and exposure

Aluminum is the most commonly available element in homes and workplaces. Aluminum is readily available for human ingestion through the use of food additives, antacids, buffered aspirin, astringents, nasal sprays, and antiperspirants; from drinking water; from automobile exhaust and tobacco smoke; and from using aluminum foil, aluminum cookware, cans, ceramics, and fireworks. Aluminum toxicity and its association with Alzheimer's disease in humans require more studies. Some data are against and some are for, because the evidences are inadequate and inconclusive to suggest aluminum as the primary cause of the disease. Prolonged periods of exposure to aluminum and dust causes coughing, wheezing, shortness of breath, memory loss, learning difficulty, loss of coordination, disorientation, mental confusion, colic, heartburn, flatulence, and headaches. Chronic exposures to alumina dust cause irritation to the eyes, skin, respiratory system, pulmonary fibrosis, and lung damage.

Toxicity and health effects

Occupational exposure to aluminum dust and fumes during welding provide suggestive evidence that there may be a relationship between chronic aluminum exposure and subclinical neurological effects, such as impairment on neurobehavioral tests for psychomotor and cognitive performance. Inhalation exposure has not been associated with overt symptoms of neurotoxicity. Prolonged exposure to high concentrations of aluminum and its accumulation causes disturbances in renal function, dialysis, and encephalopathy syndrome—a degenerative neurological syndrome characterized by the gradual loss of motor, speech, and cognitive functions.

Aluminum and carcinogenicity

The DHHS and the US EPA have not evaluated the carcinogenic potential of aluminum in humans. Aluminum has not been shown to cause cancer in animals. However, the IARC has classified aluminum under Group 1, meaning a known human carcinogen.

Aluminum hydroxide (CAS No. 21645-51-2)

Molecular formula: $Al(OH)_3$
Synonyms and trade names: Hydrated alumina; Alumina trihydrate; Aluminium hydrate;
Aluminum hydroxide

Use and exposure

Aluminum hydroxide is a white amorphous powder, incompatible with strong bases. It is
the most stable form of aluminium in normal conditions and is insoluble in water and ethanol. Aluminium hydroxide is amphoteric. It is soluble in strong acids and forms $Al(H_2O)_6^{3+}$,
alkalis, and $Al(OH)_4$. Aluminum hydroxide forms a gel on prolonged contact with water; it
absorbs acids and carbon dioxide. It is used in waterproof textiles, printing ink, glassware,
filler for paper, mordant, and in the manufacture of aluminum and lubricants. Aluminum
hydroxide is a flame retardant. Heat-dried aluminium hydroxide powder is known as
activated alumina and is used in gas purification, as a catalyst support, and an abrasive.
Pharmacologically, aluminum hydroxide is used as an antacid (as Alu-Cap, Aludrox, or
Pepsamar). The hydroxide reacts with excess acid in the stomach, reducing its acidity.

Toxicity and health effects

Exposures to aluminum hydroxide cause irritation to the skin and eyes, particularly when
wet. It causes coughing, shortness of breath, irritation of the upper respiratory tract, and
mild irritation. Aluminum hydroxide has not been reported as hazardous when ingested.

Exposure limits

No exposure limit has been established for aluminum hydroxide.

Handling and storage

Aluminum hydroxide should be kept stored in a tightly closed container, in a cool, dry,
ventilated area. It should be protected against physical damage and isolated from incompatible substances. Containers of this material may be hazardous when empty since they
retain product residues (dust, solids); observe all warnings and precautions listed for the
product.

References

Environmental Health & Safety. 2006. Aluminum hydroxide. MSDS no. A2796. Mallinckrodt Baker,
    Phillipsburg, NJ.
International Labor Organization (ILO). 1998. Aluminum hydroxide. International Programme on
    Chemical Safety (IPCS ICSC: 0373). Geneva, Switzerland (updated 2005).

Aluminum metallic powder (CAS No. 7429-90-5)

Molecular formula: **Al**
Synonyms and trade names: Aluminium powder; Aluminum; Aluminum flake; Aluminum
Metal; Metana

Use and exposure

Aluminum metallic powder is a light, silvery-white to gray, odorless powder. Aluminum metallic powder is reactive and flammable. Aluminum is normally coated with a layer of aluminum oxide unless the particles are freshly formed. There are two main types of aluminum powder: the "flake" type made by stamping the cold metal and the "granulated" type made from molten aluminum. Pyro powder is an especially fine type of "flake" powder. Aluminum powders are used in paints, pigments, protective coatings, printing inks, rocket fuel, explosives, abrasives, and ceramics; the production of inorganic and organic aluminum chemicals; and as catalysts. Pyro powder is mixed with carbon and used in the manufacture of fireworks. The coarse powder is used in aluminothermics.

Toxicity and health effects

Exposures to aluminum metallic powder have been known to cause health effects with symptoms such as irritation, redness, and pain to the eyes, coughing, shortness of breath, irritation to the respiratory tract, nausea, and vomiting in extreme cases. In prolonged periods of inhalation exposures, as in occupational situations, aluminum metallic powder is known to cause pulmonary fibrosis, numbness in fingers, and (in limited cases) brain effects. Workers with pre-existing skin disorders, eye problems, or impaired respiratory function are known to be more susceptible to the effects of aluminum metallic powder.

Aluminum metallic powder and carcinogenicity

The International Agency for Research on Cancer (IARC) has not evaluated the carcinogenicity of this chemical. The American Conference of Governmental Industrial Hygienists (ACGIH) has proposed a carcinogenicity designation of A4 (not classifiable as a human carcinogen) (aluminum metal and insoluble compounds). The National Toxicology Program (NTP) has not listed this chemical in its report on carcinogens.

Precautions

The dry powder is stable but the damp or moist bulk dust may heat spontaneously and form flammable hydrogen gas. Moist aluminum powder may ignite in air, with the formation of flammable hydrogen gas and a combustible dust. Powdered material may form explosive dust-air mixtures. Contact with water, strong acids, strong bases, or alcohols releases flammable hydrogen gas. The dry powder can react violently or explosively with many inorganic and organic chemicals.

Exposure limits

For aluminum metal dusts, the OSHA has set the PEL for aluminum metallic powder as 15 mg/m^3 (TWA) total dust and 5 mg/m^3 (TWA) and the repairable fraction for aluminum metal as Al; the ACGIH has set the TLV as 10 mg/m^3 (TWA).

Precautions

During handling and use of aluminum metallic powder, avoid formation of dust and control ignition sources. Employ grounding, venting, and explosion relief provisions in accord with accepted engineering practices in any process. Only empty the substance in an inert or non-flammable atmosphere, because disposal and emptying contents into a non-inert atmosphere where flammable vapors may be present could cause a flash fire or explosion due to electrostatic discharge.

Storage

Aluminum metallic powder should be kept stored in a tightly closed container, in a cool, dry, ventilated area, protected against physical damage and isolated from sources of heat, ignition, smoking areas, and moisture. Aluminum metallic powder should be kept away from acidic, alkaline, combustible, and oxidizing materials and separate from halogenated compounds.

Reference

Environmental Health & Safety. 2001. Aluminum powder. MSDS no. A2712. Mallinckrodt Baker, Phillipsburg, NJ.

Aluminum potassium sulfate (CAS No. 7778-80-5)

Synonyms and trade names: Sulfuric acid dipotassium salt; Sal polychrestum

Aluminum sulfate (CAS No. 10043-01-3)

Synonyms and trade names: Aluminium sulfate; Cake alum; Filter alum; Papermaker's alum; Alumogenite

Aluminum potasium sulfate dodecahydrate (CAS No. 7784-24-9)

Molecular formula: $AlK(SO_4)_2.12H_2O$
Synonyms and trade names: Alum; Aluminum potassium sulfate, Dodecahydrate; Kalinite; Potassium alum; Aluminum potassium salt

Use and exposure

Aluminum is the most commonly available element in homes and workplaces. Aluminum is readily available for human ingestion through the use of food additives, antacids, buffered aspirin, astringents, nasal sprays, and antiperspirants; from drinking water; from automobile exhaust and tobacco smoke; and from using aluminum foil, aluminum cookware, cans, ceramics, and fireworks. Workers who are at risk for toxicity are those in refineries, foundries and also welders and grinders.

Toxicity and health effects

Prolonged periods of exposure to aluminum and dust cause symptoms of toxicity that include, but are not limited to, coughing, wheezing, shortness of breath, memory loss, learning difficulty, loss of coordination, disorientation, mental confusion, colic, heartburn, flatulence, and headaches. Chronic exposures to alumina dust cause irritation to the eyes, skin, respiratory system, pulmonary fibrosis, and lung damage. Occupational exposure to aluminum dust and fumes during welding provide suggestive evidence that there may be a relationship between chronic aluminum exposure and subclinical neurological effects, such as impairment on neurobehavioral tests for psychomotor and cognitive performance. Inhalation exposure has not been associated with overt symptoms of neurotoxicity. Prolonged exposure to high concentrations of aluminum and its accumulation causes disturbances in renal function. Further, dialysis encephalopathy syndrome or dialysis dementia can result from the accumulation of aluminum in the brain. Dialysis encephalopathy is a degenerative neurological syndrome characterized by the gradual loss of motor, speech, and cognitive functions.

Today, aluminum is everywhere—under your arms (deodorants), in your teeth (toothpaste), and on your baby's skin (powder). In addition, dental amalgams, cosmetics, and cigarette filters contain aluminum. We ingest it in some drinking waters, commercial teas, cheeses, white flour, baking powder, aspirin, and table salt. We cook with it too; most pots and pans are, at least in part, made of aluminum. Unfortunately, many over-the-counter and prescription antacids for digestive difficulties contain aluminum.

The common routes of human exposure to aluminum are ingestion and inhalation. Exposures to aluminum cause different signs and symptoms of toxicity that include, but are not limited to, burning pain in the head, throbbing headache with constipation, vertigo, nausea, hair loss, eyes feel cold, dry, burning, chronic conjunctivitis, humming in the ears, facial pimples, sore and bleeding gums, abnormal cravings for chalk, charcoal, dry food, pain in arms and fingers, restless and anxious sleep and dreams, chapped skin, and brittle nails. Also, aluminum has been shown to be a risk factor in the onset of neurological diseases, notably Alzheimer's disease and other dementias. Therefore, it is important to stop contamination from all potential aluminum sources. Acute toxicity related to aluminum is extremely rare. However, aluminum toxicity is usually found among patients with impaired renal function. Reports have shown that aluminum causes conjunctivitis, eczema, upper airway irritation, pneumoconiosis, and especially with dialysis patients, neurotoxicity and osteomalacia. For years, elevated aluminum has been implicated in several brain diseases, such as Alzheimer's and Parkinson's disease, and has also been found in some senior citizens with extreme memory loss, absent-mindedness, or dementia. It is sometimes found in the hair of children diagnosed with ADHD, ADD, and those with seizures. Hyperactivity, memory disturbances, and learning disabilities may result from even mildly elevated levels of aluminum. Inhibition of neurotransmission and impaired motor coordination may also result. According to a report in the *Lancet* in 1989, many infant formulas contain aluminum. In this report, it was revealed that human breast milk contained 5–20 µg/L of aluminum, cow's milk-based formulas contained 20 times as much aluminum, and soy-based formulas contained 100 times as much. Physical symptoms of aluminum toxicity include brittle bones or osteoporosis, as aluminum is stored in the bones. Kidney malfunction may also result, as the kidneys filter aluminum. Toxicity of aluminum and its association with Alzheimer's disease in humans has not been well confirmed since some data are against while other data are inadequate and inconclusive.

Aluminum and carcinogenicity

The United States Environmental Protection Agency (US EPA) has not classified aluminum for its carcinogenicity. It has been reported that although aluminum has been classified by the IARC as Group 1, meaning a known human carcinogen, the responsible agent(s) for the carcinogenicity are probably chemicals other than aluminum.

Exposure limits

The US EPA recommends that the concentration of aluminum in drinking water not exceed 0.2 ppm. The Food and Drug Administration (FDA) has determined that aluminum cooking utensils, aluminum foil, antiperspirants, antacids, and other aluminum products are generally safe. The OSHA set a maximum concentration limit for aluminum dust in workplace air as 15 mg/m^3 for an 8-h period (TWA) and the NIOSH has recommended a limit of 10 mg/m^3 in workplace air for up to a 10-h period (TWA). For aluminum, the ACGIH has set the TLV as 10 mg/m^3 and the OSHA has set the PEL as 5 mg/m^3.

References

Agency for Toxic Substances and Disease Registry (ATSDR). 1992. Toxicological profile for aluminum. US Department of Health and Human Services (DHHS), Public Health Service, Atlanta, GA.
———. 2006. Toxicological profile for aluminum (Draft for Public Comment). US Department of Health and Human Services, Public Health Service, Atlanta, GA (updated 2007).

alpha-Alumina (CAS No. 1344-28-1)

Molecular formula: Al_2O_3
Synonyms and trade names: Alumina; Activated Alumina; Aluminum oxide

Use and exposure

alpha-Alumina is a silver-white, odorless powder, insoluble in water. It is incompatible with chlorine trifluoride, ethylene oxide, with halo carbons, halogens, combustible materials, and oxidizing materials.

Toxicity and health effects

Exposures to (inhalation) high concentrations of alumina cause poisoning with symptoms that include, but are not limited to, coughing, shortness of breath, respiratory tract irritation due to mechanical action, unpleasant deposits in the nasal passages, impaired pulmonary function, a sweet, metallic, or foul taste in the mouth, metal fume fever, an influenza-like illness, sudden thirst, fever, chills, profuse sweating, excessive urination, diarrhea, and prostration. Other symptoms include upper respiratory tract irritation accompanied by coughing, and a dryness of the mucous membranes.

Precautions

alpha-Alumina containers should be tightly closed and kept stored in a cool, dry, ventilated area. Also, the containers should be protected from physical damage since this material may be hazardous when the containers are empty as they retain dust, solids of the product.

Aminocarb (CAS No. 2032-59-9)

Molecular formula: $C_{11}H_{16}N_2O_2$
IUPAC name: 4-Dimethyamino-m-tolyl methylcarbamate; Phenol, 4-(dimethylamino)-3-methyl-methylcarbamate
Synonyms and trade names: Metacil
Toxicity class: USEPA: II; WHO: II

Use and exposure

Aminocarb is a non-systemic insecticide. It is combustible and the liquid formulations containing organic solvents may be flammable and release irritating or toxic fumes. It has been in use for the control of lepidopterous larvae and other chewing insects in cotton, field crops, and in forestry, other lepidopterous insects, aphids, and soil mollusks. The acute oral LD_{50} for rats is 50 mg/kg, while the acute dermal LD_{50} for rats is 275 mg/kg, suggesting toxicity to skin is appreciably severe compared to carbamate pesticide.

Toxicity and health effects

Occupational exposures to high concentrations of aminocarb cause poisoning with symptoms including, but not limited to, malaise, muscle weakness, dizziness, sweating, headache, salivation, nausea, vomiting, abdominal pain, diarrhea, labored breathing, miosis with blurred vision, unconsciousness, pupillary constriction, muscle cramp, excessive salivation, incoordination, muscle twitching, and slurred speech. Occupational workers severely exposed to aminocarb show inhibition of cholinesterase activity, develop disorders in the nervous system, convulsions, cardiorespiratory depression, bronchospasms, bronchorrhea pulmonary edema, respiratory failure, and death.

Aminocarb and carcinogenicity

Aminocarb has not been listed by the IARC, the NTP, or the USEPA as a human carcinogen.

Exposure limits

No TLV has been established for aminocarb.

Precautions

Aminocarb decomposes on heating and produces toxic fumes, irritating fumes, and oxides of nitrogen. During handling of aminocarb workers should wear protective clothing, gloves, face shield, breathing protection and avoid open flames.

Reference

International Chemical Safety Cards (ICSC). 1993. Aminocarb Card no. ICSC: 0097. International Programme on Chemical Safety & the Commission of the European Communities (IPCS CEC). Geneva, Switzerland.

4-Aminophenyl ether (CAS No. 101-80-4)

Molecular formula: $(H_2N_6H_4)_2O$

Synonyms and trade names: 4,4-Oxydianiline; 4,4-Diaminodiphenyl oxide; N-Phenylbenzeneamine; *para*-Aminophenyl ether; Bis(4-aminophenyl) ether; Bis(*para*-aminophenyl)ether; Diaminodiphenyl ether; 4-Aminophenyl ether; Oxydianiline; 4-Aminophenyl ether; *para*-Aminophenyl ether; Diaminodiphenyl ether; *para,para'*-Oxybis(aniline); ODA; DADPE; 4,4'-DADPE

Use and exposure

4-Aminophenyl ether is a resin used in the manufacture of a variety of industrial products. For example as, insulating varnishes, flame-retardant fibers, wire enamels, coatings, and films. It is also used for the manufacture of other fire-resistant products.

Toxicity and health effects

4-Aminophenyl ether is highly toxic to animals. There is sufficient evidence as a carcinogenicity and has caused adenomas and carcinomas in the thyroid and liver of experimental rats.

Reference

International Agency for Research on Cancer (IARC) Summaries & Evaluations. 1998. 4,4' Diaminodiphenyl ether. Vol. 29, p. 203, 1982., Lyon, France.

4-Aminopyridine (Avitrol) (CAS No. 504-24-5)

Synonyms and trade names: 4-AP, P-aminopyridine; Gamma-aminopyridine; Amino-4-pyridine; Avitroland; Avitrol 200; Fampridine
Chemical class: Pyridine compound

Use and exposure

Technical 4-aminopyridine is a white crystalline solid that contains about 98% active ingredient. It is soluble in water and slightly soluble in benzene and ether. 4-Aminopyridine formulations are classified by the US Environmental Protection Agency (US EPA) as restricted use pesticides (RUPs) and therefore should be purchased and used only by certified and trained workers and applicators. Avitrol is available as grain baits or as a powder concentrate. It is one of the most prominent avicides. It is registered with the EPA for use against red-winged blackbirds, blackbirds in agricultural fields, grackles, pigeons, and sparrows around public buildings, and various birds around livestock feeding pens. Avitrol repels birds by poisoning a few members of a flock, causing them to become hyperactive.

Toxicity and health effects

4-Aminopyridine, a pyridine compound, is an extremely effective bird poison. In agriculture, 4-AP is used as an extremely effective bird poison sold under the brand name Avitrol. It is highly toxic to all mammals including humans if dosages are exceeded, and as an experimental drug, the recommended dose data is unavailable. 4-Aminopyridine blocks the potassium channels in neurons. It should be borne in mind that this is an experimental drug and its side effects are uncertain. Dosages should be carefully regulated, as potassium is a chemical that is used extensively in other parts of the body, including heart functions.

In humans, 4-aminopyridine is rapidly absorbed into the bloodstream from the gastrointestinal tract. In the liver, it is readily broken down, or metabolized, into removable compounds excreted in urine. After intravenous and oral doses were given to humans, (90.6% and 88.5%, respectively,) 4-aminopyridine excreted in the urine. It does not to concentrate or accumulate in the skin. Birds killed with Aritrol are not poisonous to predators.

4-Aminopyridine and carcinogenicity

There is no data available on the carcinogenicity of 4-aminopyridine.

Exposure limits

There is no data available on the acceptable daily intake (ADI), permissible exposure limit (PEL), and threshold limit value (TLV) of 4-aminopyridine.

Precautions

Grain bait formulations of 4-aminopyridine are grouped in toxicity class III and must bear the signal word CAUTION. The powder concentrate formulations are grouped in toxicity class I and must bear the signal word DANGER.

References

Sax, N.I. 1984. *Dangerous Properties of Industrial Materials*, 6th ed. Van Nostrand Reinhold, New York.
US Environmental Protection Agency (US EPA). 1980. 4 Aminopyridine: Avitrol. Pesticide Registration Standard. Office of Pesticides and Toxic Substances, Washington, DC.
US National Library of Medicine (USNLM). 1995. Hazardous substances data bank. NLM, Bethesda, MD.

Amyl alcohol (CAS No. 75-85-4)

Molecular formula: $C_5H_{11}OH$
Synonyms and trade names: Pentanols; Pentyl alcohols; Fusel oil; Potato spirit

Use and exposure

Amyl alcohol is produced during the fermentation of grains, potatoes, and beets. It is also produced during the acid hydrolysis of petroleum fraction. Amyl alcohol is widely used in industry. For example, in the manufacturing of lacquers, paints, varnishes, perfumes, pharmaceuticals, plastics, rubber, explosives, hydraulic fluids, for the extraction of fats, is also used in the petroleum refinery industries.

Toxicity and health effects

Inhalation of the vapors of amyl alcohol causes tearing, pain, redness, swelling, irritation of the mucous membrane of the eyes, nose, throat, and upper respiratory tract and of the skin. Acute and long-term exposure to amyl alcohol causes nausea, vomiting, headache, vertigo, and muscular weakness. Vomiting may cause aspiration into the lungs, resulting in chemical pneumonia. After a prolonged period of exposure to amyl alcohol, workers develop dizziness, double vision, shortness of breath, delirium, and related narcotic effects. In severe cases, inhalation leads to pulmonary edema, kidney injury, effects on the heart and becomes fatal.

Occupational workers with pre-existing skin disorders, eye problems, or impaired liver, kidney, or respiratory function, may be more susceptible to the effects of amyl alcohol.

Precautions

During handling and use of amyl alcohol, persons with pre-existing skin disorders, eye problems, or impaired liver, kidney, or respiratory function, should be careful since these workers/persons are more susceptible to the effects of amyl alcohol.

Reference

Environmental Health & Safety. 2007. Amyl alcohol. MSDS no. A6372. Mallinckrodt Baker, Phillipsburg, NJ.

Aniline (CAS No. 62-53-3)

Molecular formula: $C_6H_7N/C_6H_5NH_2$
Synonyms and trade names: Aminobenzene; Aminophen; Arylamine; Benzenamine; Aniline oil; Phenylamine

Use and exposure

Aniline was first isolated from the destructive distillation of indigo in 1826 by Otto Unverdorben. Aniline is oily and, although colorless, it slowly oxidizes and turns into a kind of resin in air, giving the sample a red-brown tint. At room temperature, aniline, the simplest aromatic amine, is a clear to slightly yellow, oily liquid that darkens to a brown color on exposure to air. Like most volatile amines, it possesses the somewhat unpleasant odor of rotten fish and also has a burning aromatic taste. It has a low vapor pressure at room temperature and ignites readily, burning with a smoky flame. It does not readily evaporate at room temperature. Aniline is slightly soluble in water and mixes readily with most organic solvents. It is synthesized by catalytic hydrogenation of nitrobenzene or by ammonolysis of phenol. Aniline is incompatible with strong acids, strong oxidizers, albumin, and solutions of iron, zinc, aluminum, toluene diisocyanate, and alkalis. It ignites spontaneously in the presence of red fuming nitric acid, and with sodium.

Originally, the great commercial value of aniline was due to the readiness with which it yields, directly or indirectly, valuable dyestuffs. Currently, the largest market for aniline is in the preparation of methylene diphenyl diisocyanate (MDI), some 85% of aniline serving this market. In fact, in industry, aniline is an initiator or intermediary in the synthesis of aniline being used as a precursor to more complex chemicals. It is the starting material for many dyestuffs, known as aniline dyes. Its main application is in the manufacture of polyurethane foam, and a wide variety of products, such as MDI, agricultural chemicals, synthetic dyes, antioxidants, stabilizers for the rubber industry, varnishes, explosives, analgesics, and hydroquinone for photographic developing, and as an octane booster in gasoline. Aniline has also been detected in tobacco smoke and exposures to aniline have been reported among workers in related industrial workplaces, hazardous waste sites, and the general population through food and drinking water.

Toxicity and health effects

Exposures to aniline on inhalation, ingestion and/or through skin contact cause adverse health effects. Exposures to liquid aniline cause mild irritation to the skin and eyes. Aniline is a blood toxin and its absorption through the skin and by inhalation of its vapor results in systemic toxicity, damage to the kidney, liver, bone marrow and of methemoglobinemia. The symptoms of poisoning include, but are not limited to, drowsiness, dizziness, severe headache, nausea, tiredness, bluish discoloration of the lips and tongue, loss of appetite, irregular heart beat, mental confusion, and shock. A prolonged period of exposure to the vapor results in respiratory paralysis, convulsions, coma, and death.

Aniline and carcinogenicity

Laboratory animals exposed to aniline over a lifetime developed cancer of the spleen. The clusters of cases of bladder cancer deaths observed in workers in the aniline-dye industry have been attributed to exposure to chemicals other than aniline. The available studies in humans are inadequate to determine whether exposure to aniline can increase the risk of developing cancer in people. The US EPA classifies aniline as B2, meaning a probable human carcinogen, while the International Agency for Research on Cancer (IARC) categorized aniline as Group 3, meaning not classifiable as a human carcinogen.

Exposure limits

For aniline, the Occupational Safety and Health Administration (OSHA) has set the PEL as 5 ppm in workplace air for 8 h (TWA), and the American Conference of Governmental Industrial Hygienists (ACGIH) set the TLV as 2 ppm (TWA) skin.

Precautions

When using aniline, occupational workers should wear impervious protective clothing, including boots, gloves, laboratory coat, apron or coveralls, chemical safety goggles, and/or a full face shield as appropriate, to prevent skin contact. Workplace facilities should maintain an eye-wash fountain and quick-drench facilities. Workers should not eat, drink, or smoke in the workplace.

Storage

Aniline should be kept stored against physical damage in a cool (but not freezing), dry, well-ventilated location, away from smoking areas and fire hazard. It should be kept separated from incompatibles and the containers should be bonded and grounded for transfer to avoid static sparks.

Reference

Agency for Toxic Substances and Disease Registry (ATSDR). 2002. Aniline. Managing Hazardous Materials Incidents. Volume III – Medical Management Guidelines for Acute Chemical Exposures. US Department of Health and Human Services, Public Health Service, Atlanta, GA (updated 2008).
Environmental Health & Safety. 2008. Aniline. Mallinckrodt Baker, Phillipsburg, NJ.
International Agency for Research on Cancer (IARC). 1987. Aniline, Supplement 7, 1998.

Antimony Compounds (as Sb)

Antimony and its compounds have many industrial uses. The list of compounds includes:

(i) **Antimony acetate (CAS No. 3643-76-3)**

Molecular formula: $Sb(CH_3COO)_3$

Antimony acetate is a catalyst in the production of synthetic fibers.

(ii) **Antimony pentachloride (CAS No. 7647-18-9)**

Molecular formula: $SbCl_5$

Antimony pentachloride is a yellow to red, oily hygroscopic liquid, soluble in hydrochloric acid and chloroform; solidifies with moisture and decomposes in excess water; used as an intermediate in synthesis and dyeing and used in analytical testing for cesium and alkaloids.

(iii) **Antimony pentafluoride (CAS No. 7783-70-2)**

Molecular formula: SbF_5

Antimony pentafluoride is a moderately viscous liquid, corrosive and hygroscopic. It reacts violently with water; it is soluble in glacial acetic acid. It is used as a fluorination agent in organic synthesis.

(iv) **Antimony pentasulfide (CAS No. 1315-04-4)**

Molecular formula: Sb_2S_5

Antimony pentasulfide, also known as antimony red, is a yellow to orange powder, insoluble in water, but soluble in alkal and in concentrated hydrochloric acid; it is used as a red pigment.

(v) **Antimony potassium tartrate (CAS No. 11071-15-1)**

Molecular formula: $K_2(Sb_2(C_4H_4O_6)_2 \cdot 3H_2O$

Antimony potassium tartrate is a white crystalline powder with a sweetish taste. It is used as a mordant or fixing agent in leather and textile dyeing as well as an analytical reagent and flux additive for electroplating. It is used in making insecticides and pesticides. It was used as a parasiticide or as an emetic and expectorant.

(vi) **Antimony sulfate (CAS No. 7446-32-4)**

Molecular formula: $Sb_2(SO_4)_3$

Antimony sulfate is a white deliquescent powder. It is soluble in acids and used in organic synthesis.

(vii) **Antimony trichloride (CAS No. 10025-91-9)**

Molecular formula: $SbCl_3$

Antimony trichloride appears as orthorhombic deliquescent hygroscopic crystals and is soluble in alcohol and acetone. It reacts with moisture, forming antimony oxychloride in air. It is used as a chlorinating agent, as a fireproofing agent in textiles, in bronzing steel, and as a mordant in dyeing as well as a caustic in medicine.

(viii) **Antimony trifluoride (CAS No. 7783-56-4)**

Molecular formula: SbF_3

Antimony trifluoride appears as orthorombic deliquescent crystals and is soluble in water. It is used as a fluorination agent in organic synthesis, in dyeing, and to make porcelain and pottery.

(ix) **Antimony triiodide (CAS No. 7790-44-5)**

Molecular formula: SbI_3

(x) **Antimony trioxide (CAS No. 1327-33-9)**

Molecular formula: Sb_2O_3

Antimony trioxide appears as white rhombic crystals and is insoluble in water. It is used as a powerful reducing agent, a flame retardant for a wide range of plastics, rubbers, paper, and textiles, a catalyst in pet production, an activator in the glass industry, a flocculant in titanium dioxide production, in paints and adhesives industries, and ceramic frites.

(xi) **Antimony trisulfide (CAS No. 1345-0406)**

Molecular formula: Sb_2S_3

Antimony trisulfide appears as dark orange to black rhombic crystals and is insoluble in water, but soluble in concentrated hydrochloric acid and sulfides. It is used as a pigment and in pyrotechnics; it is used on safety matches and in vulcanizing rubber. It is used in combination with antimony oxides as a yellow pigment in glass and porcelain.

(xii) **Potassium antimonate (CAS No. 12208-13-8)**

Molecular formula: $KSbO_3$

Potassium antimonate is white powder, soluble in water. It is used as an oxidizing agent at high temperatures, as a fireproofing auxiliary, as an enamel opacifier, as a glass fining agent, and to decolorize glass tubes and fiber glass.

(xiii) **Sodium antimonate (CAS No. 15432-85-6)**

Molecular formula: $NaSbO_3$

Sodium antimonite is a white powder, soluble in water. It is used as an oxidizing agent at high temperatures, as a fireproofing auxiliary, as an enamel opacifier, as a glass fining agent, and as a decolorizer in glass tubes and fiber glass.

Reference

US Department of Labor, Occupational Safety and Health Administration (OSHA), Safety and Health Topics, 2004. Antimony and Compounds (as Sb).

Antimony (CAS No. 7440-36-0)

Molecular formula: **Sb**
Synonyms and trade names: Antimony black; Antimony black; Stibium

Use and exposure

Antimony is a silvery-white metal found in the earth's crust. It is insoluble in hot or cold water, but soluble in hot concentrated sulfuric acid and hot nitric acid, and reacts with oxidizing acids and halogens (fluorine, chlorine, or bromine). It does not react with water at room temperature, but will ignite and burn in air at higher temperatures. Ores of antimony are mined and later mixed with other metals to form antimony alloys, which are used in lead storage batteries, solder, sheet and pipe metal bearings, castings, and pewter. Antimony oxide is added to textiles and plastics to prevent them from catching fire. It is also used in paints, ceramics, and fireworks, and as enamels for plastics, metal, and glass. Antimony is alloyed with other metals, such as lead, to increase its hardness and strength; its primary use is in antimonial lead, which is used in grid metal for lead acid storage batteries. Antimony salts are used in the treatment of leishmaniasis and schistosomiasis.

Toxicity and health effects

The toxicity of antimony compounds is comparable to that of arsenic, but as antimony compounds are hardly absorbed in the gastrointestinal tract, there is less hazard of acute poisoning. In addition, antimony compounds often cause vomiting, thus being removed from the organism. Chronic poisoning may result in damage to the liver, kidneys, and even the heart and the circulatory system. The symptoms differ among the compounds. Stibine accumulates in fatty tissue. Exposures to antimony and its compounds cause poisoning and toxicity to the worker with symptoms that include, but are not limited to, irritation to eyes, skin, nose, and throat, ulceration of nasal septum and larynx, dermatitis as characterized by antimony spots, cough, dizziness, seizures, headache, anorexia, nausea, vomiting, diarrhea, stomach cramps, bloody stools, insomnia, inability to smell properly, metallic taste, cardiovascular disturbances, pulmonary edema, pharyngitis, tracheitis, heart and lung damage, pneumoconiosis, slow and shallow respiration, coma, and death. Antimony fumes and dusts inhaled by occupational workers are associated with the development of benign tumors of the lungs, dermatitis and, less commonly, effects on the heart and kidneys. Laboratory animals exposed to antimony by inhalation or ingestion exhibit effects similar to those noted in humans. Antimony can be measured in the urine, feces, and blood. To date, little is known about the environmental risks involved. Water pollution seldom occurs because of the low solubility of most compounds. Extreme caution should be taken when coming into direct contact with antimony compounds.

Antimony and carcinogenicity

There is insufficient evidence for the carcinogenicity of antimony and antimony compounds. The Department of Health and Human Services (DHHS), the IARC, and the US EPA have not classified antimony as to its human carcinogenicity. Lung cancer has been observed in some studies with laboratory rats that breathed high levels of antimony, while no human studies are available. Prolonged periods of exposure of laboratory rats to high concentrations of antimony trioxide and trisulfide increase the incidence of lung tumors. The IARC has classified antimony trioxide as Group 2B, meaning a possible human carcinogen, while the ACGIH has included antimony trioxide under Group A2, meaning a suspected human carcinogen.

Exposure limits

The US EPA has set 0.006 ppm of antimony in drinking water. The OSHA reports the limit of 0.5 mg/m^3 of antimony for an 8-h workday, and the ACGIH and the National Institute for Occupational Safety and Health (NIOSH) also recommend the same levels.

Precautions

Antimony trioxide is incompatible with bromine trifluoride, strong acids, strong bases, reducing agents, perchloric acid, and chlorinated rubber. The release of the deadly gas, stibine, and its inhalation cause adverse effects on the respiratory, gastrointestinal, and cardiovascular systems. Workers must wear impervious protective clothing, including boots, gloves, laboratory coat, apron or coveralls, as appropriate, to prevent skin contact.

References

Agency for Toxic Substances and Disease Registry (ATSDR). 1992. Toxicological profile for antimony. US Department of Health and Human Services, Public Health Service, Atlanta, GA (updated 2008).
Renes, L.E. 1953. Antimony poisoning in industry. *Arch. Ind. Hyg. Occup. Med.* 7: 99–108.
Technology Transfer Network Air Toxics web site. Antimony compounds. Hazard Summary. Created April 1992; revised January 2000 (updated 2007).

Aromatic nitro compounds

Aromatic nitro compounds are typically synthesized by the action of a mixture of nitric and sulfuric acids on a suitable organic molecule. They contain one or more nitro functional groups (–NO$_2$). Aromatic nitro compounds are highly explosive, especially when the compound contains more than one nitro group. The presence of impurities or improper handling can also trigger a violent exothermic decomposition, i.e., trinitrophenol (picric acid), trinitrotoluene (TNT), and trinitroresorcinol (styphnic acid). Chloramphenicol is a rare example of a naturally occurring nitro compound. The reduction of aromatic nitro compounds with hydrogen gas over a platinum catalyst gives aniline. (See *Ullmann's Encyclopedia of Industrial Chemistry* for more details.)

References

Nitro Compounds Aromatic., 2002. p. 127. In: *Ullmann's Encyclopedia of Industrial Chemistry*. pp. 4600., Wiley-VCH Verlag., John Wiley & sons., Hoboken, NJ.
World Health Organization (WHO). 1978. Some aromatic amines and related nitro compounds: Hair dyes, coloring agents & miscellaneous industrial chemicals. Monographs on the Evaluation of Carcinogenic Risks to Humans. IARC Monograph Vol. 16. World Health Organization, Geneva, Switzerland (updated 2008).

Aromatic phenols

Phenols, also called phenolics, are a class of chemical compounds. These compounds consist of a hydroxyl group (–OH) attached to an aromatic hydrocarbon group. The simplest of the class is n phenol (C$_6$H$_5$OH). Phenol can be made from the partial oxidation of benzene, the reduction of benzoic acid, by the cumene process, or by the Raschig process. It can also be found as a product of coal oxidation. Some phenolic compounds are germicidal and are used in formulating disinfectants. Phenolic compounds are used as disinfectants and for chemical synthesis. Others possess estrogenic or endocrine disrupting activity. The following is a list of phenolics: phenol, BHT (butylated hydroxytoluene), bisphenol A, capsaicin, cresol, estradiol, eugenol, gallic acid, guaiacol, 4-nonylphenol, orthophenyl phenol, picric acid, phenolphthalein, polyphenol, raspberry ketone, and serotonin/dopamine/adrenalin/noradrenaline.

Reference

John McMurry (ed.) and Eric E. Simanek. 2007. *Fundamentals of Organic Chemistry*, 2nd ed. Brooks/ Cole pub-co

p-Arsanilic acid (CAS No. 98-50-0)

Molecular formula: $C_6H_8AsNO_3$
Synonyms and trade names: 4-Aminobenzenearsonic acid; 4-Arsanilic acid; Atoxylic acid; Benzenearsonic acid, p-aminophenyl arsenic acid.

Use and exposure

p-Arsanilic acid is an off-white powder. It is slightly soluble in cold water. It is incompatible with oxidizing agents; on hazardous decomposition, p-arsanilic acid produces nitrogen oxides, carbon monoxide, carbon dioxide, nitrogen gas, and oxides of arsenic.

Toxicity and health effects

Occupational workers exposed to p-arsanilic acid develop poisoning. The symptoms include, but are not limited to, eye and skin irritation, chemical conjunctivitis and corneal damage, hyperpigmentation of the skin and, perkeratoses of plantar and palmar surfaces, primary irritation and sensitization, digestive tract irritation, gastrointestinal hypermotility, diarrhea, hepatitis, hepatocellular necrosis, central nervous system depression, cardiac disturbances, and liver and kidney damage. The target organs include the kidneys, central nervous system, liver, and cardiovascular system. After a prolonged period of exposure to arsenic compounds, including arsenical dust, workers are known to develop shortness of breath, nausea, chest pains, garlic odor, and impairment of peripheral circulation. The toxicological properties of p-arsanilic acid have not been fully investigated.

p-Arsanilic acid and carcinogenicity

The IARC has listed p-arsanilic acid as Group 1, meaning carcinogenic to humans.

Exposure limits

The American Conference of Industrial Hygienists (ACGIH) and the National Toxicology Program (NTP) have set no exposure limits for p-arsanilic acid:

Precautions

Occupational workers should wash thoroughly after using p-arsanilic acid. Any contaminated clothing should be washed before reuse. Work areas should have adequate ventilation and minimum dust generation and accumulation. Workers should avoid any kind of contact of p-arsanilic acid with the eyes, skin, clothing, ingestion or inhalation.

Storage

p-Arsanilic acid should be kept stored in a tightly closed container in a cool, dry, well-ventilated area. It should be kept away from incompatible materials, dust generation, excess heat, and strong oxidants.

Reference

Material Safety Data Sheet (MSDS). 2008. p-Arsanilic Acid. ACC no. 70695.

Arsenates

Arsenate minerals are minerals containing the arsenate (AsO_4^{3-}) anion group—arsenic acid, calcium arsenate, chromated copper arsenate, copper(II) arsenate, lead hydrogen arsenate, monosodium methyl arsenate, potassium arsenate, arsenates, arsenic minerals, arsenical herbicides, arsenides, and arsenites. Arsenate resembles phosphate in many respects, since arsenic and phosphorus occur in the same group.

Arsenic acid (CAS No. 7778-39-4)

Molecular formula: H_3AsO_4
Synonyms and trade names: Arsenate; Arsenic acid; Arsenic acid hemihydrate; Arsenenous acid; Arsenous acid anhydride; Orthoarsenic acid solution; Orthoarsenic acid; Desiccant L-10; Metaarsenic acid; Zorox

Use and exposure

Arsenic acid is a clear, white, semi-transparent crystal (sugar or sand-like) material. Arsenic acid is one of the most commercially important pentavalent compounds of arsenic. It is used for wood treatment/wood preservative, as a drying agent, a soil sterilant, and to make other arsenates. It is a broad-spectrum biocide, a finishing agent for glass and metal, and a reagent in the retard of dyestuffs and organic arsenic compounds. Arsenic acid decomposes on heating, producing toxic and corrosive fumes. It is a strong oxidant and reacts with combustible and reducing materials and attacks/damages metals to produce toxic and flammable arsine. It is soluble in water and forms the arsenate ion.

Toxicity and health effects

Arsenic acid is absorbed into the body by inhalation of its vapor, through the skin, and by ingestion. Arsenic acid is an eye irritant and may cause burns. Most injuries result from exposure to dusts, causing conjunctivitis, lacrimation, photophobia, and chemosis. Exposures to arsenic acid cause poisoning with symptoms of irritation of the eyes, the skin, the respiratory tract, respiratory distress, sore throat, nausea, vomiting, diarrhea, poor appetite, and stomach cramps. Arsenic acid also causes effects on the blood, cardiovascular system, gastrointestinal tract, liver, and peripheral nervous system, leading to polyneuropathy and convulsions. Repeated exposures to high concentrations of arsenic acid have been reported to cause nerve damage, with "pins and needles," numbness, and weakness of the arms and legs, and even death. Tachycardia is frequently reported following ingestion of arsenic acid salts and is contributed to by anxiety, intravascular fluid depletion, and possibly direct arsenic-induced cardiotoxicity.

Arsenic acid and carcinogenicity

The IARC has grouped arsenic acid as A1, meaning a confirmed human carcinogen.

Exposure limits

The long-term exposure limit for arsenic acid (8-h TWA) has been set as 0.1 mg/m^3. The legal airborne PEL is 0.01 mg/m^3 averaged over an 8-h work shift for arsenic and inorganic arsenic compounds. The NIOSH recommended airborne exposure limit (RAEL) for arsenic acid is 0.002 mg/m^3 for any 15-min work period for arsenic, inorganic 0.2 mg/m^3 averaged over an 8-h work shift for arsenic and soluble compounds.

Storage

Arsenic acid should be kept stored in an area without drain or sewer access. It should be separated from strong oxidants, strong bases, metals, strong reducing agents, food, and feedstuffs. Occupational workers should keep arsenic acid in a cool, well-ventilated area, away from heat. Workers should avoid storing or transporting arsenic acid with aluminium, copper, iron, or zinc.

Precautions

Occupational workers should be extremely careful and cautious during handling and use of arsenic acid.

References

Bradberry, S.M., Harrison, W.N., Beer, S.T. and Vale, J.A. 1997. Arsenic acid. UKPID Monograph. National Poisons Information Service, Birmingham, U.K.

Greenberg, C., Davies, S., McGowan, T., Schorer, A. and Drage, C. 1979. Acute respiratory failure following severe arsenic poisoning. *Chest* 76: 596–98.

Hazardous Substances Data Bank (HSDB). 1995. In: *Tomes plus*. Environmental Health and Safety Series I, Vol. 26. National Library of Medicine.

International Programme on Chemical Safety and the European Commission (IPCS-ECE). 2005. Arsenic acid. ICSC card no. 1625. IPCS, WHO, Geneva, Switzerland.

McWilliams, M.E. 1989. Accidental acute poisoning by a concentrated solution of arsenic acid from percutaneous absorption: a case report. *Vet. Hum. Toxicol.* 31: 354.

Sodium arsenate dibasic (CAS No. 7778-43-0)

Molecular formula: **AsHNa$_2$O$_4$**

Synonyms and trade names: Arsenic acid, disodium salt; Disodium arsenate; Disodium hydrogen arsenate; Sodium arsenate; Sodium orthoarsenate

Use and exposure

Arsenic acid disodium salt is not combustible. It emits irritating or toxic fumes or gases in a fire.

Toxicity and health effects

Exposures to arsenic acid disodium salt cause adverse health effects. The symptoms of poisoning include, but are not limited to, cough, headache, sore throat, weakness, respiratory distress, labored breathing, irritation to the eyes, and skin, dermatitis, mucous membranes, the respiratory tract, and pigmentation disorders. Repeated or prolonged contact with arsenic acid disodium salt causes cardiovascular disorders affects the nervous system, neuropathy, and kidneys, resulting in severe gastroenteritis, loss of fluids and electrolytes, kidney impairment, collapse, shock, and death.

References

International Programme on Chemical Safety and the European Commission (IPCS-EC). 1999. Sodium arsenate dibasic. ICSC card no. 1208. Geneva, Switzerland (updated 2004).

Arsenic (Inorganic) compounds (As+3 and As+5)

Inorganic arsenic compounds are mainly used to preserve wood. During the 1800s and the early 1900s, inorganic arsenic was commonly used as a rat poison and in the treatment of some human diseases, such as syphilis. A widely used arsenic compound, Fowler's solution (potassium arsenite) was prescribed for chronic infections, anemia, and skin diseases. Pentavalent arsenic is still used to treat advanced trypanosomiasis (a disease caused by parasites, which is rare in the United States but is common in Africa), and arsenic trioxide is used today as a treatment for promyelocytic leukemia. Copper chromate arsenate (CCA) is used to make "pressure-treated" lumber. CCA is no longer used in the United States for residential uses, however, it is still used in industrial applications. Organic arsenic compounds are used as pesticides, primarily on cotton fields and orchards. Ingesting or breathing low levels of inorganic arsenic for a long time can cause a darkening of the skin and the appearance of small "corns" or "warts" on the palms, soles, and torso. Skin contact with inorganic arsenic may cause redness and swelling. Inorganic arsenic compounds cause ulceration and perforation of the nasal septum as seen in miners, dermatitis, gastrointestinal disturbances, peripheral neuropathy, respiratory irritation, and hyperpigmentation of the skin. Acute exposure causes fever, anorexia, hepatomegaly, melanosis, ischemic heart disease, cardiac arrhythmias, and cardiovascular failure. Inorganic arsenic compounds also cause jaundice, cirrhosis, acites, enlargement of the liver (hemorrhagic necrosis and fatty degeneration), kidney damage with effects on the capillaries, tubules and glomeruli, peripheral neuropathy (sensory and motor), axonal degeneration, encephalopathy, and loss of hearing due to the effect on the auditory nerves. It is a potential occupational carcinogen.

The compounds of arsenic are many and include:

(a) **Arsenenic acid (meta-arsenic acid, $HAsO_3$ CAS No. 10102-53-1)**

(b) **Arsenenous acid ($HAsO_2$, CAS No. 13768-07-5)**

(c) **Arsenic trichloride ($AsCl_3$, CAS No. 7784-34-1):** Oily, clear liquid; decomposes in water; used in ceramics, organic chemical syntheses, and in the preparation of pharmaceuticals; light sensitive and moisture sensitive.

(d) **Arsenic trisulfide (As_2S_3, CAS No. 1303-33-9)**: An acidic compound in the form of yellow or red monoclinic crystals with a melting point at 300°C; occurs as the mineral orpiment; used as a pigment.

(e) **Arsenic acid (H_3AsO_4, CAS No. 1327-52-2, 7778-39-4)**: White, poisonous crystals, soluble in water and alcohol; used in wood treatment, finishing agent for glass and metal, manufacturing of dyestuffs and organic arsenic compounds, a soil sterilant. Also known as orthoarsenic acid.

(f) **Arsenic acid monoammonium salt (AsH_6NO_4, CAS. No. 13462-93-6)**

(g) **Arsenic disulfide (As_2S_2, CAS No. 1303-32-8)**: Deep red, lustrous monoclinic crystal; insoluble in water; used in fireworks.

(h) **Arsenic pentasulfide (As_2S_5)**: Yellow crystals; insoluble in water; readily decompose to trisulfide and sulfur; used as a pigment.

(i) **Arsenic pentoxide (As_2O_5, CAS No. 1303-28-2)**: White, deliquescent compound that decomposes by heat and is soluble in water. Same application as arsenic trioxide.

(j) **Arsenic trioxide (As_2O_3, CAS No. 1327-53-3)**: Toxic glassy, amorphous lumps or crystal compound; slightly soluble in water; octahedral crystals change to the monoclinic form by heating at 200°C; arsenious acid; occurs naturally as arsenolite and claudetite; used in some medicinal preparations but in small quantities. It is produced as a by-product of metal smelting operations. It has been estimated that as much as 70% of the world's arsenic production is used in timber treatment as copper chrome arsenate, and about 20% is used in agricultural chemicals as arsenic-containing pesticides; the remainder is used in glass, pharmaceuticals, and non-ferrous alloys.

(k) **Arsenin**: A heterocyclic organic compound composed of arsenic in a six-membered ring structure in which the carbon atoms are unsaturated, with no nitrogen atoms present.

(l) **Arseniosiderite ($Ca_3Fe_4(AsO_4)4(OH)_4 \cdot 4H_2O$)**: A yellowish-brown mineral consisting of a basic iron calcium arsenate and occurring as concretions.

(m) **Arseno compound**: A compound containing an As-As bond with the general formula (RAs)n, where R represents a functional group; structures are cyclic or long-chain polymers.

(n) **Arsenobismite ($Bi_2(AsO_4)(OH)_3$)**: A yellowish-green mineral consisting of a basic bismuth arsenate and occurring in aggregates.

(o) **Arsenoclasite ($Mn_5(AsO_4)_2(OH)_4$)**: A red mineral consisting of a basic manganese arsenate. Also spelled arsenoklasite.

(p) **Arsenolamprite (FeAsS)**: A lead gray mineral consisting of nearly pure arsenic; occurs in masses with a fibrous foliated structure.

(q) **Arsenolite (As_2O_3)**: A mineral crystallizing in the isometric system, usually occurring as a white bloom or crust. Also known as arsenic bloom.

(r) **Arsenopyrite (FeAsS)**: A white to steel-gray mineral crystallizing in the monoclinic system with pseudo-orthorhombic symmetry because of twinning; occurs in crystalline rock and is the principal ore of arsenic. Also known as mispickel.

(s) **Arsine (H_3As)**: A colorless, highly poisonous gas with an unpleasant odor.

(t) **Arsinic acid (R_2AsO_2H)**: Derived from trivalent arsenic; an example is cacodylic acid, or dimethylarsinic acid, $(CH_3)_2AsO_2H$.

(u) **Arsanilic acid (CAS No. 98-50-0)**: 4-Aminophenylarsonic acid derived from orthoarsenic acid, is an arsenical antibacterial veterinary medicine used in the prevention and treatment of swine dysentery.

(v) **Arsonic acid:** An acid derived from orthoarsenic acid, $OAs(OH)_3$; the type formula is generally considered to be $RAsO(OH)_2$; an example is para-aminobenzenearsonic acid, $NH_2C_6H_4AsO(OH)_2$.

(w) **Arsonium (AsH_4)**: A radical that may be considered analogous to the ammonium radical in that a compound such as AsH_4OH may form.

(x) **Arsphenamine (CAS No. 139-93-5)**

(y) **Calcium arsenate ($Ca_3(AsO_4)_2$, CAS No. 7778-44-1)**

(z) **Carbarsone (CAS No. 121-59-5)**

(aa) **Copper arsenate ($As_2Cu_3O_8$, CAS No. 10103-61-4)**

(bb) **Cupric arsenide ($CuHAsO_3$, CAS No. 10290-12-7)**

(cc) **Dimethylarsine ($(CH_3)_2AsH$, CAS No. 593-57-7)**

(dd) **Gallium arsenide (AsGa, CAS No. 1303-00-0)**

(ee) **Lead arsenate ($PbHAsO_4$, CAS No. 7784-40-9)**

(ff) **Magnesium arsenate ($As_2Mg_3O_8$, CAS No. 10103-50-1)**

(gg) **Manganese arsenate ($AsHMnO_4$, CAS No. 7784-38-5)**

(hh) **Methylarsine (CH_3AsH_2, CAS No. 593-52-2)**

(ii) **3-Nitro-4-hydroxy-phenyl arsonic acid (CAS No. 121-19-7)**

Also called roxarsone; used in veterinary medicine to promote growth, feed efficiency, and pigmentation, and to control swine dysentery. It is used as a synergist of primary anticoccidials.

(jj) **4-Nitrophenylarsonic acid (CAS No. 98-72-6)**

(kk) **Potassium arsenate (KH_2AsO_4, CAS No. 7784-41-0)**

(ll) **Potassium arsenite ($KH(AsO_2)_2$, CAS No. 10124-50-2)**

(mm) **Sodium arsenate (Na_3AsO_4, CAS No. 7631-89-2)**

Dibasic (CAS No. 7778-43-0); tribasic (CAS No. 13464-38-5)

(nn) **Sodium arsenite ($NaAsO_2$, CAS No. 7784-46-5)**

(oo) **Trimethylarsine ($(CH_3)_3As$, CAS No. 593-88-4)**

(pp) **Tryparsamide (CAS No. 554-72-3)**

(qq) **Zinc arsenate ($As_2O_8Zn_3$, CAS No. 13464-44-3)**

Arsenic pentoxide (CAS No. 1303-28-2)

Molecular formula: As_2O_5
Synonyms and trade names: Arsenic (V) oxide; Arsenic acid anhydride; Arsenic anhydride

Use and exposure

Arsenic pentoxide is an important commercial compound of arsenic. It is a white, shapeless (amorphous), crystalline, lumpy solid or, hygroscopic powder, and is not combustible. It emits irritating or toxic fumes (or gases) in a fire. Arsenic pentoxide is used as a solid or as a solution in the manufacture of arsenates, weed killer, metal adhesives, insecticides, fungicide wood preservatives, and colored gases, and in printing and dyeing. It can be synthesized either by burning elemental arsenic in oxygen, or by oxidizing arsenic trioxide with oxidizing agents, such as ozone, hydrogen peroxide, and nitric acid. Arsenic pentoxide decomposes to oxygen and As_2O_3 on heating, and dissolves readily in water to form arsenic acid, H_3AsO_4. Arsenic pentoxide reacts violently with bromine pentafluoride and reducing agents, and attacks many metals in the presence of water or moisture.

Toxicity and health effect

Exposures to arsenic pentoxide dust causes eye irritation, itching, burning, mild temporary redness and/or inflammation of the eye membrane (conjunctivitis), lacrimation, diplopia (temporary double vision), photophobia (abnormal sensitivity to light), vision dimness, and other transient eye damage or lesion formation (ulceration), cough, redness, sore throat, headache, dizziness, weakness, shortness of breath, and pain in chest. There may be a delay in the appearance of the symptoms of poisoning. Ingestion of arsenic pentoxide causes vomiting, abdominal pain, diarrhea, severe thirst, muscular cramps, and shock. Arsenic pentoxide causes adverse health effects to the eyes, inflammation and redness of the skin (erythroderma) with skin shedding (exfoliative dermatitis) may result from hyperkeratosis, pulmonary edema, acute respiratory distress syndrome (ARDS), and respiratory failure. Cardiovascular disturbances (heart rate, sinus tachycardia, and ventricular dysrhythmias), acute degenerative disease or dysfunction of the brain (encephalopathy) may develop and progress over several days, leading to delirium, and confusion. Severe exposures to arsenic pentoxide cause seizures, brain swelling (cerebral edema) and brain vessel bleeding (micro-hemorrhages) and damage in, peripheral nervous system, bone marrow (hematopoietic changes), liver, and lungs. Exposure far above the occupational exposure limits (OEL) may result in death.

Arsenic pentoxide and carcinogenicity

The IARC has classified arsenic pentoxide as Group 1, meaning a confirmed human carcinogen. Inorganic arsenic compounds as a generic class are carcinogenic to humans. Also, the US EPA has classified inorganic arsenic as Group A, meaning a human carcinogen.

Exposure limits

The ACGIH has set the TLV for arsenic pentoxide as 0.01 mg/m^3 (as As) TWA.

Storage

Arsenic pentoxide should be kept stored separately from strong bases, reducing substances, food, and feedstuffs and dry and safe area.

Precautions

During handling and use of arsenic pentoxide, occupational workers should use protective gloves, protective clothing, safety goggles or eye protection in combination with breathing protection if using powder, and workers should not eat, drink, or smoke in the area.

The worker should NOT wash away the spillage/waste into the sewer. The containers of arsenic pentoxide may explode when heated.

References

International Programme on Chemical Safety and the Commission of the European Communities (IPCS, CEC). 2005. Arsenic pentoxide. ICSC card no. 0377.
Technology Transfer Network Air Toxics web site. 2007. Arsenic Compounds. US EPA, Washington, DC.

Arsenic sulfide (CAS No. 1303-33-9)

Molecular formula: As_2S_3
Synonyms and trade names: Arsenic sesquisulfide; Arsenic tersulfide; Arsenic trisulfide; Arsenic yellow; Arsenious sulfide; Arsenous sulfide; Diarsenic Trisulfide; Kings Gold; Orpiment

Arsenic trioxide (CAS No. 1327-53-3)

Molecular formula: As_2O_3
Synonyms and trade names: Arsenic oxide; Arsenic (III) oxide; Arsenous trioxide; Arsenous acid; Arsenous oxide; Arsenic sesquioxide; White arsenic; Diarsenic trioxide; Crude arsenic; Arsenic (white); Arsenious oxide; Arsenic (III) trioxide; Arsenous anhydride; Arsenite; Arsenolite; Arsenous acid anhydride; Arsenous oxide anhydride; Arsodent; Claudelite; Claudetite; Arsenic oxide (3); Arsenic oxide (As_2O_3); Arsenicum album; Diarsonic trioxide; Diarsenic oxide

Use and exposure

Arsenic trioxide is an amphoteric oxide that shows a marked preponderance for its acidic properties. It dissolves readily in alkaline solutions to give arsenites. It is the starting point for the manufacture of arsenic-based pesticides (sodium arsenite, sodium arsenate, sodium cacodylate); a starting point for the manufacture of certain arsenic-based pharmaceuticals (Neosalvarsan) and veterinary products. Arsenic trioxide has several applications: (i) a decolorizing agent for glasses and enamels; (ii) a preservative for wood, in animal hides; (iii) in hydrogen recombination poison for metallurgical studies, as a termite poison; (iv) as a starting material for the preparation of elemental arsenic, arsenic alloys, and to enhance electrical junctions in semiconductors; (v) as a cytostatic in the treatment of refractory promyelocytic (M3), a subtype of acute myeloid leukemia; (vi) arsenic trioxide is also used to treat leukemia in patients who have not responded to other medications; (vii) arsenic trioxide was mixed with copper II acetate to form the extremely toxic but exceedingly vibrant pigment, known as Paris green, for use as a rodenticide in the Paris subways; and (viii) in cases of suicide and murder. Humans can be exposed to arsenicin several ways, such as ingesting small amounts present in food and water or breathing air containing arsenic, breathing sawdust or burning smoke from wood treated with arsenic, living in areas with unusually high natural levels of arsenic in rock, working in a job that involves arsenic production or use, such as copper or lead smelting, wood treating, or pesticide application.

Toxicity and health effects

Arsenic trioxide is readily absorbed by the digestive system. The toxic effects are also well known after inhalation of the dust or fumes and after skin contact. Initially, elimination is rapid (half-life of 1–2 days) by methylation to cacodylic acid and excretion in the urine, but a certain amount (30%–40% in the case of repeated exposure) is incorporated into the bones, muscles, skin, hair, and nails (all tissues rich in keratin) and eliminated over a period of weeks or months.

The first symptoms of acute arsenic poisoning by ingestion are digestive problems: vomiting, abdominal pains, and diarrhea often accompanied by bleeding. Sub-lethal doses can lead to convulsions, cardiovascular problems, inflammation of the liver and kidneys, and abnormalities in the coagulation of the blood. These are followed by the appearance of characteristic white lines (Mees stripes) on the nails and by hair loss. Lower doses lead to liver and kidney problems and to changes in the pigmentation of the skin. Cases of acute arsenic poisoning after inhalation and after skin contact with arsenic trioxide are many. The first signs are severe irritation, either of the respiratory tract or of the exposed skin, followed by longer-term neurological problems. Even dilute solutions of arsenic trioxide are dangerous on contact with the eyes. Ingesting or breathing low levels of inorganic arsenic for a long time can cause a darkening of the skin and the appearance of small "corns" or "warts" on the palms, soles, and torso. Skin contact with inorganic arsenic may cause redness and swelling. In brief, acute human exposures to arsenic in well water (typically containing more than 1200 µg/L) is known to cause many health effects, including abdominal pain, vomiting, diarrhea, muscular weakness and cramping, pain in arms and legs, skin changes or rashes, swelling of the eyelids, feet, and hands, and in serious poisoning, death.

Arsenic trioxide and carcinogenicity

The risk of lung cancer is clearly increased in certain smelter workers who inhale high levels of arsenic trioxide. The available information consistently indicates a causal relationship between skin cancer and heavy exposure to inorganic arsenic in drugs, in drinking water with a high arsenic content, or in the occupational environment. Studies on workers exposed in copper foundries in the United States, Japan, and Sweden, indicate a risk of lung cancer 6–10 times higher for the most exposed workers compared with the general population. Long-term ingestion of arsenic trioxide either in drinking water or as a medical treatment can lead to skin cancer. Reproductive problems (high incidence of miscarriage, low birth weight, congenital deformations) have also been indicated in one study of women exposed to arsenic trioxide dust as employees or neighbors of a copper foundry.

Studies of populations with high levels of arsenic in drinking water in West Bengal (India), Taiwan, Bangladesh, China, and Argentina, have shown elevated risks of cancers of the urinary tract, lung, skin, and, less consistently, cancers of the colon and liver. Some of these studies reported an increased risk of transitional cell bladder cancer at levels of arsenic below the then current US standard of 50 ppb. For example, a 2001 study from Taiwan compared the incidence of this cancer in communities with known levels of arsenic in their drinking water. The findings showed that the bladder cancer risk for residents of communities with levels above 100 ppb was over 15 times that of people living in areas with levels no higher than 10 ppb.

Several studies have shown that ingestion of inorganic arsenic can increase the risk of skin cancer and cancer of the liver, bladder, and lungs. Inhalation of inorganic arsenic can cause an increased risk of lung cancer. The DHHS and the US EPA have determined that inorganic arsenic is a known human carcinogen. The IARC has determined that inorganic arsenic is carcinogenic to humans. The IARC classifies arsenic trioxide and other compounds of arsenic as Group 1, meaning human carcinogens.

Exposure limits

The US EPA has set limits on the amount of arsenic that industrial sources can release to the environment and has restricted or cancelled many of the uses of arsenic in pesticides. The US EPA has set a limit of 0.01 ppm for arsenic in drinking water. The OSHA has set a PEL of 10 $\mu g/m^3$ in workplace air for 8-h shifts and 40-h work weeks.

References

Agency for Toxic Substances and Disease Registry (ATSDR). 2007. Toxicological Profile for Arsenic. US Department of Health and Human Services, Public Health Service, Atlanta, GA.
International Agency for Research on Cancer (IARC). 1973. Summaries and Evaluations. Arsenic and inorganic arsenic compounds, Vol. 2, 48. IARC, Lyon, France.

Arsine (CAS No. 7784-42-1)

Molecular formula: AsH_3
Synonyms and trade names: Arsenic trihydride; Arsenic hydride; Hydrogen arsenide

Use and exposure

Arsine is a highly toxic, colorless gas with a garlic odor. It is soluble in water, benzene, and chloroform. It is extremely flammable and explosive when exposed to heat, sparks, or flames. Arsine decomposes on heating and under the influence of light and moisture, producing toxic arsenic fumes. Arsine reacts with strong oxidants, causing explosion hazard and may explosively decompose on shock, friction, or concussion. Workers in the metallurgical industry involved in the production process and the maintenance of furnaces, and in the microelectronics industry get exposed to the substance. Arsine is extensively used in semiconductor industries, and in the manufacture of microchips.

Toxicity and health effects

Arsine is a highly toxic gas. It primarily targets the erythrocyte (red blood cell) and rapidly induces intravascular hemolysis. Secondary effects resulting from hemolysis include hemolytic anemia, hepatic and renal damage. The exact mechanism by which arsenic causes erythrocytes to rupture is unknown, but it is believed to be due to either oxidative damage or a reaction with sulfydryl. As stated, arsine is a potent hemolytic agent and causes acute intravascular hemolysis, rapid red blood cell destruction, and renal failure. Arsine is highly soluble in body fat and hence, can easily cross the alveolo-capillary membrane and into the red blood cell. Arsine causes chemical burns. Exposures to arsine cause headaches, malaise, weakness, dizziness, dyspnea; abdomen and back pain;

nausea, vomiting, diarrhea, bronze skin; hematuria (hemoglobin in urine), jaundice, liver enlargement, fever, anxiety, disorientation, delirium, shivering, muscular cramps, tachypnea, tachycardia, anemia, hyperkalemia, electro-cardiographic changes, burning sensations, peripheral neuropathy (focal anesthesia and paresthesia), agitation, disorientation, and hallucinations. The exposed individual and/or the occupational worker soon develops a sensation of cold and paresis in the limbs, hemoglobinuria, a garlic-like odor to the breath, multi-organ failure, and massive hemolysis and kidney failure. Toxic pulmonary edema or acute circulatory failure has been reported as the cause of death in some cases of arsine poisoning. Studies have indicated that occupational exposures to arsine cause increased rates of miscarriage among women associated with the semiconductor industry.

Arsine and carcinogenicity

Arsine and airborne arsenic compounds have been associated with carcinogenicity (Hubert et al., 1988). An increased risk of lung cancer has been reported in several epidemiological studies. Arsine is a human carcinogen. The IARC has classified arsenic and arsenic compounds as Group 1, meaning carcinogenic to humans.

Precautions

Occupational workers should be careful during handling/use of arsine. Workers should use protective gloves: neoprene, butyl rubber, PVC, polyethylene, or Teflon. Workers should also use appropriate protective equipment. If a leak occurs in a user's equipment, be certain to purge the piping with an inert gas prior to attempting repairs and evacuate all personnel from the affected area. The compressed gas cylinders should not be refilled without the express written permission of the owner.

References

Agency for Toxic Substances and Disease Registry (ATSDR). 2007. Medical Management Guidelines for Arsine. Division of Toxicology and Environmental Medicine, Atlanta, GA.
Hazardous Substances Data Bank (HSDB). 1999. US National Library of Medicine, Bethesda, MD.
National Institute of Occupational Safety and Health (NIOSH). Arsine. 2005. *Pocket Guide to Chemical Hazards*. Centers for Disease Control and Prevention, NIOSH, Atlanta, GA.
Stokinger, H.E. 1981. Arsine. In: *Patty's Industrial Hygiene and Toxicology*. Clayton, G.D., and Clayton, E. (Eds.). Wiley-Interscience, New York. pp. 1528–30.
World Health Organization (WHO). International Programme on Chemical Safety (IPCS). 2001. Arsine. ICSC card no. 0222. IPCS, WHO, Geneva, Switzerland (updated 2008).

Arsenic and Arsenic Compounds (CAS No. 7440-38-2)

Synonyms and trade names: Arsenic black; arsenicals; arsenic-75; colloidal arsenic; gray Arsenic; Metallic arsenic

(i) **Arsenic (CAS No. 7440-38-2)**

Molecular formula: **As**

(ii) **Arsenic acid (CAS No. 7778-39-4)**

Molecular formula: H_3AsO_4

(iii) **Arsenous acid (Arsenious acid) (CAS No. 10124-50-2)**
    Molecular formula: H_3AsO_3
(iv) **Arsenic trioxide (CAS No. 1327-53-3)**
    Molecular formula: As_2O_3
 (v) **Arsenic trihydride (Arsine) (CAS No. 7784-42-1)**
    Molecular formula: AsH_3
(vi) **Cadmium arsenide (CAS No. 12006-15-4)**
    Molecular formula: Cd_3As_2
(vii) **Gallium arsenide (CAS No. 1303-00-0)**
    Molecular formula: $GaAs$
(viii) **Lead arsenate (CAS No. 7778-40-9)**
    Molecular formula: $PbHAsO_4$

Synonyms and trade names: Arsenic acid, lead salt; Acid lead arsenate; Dibasic lead arsenate

Use and exposure

Lead arsenate is an odorless, white, heavy powder. It is not combustible and emits irritating or toxic fumes (or gases) in a fire. Lead arsenate enters the body by inhalation, or from contaminated food and beverages. It is used to make insecticides.

Toxicity and health effects

Exposures to lead arsenate occur in work areas and it is absorbed into the body by inhalation of its dust and by ingestion. It is irritating to the eyes, the skin, and the respiratory tract, and causes adverse effects on the gastrointestinal tract, nervous system, kidneys, liver, and blood. Accidental ingestion of lead arsenate in the workplace causes cough, sore throat, abdominal pain, diarrhea, drowsiness, headache, nausea, vomiting, muscular cramp, constipation, excitation, and disorientation in the worker.

Lead arsenate and carcinogenicity

Lead arsenate is a human carcinogen. It has been shown to cause skin, lung, and liver cancer.

Precautions

All arsenic compounds are generally regarded as carcinogens and no maximum value has been established. Arsenic compounds are mutagenic, extremely toxic by inhalation, through skin contact, and accidental ingestion/swallowing, and may become fatal.

References

Bhattacharyya, R., Jana, J., Nath, B., Sahu, S.J., Chatterjee, D. and Jacks, G. 2005. Groundwater As mobilization in the Bengal Delta Plain. The use of ferralite as a possible remedial measure: a case study. *Appl. Geochem.* 18: 1435–51.

Chakraborti, D., et al. 2003. Arsenic groundwater contamination in Middle Ganga Plain, Bihar, India: A future danger? *Environ. Health Perspect.* 111: 1194–1201.

———. 2004. Groundwater arsenic contamination and its health effects in the Ganga–Meghna–Brahmaputra plain. *J. Environ. Monit.* 6: 74N–83N.

International Programme on Chemical Safety and the Commission of the European Communities (IPCS-CEC). 1997. Lead arsenate. ICSC no. 0911. IPCS, Geneva, Switzerland (updated 2004).

Mukherjee, A.B. and Bhattacharya, P. 2001. Arsenic in groundwater in the Bengal Delta Plain: Slow poisoning in Bangladesh. *Environ. Rev.* 9: 189–220.

National Academy of Sciences (NAS). 1977. Arsenic. Medical and biologic effects of environmental pollutants: Arsenic. National Academy of Sciences, Washington, DC.

Singh, A.K. 2006. Chemistry of arsenic in groundwater of Ganges–Brahmaputra river Basin. *Curr. Sci.* 91 (No. 5). PP. 599–606. (Review article).

Sodium arsenate (CAS No. 7778-43-0)

Molecular formula: $AsHNa_2O_4$

Synonyms and trade names: Sodium orthoarsenate; Disodium hydrogen arsenate; Arsenic acid, disodium salt; Sodium arsenate dibasic

Use and exposure

Sodium arsenate is currently registered for use as ant bait. These baits are used in approximately 1% of US homes. Sodium arsenate is a pentavalent form of inorganic arsenic. It is a heptahydrate that normally exists as colorless crystals with no discernible odor. Sodium arsenate contains 24% arsenic.

Toxicity and health effects

Inorganic arsenical compounds have been classified as Class A oncogenes, demonstrating positive oncogenic effects based on sufficient human epidemiological evidence. Inorganic arsenicals are known to be acutely toxic. The symptoms that follow oral exposure include severe gastrointestinal damage resulting in vomiting and diarrhea, and general vascular collapse leading to shock, coma, and death. Muscular cramps, facial edema, and cardiovascular reactions are also known to occur following oral exposure to arsenic.

On ingestion, organic arsenic compounds cause severe health effects, including burning lips, throat constriction, abdominal pain, dysphagia, nausea, vomiting, diarrhea, convulsions, coma, and death. Irritation of the respiratory tract, skin, and eyes may result from inhalation exposures. Chronic exposure to organic arsenic compounds may result in dermatitis, anemia, leukocytopenia, or the effects associated with several forms of cancer.

Storage

It is important to keep stored organic arsenic compounds in a cool, dry, well-ventilated area in tightly sealed containers and with proper labels and identification. The storage containers of organic arsenic compounds should be protected from physical damage and stored separately from oxidizers such as perchlorates, peroxides, permanganates, chlorates, or nitrates, and strong acids such as hydrochloric, sulfuric, or nitric acid. Further, specific organic arsenic compounds must have separate storage requirements, which should be evaluated prior to storage.

Precautions

During handling and use of arsenic compounds, workers should use appropriate personal protective clothing and equipment. Users/workers must be careful during use, and maintain effective measures to prevent skin contact with organic arsenic compounds. The selection of the appropriate personal protective equipment (PPE), such as gloves, sleeves, and encapsulating suits should be based on the extent of the worker's potential exposure to organic arsenic compounds. The users and the management should periodically evaluate and determine the effectiveness of the chemical-resistant clothing in preventing dermal contact and ensuring users' safety.

References

National Library of Medicine (NLM). 1995. National Institutes of Health, Health & Human Services (NIHHS), 1995.

US Environmental Protection Agency (US EPA). 1986. Sodium arsenate. Pesticide Fact Sheet no. 114m 1986 12/86. Cornell University, Ithaca, NY.

World Health Organization (WHO). 2001. Arsenic and arsenic compounds. Environmental Health Criteria No. 224 (second edition), Geneva, Switzerland.

Sodium arsenate heptahydrate (CAS No. 10048-95-0)

Synonyms and trade names: Arsenic acid; Disodium salt; Heptahydrate; Sodium acid arsenate heptahydrate; Disodium arsenate heptahydrate; Sodium arsenate; Dibasic, 7-hydrate

Molecular formula: $Na_2HAsO_4, 7H_2O$

Data on ingredients

Arsenic Acid, Disodium salt (**CAS No. 7778-43-0**)

Use and exposure

Sodium arsenate heptahydrate is a poisonous solid/powdered solid/crystalline powder. It is odorless with a saline taste. The aqueous solution is alkaline to litmus and decomposes on heating at 100°C (212°F), rapidly becoming anhydrous. It is incompatible with acids, iron, aluminum, and zinc in the presence of water, and strong reducing agents. It is non-corrosive in the presence of glass. Sodium arsenate heptahydrate emits toxic fumes of arsenic when heated to decomposition.

Toxicity and health effects

Arsenic is highly toxic. Exposures through ingestion and inhalation cause adverse health effects that include, but are not limited to, irritation with itching, burning, and conjunctiva damage, photophobia, corneal injury, dimness of vision, diplopia, lacrimation, cold and clammy skin, low blood pressure, weakness, headache, cramps, inflammation of the mucous membranes with cough and foamy sputum, restlessness, dyspnea, cyanosis, pulmonary edema, burning in the esophagus, vomiting, and bloody diarrhea, damage to the liver and kidneys, and death from circulatory failure.

Repeated exposures to sodium arsenate heptahydrate cause bronzing of the skin, edema, dermatitis, and lesions. Repeated or prolonged inhalation of dust may cause damage to the nasal septum. Chronic exposure from inhalation or ingestion may cause hair and weight loss, a garlic odor to the breath and perspiration, excessive salivation and perspiration, central nervous system damage, hepatitis, gastrointestinal disturbances, and cardiovascular damage.

Sodium arsenate heptahydrate and carcinogenicity

Sodium arsenate heptahydrate has been classified as an A1 carcinogen (confirmed as a human carcinogen) by the ACGIH and as a proven human carcinogen by the IARC. It is mutagenic for mammalian somatic cells, and mutagenic for bacteria and/or yeast. Sodium arsenate heptahydrate has been reported as a suspected fetal toxin. Arsenic compounds are known human carcinogens.

Exposure limits

For sodium arsenate heptahydrate, the OSHA has set the PEL as 10 µg (As)/m^3 ppm (TWA); the ACGIH has set the TLV as 0.01 mg (As)/m^3 (TWA).

Storage

Workers should keep containers of sodium arsenate heptahydrate tightly closed and under lock and key in a cool, well-ventilated room/store area.

Precautions

Occupational workers should avoid ingestion or breathing the dust of sodium arsenate heptahydrate. On exposure to sodium arsenate heptahydrate, immediately flush eyes with plenty of water. Occupational workers should wear suitable protective clothing, suitable respiratory equipment, and work in proper ventilation. During use, workers should avoid contact with the skin, eyes be away from incompatibles such as acids. For any kind of accidental ingestion, workers should seek immediate medical advice.

Occupational workers and users should be aware that sodium arsenate heptahydrate is very toxic and its improper use and handling is dangerous, causing fatal injury if swallowed and/or inhaled. Sodium arsenate heptahydrate is a cancer hazard. It contains inorganic arsenic that can cause cancer. The risk of cancer from sodium arsenate heptahydrate depends on the duration and level of exposure. Because sodium arsenate heptahydrate causes irritation to the skin, the eyes, and the respiratory tract, and causes liver and kidney damage, workers should handle the chemical substance only with adequate ventilation, respiratory equipment, and under qualified supervision.

Reference

Material Safety Data Sheets (MSDS). 2009. Sodium arsenate heptahydrate. MSDS no. S2858. Environmental Health & Safety (EHS). Mallinckrodt Baker, Phillipsburg, NJ.

Asbestos (CAS No. 1332-21-4)

Synonyms and trade names: Serpentine asbestos; Amphibole asbestos; Fiber asbestos; Blue asbestos

Use and exposure

Asbestos, a naturally occurring fibrous mineral, is widespread in nature. Asbestos is non-combustible, resistant to heat, and features a low conductivity and is insoluble in water. Asbestos is classified into two groups: (i) serpentine and (ii) amphibole. The serpentine group includes: (i) chrysotile: molecular formula: $Mg_3Si_2O_5(OH)_4$; and (ii) crocidolite: molecular formula: $Na_2Fe_2(Fe,Mg)_3Si_8O_{22}(OH)_2$. The amphibole group includes (i) amosite: molecular formula: $Na_2Fe_2(Fe,Mg)_3Si_8O_{22}(OH)_2$; (ii) anthophyllite: molecular formula: $Mg_7Si_8O_{22}(OH)_2$; (iii) tremolite: molecular formula: $Ca_2Mg_5Si_8O_{22}(OH)_2$; and (iv) actinolite: molecular formula: $Ca_2(Mg,Fe)_5Si_8O_{22}(OH)_2$.

Asbestos fibers can be very small—up to 700 times smaller than a human hair. Because it is fire resistant, resists many chemicals, and is an excellent insulator, asbestos was added to a variety of building materials and other products. The most important deposits are to be found in the former Soviet Union, Canada, and South Africa. Natural emissions are produced, for example, by the weathering of serpentine rocks.

Asbestos is not a chemical element, but rather an umbrella term for two groups of minerals, namely, serpentine and amphibole asbestos. Asbestos is a fibrous, impure mineral and its color varies between pure white and green, brown or gray, depending on the iron-oxide content. Crocidolite is blue or blue asbestos with fibers between 20 and 25 nm long that are smooth and greasy to the touch. Asbestos, because of its heat resistance, suppleness, and its capacity for being incorporated into inorganic and organic binders, is used in industry for thermal insulation, fireproofing, and sealing, in addition to acting as a filler in the production of countless composites (more than 3000). Economic exploitation is centered on chrysotile asbestos from the serpentine group (95%). Five percent of amphibole asbestos is used for asbestos fiber production.

Spun asbestos is used for fireproof, chemical-resistant clothing. When mixed with cement, asbestos is cast to form panels and pipes (70%–90% of the world's production is used in the construction industry in Western Europe; acc. WHO, 1987). Asbestos has also been used as a filter material in the drinks and pharmaceutical industry and for the manufacture of brake and clutch linings for motor vehicles.

Toxicity and health effects

Various asbestos-induced illnesses are known from the industrial-medicine sector in which the size of the fibers plays a crucial role. Generally speaking, fibers with a diameter of <2 m and a length of >5 μm are considered to be hazardous to health (diameter:length = 1:3). Such a fiber size is capable of entering the lungs, accumulating and becoming encapsulated. Fibers have also been found to have a certain migration capability in the organism and the cell metabolism. Accumulation in the lungs causes sclerosis of the pulmonary alveoli, thereby impairing oxygen exchange. The inhalation of large quantities of fiber can cause asbestosis, which increases the risk of bronchial cancer. In particular, dusts <200 μm are highly toxic and are suspected of being a direct cause of tumors. Exposure to asbestos irritates the eyes and the respiratory tract. Direct penetration into damaged skin produces excessive hornification. Fibers in the lungs cause chronic bronchitis, irritation of the pleura and pleurisy. Distension of the lungs can result in lung cancer. Workplace exposure may produce periods of latency in the gastrointestinal tract lasting up to 40 years. To date, there are no known characteristic toxicology data (DVGW, 1988).

Asbestos fibers are released into the environment by natural and anthropogenic processes. The liberation of asbestos during the manufacture and processing of materials containing asbestos must be significantly reduced, especially in confined areas and at workplaces because of the great hazard to the respiratory organs, in particular the lungs. There is increasing speculation that the oral intake of asbestos can cause tumors. Therefore, there is no justification for the further use of asbestos cement in drinking water pipes. Numerous products containing substitute materials are now making it possible to dispense with the use of substances containing asbestos.

References

Material Safety Data Sheets (MSDS). 2005. Safety data for asbestos. The Physical and Theoretical Chemistry Laboratory, Oxford University, U.K.
US Environmental Protection Agency (US EPA). 2010. Asbestos: Basic information. US EPA, Washington, DC.

Atrazine (CAS No. 1912-24-9)

Molecular formula: $C_8H_{14}ClN_5$
Synonyms and trade names: 6-Chloro-N-ethyl-N'-(1-methylethyl)-1,3,5-triazin-2,4-diamine; 2-Chloro-4-ethylamino-6-isopropylamino-s-triazin; 2-Chloro-4-(ethylamine)-6-(isopropylamine)-s-triazine, Gesaprim

Use and exposure

Atrazine is a colorless, crystalline solid. Although atrazine is very stable, it is only slightly soluble in water, but soluble in N-pentane, chloroform, dimethyl sulfoxide, ethyl acetate, diethyl ether, and methanol. Atrazine is a broad-spectrum triazine herbicide and is used as a selective herbicide for weed control in corn and asparagus, in the culture of sugarcane and pineapple. Additionally, it is used as a total herbicide on roads and public places as well as on uncultivated ground in combination with amitrol, bromacil, dalapon, and growth promoters. Atrazine inhibits photosynthesis and other metabolic processes in plants. There are no natural sources of atrazine. It is produced from cyanuric acid chloride with ethylamine and isopropylamine. The reaction takes place successively in tetrachloromethane. All atrazine produced is released into the environment. The formulations include granules, water dispersible granules, liquid, suspension concentrate, wettable powder, and a combination with many other herbicides. Atrazine is compatible with various insecticides and fungicides.

Atrazine was banned in the European Union (EU) in 2004 because of its persistent groundwater contamination. In the United States, however, atrazine is one of the most widely used herbicides, with 76 million pounds of it applied each year. It is probably the most commonly used herbicide in the world.

Toxicity and health effects

The acute toxicity of atrazine for mammals is very low. In addition, there is hardly any resorption via the skin due to its low solubility. Atrazine does not irritate the skin or the eyes. It is easily resorbed after oral intake and within 24 h, more than 50% is excreted via urine. During the passage, atrazine is completely metabolized mainly by way of oxidative dealkylation of the amino group and reaction of the chlorine atom with endogenic thiolic reagents.

Exposures to atrazine cause damage to the liver, kidney, and heart in animals. It has also been shown to cause changes in blood hormone levels in animals, which affect ovulation and the ability to reproduce. In animal experiments, mutagenic or teratogenic effects have not been discovered. There is little data about the health effects of atrazine in humans. In Germany, atrazine has not been classed as toxic.

Atrazine and carcinogenicity

Available information is inadequate to definitely state whether atrazine causes cancer in humans. There are limited human and animal data that suggest there may be a link between atrazine exposure and various types of cancer. A Cancer Assessment Review Committee (CARC) sponsored by the US EPA has classified atrazine as not likely to be carcinogenic to humans. The IARC has determined that atrazine is not classifiable as to its carcinogenicity to humans. The ACGIH observed that atrazine is not classifiable as a human carcinogen.

Exposure limits

The US EPA has set a maximum limit of 0.003 mg/L of atrazine in drinking water. The OSHA has set a limit of 5 mg/m^3 of atrazine in workplace air for an 8-h workday. The US EPA has set the maximum allowed level in foods as 0.02–15 ppm.

References

Agency for Toxic Substances and Disease Registry (ATSDR). 2003. Toxicological profile for Atrazine. US Department of Health and Human Services, Public Health Service. Atlanta, GA.
Tomlin, C.D.S. (ed.). 2006. *The Pesticide Manual: A World Compendium*, 14th ed. British Crop Protection Council, Farnham, Surrey, U.K.

Azinphos-ethyl (CAS No. 2642-71-9)

Molecular formula: $C_{12}H_{16}N_3O_3PS_2$
IUPAC name: S-(3,4-dihydro-4-oxobenzo[d]-[1,2,3]-triazin-3-ylmethyl)   0,0-diethyl phosphorodithioate
Synonyms and trade names: Athyl-Gusathion; Azinos; Azinphos-etile; Azinfosethyl; Ethyl-azinophos; Ethyl Guthion; Gusathion; Guthion (ethyl); Triazotion

Use and exposure

Azinphos-ethyl forms colorless, clear crystals. It is insoluble in water, but soluble in most organic solvents. As an acaricide, azinphos-ethyl is used for the control of pests, such as, spider mites, aphids, caterpillars, potato bug, beetles, bollweevils, whiteflies, bollworms, thrips, and other biting and sucking insects. Human exposures to azinphos-ethyl occur through absorption from the gastrointestinal tract, through the intact skin, and by inhalation of fine spray mist and dusts.

Azinphos-ethyl is a non-systemic organophosphorus insecticide, which is used against a relatively broad spectrum of insects. These include lepidopterous larvae, beetles and their larvae, aphids, jassids, and spider mites on various crops. And also crops of cotton, rice, sugar and fodder beets, fruits such as apples, pears, citrus-fruit, grapes; tobacco. Many countries around the world have stopped the use of this pesticide.

Toxicity and health effects

Azinphos-ethyl as an organophosphate pesticide is very toxic to animals and humans, inhibiting the cholinesterase enzyme. Occupational workers exposed to azinphos-ethyl show symptoms of poisoning. On exposure through skin contact, inhalation of dust or spray, or accidental ingestion/swallowing, azinphos-ethyl is fatal. The early symptoms of toxicity include, but are not limited to, excessive sweating, headache, miosis, dyspnea, nausea, vomiting and diarrhea, blurred vision, muscle fasciculations, weakness, increased salivation, stomach pains, slurred speech. More severe poisoning leads to respiratory failure due to a combination of bronchorrhea, bronchoconstriction (muscarinic effects), paralysis of the respiratory muscles (nicotinic effects), and respiratory center paralysis (central effects), which all may eventually lead to shortness of breath, brain hypoxia, convulsions, and coma. Oral and dermal exposures to azinphos-ethyl cause a disturbed heart rate with chest pain. Hypotension (low blood pressure) may be observed, although hypertension (high blood pressure) is not uncommon. Exposed and poisoned occupational workers show respiratory symptoms, such as dyspnea, pulmonary edema, respiratory depression, and respiratory paralysis.

Reports have shown that azinphos-ethyl causes the inhibition of acetylcholinesterase activity in plasma, erythrocytes, the brain and sub-maxillary gland. Studies have indicated that multiple doses of azinphos-ethyl in low doses cause the plasma activity of rats and dogs to fall rapidly to a stable level, while the activity of the erythrocyte acetylcholinesterase falls more gradually.

Azinphos-ethyl and carcinogenicity

The DHHS and the IARC have not classified azinphos-ethyl as to its carcinogenicity. The US EPA also observes that azinphos-ethyl showed no evidence of carcinogenicity in rats and mice for classification.

Precautions

Azinphos-ethyl is an organophosphorus compound that inhibits cholinesterase enzymes. It is very toxic and occupational workers should be very careful during use, storage, and waste disposal of it.

References

Tomlin, C.D.S. (Ed.). *The Pesticide Manual*, 15th ed. British Crop Protection Council (BCPC). Cambridge, U.K. Black Well Scientific Publications.

US Environmental Protection Agency (US EPA). 2007. Chemical Emergency Preparedness and Prevention Azinphos ethyl, Emergency First Aid Treatment Guide. US EPA.

World Health Organization (WHO). 1994. Azinphos-ethyl. Data Sheet on Pesticides No. 72, WHO/PCS/DS/94.72.

Azinphos methyl (CAS No. 86-50-0)

Molecular formula: $C_{10}H_{12}N_3O_3PS_2$
IUPAC name: S-(3,4-dihydro-4-oxobenzo[d]-(1,2,3)- triazin-3-ylmethyl) O,O-dimethyl phosphorodithioate

Synonyms and trade names: Crysthion; Crysthyon; DBDR; Gothnion; Gusathion; Guthion; Methyl; Gusathion; Methyl Guthion; Gusathion; Guthion
Toxicity class: USEPA: I; WHO: Ib

Uses and exposures

Azinphos methyl is an organophosphate pesticide available as a white or colorless crystalline solid substance. Technical azinphos methyl is a brown waxy solid that is sparingly soluble in water, but very soluble in most organic solvents. It is incompatible with strong oxidizing agents, and is hydrolyzed by water, acids, and alkalies. Azinphos methyl is used on many crops, especially apples, pears, cherries, peaches, almonds, and cotton for the control of a variety of crop pests and fruit crop pests. These include, codling moth, light brown apple moth, spring beetle, apple leaf hopper, bryobia mite, pear and cherry slug, woolly aphid, mites, codling moths, lygus-bugs, bollworms, armyworms, boll weevils, thrips, grasshoppers, stinkbugs, spittle bugs, plum curculio, pear psylla, scale, and the European brown snail. The use of azinphos methyl is not recommended either for public health or domestic purposes. Azinphos methyl is absorbed from the gastrointestinal tract, through the intact skin, and by inhalation of fine spray mist or dusts during spray operations. It is available as an emulsifiable liquid, a flowable liquid, a ULV liquid, and wettable powder formulations.

Toxicity and health effects

Azinphos methyl is extremely toxic, the probable lethal oral dose for humans being from 5 to 50 mg/kg body weight (just seven drops). Different solvents used in the preparation and commercial formulations of azinphos methyl further change the toxicological properties. The symptoms of poisoning include, but are not limited to, excessive sweating, headache, weakness, giddiness, nausea, hypersalivation, vomiting, stomach pains, blurred vision, slurred speech, and muscle twitching, and with advanced time and poisoning, chest tightness, vomiting, cramps, convulsion, coma, loss of reflexes and loss of sphincter control, respiratory failure, unconsciousness, or death.

Reports have indicated that azinphos methyl causes no delayed neurotoxicity. Azinphos methyl has been classified as a tumorigen. After accidental or intentional exposure to azinphos methyl, occupational workers exhibit over stimulation of the nervous system, rapid twitching and paralysis of muscles, and death.

Azinphos methyl and carcinogenicity

The DHHS and the IARC have not classified azinphos methyl as to its carcinogenicity. The US EPA also observes that azinphos methyl shows no evidence of carcinogenicity in rats and mice for classification.

Storage and precautions

Azinphos methyl only should be handled by trained occupational workers/personnel wearing proper protective clothing. Azinphos methyl should always be stored and transported in clearly labeled impermeable containers under lock and key. Its storage area should be secure from access by unauthorized persons and children. Occupational workers must be properly informed that no food or drink should be stored in the storage areas.

References

Agency for Toxic Substances and Disease Registry (ATSDR). 2008. Toxicological profile for Guthion. US Department of Health and Human Services (US DHHS), Public Health Service (PHS), Atlanta, GA.

International Programme on Chemical Safety and the Commission of the European Communities, IPCS-CEC. 2001. Azinphos-methyl. ICSC Card No. 0826, WHO, Geneva, Switzerland.

Material Safety Data Sheets (MSDS). 1995. Acephate. Extension Toxicology Network (EXTOXNET), Pesticide Information Profiles. Cornell University, Ithaca, NY (updated 2001).

Routt, R.J. and Roberts, J.R. 1999. Organophosphate insecticides. Recognition and Management of Pesticide Poisonings. US Environmental Protection Agency, National Technical Information Service. EPA 735-R-98-003; pp. 55–57. Washington, DC.

Tomlin, C.D.S. (ed.). 2008. *The Pesticide Manual,* 15th ed. British Crop Protection Council (BCPC). Blackwell Scientific, Hampshire, U.K.

US Environmental Protection Agency (US EPA). 1987. Acephate. Pesticide Fact Sheet. National Technical Information Service. US EPA, Washington, DC.

———. 1991. Azinphos-methyl. Pesticide Information Profiles, US EPA., Washington, DC.

———. 2000. Azinphos-methyl. Technology Transfer Network Air Toxics web site (Revised 2007).

———. 2001. Azinphos methyl. Pesticide Registration. IRED Facts (updated June 2007).

B

- Bacillus thuringiensis
- Barium
- Barium chloride
- Barium hydroxide (Anhydrous)
- Barium hydroxide (Octahydrate)
- Barium nitrate
- Barium sulfate
- Bendiocarb
- Benomyl
- Benzaldehyde
- 1,2-Benzanthracene
- Benzene
- Benzidine
- 2,3-Benzofuran
- Benzyl acetate
- Benzyltrimethylammonium dichloroiodate
- Beryllium
- Beryllium chloride
- Bifenthrin

- 1,1'-Biphenyl
- Bis(chloromethyl) ether
- Bismuth
- Boron
- Boric acid
- Boron carbide
- Sodium tetraborate decahydrate/Borax
- Sodium tetraborate pentahydrate
- Boron nitride
- Borazole (Borazine)
- Boron trichloride
- Boron trifluoride
- Bromodichloromethane (BDCM)
- Bromomethane
- 1,3-Butadiene
- 2-Butanone
- 1-Butene
- 2-Butoxyethanol
- Sec-Butyl acetate
- n-Butyl acetate,
- Isobutyl acetate
- 1-Butyne
- Butane
- Butanethiol
- Sec-Butyl acetate
- n-Butyl acetate
- Isobutyl acetate
- tert-Butyl acetate/t-butyl acetate
- n-Butyl alcohol
- n-Butylamine

Bacillus thuringiensis (CAS No. 68038-71-1) (*B.t.* var. kurstaki)

Synonyms and trade names: Berliner (*B.t.* var. kurstaki); Dipel; Thuricide; Bactospeine; Leptox; Novabac; Victory; Certan (*B.t.* var. aizawa); Teknar (*B.t.* var. israelensis)

Use and exposure

Bacillus thuringiensis is commonly known as *B.t.* It is a microorganism that produces chemicals toxic to insects. *B.t.* was registered in the United States for use as a pesticide in 1961 and re-registered in 1998. *B.t.* occurs naturally in the environment and has been isolated from soil, insects, and plant surfaces. *B.t.* pesticides are used for food and non-food crops, greenhouses, forests, and outdoor home use. *B.t.* pesticides exist in granular, powder, dust, suspension, and flowable forms.

A number of insecticides are based on these toxins. *B.t.* is considered ideal for pest management because of its specificity to pests and because of its lack of toxicity to humans as well as natural enemies of many crop pests. There are different strains of *B.t.*, each with specific toxicity to particular types of insects. For instance, *B.t. aizawai* (*B.t.a.*) is used against wax moth larvae in honeycombs, *B.t. israelensis* (*B.t.i.*) is effective against mosquitoes, blackflies and some midges, and *B.t. kurstaki* (*B.t.k.*) controls various types of lepidopterous insects, including the gypsy moth and cabbage looper. A new strain, *B.t. san diego*, has been found to be effective against certain beetle species and the boll weevil. In order to be effective, *B.t.* must be eaten by insects in the immature, feeding stage of larvae development. *B.t.* is ineffective against adult insects. Regular monitoring of the target insect population before application of *B.t.* ensures good control of the vulnerable larval stage. More than 150 insects, mostly lepidopterous larvae, are known to be susceptible in some way to *B.t.* Death of target larvae is known to occur within a few hours to a few weeks of *B.t.* application, depending on the species of insect and the amount of *B.t.* ingested by the insect.

B.t. is moderately persistent in soil and its toxins degrade rapidly. The movement of *B.t.* is limited following pesticide application and it is unlikely to contaminate groundwater. *B.t.* is not native to water and is not likely to multiply in water.

Toxicity and health effects

The insecticidal action of *B.t.* is attributed to protein crystals produced by the bacterium. Exposures of test animals to *B.t.* using several routes did not produce any acute toxicity in birds, dogs, guinea pigs, mice, or rats. Also laboratory rats when injected with *B.t.k.*, showed no toxic or virus-like effects. No oral toxicity was found in rats, mice, or Japanese quail fed protein crystals from *B.t.* var. *israelensis*. Studies indicated that after rats ate *B.t.*, the microorganism remained in the digestive system until it was eliminated from the body. Rabbits exposed to *B.t.* showed mild skin irritation and rats showed low inhalation toxicity to *B.t.* In fact, chronic toxicity studies in dogs, guinea pigs, rats, and other species of test animals showed no evidence of adverse health effects.

The toxicity of *B.t.* is insect specific. Researches have provided valuable data and identified *B.t.* subspecies that differ in toxicity to different insects. Examples of *B.t.* subspecies and the insects they affect are *aizawai* (moths), *kurstaki* (moths), *israelensis* (mosquitoes and flies), and *tenebrionis* (beetles). Also, phytotoxicity studies (plant researches) showed *B.t.* genes in some crops (*B.t.* crops) to combat insects of corn crops, cotton, and potatoes. *B.t.* must be eaten by insects to be effective and works by interfering with digestion. Insects are most sensitive to *B.t.* when they are larvae, an immature life stage. Insects that eat *B.t.*, die from hunger or infection. It does not cause disease outbreaks in insect populations. *B.t.* may produce toxic chemicals that are released from the organism.

Bacillus thuringiensis and carcinogenicity

Different studies indicate that *B.t.* has no tumor-producing effects, mutagenic effects, birth or reproductive toxicity to animals. The NTP, the IARC, and the OSHA report no information about *B.t.* as a human carcinogen.

References

Material Safety Data Sheets (MSDS). 1994. *Bacillus thuringiensis.* Extension Toxicology Network (EXTOXNET). Oregon State University, USDA/Extension Service, Corvallis, OR.

Meister, R.T. 2004. *Crop Protection Handbook*, Meister Media Worldwide, Willoughby, OH.

Tomlin, C.D.S. (ed.). 2006. *The Pesticide Manual*, 14th ed. British Crop Protection Council. Blackwell Scientific, Cambridge, MA.

US Environmental Protection Agency (US EPA). 1986. Pesticide fact sheet for *Bacillus thuringiensis.* Fact sheet no. 93. Office of Pesticide Programs, Washington, DC.

World Health Organization (WHO). 1999. *Bacillus thuringiensis.* Environmental Health Criteria, 217. Geneva, Switzerland.

Barium (CAS No. 7440-39-3)

Molecular formula: **Ba**

Use and exposure

Barium is a silvery-white metal. It exists in nature only in ores containing mixtures of elements. The important combinations are peroxide, chloride, sulfate, carbonate, nitrate, and chlorate. The pure metal oxidizes readily and reacts with water, emitting hydrogen. It combines with other chemicals such as sulfur or carbon and oxygen to form barium compounds. Barium compounds are used by the oil and gas industries to make drilling muds. Barium attacks most metals with the formation of alloys; iron is the most resistant to alloy formation. Barium forms alloys and intermetallic compounds with lead, potassium, platinum, magnesium, silicon, zinc, aluminum, and mercury. Barium compounds exhibit close relationships with the compounds of calcium and strontium, which are also alkaline earth metals. Doctors sometimes use barium sulfate to perform medical tests and to take x-rays of the gastrointestinal tract. Twenty-five barium isotopes have been identified. ^{138}Ba is the most abundant; the others are unstable isotopes with half-lives ranging from 12.8 days for ^{140}Ba to 12 sec for ^{143}Ba. Two of these isotopes, ^{131}Ba and ^{139}Ba, are used in research as radioactive tracers. The general population is exposed to barium through air, drinking water, and food.

Toxicity and health effects

The health effects of barium compounds depend on how well the compound dissolves in water or in the stomach contents. Barium compounds that do not dissolve well, such as barium sulfate, are considered not harmful. Barium carbonate dust and barium oxide dust have been reported to be a bronchial irritant. While barium carbonate is a dermal irritant, barium oxide is a nasal irritant. Occupational workers exposed to barium dust, usually in the form of barium sulfate or carbonate, often develop a benign pneumoconiosis also called "baritosis." The effect of baritosis has been shown to be reversible and has not caused any kind of severe pulmonary adverse effect. Barium compounds that do not dissolve in water are considered safe. However, the health effects of the different barium compounds depend on the degree of their water solubility. The compounds that dissolve well in water are known to cause harmful health effects when ingested in high levels. Symptoms of poisoning include stomach irritation, brain swelling, muscle weakness, liver and kidney damage, adverse effects to the heart, increased blood pressure, changes in heart rhythm, effects on the spleen, difficulties in breathing, and swelling of the brain. Exposures to high concentrations of barium through food and drinking water

TABLE 1

Toxicity of Barium Compounds to Humans

Barium Compound	Exposure Data	Effect
Barium carbonate	Lowest lethal dose (57 mg/kg)	Death
Barium carbonate	Lowest toxic dose (29 mg/kg)	Flaccid paralysis without anesthesia; paresthesia; muscle weakness
Barium chloride	Lowest lethal dose (11.4 mg/kg)	Death
Barium polysulfide	Lowest toxic dose (226 mg/kg)	Flaccid paralysis without anesthesia; muscle weakness; dyspnea

cause gastrointestinal disturbances. Barium causes vomiting, abdominal cramps, diarrhea, difficulties in breathing, increased or decreased blood pressure, numbness around the face, and muscle weakness, changes in heart rhythm or paralysis, and possibly death. Animals exposed to barium over long periods showed kidney damage, decreased body weight, and fatal injury. Ingestion of large amounts of barium chloride (2 and 4 g) causes fatal injury, because barium ions paralyze the heart. Acute poisoning with barium causes nausea and diarrhea, cardiac problems, and muscular spasms, as well as cardiac arrest. Thus, barium, at concentrations normally found in our environment, does not pose any significant risk for the general population. However, for specific subpopulations and under conditions of high barium exposure, the potential for adverse health effects should be taken into account.

Barium and carcinogenicity

Barium has not been classified as a carcinogen by the DHHS or the IARC. Similarly, the US EPA reported that barium on ingestion or inhalation has not caused any carcinogenic effects to humans. The IARC indicated that barium chromate (VI) is the only barium compound for which there is sufficient evidence that it is a human carcinogen.

Exposure limits

For barium, the US EPA has set a limit of 2 ppm (2.0 mg/L) in drinking water. The OSHA has set the PEL as 0.5 mg m^3 TWA in workplace air. The NIOSH has set the REL limit as 500 mg/m^3 for soluble barium compounds, 10 mg/m^3 for total barium dust, and 5 mg/m^3 for barium sulfate. The OSHA has set a limit for barium sulfate dust as 15 mg/m^3 of total dust and 5 mg/m^3 as desirable fraction.

Precautions

Barium is a strong reducing agent and reacts violently with oxidants, acids, and halogenated solvents. It reacts with water, forming flammable/explosive gas. Barium is known to spontaneously ignite on contact with air, if in powder form, causing fire and explosion hazard.

References

Agency for Toxic Substances and Disease Registry (ATSDR). 2007. Toxicological Profile for Barium and Compounds. U.S. Department of Public Health and Human Services, Public Health Service, Atlanta, GA.

Beliles, R.P. 1994. The Metals. In: *Patty's Industrial Hygiene and Toxicology*, 4th ed. Clayton, G.D. and Clayton, F.E. (eds.). Wiley, J. New York. pp. 1925–29.

World Health Organization (WHO). 1990. Barium. Environmental Health Criteria no. 107. WHO, Geneva, Switzerland.

World Health Organization (WHO) International Programme on Chemical Safety (IPCS). 1991. Barium. Health and Safety Guide No. 46. WHO, IPCS, Geneva, Switzerland.

Barium chloride (Anhydrous) (CAS No. 10361-37-2)

Molecular formula: **Ba**

Barium chloride (Dihydrate) (CAS No. 10326-27-9)

Molecular formula: $BaCl_2.2H_2O$
Synonyms: Barium chloride

Use and exposure

Barium chloride is a white/colorless solid, stable under ordinary conditions of use and storage. It is incompatible with bromine trifluoride, 2-furan percarboxylic acid (anhydrous).

Toxicity and health effects

Exposures to barium chloride cause sore throat, coughing, and labored breathing, and become harmful and fatal if swallowed or inhaled. Prolonged exposures cause irritation to the skin, eyes, and respiratory tract, and involve the heart, respiratory system, and the CNS. An accidental ingestion of barium chloride causes severe gastroenteritis, abdominal pain, vomiting, diarrhea, tremors, faintness, paralysis of arms and legs, and a slow or irregular heartbeat. In severe cases, barium chloride may cause collapse and death from respiratory failure.

Exposure limits

The OSHA has set the PEL for soluble barium compounds as 0.5 mg (Ba)/m^3; the ACGIH has set the TLV as 0.5 mg (Ba)/m^3 and the estimated lethal dose in humans.

Barium chloride and carcinogenicity

Barium chloride is grouped by the IARC and the NTP as A4, meaning not classifiable as a human carcinogen. However, barium chromate (VI) is the only barium compound for which there is sufficient evidence that it is a human carcinogen.

Storage

Barium chloride should be kept stored in a tightly closed container. Protect from physical damage. Store in a cool, dry, ventilated area away from sources of heat, moisture, and incompatibilities. Containers of this material may be hazardous when empty since they retain product residues (dust, solids); observe all warnings and precautions listed for the product.

Precautions

Occupational workers with pre-existing skin disorders or impaired respiratory function may be more susceptible to the effects of barium chloride. During handling and use of barium chloride, it is important to avoid contact of the substance with the eyes, skin, and clothing, avoid ingestion and inhalation. After handling/use, workers should wash the exposed parts such as, the eyes and skin, and change clothes.

References

International Agency for Research on Cancer (IARC). 1980. *Some Metals and Metallic Compounds*. International Agency for Research on Cancer, Lyon. p. 205 (IARC Monographs on the Evaluation of the Carcinogenic Risk of Chemicals to Humans, Vol. 23).

Material Safety Data Sheets (MSDS). 2008. Barium chloride. Environmental Health & Safety (EHS). MSDS no. B0372. Mallinckrodt Baker, Phillipsburg, NJ.

Barium hydroxide (Anhydrous) (CAS No. 17194-00-2)

Barium hydroxide (Octahydrate) (CAS No. 12230-71-6)

Molecular formula: **Ba (OH)$_2$**
Synonyms and trade names: Barium hydroxide octahydrate; Barium hydrate; Barium hydroxide; 8-Hydrate

Use and exposure

Barium hydroxide is available as colorless or white crystals. It is odorless and soluble in water. It is stable under ordinary conditions of use and storage. It is incompatible with acids, oxidizers, and chlorinated rubber. Barium hydroxide is corrosive to metals such as zinc. It is very alkaline and rapidly absorbs carbon dioxide from air and becomes completely insoluble in water. Barium hydroxide is used in analytical chemistry for the titration of weak acids and is used in organic synthesis as a strong base. Barium hydroxide decomposes to barium oxide when heated to 800°C.

Toxicity and health effects

Exposures to barium hydroxide by inhalation in the form of dust cause adverse health effects and toxicity. Barium hydroxide is corrosive and toxic. Barium hydroxide presents the same health hazards as other strong bases and other water-soluble barium compounds. The signs and symptoms of toxicity include, but are not limited to, irritation of the nose, throat, and respiratory tract. Workers without proper protection suffer irritation of the eyes and skin, skin burns, sore throat, coughing, and shortness of breath. Repeated exposures by ingestion to high concentrations of barium hydroxide cause systemic poisoning with symptoms of severe irritation of the gastrointestinal tract, tightness in the muscles of the face and neck, vomiting, diarrhea, abdominal pain, muscular tremors, anxiety, weakness, labored breathing, cardiac irregularity, convulsions, cardiac and respiratory failure, and death. Exposures to very high doses (1–15 g) cause kidney damage and death within hours or a few days. Occupational workers who suffer with pre-existing health disorders of the skin and nervous system and impaired respiratory or kidney function have been shown to be more susceptible to the effects of barium hydroxide.

Barium hydroxide and carcinogenicity

Barium hydroxide is grouped as A4, meaning not classifiable as a human carcinogen.

Exposure limits

The US OSHA has set the PEL for the soluble barium compounds as 0.5 mg (Ba)/m^3 at the workplace air and the ACGIH has set the TLV as 0.5 mg (Ba)/m^3.

Storage

Barium hydroxide should be kept stored in a tightly closed container, in a cool, dry, ventilated area. The chemical substance should be protected against physical damage, in isolation and away from incompatible chemical substances. The containers of barium hydroxide may be hazardous when empty because they retain residues as dust and solids.

Precautions

Occupational workers should be careful during handling of barium hydroxide. Workers should wear impervious protective clothing, including boots, gloves, laboratory coat, apron, or coveralls, as appropriate, to prevent skin contact. Workers should use chemical safety goggles or full-face shields. The workplace should maintain an eye-wash fountain and quick-drench facilities.

Reference

Material Safety Data Sheets (MSDS). 2007. Barium hydroxide. Environmental Health & Safety (EHS). MSDS no. B0372. Mallinckrodt Baker, Phillipsburg, NJ.

Barium nitrate (as Ba) (CAS No. 10022-31-8)

Molecular formula: **Ba (NO$_3$)$_2$**
Synonyms and trade names: Barium (II) nitrate; Barium dinitrate; Nitric acid barium salt

Use and exposure

Barium nitrate is a stable, strong oxidizer. It is incompatible with combustible material, reducing agents, acids, acid anhydrides, and moisture sensitive substance. It is hazardous as magnesium plus barium oxide plus zinc, aluminum and magnesium alloys, combustibles (paper, oil, wood), acids, and oxidizers. Mixtures with finely divided aluminum-magnesium alloys are easily ignitable and extremely sensitive to friction or impact.

Barium nitrate on contact with combustible materials will ignite. Barium nitrate mixed with aluminum powder, a formula for flash powder is highly explosive. However, barium nitrate is non-corrosive in the presence of glass. It is used in military thermite grenades, in the manufacturing process of barium oxide, the vacuum tube industry, and for green fire in pyrotechnics.

Toxicity and health effects

Exposures to barium nitrate by ingestion or inhalation cause poisoning. The symptoms include, but are not limited to, ringing of the ears, dizziness, irregular and elevated blood pressure, blurred vision, irritation to the respiratory system, tightness of muscles (especially in the face and neck), vomiting, diarrhea, abdominal pain, muscular tremors, anxiety, weakness, labored breathing, cardiac irregularity, and convulsions. Poisoned workers also suffer from kidney damage, cardiac or respiratory failure, tremors, convulsions, coma, and possibly death. Prolonged periods of exposure to barium nitrate is known to cause damage of the liver (anemia and possibly methemoglobinemia), spleen, kidney, bone marrow, and the CNS.

Barium nitrate and carcinogenicity

Barium nitrate has not been listed by the NTP, the IARC, or the OSHA as a human carcinogen.

Exposure limits

For barium nitrate, the OSHA has set the PEL as 0.5 mg m^{-3} TWA and the ACGIH has set the TLV as 0.5 mg/m^3 (Ba).

Storage

Barium nitrate should be kept stored in a tightly closed container, in a cool, dry, ventilated area, protected against physical damage. It should be separated from heat, sources of ignition, incompatible substances, combustibles, and organic or other readily oxidizable materials. Barium nitrate should not be stored on wood floors or with food and beverages.

Precautions

After accidental exposures to barium nitrate by ingestion, swallow, or inhalation, workers should induce vomiting immediately as directed by medical personnel. Never give anything by mouth to an unconscious person. Get medical attention immediately.

Reference

Material Safety Data Sheets (MSDS). 2007. Barium nitrate. MSDS no. B0432. Environmental Health & Safety. Mallinckrodt Baker, Phillipsburg, NJ.

Barium sulfate (CAS No. 7727-43-7)

Molecular formula: $BaSO_4$
Synonyms and trade names: Barite; Barium sulfate

Use and exposure

Barium sulfate is available as white powder or polymorphous crystals. It is stable, odorless, insoluble, or negligibly soluble in water, and may produce sulfur oxides on burning. It occurs naturally as mineral barite (barytes). It has wide use as an inert filler pigment extender in paints, primers, inks, plastics, floor tiles, paper coatings, polymer fibers, and rubber. It is used as the semi-transparent base (lake) for organic pigments and as a thixo-tropic weighting mud in oil well drilling. Barium sulfate is radio-opaque and is used as bar-ium meal in medical x-ray diagnosis. Barium sulfate is a contrast agent that is used to help x-ray diagnosis of problems in areas of the upper GI tract, like the esophagus, the stomach, and/or the small intestine. Barium sulfate is the raw material for the manufacture of litho-pone, a white pigment, and is used in the manufacture of photographic paper, wallpaper, and glassmaking; in battery plate expanders; and in heavy concrete for radiation shield.

Toxicity and health effects

Exposures to barium sulfate cause irritation to the eyes, lachrimation; redness, scaling, and itching are characteristics of skin inflammation. Although barium sulfate has been identi-fied as a non-toxic dust, long-term inhalation of dust in high concentrations has caused benign pneumoconiosis (baritosis), deposition of dust in the lungs in sufficient quantities to produce adverse effects. This produces a radiological picture even though symptoms and abnormal signs may not be present. The Fumes of barium are respiratory irritants and over-exposure to dusts and fumes is known to cause rhinitis, frontal headache, wheezing, laryngeal spasm, salivation, and anorexia. Long-term effects include nervous disorders and adverse effects on the heart, circulatory system, and musculature. Heavy exposures to barium sulfate may cause benign pneumoconi in exposed workers. However, there are no reports indicating that barium sulfate has potential occupational hazards or carcinogenicity.

Exposure limits

The exposure limits for barium sulfate has been determined as follows: the OSHA PEL as 15 mg/m^3 total dust, 5 mg/m^3 respirable dust; the ACGIH TLV as 10 mg/m^3 (TWA).

Storage

Barium sulfate should be kept stored in its original sealed glass container with proper security when not in use. Barium sulfate should be stored in a cool, dry, ventilated area, away from incompatible materials and foodstuff containers. The containers need to be protected against physical damage and checked regularly for leaks.

Precautions

Barium sulfate should only be used by or under the direct supervision of a qualified super-visor or doctor. During use and handling of barium sulfate, workers should avoid contact of the chemical substance with the skin and should not breathe dust. After use, the work-ers should wash their hands with soap and water. While dispensing, care is needed to label barium sulfate properly and correctly, to avoid confusion with poisonous barium sulfide, sulfite, or carbonate.

Reference

Agency for Toxic Substances and Disease Registry (ATSDR). 2007. Toxicological Profile for Barium and Compounds. US Public Health Service, US Department of Health and Human Service, Atlanta, GA.

Bendiocarb (CAS No. 22781-23-3)

Molecular formula: $C_{11}H_{13}NO_4$
IUPAC name: 2,3-Isopropylidenedioxyphenyl methylcarbamate
Synonyms and trade names: Ficam; Garvox; Seedox; Dycarb; Multamat; Niomil; Rotate; Seedox; Tattoo; Turcam
Toxicity class: USEPA: II; WHO: II

Use and exposure

Bendiocarb is an odorless, white crystalline solid. Some of the formulations of bendiocarb are classified as a GUP, while Turcam and its 2.5 G formulation have been classified as a RUP. Bendiocarb is stable under normal temperatures and pressures, but should not be mixed with alkaline preparations. Thermal decomposition products may include toxic oxides of nitrogen. It is non-corrosive. Bendiocarb as a carbamate insecticide is effective against a wide range of insects that cause nuisance and act as disease vectors. It is used to control mosquitoes, flies, wasps, ants, fleas, cockroaches, silverfish, ticks, and other pests in homes, industrial plants, and food storage sites. In agriculture, it is used against a variety of insects, especially those in the soil. Bendiocarb is also used as a seed treatment on sugar beets and maize and against snails and slugs. Pesticides containing bendiocarb are formulated as dusts, granules, ultra-low volume sprays, and as wettable powders.

Toxicity and health effects

Bendiocarb is moderately toxic if ingested or absorbed through the skin. Absorption through the skin is the most likely route of exposure. Bendiocarb is absorbed through all the normal routes (oral, dermal, and inhalation) of exposure, but dermal absorption is especially rapid. Carbamates are generally excreted rapidly and do not accumulate in mammalian tissue. If exposure does not continue, cholinesterase inhibition and its symptoms reverse rapidly. In non-fatal cases, the illness generally lasts less than 24 h. Bendiocarb is moderately toxic to birds. The LD_{50} in mallard ducks and quail is 3.1 and 19 mg/kg, respectively. Bendiocarb is moderately to highly toxic to fish. The LC_{50} (96 h) for bendiocarb in rainbow trout is 1.55 mg/L.

Bendiocarb is a mild irritant to the skin and eyes. Symptoms of bendiocarb poisoning include, but are not limited to, weakness, blurred vision, headache, nausea, abdominal cramps, chest discomfort, constriction of pupils, sweating, muscle tremors, and decreased pulse. Prolonged exposures to high concentrations of bendiocarb cause severe poisoning with symptoms of twitching, giddiness, confusion, muscle incoordination, slurred speech, low blood pressure, heart irregularities, and loss of reflexes may also be experienced. Death can result from respiratory arrest, paralysis of the muscles of the respiratory system, intense constriction of the openings of the lung, or all three. In one case of exposure while applying bendiocarb, the victim experienced symptoms of severe headache, vomiting, and excessive salivation, and his cholinesterase level was depressed by 63%. He recovered from these symptoms in less than 3 h with no medical treatment and his cholinesterase level returned to normal within 24 h. In another case, poisoning occurred when an applicator, who was not wearing protective equipment, attempted to clean contaminated equipment. The victim experienced nausea, vomiting, incoordination, pain in his arms, hands, and legs, muscle spasms, and breathing difficulties. Bendiocarb is readily absorbed by the gastrointestinal tract and is rapidly metabolized.

A two-year study with rats exposed to high doses of bendiocarb (10 mg/kg/day) showed a wide range of adverse effects in organ weights, blood, and urine characteristics, as well as an increased incidence of stomach and eye lesions. In a three-generation study with rats, fertility and reproduction were not affected by bendiocarb at dietary doses up to 12.5 mg/kg/day. Very high doses (40 mg/kg/day) during prenatal and postnatal periods caused toxic effects to rat dams and reduced pup weight and survival rates. No effects were seen at 20 mg/kg/day. Thus, no reproductive effects are likely to occur in humans at expected exposure levels.

Bendiocarb and carcinogenicity

Bendiocarb was not carcinogenic in two-year studies of rats and mice. Similarly, it has not caused any teratogenic effects in the offspring of rats and rabbits during gestation. Numerous studies show that bendiocarb is not mutagenic.

Exposure limits

The ADI for bendiocarb has been set as 0.004 mg/kg/day. Published data regarding recommended limits/levels of bendiocarb in drinking water, surface water, foods, or other items in daily diets are not available from the OSHA, the NIOSH, or the ACGIH.

Precautions

The preliminary risk assessment of the US EPA showed that applicators, including home owners, were at risk when mixing or applying the pesticide. Thus, bendiocarb is no longer sold in the United States. All bendiocarb-containing products in the United States recently had their registrations cancelled due to concerns over exposure of those applying the products. In the UK, however, bendiocarb is still registered as a biocide and amateurs can use it even though it lacks proper warning labels. In view of this, bendiocarb should be purchased and used only by certified and trained applicators. Like all RUPs, management must be done only by trained and certified workers.

Reference

Kidd, H. and James, D.R. (eds.). 1991. *The Agrochemicals Handbook*. Cambridge, U.K.

Benomyl (CAS No. 17804-35-2)

Molecular formula: $C_{14}H_{18}N_4O_3$
Synonyms and trade names: Benlate; Tersan; Fungicide 1991; Fundazol
Chemical name: Carbamic acid, ((1-(butylamino)carbonyl)-1H-benzimidazol-2-yl), methyl ester

Use and exposure

Benomyl, a tan-colored crystalline solid/powder, is a systemic fungicide with a characteristic odor. It belongs to the benzimidazole family. Benomyl decomposes at high temperatures. Benomyl is essentially insoluble in water. It is stable under normal storage conditions, but will decompose to carbendazim in water. On decomposition by heat, benomyl produces toxic fumes including nitrogen oxides. Benomyl is a systemic and broadspectrum fungicide that is currently registered for use in more than 50 countries on more than 70 crops for the control of diseases in fruit trees, nut crops, vegetables, cereals, tropical crops and ornamentals, turf, and many field crops. Benomyl is marketed as a wettable powder and as a dry flowable formulation (dispersible granules).

Toxicity and health effects

Benomyl has a very low acute toxicity. It causes contact dermatitis and dermal sensitization in some farm workers. It is only mildly irritating to the skin and eyes, and prolonged contact may cause skin sensitization. Animal tests show that this substance possibly causes toxic effects on human reproduction and malformations in human babies. Benomyl poisoning in the general population has not been reported in the scientific literature. Recent data used to estimate dietary exposure based on food consumption patterns within the United States indicate exposures well below the no observed effects levels (NOELs) in animal toxicity tests. Also, no inadvertent poisoning of agricultural or forestry workers has been documented. The primary toxic effects of benomyl under these conditions of exposure are dermal sensitization and contact dermatitis, but these effects can be significantly reduced by limited exposure. Benomyl is rapidly converted to carbendazim in various environmental compartments.

Storage

Benomyl should be kept stored in a well-ventilated area. Keep container tightly closed. Do not store or consume food, drink, or tobacco in areas where they may be contaminated with this material. Workers should never allow benomyl to get wet during storage. This may lead to certain chemical changes that could increase its toxicity (lacrimation because of the formation of butyl isocyanate) and reduce the effectiveness of benomyl as a fungicide.

Precautions

Liquid formulations of benomyl containing organic solvents may be flammable. When using benomyl, workers should observe the rules about treatment, storage, transportation, and disposal must follow local regulations. During handling of large quantities (2 kg bags or greater) of solid formulations of benomyl, workers should use a dust mask and protective clothing.

References

World Health Organization (WHO). 1990. Benomyl. Health and Safety Guide 81. WHO, Geneva, Switzerland.
———. 1993. Benomyl. International Programme on Chemical Safety (IPCS). Health and Safety Guide No. 81. IPCS, Geneva, Switzerland.

Benzaldehyde (CAS No. 100-52-7)

Molecular formula: C_7H_6O
Synonyms and trade names: Artificial almond oil; Benzoic aldehyde; Benzene carbaldehyde; Benzenecarbonal; Benzene carboxaldehyde; Benzenemethylal; Phenylmethanal; Synthetic bitter almond oil

Use and exposure

Benzaldehyde is a colorless to yellow, oily liquid with an odor of bitter almonds. Benzaldehyde is commercially available in two grades: (i) pure benzaldehyde and (ii) and double-distilled benzaldehyde. The latter has applications in the pharmaceutical, perfume, and flavor industries. Benzaldehyde may contain trace amounts of chlorine, water, benzoic acid, benzyl chloride, benzyl alcohol, and/or nitrobenzene. Benzaldehyde is ignited relatively easily on contact with hot surfaces. This has been attributed to the property of very low auto-ignition temperature. Benzaldehyde also undergoes autoxidation in air and is liable to self-heat. Benzaldehyde exists in nature, occurring in combined and uncombined forms in many plants. Benzaldehyde is also the main constituent of the essential oils obtained by pressing the kernels of peaches, cherries, apricots, and other fruits. Benzaldehyde is released into the environment in emissions from combustion processes, such as gasoline and diesel engines, incinerators, and wood burning. It is formed in the atmosphere through photochemical oxidation of toluene and other aromatic hydrocarbons. Benzaldehyde is corrosive to gray and ductile cast iron (10% solution), and all concentrations of lead. However, pure benzaldehyde is not corrosive to cast iron. Benzaldehyde does not attack most of the common metals, like stainless steels, aluminum, aluminum bronze, nickel and nickel-base alloys, bronze, naval brass, tantalum, titanium, and zirconium. On decomposition, benzaldehyde releases peroxybenzoic acid and benzoic acid.

Benzaldehyde is used in perfumes, soaps, foods, drinks, and other products; as a solvent for oils, resins, some cellulose ethers, cellulose acetate, and cellulose nitrate. The uses of benzaldehyde in industries are extensive. For instance, in the production of derivatives that are employed in the perfume and flavor industries, like cinnamaldehyde, cinnamyl alcohol, cinnamic acid, benzylacetone, and benzyl benzoate, in the production of triphenylmethane dyes and the acridine dye, benzoflavin; as an intermediate in the pharmaceutical industry, for instance, to make chloramphenicol, ephedrin, and ampicillin, as an intermediate to make benzoin, benzylamine, benzyl alcohol, mandelic acid, and 4-phenyl-3-buten-2-one (benzylideneacetone), in photochemistry, as a corrosion inhibitor and dyeing auxiliary, in the electroplating industry, and in the production of agricultural chemicals.

Toxicity and health effects

Exposures to the vapor of benzaldehyde cause irritation to the upper respiratory tract, intolerable irritation of the nose and throat, headache, nausea, dizziness, drowsiness, and confusion. It is a CNS depressant. Exposures to benzaldehyde cause moderate to severe eye irritation and prolonged periods of exposure cause corrosive effects to the skin, like burns, scarring, and skin injury, fatigue, headache, nausea, dizziness, and loss of coordination. At higher concentrations, benzaldehyde produces more severe effects, such as sore throat, abdominal pain, nausea, CNS depression, convulsions, and respiratory failure. The estimated lethal dose of benzaldehyde has been reported as 2 oz. There is no data on humans and the information and conclusions are based on evidence obtained from animal studies.

Benzaldehyde and carcinogenicity

Benzaldehyde has not been listed as a human carcinogen by the regulatory agencies, namely, the OSHA, the NIOSH, the ACGIH, the NTP, or the IARC.

Storage

Benzaldehyde should be kept stored in a tightly closed container and protected against physical damage. Storage of the chemical substance outside or in a detached area is preferred, whereas inside storage should be in a standard flammable liquids storage room or cabinet. Benzaldehyde should be kept separated from oxidizing materials. Also, storage and use areas should be no smoking areas. Containers of this material may be hazardous when empty since they retain product residues (vapors, liquid); observe all warnings and precautions listed for the product.

Precautions

Workers should be careful when using benzaldehyde because there is a risk of spontaneous combustion. It may ignite spontaneously if it is absorbed onto rags, cleaning cloths, clothing, sawdust, diatomaceous earth (kieselguhr), activated charcoal, or other materials with large surface areas in workplaces. Workers should avoid handling the chemical substance and should not cut, puncture, or weld on or near the container. Exposure of benzaldehyde to air, light, heat, hot surfaces such as hot pipes, sparks, open flames, and other ignition sources should be avoided. Workers should wear proper personal protective clothing and equipment.

References

Brühne, F., et al. 2005. Benzaldehyde. In: *Ullmann's Encyclopedia of Industrial Chemistry*, 7th ed. Wiley, J. New York.

Canadian Centre for Occupational Health & Safety (CCOHS). 2007. Benzaldehyde. Chemical Information. no. 232. CCOHS, Canada.

Lewis, R.J. Sr. (ed.). 2002. Benzaldehyde. In: *Hawley's Condensed Chemical Dictionary*, pp 1379., 14th ed. Wiley, J. New York.

Material Safety Data Sheets (MSDS). 1999. Benzaldehyde. MSDS no. B0696. Mallinckrodt Baker, Phillipsburg, NJ.

1,2-Benzanthracene (CAS No. 56-55-3)

Molecular formula: $C_{18}H_{12}$
Synonyms and trade names: 1,2-Benz(a)anthracene; Benzanthrene; Benzo(a)anthracene; Benzo(b)phenanthrene; 2,3-Benzophenanthrene; Naphthanthracene; Tetraphene

Use and exposure

1,2-Benzanthracene is available as colorless to yellow brown fluorescent flakes or powder. It is stable, combustible, and incompatible with strong oxidizing agents. On decomposition, 1,2-benzanthracene releases carbon monoxide, carbon dioxide, acrid smoke, and fumes. During work, 1,2-benzanthracene can be absorbed into the body of occupational workers by inhalation, through the skin, and by ingestion. Exposures may cause irritation to the eyes, skin, and respiratory tract.

Toxicity and health effects

Exposures to 1,2-benzanthracene is known to cause kidney damage. However, published data on the neurotoxicity, teratogenicity, reproductive toxicity, and mutagenicity of 1,2-benzanthracene is not available.

1,2-Benzanthracene and carcinogenicity

Benz[*a*]anthracene has been shown to be carcinogenic to experimental animals. 1,2-Benzanthracene has been listed and grouped as an A2, meaning a suspected human carcinogen by the ACGIH and the NTP, and the IARC classified it as Group 2A, meaning a probable human carcinogen.

Exposure limits

The OSHA has not listed the exposure limits of 1,2-benzanthracene.

Precautions

Workers should wash thoroughly after using and handling 1,2-benzanthracene. Use only in a well-ventilated area. Minimize dust generation and accumulation. Avoid contact with the eyes, skin, and clothing. Keep container tightly closed. Avoid ingestion and inhalation.

Storage

Store in a cool, dry, well-ventilated area away from incompatible substances. Keep containers tightly closed.

Reference

International Chemical Safety Cards. Benz(a)anthracene. ICSC card no. 0385.
Material Safety Data Sheet (MSDS). 1999. 1,2-Benzanthracene. ACC no. 50930 (updated 2007).

Benzene (CAS No. 71-43-2)

Molecular formula: C_6H_6
Synonyms and trade names: Benzine; Benzol; Aromatic hydrocarbon

Use and exposure

Benzene is a colorless, flammable liquid with a pleasant odor. It is used as a solvent in many areas of industries, such as rubber and shoe manufacturing, and in the production of other important substances, such as styrene, phenol, and cyclohexane. It is essential in the manufacture of detergents, pesticides, solvents, and paint removers. It is present in fuels such as gasoline up to the level of 5%.

Toxicity and health effects

Exposure to low concentrations of benzene vapor or liquid causes dizziness, light-headedness, headache, loss of appetite, stomach upset, and irritation to the nose and throat. Prolonged exposure to high concentrations of benzene leads to functional irregularities in the heart beat and in severe cases to death. Benzene is a known carcinogen to humans. It causes leukemia and blood disorders such as aplastic anemia. The major types of leukemia related to benzene exposure are (i) acute myelogenous leukemia (AML); (ii) acute lymphocytic leukemia (ALL); (iii) chronic myelogenous leukemia, also called chronic myeloid leukemia (CML); (iv) chronic lymphocytic leukemia (CLL), and hairy cell leukemia (HCL). Occupational exposure to benzene is frequent, such as in road-tanker drivers and Chinese glue- and shoe-making factory workers. Exposure to benzene has been linked with the development of rarer forms of leukemia, such as AML and ALL. It has also been linked to lymphoma and rare blood diseases.

Acute myelogenous leukemia

AML (acute myeloid leukemia or acute non-lymphocytic leukemia) is a blood cancer that develops in specific types of white blood cells (granulocytes or monocytes). White blood cells are used by the body to fight infections. These blood stem cells originate in a person's bone marrow. With the development of ALL, the normal development of white blood cells is disturbed and they do not grow properly. Possibly due to some sort of change or damage to their genetic material or DNA, the cells are prevented from growing beyond a certain point. This disturbs their development and affects the differentiation process of cells into functional types of white cells.

Acute lymphocytic leukemia

ALL (acute lymphocytic leukemia) is a malignant cancer that develops in a person's white blood cells, called lymphocytes. White blood cells are used by the body to ward off disease and infection. ALL is rare among adults, but is the prevalent form of leukemia in children. Nearly 85% of leukemia in children is ALL. In adults, the disease may be related to genetics or exposure to solvents containing benzene. In people that develop ALL (and other types of acute leukemia), white blood cells do not grow properly. Because of some change or damage to their genetic material or DNA that scientists do not fully understand, the cells are prevented from growing beyond a certain point early in their development, and they cannot differentiate into functional types of white cells. Long-term exposure to benzene increases the risks of getting cancer, however cancer linked to benzene has been discovered in people exposed for less than 5 years. Workers exposed for decades are at increased risk for these rare forms of leukemia and long-term exposure may also adversely impact on bone marrow and blood production. Still other workers have been diagnosed with aplastic anemia, a group of disorders that prevent the bone marrow from producing all three types of blood cells: red blood cells, white blood cells, and platelets.

Exposure limits

The US EPA has set the maximum permissible level of benzene in drinking water at 5 ppb. The OSHA has set a limit of 1 ppm of benzene in workplace air for 8 h (TWA).

Benzene and carcinogenicity

The NIOSH recommends that benzene is treated as a potential human carcinogen.

Exposure limits

The National Institute of Occupational Safety and Health (NIOSH) recommends that occupational exposures to carcinogens be limited to the lowest feasible concentration: 0.1 ppm for 10 h (TWA); 1 ppm for 15 min short-term exposure limit (STEL). The TLV for benzene are 0.5 ppm for 8 h (TWA); 2.5 ppm for 15 min STEL on skin. More information on the toxicity and health effects of benzene among occupational workers may be found in the literature.

References

Agency for Toxic Substances and Disease Registry (ATSDR). 2007. Toxicological Profile for Benzene. US Public Health Service, US Department of Health and Human Service, Atlanta, GA.

Dikshith, T.S.S. (ed.). 2009. *Safe Use of Chemicals – A Practical Guide.* CRC Press, Boca Raton, FL.

Snyder, R., Witz, G. and Goldstein, B.D. 1993. The toxicology of benzene. *Environ. Health Perspect.* 100: 293–306.

Benzidine (CAS No. 92-87-5)

Molecular formula: $C_{12}H_{12}N_2$
Synonyms and trade names: 4,4'-Diaminophenyl; 4,4'Diphenylenediamine; 4,4'-Biphenyldiamine; 4,4'-Biphenylenediamine; p-Benzidine; 1,1'-Biphenyl-4; 4'-diamine; p-Diaminodiphenyl; p,p'-Bianiline; 4,4'-Bianiline; p,p'-Diaminobiphenyl; Fast corinth base b; Benzidine base

Use and exposure

Benzidine is a white, grayish-yellow, or slightly reddish crystalline solid or powder. The major use for benzidine is in the production of dyes, especially azo dyes in the leather, textile, and paper industries, as a synthetic precursor in the preparation and manufacture of dyestuffs. It is also used in the manufacture of dyes and rubber, as a reagent, and as a stain in microscopy. It is slightly soluble and slowly changes from a solid to a gas.

Toxicity and health effects

Exposure to benzidine causes irritation to the eyes. Laboratory animals exposed to benzidine at as low as 0.01% to 0.08% in food showed adverse health effects, such as organ weight decrease in the liver, kidney, and body weight, and an increase in spleen weight, swelling of the liver, and blood in the urine. Exposure may cause an increase in urination, blood in the urine, and urinary tract tumors. Benzidine is considered acutely toxic to humans by ingestion, with an estimated oral lethal dose of between 50 and 500 mg/kg. The symptoms of acute ingestion exposure include cyanosis, headache, mental confusion, nausea, and vertigo. Dermal exposure may cause skin rashes and irritation. Prolonged exposure to benzidine causes bladder injury in humans.

Benzidine and carcinogenicity

Laboratory studies of animals exposed to benzidine via oral, inhalation, and injection developed various tumor types at multiple sites. Epidemiologic studies have indicated that occupational exposure to benzidine causes an increased risk of bladder cancer. The US EPA has classified and listed benzidine as Group A, meaning a human carcinogen.

Precautions

At high temperatures, benzidine breaks down and releases highly poisonous fumes. During use and handling, workers should wear butyl rubber gloves, goggles, and full body plastic coveralls and ensure that no skin is exposed.

Storage

Benzidine should be kept stored in a cool, well-ventilated area, in closed, sealed containers and out of sunlight and away from heat.

References

Agency for Toxic Substances and Disease Registry (ATSDR). 1989. Toxicological Profile for Benzidine (Draft). US Public Health Service, US Department of Health and Human Services, Atlanta, GA.
———. 1995. Toxicological Profile for Benzidine. US Public Health Service, US Department of Health and Human Services, Atlanta, GA.
Technology Transfer Network Air Toxics web site. Benzidine. Hazard Summary–1992 (revised 2000 and updated 2007).
US Environmental Protection Agency (US EPA). 1999. Benzidine. Integrated Risk Information System (IRIS). National Center for Environmental Assessment, Office of Research and Development, Washington, DC.

2,3-Benzofuran (CAS No. 271-89-6)

Molecular formula: C_8H_6O
Synonyms and trade names: Benzofurfuran; NCI-C56166; Coumarone; Cumarone; 1-Oxindene; Benzo(b) furan

Use and exposure

2,3-Benzofuran is a colorless, sweet-smelling, oily liquid made by processing coal into coal oil. It may also be formed during other uses of coal or oil. 2,3-Benzofuran is not used for any commercial purposes, but the part of the coal oil that contains 2,3-benzofuran is made into a plastic called coumarone-indene resin. This resin resists corrosion and is used to make paints and varnishes. The resin also provides water resistance and is used in coatings on paper products and fabrics. It is used as an adhesive in food containers and some asphalt floor tiles. The resin has been approved for use in food packages and as a coating on citrus fruits. We do not know how often the resin is used or whether any 2,3-benzofuran in the coating or packaging gets into the food. 2,3-Benzofuran may enter the air, water, and soil during its manufacture, use, or storage at hazardous waste sites. Breathing contaminated air or touching the chemical in the workplace is the source of human exposure.

Toxicity and health effects

Rats and mice that ingested high levels of 2,3-benzofuran over a short time had liver and kidney damage. Those exposed over a long time to moderate levels had liver, kidney, lung, and stomach damage. In one study, the ability of animals to reproduce was not affected. We do not know if people will experience health effects similar to those seen in animals.

Very little is known about the possible harmful effects of 2,3-benzofuran to human health. There are no studies that have looked at the effects in people from exposures to air, water, or food, or through skin contact.

2,3-Benzofuran and carcinogenicity

Laboratory mice and rats exposed to 2,3-benzofuran for long periods of time developed cancer of the kidneys, lungs, liver, or stomach. There are no studies on 2,3-benzofuran's potential to cause cancer in humans. The DHHS has not classified 2,3-benzofuran as a human carcinogen. The IARC and the US EPA have also not classified 2,3-benzofuran as to its human carcinogenicity.

Reference

Agency for Toxic Substances and Disease Registry (ATSDR). 1992. Toxicological Profile for 2,3-Benzofuran. US Department of Health and Human Services, Public Health Service, Atlanta, GA.

Benzyl acetate (CAS No. 140-11-4)

Molecular formula: $C_9H_{10}O_2$
Synonyms and trade names: Acetic acid benzyl ester; Acetic acid phenylmethyl ester; Benzyl ethanoate

Use and exposure

Benzyl acetate is a colorless liquid with a fruity odor. On burning and decomposition, it produces irritating fumes. Benzyl acetate reacts with strong oxidants causing fire and explosion hazard.

Toxicity and health effects

Exposures to benzyl acetate cause adverse health effects. The symptoms of toxicity and poisoning include irritation to the skin, eyes, burning sensation, confusion, dizziness, drowsiness, labored breathing, sore throat, nausea, vomiting, and diarrhea. Benzyl acetate also causes adverse health effects to the respiratory tract and the CNS system with neurological effects.

Benzyl acetate and carcinogenicity

Exposures to benzyl acetate have not caused any carcinogenic effects in humans. The ACGIH classified benzyl acetate as Group 4, meaning not classifiable as a human carcinogen, and the IARC classified acetate as a Group 3 carcinogen.

Exposure limits

The TLV of benzyl acetate has been set as 10 ppm (TWA).

Storage

Benzyl acetate should be kept stored in a cool, dry place with the container closed when not in use.

Precautions

Exposures to benzyl acetate far above the OEL may result in unconsciousness. After handling and using benzyl acetate, workers should wash thoroughly and remove contaminated clothing, washing it before reuse. Workers should avoid any kind of contact of benzyl acetate with the eyes, skin, ingestion, and inhalation. Workers should wear safety glasses and chemical goggles to avoid splashing of the chemical substance during work, and wear appropriate protective gloves and clothing to prevent skin exposure.

Reference

Material Safety Data Sheet (MSDS). 1997. Benzyl acetate. International Programme on Chemical Safety and the European Commission (IPCS-EC). 1999. Benzyl acetate. ICSC card no. 1331. IPCS, Geneva, Switzerland (updated 2004).

Benzyltrimethylammonium dichloroiodate (CAS No. 114971-52-7)

Molecular formula: $C_6H_5CH_2N(ICl_2)(CH_3)_3$

Use and exposure

Benzyltrimethylammonium dichloroiodate is a slightly brown solid and hygroscopic.

Toxicity and health effects

Exposures to benzyltrimethylammonium dichloroiodate cause irritation to the eyes, skin, and respiratory tract. Accidental ingestion causes gastrointestinal irritation with nausea, vomiting, and diarrhea. Sparse information on chronic toxicity is available in literature.

Beryllium (CAS No. 7440-41-7)

Molecular formula: Be

Use and exposure

Beryllium is a brittle, steel-gray metal found as a component of coal, oil, certain rock minerals, volcanic dust, and soil. It reacts with strong acids and strong bases forming flammable/explosive gas. It has several applications in the aerospace, nuclear, and manufacturing industries. In addition, beryllium is amazingly versatile as a metal alloy where it is used in dental appliances, golf clubs, non-sparking tools, wheel chairs, and electronic devices. Beryllium is used in alloys with a number of metals, including steel, nickel, magnesium, zinc, and aluminum, the most widely used alloy being beryllium-copper—properly called "a bronze"—which has a high tensile strength and a capacity for being hardened by heat treatment. Beryllium bronzes are used in non-spark tools, electrical switch parts, and watch springs.

One of the largest uses of the metal is as a moderator of thermal neutrons in nuclear reactors and as a reflector to reduce the leakage of neutrons from the reactor core. A mixed uranium-beryllium source is often used as a neutron source. As a foil, beryllium is used as window material in x-ray tubes. Its lightness, high elastic modulus, and heat stability make it an attractive material for the aircraft and aerospace industry. Beryllium ores are used to make special ceramics for electrical and high-technology applications. Beryllium alloys are used in automobiles, computers, sports equipment (golf clubs and bicycle frames), and dental bridges. It used in nuclear reactors as a reflector or moderator because it has a low thermal neutron absorption cross section. It is used in gyroscopes, computer parts, and instruments where lightness, stiffness, and dimensional stability are required. The oxide has a very high melting point and is also used in nuclear work and ceramic applications.

Normally, the general population is exposed to low levels of beryllium in air, food, and water. People working in industries where beryllium is mined, processed, machined, or converted into metal, alloys, and other chemicals, may be exposed to high levels of beryllium. People living near these industries may also be exposed to higher than normal levels of beryllium in air. People living near uncontrolled hazardous waste sites may be exposed to higher than normal levels of beryllium.

Toxicity and health effects

Beryllium and its compounds are highly toxic substances. Beryllium can affect all organ systems, although the primary organ involved is the lung. Beryllium causes systemic disease by inhalation and can distribute itself widely throughout the body after absorption from the lungs. The signs and symptoms of chronic beryllium poisoning include, cough, chest pain, fatigue, dyspnea, anorexia, cyanosis, cubbing, hepatomegaly, splenomegaly with complications of cardiac failure, renal stone, and pneumothorax. Little beryllium is absorbed from the gastrointestinal tract. Beryllium can cause skin irritation and its traumatic introduction into subcutaneous tissue can cause local irritation and granuloma formation. Beryllium is a potent inhibitor of various enzymes of phosphate metabolism, particularly of alkaline phosphatase. The health hazards of beryllium are almost exclusively confined to inhalation exposure and skin contact.

Beryllium and its salts are toxic and should be handled with the greatest of care. Beryllium and its compounds should not be tasted to verify its sweetish nature. Ingestion and breathing of beryllium is harmful. Acute exposures to high levels of beryllium cause mild inflammation of the nasal mucous membranes and pharynx, rhinitis and pharyngitis, tracheo-bronchitis, and pneumonitis. The symptoms of acute pneumonitis are cough, respiratory distress, substernal discomfort or pain, loss of appetite, weakness, tiredness, chest pain, and cyanosis.

Beryllium can be very harmful when humans breathe it in, because it can damage the lungs and cause pneumonia. The most commonly known effect of beryllium is called berylliosis, a dangerous and persistent lung disorder that can also damage other organs, such as the heart. In about 20% of all cases, people die of this disease. Breathing in beryllium in the workplace causes berylliosis. People that have weakened immune systems are most susceptible to this disease. Beryllium can also cause allergic reactions with people that are hypersensitive to this chemical and cause chronic beryllium disease (CBD). The symptoms are weakness, tiredness, and breathing problems. Some people that suffer from CBD will develop anorexia and blueness of hands and feet. Sometimes, people can even be in such a serious condition that CBD can cause their death. Next to causing berylliosis and CBD, beryllium can also increase the chances of developing cancer and DNA damage. Chronic beryllium disease is a pulmonary and systemic granulomatous disease caused by inhalation of beryllium. The latency of the disease can be from 1 to 30 years, most commonly occurring 10–15 years after first exposure. From the reported use pattern of beryllium, it can be deduced that toxicologically relevant exposure to beryllium is largely confined to the workplace. Only a few exposure situations have been reported for the general population.

Beryllium and carcinogenicity

Epidemiological studies have indicated that beryllium and its compounds could be carcinogenic to humans. Based on sufficient evidence for animals, but inadequate evidence for humans, the US EPA grouped beryllium as B2, meaning probable human carcinogen and the IARC observed that there is sufficient evidence in animals and humans for the carcinogenicity of beryllium and beryllium compounds and classified it as Group 1, meaning beryllium compounds are carcinogenic to humans. Similarly, the DHHS and the IARC have determined that beryllium is a human carcinogen.

Exposure limits

The OSHA has set a limit of 0.002 mg/m^3 (2 μg/m^3) of workroom air (8-h TWA). By contrast, the NIOSH has set a REL at 0.5 μg/m^3 for beryllium.

References

Agency for Toxic Substances and Disease Registry (ATSDR). 2002. Toxicological Profile for Beryllium. US Department of Health and Human Services, Public Health Service, Atlanta, GA (updated 2007).

Budavari, S., O'Neil, M.J., Smith, A. and Heckelman, P.E. (eds.). 1989. Beryllium. In: *The Merck Index*. Merck, Rahway, NJ, p. 181.

Groth, D.H., Kommiheni, C. and MacKay, G.R. 1980. Carcinogenicity of beryllium hydroxide and alloys. *Environ. Res.* 21: 63–84.

Haley, P.J., Finch, G.L., Hoover, M.D. and Cuddihy, R.G. 1990. The acute toxicity of inhaled beryllium metal in rats. *Fundam. Appl. Toxicol.* 15: 767–78.

Hall, T.C., Wood, C.H., Stoeckle, J.D. and Tepper, L.B. 1959. Case data from the beryllium registry. *Am. Med. Assoc. Arch. Ind. Health* 19: 100–3.

International Agency for Research on Cancer (IARC). 1997. Summaries & Evaluations, Beryllium and Beryllium compounds. IARC, World Health Organization, Geneva, Switzerland.

International Programme on Chemical Safety (IPCS). 1990. Beryllium. Environmental health criteria, 106. World Health Organization, Geneva, Switzerland.

International Programme on Chemical Safety & Commission of the European Communities (IPCS CEC). 1994. Beryllium. ICSC card no. 0226. IPCS, Geneva, Switzerland.

US Environmental Protection Agency (US EPA). 1987. Health Assessment Document for Beryllium. Prepared by the Office of Health and environmental Assessment, Environmental Criteria and Assessment Office, External Review Draft. EPA 600-8-84-026B, 1986.

———. 1998. Beryllium and Compounds. The Risk Assessment Information System (IRIS) (updated 2007).

Beryllium chloride (CAS No. 7787-47-5)

Molecular formula: **BeCl$_2$**

Use and exposure

Beryllium chloride is available as colorless to yellow crystals. It decomposes rapidly on contact with water producing hydrogen chloride, and attacks many metals in the presence of water. Beryllium chloride emits irritating or toxic fumes (or gases) in a fire.

Toxicity and health effects

Exposures to beryllium chloride cause redness, pain and blurred vision, nausea, vomiting, and abdominal pain. Inhalation of beryllium chloride causes cough, sore throat, shortness of breath.

Beryllium chloride and carcinogenicity

The ACGIH reported and classified beryllium chloride as A1, meaning confirmed human carcinogen.

Exposure limits

The TLV of beryllium chloride (as Be) has been set at 0.002 mg/m^3 as TWA, 0.01 mg/m^3 as STEL.

Reference

International Programme on Chemical Safety and the Commission of the European Communities (IPCS-CEC). 2000. Beryllium Chloride. ICSC card no. 1354. IPCS, Geneva, Switzerland (updated 2004).

Beryllium compounds

Reference

US Environmental Protection Agency (US EPA). 2000. Beryllium Compounds. Technology Transfer Network. Air Toxics web site (updated 2007).

Bifenthrin (CAS No. 82657-04-3)

Molecular formula: $C_{23}H_{22}ClF_3O_2$
Synonyms and trade names: Bifenthrin; Bifenthrin (ANSI); Bifenthrin (Talstar); Bifenthrin (trans isomer); Bifentrin; Bifentrina; Biflex; Biphenthrin; Brigade; Capture; Cyclopropanecarboxylic acid; 3-(2-Chloro-3,3,3-trifluoro-1-propenyl)-2,2-dimethyl-; (2-Methyl{1,1'-biphenyl}-3-yl)methyl ester; FMC 54800

Use and exposure

Bifenthrin is a member of the pyrethroid chemical class. It is an insecticide and acaricide. Bifenthrin is off-white to pale tan waxy solid granules with a faint, musty odor and slightly sweet smell. Bifenthrin is soluble in methylene chloride, acetone, chloroform, ether, and toluene, and is slightly soluble in heptane and methanol. It is slightly combustible and support combustion at elevated temperatures. Thermal decomposition and burning may form toxic by-products, such as carbon monoxide, carbon dioxide, hydrogen chloride, and hydrogen fluoride. Bifentrin treatment affects the nervous system and causes paralysis in insects.

Toxicity and health effects

Bifenthrin is moderately toxic to species of mammals when ingested. Exposures to large doses of bifenthrin cause poisoning with symptoms that include, but are not limited to, incoordination, tremor, salivation, vomiting, diarrhea, and irritability to sound and touch. Exposures to bifenthrin through skin absorption and/or inhalation of dust cause adverse health effects. On contact with bifenthrin, occupational workers develop adverse health effects that include skin sensations, rashes, numbness, and a burning and tingling type of effect. As a pyrethroid poison, bifenthrin disturbs the electrical impulses in nerves, over-stimulating nerve cells, causing tremors and eventually causing paralysis. The skin-related health effects were found to be reversible and subside after a brief period of time and stoppage of further exposures to bifenthrin. Although bifenthrin causes no inflammation or irritation on human skin, it can cause a tingling sensation that lasts about 12 h. Bifenthrin has caused no symptoms of irritation to rabbits' eyes. The US EPA has classified bifenthrin as toxicity class II, meaning moderately toxic.

Bifenthrin and carcinogenicity

In a 2-year study of rats exposed to bifenthrin (10 mg/kg/day), bifenthrin showed no evidence of cancer. However, significantly higher doses of bifenthrin caused an increased tumor incidence in the urinary bladder and lung of laboratory mice. The US EPA classified bifenthrin as a class C carcinogen, meaning a possible human carcinogen. Bifenthrin has not been listed as a carcinogen by the NTP, the IARC, the OSHA, or the ACGIH.

Exposure limits

The ADI has been set at 0.015 mg/kg. There is no data on the TLV or the PEL. The no observed effect level (NOEL) has been set at 2.5 mg/kg/day (rat); 1.5 mg/kg/day (dog).

Storage

Bifenthrin should be kept stored in a cool, dry, well-ventilated place away from heat, open flame or hot surfaces. It should only be stored in its original containers and should not be contaminated with other pesticides, fertilizers, water, food, or feed by storage or disposal.

References

US EPA Office of Pesticides and Toxic Substances. 1988. Fact Sheet No. 177 Bifenthrin. US EPA, Washington, DC.

Material Safety Data Sheets (MSDS). 1995. Bifenthrin. Extension Toxicology Network, Pesticide Information Profiles. Oregon State University, USDA/Extension Service, Corvallis, OR.

Kidd, H. and James, D.R. (eds.). 1991. *The Agrochemicals Handbook*, 3rd ed. Royal Society of Chemistry Information Services, Cambridge, U.K. (as updated).

World Health Organization. 1989. Allethrins: Allethrin, D-allethrin, Bioallethrin, S-bioallethrin. Environmental Health Criteria 87. International Programme on Chemical Safety, Geneva, Switzerland.

1,1'-Biphenyl (CAS No. 92-52-4)

Molecular formula: $C_{12}H_{10}$
Synonyms and trade names: Diphenol; 1,1'Biphenyl; Phenylbenzene

Use and exposure

1,1'-Biphenyl is a clear colorless liquid with a pleasant odor and stable organic compound. It is combustible at high temperatures, producing carbon dioxide and water when combustion is complete. Partial combustion produces carbon monoxide, smoke, soot, and low molecular weight hydrocarbons. It is used extensively in the production of heat-transfer fluids, for example, as an intermediate for polychlorinated biphenyls, and dye carriers for textile dyeing. It is also used as a mold retardant in citrus fruit wrappers, in the formation of plastics, optical brighteners, and hydraulic fluids.

Toxicity and health effects

Exposures to 1,1'-biphenyl cause acute effects with symptoms that include, but are not limited to, polyuria, accelerated breathing, lacrimation, anorexia, weight loss, muscular weakness, coma, fatty liver cell degeneration, and severe nephrotic lesions. Exposure to biphenyl fumes for short periods of time causes nausea, vomiting, irritation of the eyes and respiratory tract, and bronchitis. Breathing small amounts of 1,1'-biphenyl over long periods of time causes damage to the liver and the nervous system of exposed workers. Breathing the mists, vapors, or fumes may irritate the nose, throat, and lungs. Depending on the concentration and duration of exposure, the symptoms include, but are not limited to, sore throat, coughing, labored breathing, sneezing, a burning sensation, and the effects of CNS depression. Symptoms may include headache, excitation, euphoria, dizziness, incoordination, drowsiness, light-headedness, blurred vision, fatigue, tremors, convulsions, loss of consciousness, coma, respiratory arrest, and death, depending on the concentration and duration of exposure.

Precautions

Occupational workers should be careful during use and chemical management because the finely dispersed particles of 1,1'-biphenyl form explosive mixtures in air.

Exposure limits

The OSHA recommends the PEL for biphenyl as 0.2 ppm (TWA). Similarly, the ACGIH recommends the TLV as 0.2 ppm (TWA).

Storage

1,1'-Biphenyl should be kept stored in tightly closed containers in a cool, dry, isolated, and properly ventilated area, away from heat, sources of ignition, incompatibles, and contact with strong oxidizers.

Bis(chloromethyl) ether (CAS No. 542-88-1)

Molecular formula: $C_2H_4Cl_2O$
Synonyms and trade names: BCME; Chloro(chloromethoxy)methane; Dichloromethyl ether; Dimethyl-1,1'-dichloroether

Use and exposure

Bis(chloromethyl) ether is a clear liquid with a strong unpleasant odor. It does not occur naturally. It dissolves easily in water, but degrades rapidly and readily evaporates into air. During earlier years, bis(chloromethyl) was used to make several types of polymers, resins, and textiles, but its use is now highly restricted. Only small quantities of bis(chloromethyl) ether are produced in the United States. The small quantities produced are only used in enclosed systems to make other chemicals. However, small quantities of bis(chloromethyl) ether may be formed as an impurity during the production of chloromethyl methyl ether. Along with other chemicals, rain, and sunlight, it undergoes chemical reactions and breaks down as formaldehyde and hydrochloric acid.

Toxicity and health effects

Exposures to bis(chloromethyl) ether cause irritation to the skin, eyes, throat, and lungs, and in cases of severe exposures cause damage to the lungs (swelling and bleeding) and death. Breathing low concentrations will cause coughing and nose and throat irritation.

Bis(chloromethyl) ether carcinogenicity

After subcutaneous administration, chloromethyl methyl ether produced local sarcomas in mice and was found to be an initiator of mouse skin tumors. Reports have indicated that bis(chloromethyl) ether causes lung cancer and other tumors in occupational workers. The DHHS observe that bis(chloromethyl) ether is a known human carcinogen. Bis(chloromethyl)ether and chloromethyl methyl ether (technical grade) are classified as Group 1, meaning human carcinogens.

Exposure limits

The US EPA recommends that the levels of bis(chloromethyl) ether in drinking water and fish in lakes and streams should be limited to 0.0000038 ppb to prevent possible health effects. Also, the OSHA has set a limit of 1 ppb as the highest acceptable level in workplace air.

References

Agency for Toxic Substances and Disease Registry (ATSDR). 1999. Toxicological Profile for Bis(chloromethyl) Ether. US Department of Health and Human Services, Public Health Service, Atlanta, GA (updated 2008).

Sakabe, H. 1973. Lung cancer due to exposure to bis(chloromethyl)ether. *Ind. Health* 11: 145–48.

Weiss, W. and Figueroa, W.G. 1976. The characteristics of lung cancer due to chloromethyl ethers. *J. Occup. Med.* 18, 623–27.

Weiss, W., Moser, R. and Auerbach, O. 1979. Lung cancer in chloromethyl ether workers. *Am. Rev. Respir. Dis.* 120: 1031–37.

Bismuth (CAS No. 7440-69-9)

Molecular formula: **Bi**

Use and exposure

Bismuth is a white, crystalline, brittle metal with a pinkish tinge. Bismuth is the most diamagnetic of all metals, and its thermal conductivity is lower than any metal. It occurs naturally in the metallic state and in minerals such as bismite. The most important ores are bismuthinite or bismuth glance and bismite, and countries such as Peru,

Japan, Mexico, Bolivia, and Canada are major producers of bismuth. It is found as crystals in the sulfide ores of nickel, cobalt, silver, and tin. Bismuth is mainly produced as a by-product from lead and copper smelting. It is insoluble in hot or cold water. Bismuth explodes if mixed with chloric or perchloric acid. Molten bismuth explodes and bismuth powder glows red-hot on contact with concentrated nitric acid. It is flammable in powder form.

Bismuth is used in the manufacture of low melting solders and fusible alloys, as key components of thermoelectric safety appliances, such as automatic shut offs for gas and electric water heating systems and safety plugs in compressed gas cylinders, in the production of shot and shotguns, in pharmaceuticals, in the manufacture of acrylonitrile, the starting material for synthetic fibers and rubbers. Bismuth oxychloride is sometimes used in cosmetics. Also bismuth subnitrate and bismuth subcarbonate are used in medicine. Bismuth subsalicylate is used as an antidiarrheal and to treat other gastrointestinal diseases.

Toxicity and health effects

Exposures to bismuth salts are associated primarily by ingestion. Bismuth is known to cause adverse health effects. The symptoms include, but are not limited to, irritation of the eyes, skin, respiratory tract, lungs, foul breath, metallic taste, and gingivitis. On ingestion, bismuth causes nausea, loss of appetite, weight, malaise, albuminuria, diarrhea, skin reactions, stomatitis, headache, fever, sleeplessness, depression, rheumatic pain, and a black line may form on gums in the mouth due to deposition of bismuth sulfide. Prolonged exposure to bismuth causes mild but deleterious effects on the kidneys and high concentrations of bismuth result in fatalities. Occupational exposures to bismuth occur during the manufacture of cosmetics, industrial chemicals, and pharmaceuticals. Acute exposure with over dosage of bismuth-containing drugs causes anorexia, nausea, vomiting, abdominal pain, and possibly a dry mouth and thirst. Bismuth also causes neurotoxicity. Bismuth pentafluoride is highly toxic and causes irritation to the skin, eyes, and respiratory tract, while bismuth subnitrate causes blurred vision.

Bismuth and carcinogenicity

There are no reports about the carcinogenicity of bismuth to humans and bismuth has not been grouped as a human carcinogen.

References

Beliles, R.P. 1994. The metals: bismuth. In: Clayton, G.D. and Clayton, F.E. (eds.), *Patty's Industrial Hygiene and Toxicology*, Vol. 2, 4th ed. Wiley, J. New York. pp. 1948–54.
Bradberry, S.M. 1996. International Program on Chemical Safety (IPCS). INTOC Data Bank CEHM Bismuth UKPID Monograph. National Poisons Information Service, Birmingham, U.K.
Budavari, S. (ed.). 1989. *The Merck Index. An Encyclopedia of Chemicals, Drugs and Biologicals*, 11th ed. Merck, Whitehouse Station, NJ.
Czerwinski, A.W. and Ginn, H.E. 1964. Bismuth nephrotoxicity. *Am. J. Med.* 37: 969–75.
Grant, W.M. and Schuman, J.S. 1993. *Toxicology of the Eye*, 4th ed. Charles C Thomas, Springfield, IL.
Hudson, M., Ashley, N. and Mowat, G. 1989. Reversible toxicity in poisoning with colloidal bismuth subcitrate. *Br. Med. J.* 299: 59.

Huwez, F., Pall, A., Lyons, D. and Stewart, M.J. 1992. Acute renal failure after overdose of colloidal bismuth subcitrate. *Lancet* 340: 1298.

James, J.A. 1968. Acute renal failure due to a bismuth preparation. *Calif. Med.* 109: 317–19.

Winship, K.A. 1983. Toxicity of bismuth salts. *Adverse Drug React. Acute Poisoning Rev.* 2: 103–21.

Boron (CAS no. 7440-42-8)

Molecular formula: **B**

Use and exposure

In 1808, Sir Humphry Davy and J. L. Gay-Lussac discovered boron. It is a trivalent, non-metallic element that occurs abundantly in the evaporite ores, borax and ulexite. Boron is never found as a free element on Earth. Boron as a crystalline is a very hard, black material with a high melting point, and exists in many polymorphs. Boron has several forms, the most common form being amorphous boron, a dark powder, non-reactive to oxygen, water, acids, and alkalis. It reacts with metals to form borides. Boron is an essential plant micronutrient. Sodium borate is used in biochemical and chemical laboratories to make buffers. Boric acid is produced mainly from borate minerals by the reaction with sulfuric acid. Boric acid is an important compound used in textile products.

Compounds of boron are used in organic synthesis, in the manufacture of special types of glasses, and as wood preservatives. Boron filaments are used for advanced aerospace structures owing to their high strength and light weight. It is used as an antiseptic for minor burns or cuts and is sometimes used in dressings. Boric acid was first registered in the United States in 1948 as an insecticide for control of cockroaches, termites, fire ants, fleas, silverfish, and many other insects. It acts as a stomach poison affecting the insects' metabolism, and the dry powder is abrasive to the insects' exoskeleton. Boric acid is generally considered to be safe for use in household kitchens to control cockroaches and ants. The important use of metallic boron is as boron fiber. Borate-containing minerals are mined and processed to produce borates for several industrial uses, i.e., glass and ceramics, soaps and detergents, fire retardants and pesticides.

The fibers are used to reinforce the fuselage of fighter aircraft, e.g., the B-1 bomber. The fibers are produced by vapor deposition of boron on a tungsten filament. Pyrex is a brand name for glassware, introduced by Corning Incorporated in 1915. Originally, Pyrex was made from thermal shock-resistant borosilicate glass. The common borate compounds include boric acid, sodium tetraborates (Borax), and boron oxide.

Toxicity and health effects

Boron has been studied extensively for its nutritional importance in animals and humans. There is a growing body of evidence that boron may be an essential element in animals and humans. Many nutritionists believe that people would benefit from more boron and many popular multivitamins, such as centrum, in the diet. The adverse health effects of boron on humans is limited. However, ingestion/inhalation causes irritation to the mucous membrane and boron poisoning.

Short-term exposures to boron in work areas are known to cause irritation of the eye, the upper respiratory tract, and the naso-pharynx, but the irritation disappears with the stoppage of further exposure. Ingestion of large amounts of boron (about 30 g of boric acid) over short periods of time is known to affect the stomach, intestines, liver, kidney, and brain and can eventually lead to death in exposed people.

Boron and carcinogenicity

Boron has not been listed as a human carcinogen by the IARC, the NTP, or the OSHA.

Exposure limit

For boron, the US EPA has determined the limit in drinking water as 4 mg/L/day. The OSHA has set a legal limit of 15 mg/m^3 for boron oxide in air averaged over an 8-h workday.

Precautions

Elemental boron is non-toxic and common boron compounds, such as borates and boric acid, have low toxicity (approximately similar to table salt with the lethal dose being 2–3 g/kg) and do not require special precautions while handling. Some of the more exotic boron hydrogen compounds, however, *are* toxic as well as highly flammable and do require special care when handling.

Reference

Agency for Toxic Substances and Disease Registry (ATSDR). 2007. Toxicological Profile for Boron. (Draft for Public Comment.) US Department of Health and Human Services, Public Health Service, Atlanta, GA.

Boron compounds

There are several commercially important borates, including borax, boric acid, sodium perborate, and many boron compounds. The following list includes a few selected compounds.

(i) Boric acid (CAS No. 10043-35-3)

Molecular formula: H_3BO_3
Synonyms and trade names: ortho-Boric acid; Boracic acid; Borofax, boric acid; Hydrogen orthoborate

Boric acid is white powder or granules and odorless. It is incompatible with potassium, acetic anhydride, alkalis, carbonates, and hydroxides. Boric acid has uses in the production of textile fiberglass, flat panel displays, and eye drops. Boric acid is recognized for its application as a pH buffer and as a moderate antiseptic agent and emulsifier.

(ii) Boron carbide (CAS No. 12069-32-8)

Molecular formula: B_4C

Boron carbide is used as a ceramic material and is used to make armor materials, especially bulletproof vests for soldiers and police officers.

(iii) Sodium tetraborate decahydrate/Borax (CAS No. 1330-43-4 (anhydrous); 1303-96-4 (decahydrate))

Molecular formula: $Na_2B_4O_7, 10H_2O$
  Borax is used in the production of adhesives and in anticorrosion systems.

(iv) Sodium tetraborate pentahydrate (CAS No. 12179-04-3)

Molecular formula: $Na_2B_4O_7, 5H_2O$
Synonyms and trade names: Sodium tetraborate pentahydrate; Disodium tetraborate Pentahydrate; Borax pentahydrate
  Used in large amounts in making insulating fiberglass and sodium perborate bleach.

(v) Boron nitride (CAS No. 10043-11-5)

Molecular formula: BN
  Boron nitride is a material in which the extra electron of nitrogen (with respect to carbon) enables it to form structures that are isoelectronic with carbon allotropes.

(vi) Borazole/Borazine (CAS No. 6569-51-3)

Molecular formula: $B_3N_3H_6$
  Borazole (Borazine) is a colorless liquid with an aromatic smell. It decomposes with water to form boric acid, ammonia, and hydrogen. The reaction product of boron and ammonia at high temperatures is known as inorganic benzene.

Boron trichloride (CAS No. 10294-34-5)

Molecular formula: BCl_3
Synonyms and trade names: Borane, trichloro-; Boron chloride; Trichloroborane; Trichloroboron

Use and exposure

Boron trichloride is a colorless gas with a pungent odor. Boron trichloride reacts violently with water, and on decomposition and hydrolysis yields hydrochloric and boric acid. It has a pungent, highly irritating odor. Occupational exposure to boron and boron compounds can occur in industries that produce special glass, washing powder, soap and cosmetics, leather, cement, etc.

Toxicity and health effects

The fumes of boron trichloride irritate the eyes and mucous membranes. On contact, it can cause severe skin burns, severe pain, redness, necrosis, and scarring. It is corrosive to metals and living tissues. On inhalation, boron trichloride causes chemical pneumonitis and pulmonary edema—a result of exposure to the lower respiratory tract and deep lung. Occupational workers exposed to boron trichloride show symptoms such as tearing of the eyes, coughing, labored breathing, excessive salivary and sputum formation leading to pulmonary malfunction. Boron trichloride is a strong irritant to tissues and its fumes are corrosive and highly toxic. Boron affects the CNS causing depression of circulation as well as shock and coma.

The signs and symptoms of acute exposure to boron trichloride include salivation, intense thirst, difficulty in swallowing, chills, pain, shock, and oral, esophageal, and stomach burns. Ingestion of boron trichloride in work areas leads to circulatory collapse of the worker. On acute inhalation, boron trichloride result in sneezing, hoarseness, choking, laryngitis, and respiratory tract irritation along with bleeding of the nose and gums, ulceration of the nasal and oral mucosa, bronchitis, pneumonia, dyspnea, chest pain, and pulmonary edema.

Boron trichloride and carcinogenicity

There are no reports indicating that boron trichloride is a human carcinogen.

Precautions

Boron trichloride vigorously attacks elastomers and packing materials, natural and synthetic rubbers. It also reacts energetically with nitrogen dioxide/dinitrogen tetraoxide, aniline, phosphine, triethylsilane, or fat and grease. It reacts exothermically with chemical bases such as amines, amides, and inorganic hydroxides. Occupational workers should use gloves of neoprene or butyl rubber, PVC or polyethylene, safety goggles, or glasses and face shield, and safety shoes.

Storage

Boron trichloride cylinders should be protected from physical damage. The cylinders should be stored upright and firmly secured to prevent falling or being knocked over, in a cool, dry, well-ventilated area of non-combustible construction away from heavily trafficked areas and emergency exits.

References

Material Safety Data Sheet (MSDS). 2002. Boron Trichloride (updated 2006). Mallinckrodt Baker, Phillipsburg, NJ.
National Institute for Occupational Safety and Health (NIOSH). 1994. International Programme on Chemical Safety (IPCS) and The Commission of the European Communities (CEC). Boron Trichloride. ICSC card no. 0616. Atlanta, GA.

Boron trifluoride (CAS No. 7637-07-2)

Molecular formula: BF_3
Synonyms and trade names: Boron fluoride, Triflroborane

Use and exposure

Boron trifluoride is a colorless gas with an acrid suffocating odor. It forms thick acidic fumes in moist air. Dry boron trifluoride is used with mild steel, copper, copper-zinc and copper-silicon alloys, and nickel. Moist gas is corrosive to most metallic materials and some plastics. Therefore, Kel-F® and Teflon® are the preferred gasketing materials.

Mercury containing manometers should not be used because boron trifluoride is soluble in mercury. It decomposes in hot water yielding hydrogen fluoride.

Toxicity and health effects

Exposures to boron trifluoride cause irritating effects, painful burns, lesions, loss of vision, stinging of the skin, irritation of the upper respiratory tract, and cough. Higher concentrations may cause inflammation and congestion of the lungs. Occupational exposures to high concentrations of boron trifluoride result in burns to the mucous membranes. Even at low concentrations (as low as 50 ppm), boron trifluoride causes cardiac collapse, pulmonary edema, and chemical pneumonitis.

Boron trifluoride and carcinogenicity

There are no reports by the NTP, the IARC, or the OSHA indicating that boron trifluoride is a human carcinogen.

Exposure limit

The OSHA and the ACGIH have set the PEL and the TLV for boron trifluoride at 1 ppm, respectively.

Precautions

Exposures to boron trifluoride in occupational work areas cause irritating effects, painful burns, lesions, and loss of vision. Workers with potential exposure to boron trifluoride should not wear contact lenses. Prompt medical attention is mandatory in all cases of overexposure to boron trifluoride and the rescue personnel should be equipped with proper protectives. Occupational workers should handle/use boron trifluoride only in well-ventilated areas. The valve protection caps must remain in place. Workers should not drag, slide, or roll the cylinders, and use a suitable hand truck for cylinder movement. Compressed gas cylinders shall not be refilled without the express written permission of the owner. Boron trifluoride is listed as an extremely hazardous substance (EHS).

The cylinder should not be heated by any means to increase the discharge rate of the product from the cylinder. The cylinder of boron trifluoride should be kept stored in a cool, dry, well-ventilated area of non-combustible construction away from heavily trafficked areas and emergency exits.

Reference

Material Safety Data Sheet (MSDS). 2001. Boron Trifluoride. http://www.voltaix.com/msds/msds-boron_trifluoride_bf3.htm#sec2 (updated 2008).

Bromodichloromethane (CAS No. 75-27-4)

Molecular formula: $CHBrCl_2$
Synonyms and trade names: Dichlorobromomethane; Dichloromonobromomethane; Methane, bromodichloro-; Monobromodichloromethane

Use and exposure

Bromodichloromethane (BDCM) is a colorless, heavy, non-burnable/non-flammable liquid. It was formerly used as a flame retardant, a solvent for fats and waxes and because of its high density for mineral separation. Now, it is only used as a reagent or intermediate in organic chemistry. Small amounts of BDCM are also made in chemical plants for use in laboratories or in making other chemicals. On contact with hot surfaces or flames, BDMC decomposes forming toxic and corrosive gases, including hydrogen bromide and hydrogen chloride and reacts with strong bases, strong oxidants, and magnesium. BDMC is found in chlorinated drinking water as a consequence of the reaction between chlorine, added during water treatment.

Toxicity and health effects

On ingestion, BDMC causes damage to the kidneys, liver, and impaired functions.

Bromodichloromethane and carcinogenicity

There is sufficient evidence for the carcinogenicity of bromodichloromethane in experimental animals, but the evidence is inadequate in humans. The IARC has classified bromodichloromethane as Group 2B, meaning a possible human carcinogen.

Exposure limits

For bromodichloromethane, the US EPA has set a maximum contaminant level (MCL) of 0.1 ppm.

References

Agency for Toxic Substances and Disease Registry (ATSDR). 1989. Toxicological Profile for Bromodichloromethane. US Department of Health and Human Services, Public Health Service, Atlanta, GA.

International Programme on Chemical Safety and the European Commission (IPCS EC). 2006. Bromodichloromethane. ICSC card no. 0393 (updated 2008).

Bromomethane (CAS No. 74-83-9)

Molecular formula: CH_3Br
Synonyms and trade names: Brom-o-Gas; Celfume; Curafume; Dowfume MC-33; Embafume; Halon 1001; Haltox; MB; Iscobrome; Kayafume; MBX; MeBr; Metfume; Methogas; Methyl bromide; Monobromomethane; Pestmaster; Profume; Terabol; Terr-o-Gas; Zytox

Use and exposure

Bromomethane is a highly toxic compound and the US EPA has grouped it as a toxicity class I chemical substance. Bromomethane is a colorless gas or volatile liquid that is usually odorless, but has a sweet, chloroform-like odor at high concentrations and is easily miscible with ethanol, ether, aromatic carbon disulfide, and ketones. It decomposes on heating and, on burning, produces highly toxic and irritating fumes, bromides, carbon oxybromide, carbon dioxide, and monoxide. It is also used as a general purpose fumigant to kill a variety of pests, including rats and insects, and a gas soil fumigant against insects, termites, rodents, weeds, nematodes, and soil-borne diseases.

Bromomethane is used for post-harvest fumigation of foods, such as cereals, spices, dried fruits, nuts, fresh fruits, and vegetables. Although bromomethane is on the list of banned ozone-depleting chemical substances of the Montreal Protocol, in 2005 and 2006 it was granted a critical use exemption (under the Montreal Protocol). Bromomethane is an RUP and should be purchased and used only by certified applicators.

Toxicity and health effects

Exposures to bromomethane in high concentrations cause headaches, burns the skin, itching, redness, blisters, dizziness, nausea, vomiting, and weakness. Prolonged periods of exposure cause mental excitement, muscle tremors, seizures, bronchitis and pneumonia, numbness, tremor, speech defects, damage to the nervous system, lung, nasal mucosa, kidney, convulsions, respiratory paralysis, coma, and death. Human exposure to bromomethane is predominantly occupational, during manufacture and occupational field fumigation.

Bromomethane and carcinogenicity

The US EPA has classified bromomethane as Group D, meaning inadequate evidence as a human carcinogen.

Exposure limits

The ADI for bromomethane is set as 1.0 mg/kg/day (as bromide ion). The OSHA has set the average limit level of bromomethane in workplace air at 20 ppm for an 8-h workday over a 40-h week.

Precautions

Occupational workers should use appropriate ventilation during production and formulation of bromomethane at the workplace. The ventilation must be sufficient to maintain the levels of bromomethane below the prescribed OEL. Local exhaust ventilation at source or vapor extraction may also be used. Gloves or rubber boots should not be used as the liquid or concentrated vapor may be trapped inside them.

Storage

Handling and storage of bromomethane cylinders must meet the specifications laid down by the regulatory authorities. The cylinders must undergo the required and periodic tests.

References

Agency for Toxic Substances and Disease Registry (ATSDR). 1995. Toxicological Profile for Bromomethane. US Department of Health and Human Services, Public Health Service, Atlanta, GA (updated 2008).

Kidd, H. and James, D.R. (eds.). 1991. *The Agrochemicals Handbook*, 3rd ed. Royal Society of Chemistry Information Services, Cambridge, U.K.

Pesticide Information Profiles (PIP). 1996. Methyl bromide; Bromomethane. Extension Toxicology Network, EXTOXNET. PIP. Oregon State University, Corvallis, OR.

Technology Transfer Network. Air Toxics web site. 2000. Methyl Bromide (Bromomethane). Hazard Summary, created in April 1992; revised in January 2000 (updated 2007).

Butadienes

1,3-Butadiene (CAS No. 106-99-0)

Molecular formula: C_4H_6
Synonyms and trade names: Biethylene; Bivinyl; Butadiene; Divinyl; Erythrene; Vinylethylene

Use and exposure

1,3-Butadiene is a simple conjugated diene. It is a colorless gas with a mild aromatic or gasoline-like odor and incompatible with phenol, chlorine dioxide, copper, and crotonaldehyde. The gas is heavier than air and may travel along the ground; distant ignition is possible. It is an important industrial chemical used as a monomer in the production of synthetic rubber. Most butadiene is polymerized to produce synthetic rubber. While polybutadiene itself is a very soft, almost liquid material, polymers prepared from mixtures of butadiene with styrene or acrylonitrile, such as ABS, are both tough and elastic. Styrene-butadiene rubber is the material most commonly used for the production of automobile tires. Smaller amounts of butadiene are used to make nylon via the intermediate adiponitrile, other synthetic rubber materials such as chloroprene, and the solvent sulfolane. Butadiene is used in the industrial production of cyclododecatriene via a trimerization reaction.

Toxicity and health effects

The symptoms of poisoning include distorted blurred vision, vertigo, general tiredness, decreased blood pressure, headache, and nausea. Exposures to very high concentrations of 1,3-butadiene are known to cause CNS depression, decreased pulse rate, drowsiness, fatigue, vertigo, ataxia, unconsciousness, coma, respiratory paralysis, and death. Several studies show butadiene exposure increases the risk of cardiovascular diseases and cancer. There is a lack of human data on the effects butadiene. Animal studies have shown breathing butadiene during pregnancy can increase the number of birth defects.

1,3-Butadiene and carcinogenicity

Studies on laboratory animals have indicated that 1,3-butadiene causes carcinogenic effects, and has effects on the bone marrow, resulting in leukemia. While these data reveal important implications for the risk of human exposure to butadiene, more data are necessary to draw more conclusive risk assessments. Also, it has been reported that butadiene has a higher sensitivity to women over men when exposed to the chemical. The IARC, based on the human epidemiological evidence, observed that evidences for carcinogenicity are limited and categorized 1,3-butadiene as Group 2A, meaning a probable human carcinogen.

Exposure limits

The ACGIH set the TLV limit as 2 ppm (TWA), the OSHA set the PEL as 1 ppm (TWA), and the NIOSH set the IDLH as Ca 2000 ppm.

References

International Programme on Chemical Safety and the Commission of the European Communities (IPCS CEC). 1994. 1,3-Butadiene. ICSC card no. 0017 (updated 2000).

National Institute for Occupational Safety and Health (NIOSH). 2005. 1,3-Butadiene. *Pocket Guide to Chemical Hazards*. Centers for Disease Control and Prevention, Atlanta, GA.

2-Butanone (CAS No. 78-93-3)

Molecular formula: C_4H_8O

Synonyms and trade names: Ethyl methyl ketone; 2-Butanone; MEK; Methyl acetone; Methyl ethyl ketone (MEK); Methylpropanone

Use and exposure

2-Butanone is a stable, highly flammable chemical. It is incompatible with oxidizing agents, bases, and strong reducing agents. It is a colorless liquid with a sharp, sweet odor. 2-Butanone is produced in large quantities. It is used as a solvent and nearly half of its use is in paints and other coatings because it quickly evaporates into the air and it dissolves many substances. It is also used in glues and as a cleaning agent.

Toxicity and health effects

Occupational workers are exposed to 2-butanone by breathing contaminated air in workplaces associated with the production or use of paints, glues, coatings, or cleaning agents. Prolonged exposures to 2-butanone cause symptoms of poisoning such as cough, dizziness, drowsiness, headache, nausea, vomiting, dermatitis, irritation of the nose, throat, skin, and eyes and at very high levels cause drooping eyelids, uncoordinated muscle movements, loss of consciousness, and birth defects. Chronic inhalation studies in animals have reported slight neurological, liver, kidney, and respiratory effects. However, information on the chronic (long-term) effects of 2-butanone (methyl ethyl ketone) in humans is limited.

2-Butanone and carcinogenicity

The DHHS, the IARC, and the US EPA have not classified 2-butanone as a human carcinogen.

Exposure limits

For 2-butanone, the OSHA, the ACGIH, and the NIOSH have set an OEL of 200 ppm in workplace air for an 8-h workday, 40-h workweek.

Precautions

2-Butanone vapor and air mixtures are explosive. It reacts violently with strong oxidants and inorganic acids causing fire and explosion hazard.

Storage

2-Butanone should be protected from moisture.

References

Agency for Toxic Substances and Disease Registry (ATSDR). 1992. Toxicological Profile for 2-Butanone. US Department of Health and Human Services, Public Health Service, Atlanta, GA (updated 2008).

International Programme on Chemical Safety and the Commission of the European Communities (IPCS CEC). 1998. Methyl ethyl ketone. ICSC card no. 0179. ICSC, Geneva, Switzerland.

1-Butene (CAS no. 106-98-9)

Molecular formula: C_4H_8
Synonyms and trade names: 1-Butylene; Alpha-butene; Alpha-butylene; Butene-1; N-Butene; Normal Butene; Ethylethylene

Use and exposure

1-Butene is a colorless, extremely flammable liquefied gas with an aromatic odor. It is insoluble in water and is an isomer of butane. It is highly flammable and readily forms explosive mixtures with air. 1-Butene of high purity is made by cracking naphtha and separating it from other products by an extra-high purity distillation column. However, 1-butene is incompatible with metal salts, fluorine, nitrogen oxides, boron trifluoride, halogen acids, halogens, and strong oxidizing agents. It is an important organic compound in the production of several industrial materials, i.e., linear low density polyethylene (LLDPE), a more flexible and resilient polyethylene, a range of polypropylene resins, and in the production of polybutene, butylene oxide and in the C4 solvents, secondary butyl alcohol (SBA) and methyl ethyl ketone (MEK). The vapor of 1-butene is heavier than air and may travel long distances to an ignition source and flash back.

Toxicity and health effects

Exposures to 1-butene cause the effects of an asphyxiant and/or an anesthetic (at high concentrations). Workers exposed to 1-butene develop eye irritation.

Precautions

When working with 1-butene, occupational workers should wear proper protectives, preferably a NIOSH-approved full-face positive pressure supplied-air respirator or a self-contained breathing apparatus (SCBA). Workers should not wear contact lenses.

References

Material Safety Data Sheet. www.shell.com/chemicals/msds.
Flick, E.W. 1998. Industrial Solvents Handbook. 5th Edition, New York, William Andrew Publishing, Noyes.

2-Butoxyethanol (CAS No. 111-76-2)

Molecular formula: $C_6H_{14}O_2$

Synonyms and trade names: Ethylene glycol monobutyl ether; Monobutyl glycol ether; 2-Butoxy-1-ethanol; Ethylene glycol butyl ether; Ethylene glycol n-butyl ether; Butyl cellusolve; Butyl glycol; Butyl oxitol

Use and exposure

2-Butoxyethanol is a clear colorless liquid with an ether-like smell and belongs to the family of glycol ether/alkoxy alcohol. 2-Butoxyethanol is miscible in water and soluble in most organic solvents. 2-Butoxyethanol does not occur naturally. It is usually produced by reacting ethylene oxide with butyl alcohol. It is used as a solvent for nitrocellulose, natural and synthetic resins, soluble oils, in surface coatings, spray lacquers, enamels, varnishes, and latex paints, as an ingredient in paint thinners, quick-dry lacquers, latex paint, and strippers, varnish removers, and herbicides. 2-Butoxyethanol is also used in textile dyeing and printing, in the treatment of leather, in the production of plasticizers, as a stabilizer in metal cleaners and household cleaners, and in hydraulic fluids, insecticides, herbicides, and rust removers. It is also used as an ingredient in liquid soaps, cosmetics, industrial and household cleaners, dry-cleaning compounds, and as an ingredient in silicon caulks, cutting oils, and hydraulic fluids. 2-Butoxyethanol is a fire hazard when exposed to heat, sparks, or open flames.

Toxicity and health effects

2-Butoxyethanol is present in a variety of consumer products, including cleaning agents and surface coatings, such as paints, lacquers, and varnishes. 2-Butoxyethanol is readily absorbed following inhalation, oral, and dermal exposure. 2-Butoxyethanol is released into air or water by different industrial activities and facilities that manufacture, process, or use the chemical. Exposure to 2-butoxyethanol causes irritating effects to the eyes and skin, but it has not induced skin sensitization in guinea pigs. Information on the human health effects associated with exposure to 2-butoxyethanol is limited. However, case studies of individuals who had attempted suicide by ingesting 2-butoxyethanol-containing cleaning solutions suffered poisoning with symptoms such as hemoglobinuria, erythropenia, and hypotension, metabolic acidosis, shock, non-cardiogenic pulmonary edema, and albuminuria, hepatic disorders and hematuria.

2-Butoxyethanol and carcinogenicity

The Department of Health and Human Services (DHSS), the IARC, and the US EPA have not classified 2-butoxyethanol and 2-butoxyethanol acetate as to their human carcinogenicity. There is no available data on the carcinogenicity of 2-butoxyethanol and 2-butoxyethanol acetate in people or animals. On the basis of inadequate data, the IARC has observed that 2-butoxyethanol is grouped as 3, meaning not classifiable as a human carcinogen.

Exposure limits

For 2-butoxyethanol, the OSHA has set an exposure limit of 50 ppm in workplace air for an 8-h workday, 40-h workweek.

Precautions

Occupational workers should use minimal quantities of 2-butoxyethanol in designated areas with adequate ventilation and away from sources of heat or sparks. Whenever possible, fire-resistant containers should be used. Wear appropriate protective equipment to prevent skin and eye contact.

Storage

2-Butoxyethanol should be kept stored in tightly closed, grounded containers in a cool area with adequate ventilation, away from normal work areas and sources of heat and sparks, and electrical equipment. At the storage and handling area, workers should use solvent-resistant materials.

References

Agency for Toxic Substances and Disease Registry (ATSDR). 1998. Toxicological profile for 2-butoxy-ethanol and 2-butoxyethanol acetate. US Department of Health and Human Services, Public Health Service Atlanta, GA.

Carpenter, C.P., Pozzani, U.C., Woil, C.S., Nair, J.H., Keck, G.A. and Smyth, H.F. Jr. 1956. The toxicity of butyl cellosolve solvent. *Am. Med. Assoc. Arch. Ind. Health* 14: 114–31.

Dean, B.S. and Krenzelok, E.P. 1992. Clinical evaluation of pediatric ethylene glycol monobutyl ether poisonings. *J. Toxicol. Clin. Toxicol.* 30 (4): 557–63.

International Programme on Chemical Safety (IPCS). 1993. 2-Butoxyethanol. IPCS card no. 0059. Geneva, Switzerland.

———. 2006. Butoxyethanol, 2-. IARC Summary and Evaluation, IARC Monographs on the Evaluation of Carcinogenic Risks to Humans. Volume 88. IARC, Geneva, Switzerland.

World Health Organization (WHO). 1998. Concise International Chemical Assessment Document (CICAD) No. 10. 2-Butoxyethanol. WHO, Geneva. Switzerland.

Butyl acetates

sec-Butyl acetate (CAS No. 105-46-4)

Molecular formula: $C_6H_{12}O_2$
Synonyms and trade names: 1-Methylpropyl acetate; Acetic acid, 2-butyl ester;

Use and exposure

Sec-Butyl acetate is a colorless, highly flammable liquid with a pleasant odor. The vapor mixes well with air and become explosive mixtures. It reacts with strong oxidants, strong bases, strong acids, and nitrates, causing fire and explosion hazard.

Toxicity and health effects

During prolonged occupational exposures, sec-butyl acetate causes health effects. The symptoms of toxicity include irritation to the skin and eyes. Exposures to high concentrations of sec-butyl acetate cause irritation to the nose and throat, causing coughing and respiratory distress, headache, nausea, vomiting, dizziness, drowsiness, and coma. Occupational workers after prolonged exposures to the vapor of sec-butyl acetate showed severe irritation to eyes, headache, drowsiness, dryness in the upper respiratory system, skin, narcosis, effects on the CNS, and lowering of consciousness.

Exposure limits

For sec-butyl acetate, the legal airborne PEL is 200 ppm averaged over an 8-h work shift. Similarly, the NIOSH and the ACGIH also recommend the airborne exposure limit as 200 ppm (TWA) over a 10-h/8-h work shift, respectively.

Precautions

During handling of sec-butyl acetate, occupational workers should avoid use of open flames, sparks, remove all ignition sources, and smoking. The vapor of sec-butyl acetate mixes well with air, and easily forms explosive mixtures. Workers should avoid using compressed air for filling, discharging, or handling.

Reference

International Programme on Chemical Safety and the Commission of the European Communities (IPCS-CEC). 1999. sec-Butyl acetate. ICSC card no. 0840. IPCS, Geneva, Switzerland (updated 2003).

n-Butyl acetate (CAS No. 123-86-4)

Molecular formula: $C_6H_{12}O_2$
Synonyms and trade names: Acetic acid, butyl ester; 1-Butyl acetate; Butyl ethanoate; IMIS12: 0440

Use and toxicity

n-Butyl acetate, also known as butyl ethanoate, is an organic compound commonly used as a solvent in the production of lacquers and other products. It is also used as a synthetic fruit flavoring in foods such as candy, ice cream, cheeses, and baked goods. Butyl acetate is found in many types of fruit, where along with other chemicals it imparts characteristic flavors. Apples, especially of the Red Delicious variety, are flavored in part by this chemical. It is a colorless flammable liquid with a sweet smell of banana.

Isobutyl acetate (CAS No. 110-19-0)

Molecular formula: $C_6H_{12}O_2$
Synonyms and trade names: Acetic acid, isobutyl ester; Acetic acid, 2-methylpropyl ester; 2-methylpropyl acetate; IMIS21: 1534

Use and toxicity

Isobutyl acetate, also known as 2-methylpropyl ethanoate (IUPAC name) or β-methylpropyl acetate, is a common solvent. It is produced from the esterification of isobutanol with acetic acid. It is used as a solvent for lacquer and nitrocellulose. Like many esters, it has a fruity or floral smell at low concentrations and occurs naturally in raspberries, pears, and other plants. At higher concentrations, the odor can be unpleasant and may cause symptoms of CNS depression, such as nausea, dizziness, and headache.

1-Butyne (CAS No. 107-00-6)

Molecular formula: C_4H_6
Synonyms and trade names: Butyne-1; Ethylacetylene; Ethylethyne

Use and exposure

1-Butyne is a clear, colorless gas with a characteristic acetylenic odor. It is an extremely flammable and reactive alkyune and may cause frostbite. It is insoluble in water. It is a specialty gas mixture used in organic synthesis of compounds and instrument calibration; it has no specific or significant industrial application. On combustion, 1-butyne releases carbon monoxide.

Toxicity and health effects

1-Butyne probably has some anesthetic activity and can act as a simple asphyxiant. The symptoms of poisoning include rapid respiration, muscular incoordination, fatigue, dizziness, nausea, vomiting, unconsciousness, and death. The toxicity of 1-butyne has not been thoroughly investigated and the published information is very scanty.

Storage

Cylinders of 1-butyne should be stored in well-ventilated areas, away from heat, flame, and sparks and with proper valve protection caps on the cylinders when not in use. Non-refillable cylinders must be disposed of in accordance with federal, state, and local regulations.

Precautions

Occupational workers should be very careful during handling/using 1-butyne cylinders, which may rupture under fire conditions, thereby forming an explosive mixture in air. Also, the vapors of 1-butyne may travel a considerable distance to the source of ignition and flash back. Workers should secure the cylinder when using to protect them from falling and a suitable hand truck should be used to move the cylinders. The workplace should have adequate general and local exhaust ventilation to maintain concentrations below flammable limits and to avoid asphyxiation. Workers should have proper protective gloves and safety shoes to prevent contact with cold equipment. Workers should be cautious to protect 1-butyne containers from physical damage, avoid defacing the cylinders or labels, and the cylinders should be refilled by qualified producers of compressed gas.

Butane (CAS No. 106-97-8)

Molecular formula: C_4H_{10}
Synonyms and trade names: n-Butane; Butyl hydride; Methylethylmethane

Use and exposure

The main sources of butane are crude oil refining and natural gas processing. It is usually blended into motor vehicle gasoline to increase the fuel's volatility and to make engine starting easier. Butane contains mixtures of methane, ethane, propane, iso-butane, and n-butane and is a colorless aliphatic hydrocarbon gas with a gasoline-like odor.

Butane is a component of liquefied petroleum gas (LPG) and as such is used in a wide variety of fuel applications for both recreational and leisure use, including heating and air-conditioning, refrigeration, cooking, and in lighters. Butane is commonly used alone or in mixtures as a propellant in aerosol consumer products, such as hair sprays, deodorants and antiperspirants, shaving creams, edible oil and dairy products, cleaners, pesticides and coatings (e.g., automobile or household spray paint). Butane is used as a chemical intermediate in the production of maleic anhydride, ethylene, methyl tert-butyl ether (MTBE), synthetic rubber, and acetic acid and its by-products.

Butane is a simple asphyxiant with explosive and flammable potential. It is also widely used as a substance of abuse. The main target organs are in the central nervous and cardiovascular systems. Butane is found in aerosols, lighter fuel and refills, small blow torches, and camping stoves. Pure grades of butane are used in calibrating instruments and as a food additive. It is widely available. Misuse and adulteration of butane is a common commercial practice.

Toxicity and health effects

Exposures to butane cause excitation, blurred vision, slurred speech, nausea, vomiting, coughing, sneezing, and increased salivation. With increased periods of exposure to high concentrations of butane, the signs and symptoms of toxicity become more severe. For instance, the exposed worker demonstrates confusion, perceptual distortion, hallucinations (ecstatic or terrifying), delusions, behavioral changes, tinnitus, and ataxia. Workers exposed to larger doses of butane suffer from nystagmus, dysarthria, tachycardia, central depression of the CNS, drowsiness, coma, and sudden death. It has been reported that poisoned individuals show anoxia, vagal inhibition of the heart, respiratory depression, cardiac arrhythmias, and trauma.

References

Bauman, J.E., Dean, B.S. and Krenzelok, E.P. 1991. Myocardial infarction and neurodevastation following butane inhalation. *Vet. Hum Toxicol.* 4: 150.

Ramsey, J., Anderson, H.R., Bloor, K. and Flanagan, R.J. 1989. Mechanism of sudden death associated with volatile substance abuse. *Human Toxicol.* 8: 261–69.

Shepherd, R.T. 1989. Mechanism of sudden death associated with volatile substance abuse. *Human Toxicol.* 8: 287–92.

World Health Organization (WHO). Butane. Poisons Information Monograph 945. International Programme on Chemical Safety. IPCS, WHO, Geneva, Switzerland.

Butanethiol (CAS No. 109-79-5)

Molecular formula: $C_4H_{10}S$
Synonyms and trade names: Butyl mercaptan; Butanethiol; 1-Butanethiol; *n*-Butyl mercaptan; Thiobutyl alcohol; 1-Mercaptobutane; Butyl sulfhydrate

Use and exposure

Butanethiol is also known as butyl mercaptan. It is a volatile, highly flammable clear to yellowish liquid with an extremely foul-smelling, strong, garlic-, cabbage- or skunk-like odor. It is slightly oily in nature. Butanethiol is used as a chemical intermediate in the production of insecticides and herbicides. It is also used as a gas odorant.

Toxicity and health effects

Exposures to butanethiol cause adverse health effects and poisoning. The symptoms of poisoning in exposed workers include, but are not limited to, asthenia, muscular weakness, malaise, sweating, nausea, vomiting, headache, restlessness, increased respiration, incoordination, muscular weakness, skeletal muscle paralysis in most cases, heavy to mild cyanosis, lethargy and/or sedation, respiratory depression followed by coma. Severe cases of poisoning lead to death of the exposed worker.

In laboratory studies, animals given intraperitoneal and oral exposures of butanethiol survived near-lethal single doses. However, after 20 days post-treatment, animals showed pathological changes involving liver and kidney damage. The pathomorphological changes in the liver included cloudy swelling, fatty degeneration, and necrosis. Pathological changes in the kidneys included cloudy swelling, and the lungs displayed capillary engorgement, patchy edema, and occasional hemorrhage.

Precautions

Butanethiol decomposes an heading and release toxic fumes. It reacts with strong acids, bases and strong oxidants. Workers during handling of butanethiol should be very careful, should avoid open flames, sparkings, and smoking. Chemical vapor/air mixtures cause explosion ha3 and workers should wear protective cloathing, safety goggles and breathing apparatus and should work in good ventillation.

Reference

Material Safety Data Sheet (MSDS). Safety data for 1-Butanethiol. Department of Physical Chemistry, Oxford University, Oxford, U.K.

sec-Butyl acetate (CAS No. 105-46-4)

Molecular formula: $C_6H_{12}O_2$
Synonyms and trade names: Acetic acid; 1-Methylpropylacetate; 1-Methylpropyl acetate; Acetic acid, 2-butyl ester; Acetic acid, sec-butyl ester; sec-Butylacetate; Butyl acetate; sec-

Use and exposure

Sec-Butyl acetate is a colorless liquid with a pleasant odor. The vapor mixes well with air, and become explosive mixtures. It reacts with strong oxidants, strong bases, strong acids, and nitrates, causing fire and explosion hazard.

Toxicity and health effects

During prolonged occupational exposure, sec-butyl acetate causes health effects. The symptoms of toxicity include irritation to the skin and eyes. Exposures to high concentrations of sec-butyl acetate irritate the nose and throat causing coughing and respiratory distress, headache, nausea, vomiting, dizziness, drowsiness, and coma. After prolonged exposures to sec-butyl acetate, occupational workers show symptoms of severe irritation to the eyes, headache, drowsiness, dryness in the upper respiratory system, skin, and narcosis.

Exposure limits

The legal airborne PEL is 200 ppm averaged over an 8-h work shift. The NIOSH recommended airborne exposure limit is 200 ppm averaged over a 10-h work shift (TWA). The ACGIH recommended airborne exposure limit is 200 ppm averaged over an 8-h work shift.

References

International Programme on Chemical Safety and the Commission of the European Communities. 1999. IPCS, CEC ICSC Card No. 0840. International Labor Organization (ILO), Geneva, Switzerland.
National Institute for Occupational Safety and Health (NIOSH). 2005. sec-Butyl acetate. Centers for Disease Control and Prevention, Atlanta, GA.

n-Butyl acetate (CAS No. 123-86-4)

Molecular formula: $CH_3COO (CH_2)_3 CH_3$
Synonyms and trade names: Butyl ethanoate; n-Butyl acetate; n-Butyl ethanoate; Acetic acid butyl ester; Ethanoic acid butyl ester

Use and exposure

n-Butyl acetate is a clear, colorless, flammable liquid and vapor. It has a mild, fruity odor and is slightly soluble in water. n-Butyl acetate is incompatible with strong oxidizing agents, strong acids, and strong bases.

Toxicity and health effects

Exposures to n-butyl acetate cause harmful effects that include, but are not limited to, coughing and shortness of breath. High concentrations have a narcotic effect, with symptoms such as sore throat, abdominal pain, nausea, vomiting, and diarrhea. High concentrations of n-butyl acetate cause severe poisoning. Prolonged periods of exposure cause adverse effects to the lungs, the nervous system, and the mucous membranes. Repeated skin contact causes skin dryness or cracking, and dermatitis.

n-Butyl acetate and carcinogenicity

There is no available information about the carcinogenic, mutagenic, teratogenic, and developmental toxicity of n-butyl acetate.

Exposure limits

n-Butyl acetate has the limits of TLV/TWA 150 ppm, STEL 200 ppm, and PEL 150 ppm.

Precautions

On exposure to n-butyl acetate, immediately wash with plenty of water, also under the eyelids, for at least 15 min. Remove contact lenses. n-Butyl acetate is flammable in the presence of open flames, sparks, oxidizing materials, acids, and alkalis. It poses explosion risk in the presence of mechanical impact. For health safety, management authorities should provide exhaust ventilation facilities at the workplace to keep the airborne concentrations of vapors of n-butyl acetate below TLV.

Storage

n-Butyl acetate should be kept stored in a segregated and approved area. Workers should keep the container in a cool, well-ventilated area, closed tightly, and sealed until ready for use. Workers should avoid all possible sources of ignition/spark at the workplace.

Iso Butyl acetate (CAS no. 110-19-0)

Molecular formula: $C_6H_{12}O_2$
Synonyms and trade names: 2-Methylpropyl acetate

Use and exposure

Isobutyl acetate is a colorless liquid with a fruit flavor. Isobutyl acetate is moisture sensitive, incompatible with ignition sources, moisture, excess heat, strong oxidizing agents, and strong bases; on decomposition, it releases carbon monoxide and carbon dioxide. It is used for nitrification fiber and paint solvents, chemical reagents, and modulation spices.

Toxicity and health effects

Exposures to isobutyl acetate cause anesthetic effects, headaches, inflammation of the eye characterized by redness, watering, and itching, dizziness, nausea, vomiting and a large dose may cause coma. Exposures to the skin over a longer time cause irritation and inflammation characterized by itching, scaling, reddening and/or occasionally blistering. Exposures by ingestion and/or inhalation of isobutyl acetate cause irritation to the digestive tract and respiratory tract. The target organs of isobutyl acetate have been reported as the CNS and skin.

Isobutyl acetate and carcinogenicity

Isobutyl acetate has not been listed by the ACGIH, the IARC, the NTP, or the CA as a mutagen or teratogen.

Exposure limits

The ACGIH: 150 ppm TWA; the NIOSH: 150 ppm TWA; the IDLH: 700 mg/m³ TWA, 1300 ppm; the OSHA: PELs 50 ppm TWA, 700 mg/m³ TWA.

tert-Butyl acetate (CAS No. 540-88-5)

Molecular formula: $C_6H_{12}O_2$

Synonyms and trade names: t-Butyl acetate of acetic acid; tert-Butyl ester of acetic acid; 1,1-Dimethyl ethyl ester; Texaco Lead Appreciator; Tertbutyl acetate

Use and exposure

tert-Butyl acetate or t-butyl acetate is a colorless flammable liquid with a camphor- or blueberry-like smell. It is used as a solvent in the production of lacquers, enamels, inks, adhesives, thinners, and industrial cleaners. It is also used as an additive to improve the antiknock properties of motor fuels. tert-Butyl acetate has three isomers: n-butyl acetate, isobutyl acetate, and sec-butyl acetate.

Toxicity and health effects

Exposure to tert-butyl acetate causes eye, skin, and respiratory irritation in workers. By analogy with the effects of exposure to similar esters, tert-butyl acetate may act as a CNS depressant at high concentrations. The signs and symptoms of acute exposure to tert-butyl acetate include, but are not limited to, itchy or inflamed eyes and irritation of the nose and upper respiratory tract. Exposures to tert-butyl acetate at high concentrations may cause headache, drowsiness, and other narcotic effects.

Exposure limits

For tert-butyl acetate, the NIOSH has set the REL as 200 ppm (950 mg/m^3) TWA and the US OSHA has set the PEL as 200 ppm (950 mg/m^3) TWA. The IDLH concentration of tert-butyl acetate is reported as 500 ppm.

Storage

tert-Butyl acetate should be stored in a cool, dry, well-ventilated area in tightly sealed containers that are labeled in accordance with regulatory standards. Containers of tert-butyl acetate should be protected from physical damage and should be stored separately from nitrates, strong oxidizers, strong acids, strong alkalis, heat, sparks, and open flame. Because containers that formerly contained tert-butyl acetate may still hold product residues, they should be handled appropriately.

Precautions

If exposures to tert-butyl acetate or a solution in work areas gets into the eyes, immediately flush the eyes with large amounts of water for a minimum of 15 min, lifting the lower and upper lids occasionally. Always use safety goggles or eye protection in combination with breathing protection.

References

International Labour Organization or the World Health Organization. 2002. Butyl Acetates. CICAD no. 64. Geneva, Switzerland.

National Institute for Occupational Safety and Health (NIOSH). 2005. tert-Butyl Acetate. NIOSH, Centers for Disease Control and Prevention, Atlanta, GA.

Proctor, N.H., Hughes, J.P. and Fischman, M.L. 1988. *Chemical Hazards of the Workplace.* Lippincott, J.B. Philadelphia, PA.

US Department of Labor, Occupational Safety and Health Administration (OSHA). 1999. tert-Butyl Acetate. www.OSHA.gov.

n-Butyl alcohol (CAS No. 71-36-3)

Molecular formula: $C_4H_{10}O$

Synonyms and trade names: n-Butanol; Butyl hydroxide; n-Propylcarbinol; Butyric hydroxybutane; Butanol; n-Butanol, Butan-1-ol; Hemostyp; 1-Hydroxybutane; primary Butyl alcohol; n-Butyl alcohol; Propyl methanol; Butyl hydroxide

Use and exposure

n-Butyl alcohol is a colorless flammable liquid with a strong alcoholic odor. n-Butyl alcohol is a highly refractive liquid and burns with a strongly luminous flame. It is incompatible with strong acids, strong oxidizing agents, aluminium, acid chlorides, acid anhydrides, copper, and copper alloys. n-Butyl alcohol has extensive use in a large number of industries. For instance, it is used as a solvent in industries associated with the manufacturing of paints, varnishes, synthetic resins, gums, pharmaceuticals, vegetable oils, dyes, and alkaloids. n-Butyl alcohol is used in the manufacture of artificial leather, rubber, and plastic cements, shellac, raincoats, perfumes, and photographic films.

Toxicity and health effects

Exposures to n-butyl alcohol by inhalation, ingestion, and/or skin absorption are harmful. n-Butyl alcohol is an irritant, with a narcotic effect and a CNS depressant. Butyl alcohols have been reported to cause poisoning with symptoms that include, but are not limited to, irritation to the eyes, nose, throat, and the respiratory system. Prolonged exposure results in symptoms of headache, vertigo, drowsiness, corneal inflammation, blurred vision, photophobia, and cracked skin. It is advised that workers coming in contact with n-butyl alcohol should use protective clothing and barrier creams. Occupational workers with pre-existing skin disorders or eye problems, or impaired liver, kidney or respiratory function may be more susceptible to the effects of the substance.

Exposure limits

The typical PEL of n-butyl alcohol is 50–100 ppm (OSHA); ACGIH TLV: 20 ppm (TWA).

Storage

Store n-butyl alcohol in a cool, dry, well-ventilated location, away from smoking areas. Fire hazard may be acute. Outside or detached storage is preferred. Separate from incompatibles. Containers should be bonded and grounded for transfer to avoid static sparks.

Reference

Environmental Health and Safety. 2008. n-Butyl Alcohol. MSDS no. B5860. Mallinckrodt Baker, Phillipsburg, NJ.

n-Butylamine (CAS No. 109-73-9)

Molecular formula: $C_4H_{11}N$

Synonyms and trade names: 1-Aminobutane; Butylamine; Monobutylamine; 1-Butanamine; mono-n-Butylamine; Norvalamine

Use and exposure

n-Butylamine is one of the four isomeric amines of butane, the others being sec-butyl-amine, tert-butylamine, and isobutylamine. It is a highly flammable, colorless to yellow liquid. It is stable and/but incompatible with oxidizing agents, aluminium, copper, copper alloys, and acids. n-Butylamine is used in the manufacture of pesticides (such as thiocar-bazides), pharmaceuticals, and emulsifiers. It is also a precursor for the manufacture of N,N′-dibutylthiourea, a rubber vulcanization accelerator, and n-butylbenzenesulfonamide, a plasticizer of nylon.

Toxicity and health effects

Exposures to butylamine (inhalation, ingestion, and through skin contact) are harmful. It is very destructive to the mucous membranes and causes, redness, severe deep burns, and loss of vision. Symptoms include, but are not limited to, sore throat, cough, burning sensation, headache, flushing of the face, vomiting, dizziness, abdominal pain, diarrhea, nausea, shock or collapse, shortness of breath, labored breathing, depression, convul-sions, narcosis, and possibly unconsciousness. Exposure of this nature is unlikely, how-ever, because of the irritating properties of the vapor. On catching fire, butylamine gives off irritating or toxic fumes (or gases). Repeated or prolonged contact with the skin may cause dermatitis. The vapor is corrosive to the eyes, skin, and respiratory tract, and causes lung edema.

Exposure limits

The exposure limits of n-butylamine have been set as ACGIH (1997) TLV: 5 ppm; 15 mg/m^3 (ceiling values) (skin); OSHA PEL: C 5 ppm (15 mg/m^3) skin; NIOSH REL: C 5 ppm (15 mg/m^3) skin; NIOSH IDLH: 300 ppm.

Precautions

n-Butylamine decomposes on heating or burning, producing toxic fumes including nitrogen oxides. The substance is a weak base and reacts with strong oxidants and acids causing fire and explosion hazard. Some metals in the presence of water and in contact with n-butylamine undergo deformation. Persons with pre-existing skin disorders, eye problems, or impaired respiratory function may be more susceptible to the effects of n-butylamine.

Storage

n-Butylamine should be protected against physical damage. Store in a cool, dry, well-ventilated location, away from any area where the fire hazard may be acute. Outside or detached storage is preferred. Separate from incompatibles. Containers should be bonded and grounded for transfer to avoid static sparks.

Reference

International Programme on Chemical Safety and the Commission of the European Communities (IPCS-CEC). 1998. n-Butylamine. Card no. ICSC: 0374. IPCS, Geneva, Switzerland.

C

- Cadmium
- Cadmium acetate
- Cadmium chloride
- Cadmium chloride (Hemipentahydrate)
- Cadmium fume (as Cd)
- Cadmium iodide
- Cadmium monosulfide
- Cadmium monoxide
- Cadmium oxide
- Cadmium sulfate
- Calcium cyanide
- Carbaryl
- Carbofuran
- Carbon disulfide
- Carbon monoxide
- Carbophenothion
- Cefadroxil
- Cefixime
- Chlordane
- Chlorine
- Chlorofluorocarbons
- Chlorpyrifos
- Chlorpyrifos-methyl
- Cobalt metal, dust and fume (as Co)
- Copper
- Copper dust, fume, and mist (as Cu)

- Copper cyanide
- Copper fume (as Cu)
- Copper sulfate
- Coumaphos
- Cyanide
- Cyanide compounds
- Cyanogen
- Cyanogen bromide
- Cyanogen chloride
- Cyclohexane
- Hydrogen cyanide
- p-Chloronitrobenzene
- Potassium cyanide
- Potassium silver cyanide
- Sodium cyanide

Cadmium (CAS No. 7440-43-9)

Molecular formula: **Cd**

Use and exposure

Cadmium is a gray-white, soft, malleable, lustrous metal or a grayish-white powder. It is insoluble in cold water, hot water, methanol, diethyl ether, and n-octanol. It is stable, incompatible with strong oxidizing agents, nitrates, nitric acid, selenium, zinc, and the powdered metal may be pyrophoric and flammable. Cadmium is associated with occupations such as industrial processes, metal plating, production of nickel-cadmium batteries, pigments, plastics, and other synthetics. Cadmium metal is produced as a by-product from the extraction, smelting, and refining of the non-ferrous metals zinc, lead, and copper. In view of its unique properties, cadmium metal and cadmium compounds are used as pigments, stabilizers, coatings, specialty alloys, and as electronic compounds.

Toxicity and health effects

Cadmium is hazardous in case of ingestion and inhalation, slightly hazardous in case of skin contact, causing allergic reactions, irritant to the skin and eyes, and a sensitizer. Human exposures to cadmium cause nausea, vomiting, abdominal cramping, diarrhea, increased salivation, hemorrhagic gastroenteritis, headache, dizziness, cough, dyspnea, chills (metal fume fever), alopecia, anemia, arthritis, cirrhosis of the liver, renal cortical necrosis, and cardiomyopathy. Acute inhalation of cadmium causes nasopharyngeal irritation, chest pain, enlarged heart, pulmonary edema, pulmonary fibrosis, emphysema, bronchiolitis, alveolitis, and renal cortical necrosis, particularly proximal tubule cell necrosis. Prolonged period of exposure to high concentrations of cadmium causes adverse effects to the skeletal system, arthritis, cardiovascular system/hypertension, and severe over-exposure can result in death.

Cadmium and carcinogenicity

The carcinogenicity of cadmium is still an area of controversy. While some jurisdictions have classified cadmium as a known human carcinogen, others have indicated that it is a possible or probable human carcinogen. The DHHS has determined that cadmium and cadmium compounds are known human carcinogens. The observation is based on sufficient evidence of carcinogenicity in humans, including epidemiological and mechanistic information that indicate a causal relationship between exposure to cadmium and cadmium compounds and human cancer. The ACGIH has classified cadmium as A2, meaning suspected human carcinogen. Laboratory animals exposed to cadmium develop cancer of the lung, prostate, and testes, hematopoietic system, liver, and pancreas. On exposures to cadmium, occupational workers developed tumors of the lung and prostate.

Exposure limits

The OSHA set the PEL of 5 μg/m³ (dust and fume) at the workplace for an 8-h workday. The US FDA set a 0.005 ppm limit of cadmium concentration in bottled drinking water, and the ACGIH has set the TLV as 0.2 mg/m³ (dust and salts).

Storage

Cadmium should be kept stored in a tightly closed container in a cool place. It should be kept stored in a separate locked safety storage cabinet.

Precautions

On exposures to cadmium, wash the skin immediately with plenty of water and a non-abrasive soap. Workers should cover the exposed skin with an emollient.

References

Agency for Toxic Substances and Disease Registry (ATSDR). 2008. Toxicological Profile for Cadmium (Draft for Public Comment). US Department of Health and Human Services, Public Health Service, Atlanta, GA.

International Programme on Chemical Safety and the Commission of the European Communities (IPCS, CEC). 1999. Cadmium. ICSC card no. ICSC: 0020. IPCS-CEC, Geneva, Switzerland (updated 2005).

World Health Organization (WHO). 1992. Cadmium. Environmental Health Criteria No. 135. International Programme on Chemical Safety (IPCS), WHO, Geneva, Switzerland.

Cadmium acetate (CAS No. 543-90-8)

Molecular formula: $C_4H_6CdO_4$
Synonyms and trade names: Acetic acid, cadmium salt; Bis(acetoxy) cadmium; Cadmium(II) acetate; Cadmium diacetate; Cadmium ethanoate

Use and exposure

Cadmium acetate is colorless crystal with a characteristic odor. It is not combustible, but it decomposes on heating, producing toxic fumes of cadmium oxide. It is incompatible with oxidizing agents, metals, hydrogen azide, zinc, selenium, and tellurium. Occupational exposure to cadmium and cadmium compounds occurs in workplaces mainly in the form of airborne dust and fumes. Occupations and workplaces include cadmium production and refining, nickel-cadmium battery manufacture, cadmium pigment manufacture and formulation, cadmium alloy production, mechanical plating, zinc smelting, soldering, and polyvinylchloride compounding. Cadmium and compounds enter the body mainly by inhalation and by ingestion.

Toxicity and health effects

Exposures to cadmium acetate cause cough, skin redness, abdominal pain, nausea, vomiting, salivation, choking, dizziness, and diarrhea. On catching fire, cadmium acetate gives off irritating or toxic metal oxide fumes. Inhalation of dust produces perforation of the nasal septum, loss of smell, irritation, headache, metallic taste, and cough. Prolonged exposures to cadmium acetate may produce shortness of breath, chest pain, and flu-like symptoms, chills, weakness, fever, muscular pain, pulmonary edema, liver and kidney damage and death. Cadmium acetate may have effects on the kidneys and bones, leading to kidney impairment and osteoporosis (bone weakness), and liver damage. Accidental ingestion or inhalation of cadmium acetate may be fatal to workers.

Cadmium acetate and carcinogenicity

There is sufficient evidence in humans for the carcinogenicity of cadmium and cadmium compounds. The IARC has classified cadmium and cadmium compounds as Group 1, meaning human carcinogens. Cadmium and cadmium compounds are highly toxic and experimental carcinogens. Cadmium acetate has been listed by the NTP as A2, meaning a suspected human carcinogen.

Exposure limits

The OSHA has set the PEL as 5 $\mu g/m^3$ of cadmium (TWA), and the ACGIH has set the TLV as 0.01 mg/m^3 total dust, 0.002 mg/m^3 for cadmium and compounds.

Precautions

During use and handling of cadmium acetate, occupational workers should be careful. Workers should use protective gloves and immediately remove contaminated clothing and shoes. The workplace should provide an eye-wash fountain and quick-drench facilities. During use of cadmium acetate, workers should avoid heat, flame, ignition sources, dust, and incompatibles.

References

International Agency for Research on Cancer (IARC). 1997. Cadmium and cadmium compounds. Summaries & Evaluations. IARC, Lyon, Paris.

Material Safety Data Sheet (MSDS). 2008. Cadmium Acetate. MSDS no. C0077. Environmental Health & Safety. Mallinckrodt Baker, Phillipsburg, NJ.

National Institute for Occupational Safety and Health (NIOSH). 1994. Cadmium acetate. ICSC: 1075. NIOSH, Centers for Disease Control and Prevention, Atlanta, GA (updated 2007).

Cadmium chloride (Hemipentahydrate) (CAS No.7790-78-5)

Cadmium chloride (Anhydrous) (CAS No. 10108-64-2)

Molecular formula: $CdCl_2$

Synonyms and trade names: Cadmium dichloride; Cadmium chloride, hydrate (2:5); Cadmium chloride, hemipentahydrate

Use and exposure

Cadmium chloride is colorless and odorless crystal. It is used for the preparation of cadmium sulfide, used as "Cadmium Yellow," a brilliant-yellow pigment, which is stable to heat and sulfide fumes. Cadmium chloride has a high solubility in water, non-combustible solid, but the dust can be a moderate fire hazard when exposed to heat or flame, or when reacted with oxidizing agents. It is incompatible with bromine trifluoride, potassium oxidizers, zinc, selenium, tellurium, and hydrogen azide.

Toxicity and health effects

Exposures to cadmium chloride and cadmium salts cause headache, chills, diarrhea, abdominal pains, choking, dizziness, sweating, nausea, and muscular pain, irritations to the mucous membranes and upper respiratory tract, cough, shortness of breath, pulmonary fibrosis, emphysema, perforation of the nasal septum, loss of smell, chest pain, and flu-like symptoms, pulmonary edema, liver and kidney damage, and death.

Cadmium chloride and carcinogenicity

Cadmium chloride has been listed by the IARC as category 1, meaning a known human carcinogen, and the NTP has listed it as a known carcinogen. It is well known that cadmium and cadmium compounds are potential carcinogens, and care is required during use.

Precautions

During use of cadmium chloride, occupational workers should use proper protectives and avoid all kinds of direct contact with the body. Workers should avoid generation of cadmium dust in the workplace and use the chemical with adequate ventilation.

Reference

Material Safety Data Sheet (MSDS). 2008. Cadmium chloride. MSDS no. C0105. Environmental Health & Safety. Mallinckrodt Baker, Phillipsburg, NJ.

Cadmium fume (as Cd) (CAS No. 1306-19-0)

Use and exposure

Cadmium fume (as Cd) is finely divided solid particles dispersed in air.

Toxicity and health effects

Exposure to cadmium fume causes adverse health effects among occupational workers. The symptoms of toxicity and poisoning include, but are not limited to, pulmonary edema, dyspnea, coughing, tight chest, substernal pain, headaches, chills, muscle aches, nausea, vomiting, diarrhea, emphysema, proteinuria, anosmia, and mild anemia. Exposures also cause kidney and lung damage.

Cadmium iodide (CAS No. 10102-68-8)

Molecular formula: CdI_2

Use and exposure

Cadmium iodide is a crystalline solid.

Toxicity and health effects

Exposures to cadmium iodide cause adverse health effects on the skin, blood, kidneys, and lungs. It causes irritation to the eyes and skin, and corrosive effects to the skin and is extremely hazardous in case of ingestion.

Cadmium iodide and carcinogenicity

The ACGIH and the NTP classified cadmium iodide as A2, meaning a suspected human carcinogen and as Group 2, meaning a reasonably anticipated human carcinogen, respectively.

Cadmium monoxide (CAS No. 1306-19-0)

Molecular formula: CdO

Cadmium monosulfide (CAS No. 1306-23-6)

Molecular formula: CdS

Cadmium oxide (CAS No. 1306-19-0)

Molecular formula: CdO
Synonyms and trade names: Cadmium fume

Use and exposure

Cadmium oxide is dark brown powder or crystals and solubility in water is negligible. It is incompatible with magnesium.

Toxicity and health effects

Exposures to cadmium oxide cause poisoning and may be fatal if inhaled or swallowed. It is a severe respiratory irritant, an experimental carcinogen in animals, and a probable human carcinogen. Chronic exposure may cause irreversible lung injury, kidney damage, and other serious effects. The typical PEL is set as 0.3 mg/m^3 and the typical TLV as 0.05 mg/m^3 (TWA).

Cadmium sulfate (CAS No. 10124-36-4)

Molecular formula: $CdSO_4$; $CdSO_4 \cdot H_2O$ (monohydrate); $3CdSO_4 \cdot 8H_2O$ (octahydrate)
Synonyms and trade names: Sulfuric acid, cadmium salt (1:1); Niedermayrite

Toxicity and health effects

Exposures to cadmium salts by absorption are most efficient via the respiratory tract. The symptoms of health effects include, but are not limited to, irritation, headache, metallic taste, and/or cough. Severe exposures cause shortness of breath, chest pain, and flu-like symptoms with weakness, fever, headache, chills, sweating, nausea and muscular pain, pulmonary edema, liver and kidney damage, and death. Prolonged exposures, even at relatively low concentrations, may result in kidney damage, anemia, pulmonary fibrosis, emphysema, perforation of the nasal septum, loss of smell, male reproductive effects, and an increased risk of cancer of the lung and of the prostate. Decrease in bone density, renal stones, and other evidence of disturbed calcium metabolism have been reported.

Cadmium compounds and carcinogenicity

Cadmium compounds have been listed as A2, meaning suspected human carcinogens.

Precautions

Cadmium compounds cause more health disorders to occupational workers and persons with pre-existing skin disorders, eye problems, blood disorders, prostate problems, or impaired liver, kidney, or respiratory function. These workers are more susceptible to the effects of cadmium salts. On contact with cadmium, exposed workers should wash the skin and eyes immediately with plenty of water.

Carbaryl (CAS No. 63-25-2l)

Molecular formula: $C_{12}H_{11}NO_2$
IUPAC name: 1-Napthyl methylcarbamate
Synonyms and trade names: Bugmaster; Carbamec; Carbamine; Crunch; Denapon; Dicarbam; Hexavin; Karbaspray; Rayvon; Septene; Sevin; Tercyl; Torndao; Thinsec; Tricarnam
Toxicity class: US EPA: I; WHO: II

Use and exposure

Carbaryl is a colorless to light tan or white or gray, solid crystals depending on the purity of the compound. The crystals are essentially odorless, and stable to heat, light, and acids, but are not stable under alkaline conditions. It is non-corrosive to metals, packaging materials, and application equipment. Carbaryl is classified as a GUP. It is sparingly soluble in water, but soluble in dimethylformamide, DMSO, acetone, cyclohexanone, isopropanol, and xylene. Carbaryl is a wide-spectrum carbamate insecticide that controls over 100 species of insects on citrus, fruit, cotton, forests, lawns, nuts, ornamentals, shade trees, and other crops, as well as on poultry, livestock, and pets. It is also used as a molluscicide and an acaricide. Carbaryl works whether it is ingested into the stomach of the pest or absorbed through direct contact. It is available as bait, dusts, wettable powders, granules, dispersions, and suspensions.

Toxicity and health effects

Exposures to carbaryl cause a moderate to very toxic health disorder among workers. Carbaryl produces adverse effects in humans by skin contact, inhalation, or ingestion. The symptoms of acute toxicity are typical of the other carbamates. Direct contact of the skin or eyes with moderate levels of this pesticide can cause burns. Inhalation or ingestion of very large amounts can be toxic to the nervous and respiratory systems, resulting in nausea, stomach cramps, diarrhea, and excessive salivation. Exposures to high concentrations of carbaryl causes poisoning with symptoms such as excessive sweating, headache, weakness, giddiness, nausea, vomiting, stomach pains, blurred vision, slurred speech, muscle twitching, incoordination, and convulsions. The effects of carbaryl on the nervous system of rats, chickens, monkeys, and humans are primarily related to the inhibition of AChE that under normal situations is transitory. The only documented fatality from carbaryl was through intentional ingestion.

Laboratory studies have indicated that the acute oral toxicity (LD50) of carbaryl ranges from 250 to 850 mg/kg in rats, and from 100 to 650 mg/kg in mice. The inhalation toxicity (LC50) in rats is greater than 206 mg/L. Low doses of carbaryl cause minor skin and eye irritation in rabbits. The acute dermal toxicity (LD50) of carbaryl to rabbits is measured as greater than 2000 mg/kg. In a 90-day feeding study, carbaryl did not cause any significant adverse effects in rats. Carbaryl in high doses has caused no reproductive or fetal effects in a long-term feeding study of rats.

Ingestion of carbaryl affected the lungs, kidneys, and liver of experimental animals. Inhalation of carbaryl caused adverse effect to the lungs. High doses of carbaryl for a prolonged period caused nerve damage in rats and pigs. Several studies indicate that carbaryl can affect the immune system in animals and insects.

The evidence for teratogenic effects due to chronic exposure is minimal in test animals. Birth defects in rabbit and guinea pig offspring occurred only at dosage levels that were highly toxic to the mother.

Carbaryl and carcinogenicity

Carbaryl has been shown to affect cell division and chromosomes in rats. However, numerous studies indicate that carbaryl poses only a slight mutagenic risk. There is a possibility that carbaryl may react in the human stomach to form a more mutagenic compound, but this has not been demonstrated. The available information thus suggests the lack of evidence on the mutagenic potential of carbaryl and as a human mutagen. Long-term and lifetime studies in mice and rats exposed to technical-grade carbaryl did not show production of tumors. While N-nitrosocarbaryl, a possible by-product, has been shown to be carcinogenic in rats at high doses, this product has not been detected. The US EPA has not classified carbaryl as a human carcinogen. The International Programme on Chemical Safety (IPCS) has classified carbaryl as Group 3, meaning not classifiable as a human carcinogen.

Exposure limits

The ADI for carbaryl has been set as 0.01 mg/kg/day.

References

International Programme on Chemical Safety (IPCS). 1992. Carbaryl. Health and Safety. Guide No. 78. IPCS, World Health Organization, Geneva, Switzerland.

Material Safety Data Sheet (MSDS). 1993. Carbaryl. Extension Toxicology Network, Oregon State University, USDA/Extension Service, Corvallis, OR.

US Environmental Protection Agency (US EPA). 1992. Carbaryl. Technology Transfer Network, Air Toxics web site (revised 2000, updated 2007).

Carbofuran (CAS No. 1563-66-2)

Molecular formula: $C_{12}H_{15}NO_3$
IUPAC name: 2,3-Dihydro-2,2-dimethylbenzofuran-7-yl methylcarbamate
Synonyms and trade names: Agrofuran; Carbodan; Carbosip; Cekufuran; Chinufur; Furacarb; Furadan; Terrafuran
Toxicity class: US EPA: I (Formulation, Furadan 4F), II (Furadan G); WHO: Ib

Use and exposure

Carbofuran is a broad-spectrum carbamate insecticide and nematicide. It is an odorless, white crystalline solid. On heating, it breaks down and can release toxic fumes, and irritating or poisonous gases. It is sparingly soluble in water, but very soluble in acetone, acetonitrile, benzene, and cyclohexone. The liquid formulations of carbofuran are classified as RUPs because of their acute oral and inhalation toxicity to humans. Granular formulations are also classified as an RUP. In fact, carbofuran was first registered in the United States in 1969 and classified as an RUP. Exposure to heat breaks down carbofuran, with the release of toxic fumes. Carbofuran is used for the control of soil-dwelling and foliar-feeding insects. It is also used for the control of aphids, thrips, and nematodes that attack vegetables, ornamental plants, crops of sunflower, potatoes, peanuts, soybeans, sugar cane, cotton, rice, and a variety of other crops.

Toxicity and health effects

The acute oral LD50 of carbofuran to male and female rats is about 8 mg/kg, while the acute dermal LD50 for rats is more than 3000 mg/kg. Carbofuran is mildly irritating to the eyes and skin of rabbits. The acute inhalation toxicity (LC50, 4 h) is 0.075 mg/L for rats. As with other carbamate compounds, carbofuran's cholinesterase-inhibiting effect is short term and reversible.

The symptoms of carbofuran poisoning include, but are not limited to, nausea, vomiting, abdominal cramps, sweating, diarrhea, excessive salivation, weakness, imbalance, blurring of vision, breathing difficulty, increased blood pressure, and incontinence. Death may result at high doses from respiratory system failure associated with carbofuran exposure. Complete recovery from an acute poisoning by carbofuran, with no long-term health effects, is possible if exposure ceases and the victim has time to regain his or her normal level of cholinesterase and to recover from symptoms. Reports have indicated that risks from exposure to carbofuran are especially high among occupational workers and general public suffering with asthma, diabetes, cardiovascular disease, gastrointestinal or urogenital tracts disturbances. The available studies indicate carbofuran is unlikely to cause reproductive effects in humans at expected exposure levels. Studies indicate carbofuran is not teratogenic. No significant teratogenic effects have been found in the offspring of rats given carbofuran (3 mg/kg/day) on days 5 to 19 of gestation.

Carbofuran and carcinogenicity

Carbofuran and its carcinogenicity has not been listed by the NTP, the IARC, the OSHA, or the ACGIH. Published reports have also indicated that carbofuran does not pose a risk of cancer to animals and humans.

Carbofuran and bird mortalities

Carbofuran has been associated with the deaths of millions of wild birds since its introduction in 1967. For instance, Bald and Golden Eagles, Red-tailed Hawks, and migratory songbirds, as well as other wildlife have been reported to have suffered heavy poisonings. Reports have shown that carbofuran has a high potential for groundwater contamination, and has been detected in aquifers and surface waters. Also, reports have indicated that carbofuran has caused mortality in a large number of birds from direct spraying, ingestion of granules or contaminated drinking water, and from the consumption of contaminated prey. After a prolonged review of ecological and human health risks associated with carbofuran, all uses of carbofuran were found to be ineligible for re-registration and the USA EPA cancelled the registration of most uses of carbofuran.

Exposure limits

The TLV for carbofuran has been set as 0.1 mg/m^3 (8-h, TWA) and the ADI of carbofuran has been set as 0.01 mg/kg/day.

Precautions

During use/handling of carbofuran, workers should wear coveralls or a long-sleeved uniform, head covering, and chemical protective gloves made of materials such as rubber, neoprene, or nitrile. Occupational workers should know that areas treated with carbofuran are hazardous. The runoff of carbofuran material and the fire control releases irritating or poisonous gases. It is advisable that workers should enter storehouses or carbofuran-treated close spaces with caution.

Storage

Carbofuran should be stored in a cool, dry, well-ventilated place, in their original containers only. It should not be kept stored or used near heat, open flame, or hot surfaces.

References

Hayes, W., Jr. 1982. *Pesticides Studied in Man*. Williams & Wilkins, Baltimore, MD.
Material Safety Data Sheet (MSDS). 1993. Carbofuran. Extension Toxicology Network, USDA/ Extension Service, Oregon State University, Corvallis, OR.
Meister, R.T. (ed.). 1991. *Farm Chemicals Handbook*. Meister, Willoughby, OH.
US Environmental Protection Agency (US EPA). 1989. Health Advisory Summary for Carbofuran. US EPA, Washington, DC.

Carbon disulfide (CAS No. 75-15-0)

Molecular formula: **CS$_2$**
Synonyms and trade names: Carbon bisulfide; Dithiocarbonic anhydride

Use and exposure

Pure carbon disulfide is a colorless liquid with a pleasant odor similar to that of chloroform, while impure carbon disulfide is a yellowish liquid with an unpleasant odor, like that of rotting radishes. Exposure to carbon disulfide occurs in industrial workplaces. Industries associated with coal gasification plants release more carbon disulfide, carbonyl sulfide, and hydrogen sulfide. Carbon disulfide is used in large quantities as an industrial chemical for the production of viscose rayon fibers. In fact, the major source of environmental pollution by carbon disulfide both indoors and outdoors is caused by emissions released into the air from viscose plants.

Toxicity and health effects

Laboratory animals exposed to carbon disulfide experienced deleterious health effects, i.e., developmental effects, skeletal and visceral malformations, embryotoxicity, and functional and behavioral disturbances. Studies have also shown that animals exposed to carbon disulfide indicate destruction of the myelin sheath and axonal changes in both central and peripheral neurons along with changes in the cortex, basal ganglia, thalamus, brain stern, and spinal cord. Neuropathy and myelopathy were extensively studied in rats and rabbits. In the muscle fibers, atrophy of the denervation type occurred secondary to polyneuropathy. Studies have also shown that carbon disulfide causes vascular changes in various organs of animals as well as myocardial lesions. Occupational workers exposed to carbon disulfide showed symptoms of irritability, anger, mood changes, manic delirium and hallucinations, paranoic ideas, loss of appetite, gastrointestinal disturbances, and reproductive disorders. The slowing down of nerve conduction velocity in the sciatic nerves preceded clinical symptoms. Studies have indicated that carbon disulfide can affect the normal functions of the brain, liver, and heart. Occupational workers exposed to high concentrations of carbon disulfide have suffered with skin burns when the chemical accidentally touched people's skin.

Carbon disulfide and carcinogenicity

The US EPA and the IARC have not classified carbon disulfide as a human carcinogen. There are no confirmed reports indicating that carbon disulfide is carcinogenic to animals and humans.

Exposure limits

The OSHA has set a limit of 20 ppm of carbon disulfide for an 8-h workday (TWA) while the NIOSH has set a limit of 1 ppm for a 10-h workday (TWA).

Precautions

During handling of carbon disulfide, occupational workers require proper clothing, eye protection, and respiratory protection. Workers should use the chemical under trained management. On contact with the eyes, immediately flush with large amounts of water. On skin contact, the worker should quickly remove contaminated clothing and immediately call for medical attention.

References

Agency for Toxic Substances and Disease Registry (ATSDR). 1996. Toxicological Profile for Carbon Disulfide. US Department of Health and Human Services, Public Health Service, Atlanta, GA (updated 2007).
US Environmental Protection Agency (US EPA). 1992. Carbon disulfide hazard summary. Technology Transfer Network Air Toxics web site (revised 2000, updated 2007).
World Health Organization (WHO). 2000. Carbon disulfide. WHO Regional Office for Europe, Copenhagen, Denmark.

Carbon monoxide (CAS No. 630-08-0)

Molecular formula: **CO**
Synonyms and trade names: Carbonic oxide; Flue gas; CO; Carbon oxide

Use and exposure

Carbon monoxide (CO) is a colorless, odorless, tasteless gas that is extremely hazardous. CO can be formed from incomplete burning of gasoline, wood, kerosene, or other fuels. CO is also found in cigarette smoke and vehicle exhaust. In homes, CO can build up from a poorly vented or malfunctioning heater, furnace, range, or any appliance that runs on natural gas or oil. The presence of CO is very common inside and outside workplaces, around heaters: improper use of gas or kerosene-fired heaters, gas-fired central heating equipment combined with improper venting or chimney, due to blocked heating flues, improper flue vent connector, or hood installation, inadequate combustion air, from exhaust, and gas-fired water heaters.

Toxicity and health effects

CO is a highly toxic gas often called a chemical asphyxiant. When inhaled, it combines with hemoglobin more readily than oxygen does, displacing oxygen from hemoglobin, thereby interfering with oxygen transport by the blood. The early symptoms of CO poisoning include, headaches, nausea, and fatigue, which are often mistaken for the flu because CO is not detected in the home. Prolonged exposure to CO causes deleterious health effects, brain damage, and eventually death. CO poisoning can happen to anyone, anytime, almost anywhere. CO poisoning is often confused with the flu. Depending on the period of exposure and concentration of CO, poisoning may be severe, moderate, or mild. (i) Extreme exposures cause confusion, drowsiness, rapid breathing or pulse rate, vision problems, chest pain, convulsions, seizures, loss of consciousness, cardiorespiratory failure, and death. (ii) Moderate exposures cause severe throbbing headache, drowsiness, confusion, vomiting, and fast heart rate. (iii) Mild exposures cause slight headache, nausea, fatigue (often described as "flu-like" symptoms).

The symptoms of CO poisoning include, but are not limited to, drowsiness, nausea, tiredness, vomiting, headaches, dizziness, visual changes, abdominal pain, chest pains, memory and walking problems, brain damage and, in severe cases, death. Exposure to high concentrations of CO causes severe headache, weakness, dizziness, irregular heart beat, seizures, coma, respiratory failure, and unconsciousness.

The toxicity of CO results from its very tight binding to hemoglobin, the species that carries oxygen from the lungs to the bodily tissues. For hemoglobin to work, it can't bind oxygen very tightly (otherwise it couldn't release it at its destination). Unfortunately, CO binds to hemoglobin 200 times more tightly than oxygen. Carboxyhemoglobin (the molecule formed when CO binds to hemoglobin) does not perform oxygen transport, and it rapidly builds up. In essence, victims are slowly suffocated because their hemoglobin is consumed. The fatal concentration of CO depends on the length of the air exposure. CO also causes a decrease in the supply of oxygen to the heart and induces myocardial hypoxia. Levels above 300 ppm for more than 1–2 h can lead to death, and exposure to 800 ppm (0.08%) can be fatal after 1 h (Table 3). It is alarming to note that each year more than 500 Americans die from carbon monoxide poisoning and more than 2000 commit suicide by intentionally poisoning themselves.

Exposure limits

The OSHA has set the PEL for carbon monoxide as 50 ppm for an 8-h period (TWA) and the NIOSH has set a standard of 35 ppm.

Carbon monoxide poisoning, prevention, occupational safety

- Install a CO alarm on each level of your home.
- Home heating systems, chimneys, and flues must be inspected and cleaned by a qualified technician every year. Keep chimneys clear of bird and squirrel nests, leaves, and residue to ensure proper ventilation.
- Make sure that the furnace and other appliances, such as gas ovens, ranges, and cooktops are inspected for adequate ventilation.
- Do not burn charcoal inside the house even in the fireplace.
- Do not operate gasoline-powered engines in confined areas, such as garages or basements. Do not leave your car, mower, or other vehicle running in an attached garage, even with the door open.
- Do not block or seal shut exhaust flues or ducts for appliances such as water heaters, ranges, and clothes dryers.

References

Agency for Toxic Substances and Disease Registry (ATSDR). 2002. Managing Hazardous Materials Incidents. Volume III – Medical Management Guidelines for Acute Chemical exposures. Chlorine. US Department of Health and Human Services, Public Health Service, Atlanta, GA (updated 2007).

Department of Health. 2004. Carbon monoxide. Chemical Fact Sheets, Department of Health, Madison, WI.

Department of Public Health. 2004. Carbon monoxide poisoning. Department of Public Health Rev. Des Moines, IA.

Henry, C.R., Satran, D., Lindgren, B., Adkinson, C., Caren, I., Nicholson, R.N. and Henry, T.D. 2006. Myocardial injury and long-term mortality following moderate to severe carbon monoxide poisoning. *JAMA* 295: 398–402.

Phin, N. 2005. Carbon monoxide poisoning (acute). *Clin. Evid.* 13: 1732–43.

Satran, D., Henry, C.R., Adkinson, C., Nicholson, C.I., Yiscah Bracha, R.N. and Henry, T.D. 2005. Cardiovascular manifestations of moderate to severe carbon monoxide poisoning. *J. Am. Coll. Cardiol.* 45: 1513–16.

World Health Organization (WHO). 1999. Carbon monoxide. Environmental Health Criteria No. 213. WHO, Geneva, Switzerland.

Carbophenothion (CAS No. 786-19-6)

Molecular formula: $C_{11}H_{16}ClO_2PS_3$
IUPAC name: S-4-chlorophenylthiomethyl O,O-diethyl phosphorodithioate
Synonyms and trade names: Acarithion; Dagadip; Endyl; Garrathion; Hexathion (3); Lethox (2); Nephocarp; Trithion
Toxicity class: US EPA: I; WHO: Ib

Use and exposure

Carbophenothion is an off-white to light amber colored liquid with a mild mercaptan-like odor. It is slightly soluble in water, and miscible with many organic solvents, such as hydrocarbons, alcohols, ketones, and esters. The US EPA has classified it as an RUP. Carbophenothion is used as a non-systemic insecticide and acaricide for pre-harvest treatments on deciduous, citrus and small fruits, field crops, and vegetables, and for the control of aphids, mites, suckers, and other pests on fruit, nuts, vegetables, sorghum, and maize.

Toxicity and health effects

Carbophenothion is highly toxic through ingestion and skin absorption. Carbophenothion affects the nervous system by inhibiting ChE activity. On heating or burning, carbophenothion undergoes decomposition and produces toxic fumes such as phosphorus oxides, sulfur oxides, and hydrogen chloride. Exposures to carbophenothion cause poisoning with symptoms such as headache, blurred vision, weakness, nausea, discomfort in the chest, abdominal cramps, vomiting, diarrhea, salivation, sweating, and pinpoint pupils. It is highly toxic when eaten and nearly as toxic when absorbed through the skin. Large single doses of carbophenothion potentiate the toxicity of malathion.

Carbophenothion and carcinogenicity

Long-term feeding studies of laboratory rats with carbophenothion (4 mg/kg/day) for a period of two years did not produce any tumors. Thus, carbophenothion as a human carcinogen is not established.

References

Hayes, W.J. Jr. 1971. *Clinical Handbook on Economic Poisons*. US EPA Pesticides Programs, Public Health Service Publication 476.

Hayes, W.J. Jr. and Laws, E.R. Jr. (eds.). 1991. *Handbook of Pesticide Toxicology*. Volume 2. *Classes of Pesticides*. Academic Press, New York. p. 106.

Hill, E.F. and Camardese, M.B. 1986. Lethal Dietary Toxicities of Environmental Contaminants to Coturnix, Technical Report No. 2. US Department of Interior, Fish and Wildlife Service, Washington, DC.

International Programme on Chemical Safety and the Commission of the European Communities. 1997. Carbophenothion (IPCS, CEC). ICSC card no. 0410. WHO, Geneva, Switzerland (updated 2005).

Material Safety Data Sheets (MSDS). 1995. Carbophenothion. Extension Toxicology Network, Pesticide Information Profiles. Oregon State University, USDA/Extension Service, Corvallis, OR.

Cefadroxil (CAS No. 50370-12-2)

Molecular formula: $C_{16}H_{17}N_3O_5S$

Use and exposure

Cefadroxil is a light orange colored powder. It is a cephalosporin antibiotic and is used for the treatment of bacterial infections. Cefadroxil is stable under recommended storage conditions. Cefadroxil is not considered hazardous when handled under normal medical conditions and with good housekeeping.

Toxicity and health effects

Exposures to cefadroxil cause certain common side effects. These include nausea, vomiting, stomach disorders, rashes, itching, unusual tiredness or weakness, yellowing of the skin or eyes, red, swollen, or blistered skin, unusual bruising or bleeding, sore throat, respiratory distress, tightness in the chest, swelling of the mouth, face, lips, or tongue, decreased urination, dark urine, vaginal itching, odor, or discharge, fever, chills, joint pain, and seizures. Prolonged or long-term use of cefadroxil should be avoided.

Precautions

Cefadroxil should be used only under proper medical health care since it has properties of penicillin allergy, renal function, gastrointestinal tract damage. Pregnant women and breast-feeding women should avoid exposure to cefadroxil.

Cefixime (CAS No. 79350-37-1)

Molecular formula: $C_{16}H_{15}N_5O_7S_2$

Use and exposure

Cefixime is an oral third generation cephalosporin antibiotic. It was sold under the trade name Suprax in the United States until 2003. The oral suspension form of "Suprax" was re-launched. Cefixime is prescribed for bacterial infections of the chest, ears, urinary tract, and throat (tonsilitis and pharyngitis), and for uncomplicated gonorrhea, upper and lower respiratory tract infections, acute otitis media, and gonococcal urethritis.

Toxicity and health effects

Exposures to cefixime may cause side effects that include, but are not limited to, stomach and abdominal pain, diarrhea, vomiting, mild skin rash, headache, fever, urticaria, pruritis, eosinophilia, leucopenia, anaphylaxis, superinfection, hemolytic anemia, dyspepsia, and flatulence. Also, cefixime may cause transient elevation of SGOT, SGPT, alkaline phosphatase, BUN, and creatinine. Reports have indicated that cefixime is mainly excreted unchanged in the bile and urine.

Precautions

Users of cefixime should be careful in health conditions such as neonates, pregnancy, lactation, and renal failure. Users should follow suitable and appropriate health care measures to control superinfection (if it happens during therapy). It is contraindicated in patients with a known allergy to penicillin or any other ingredients of Taximax. Taxim-O is contraindicated in patients with a known allergy to the cephalosporin group of antibiotics.

Reference

National Institutes of Health (NIH). 2003. Department of Health and Human Services (DHHS), Bethesda, MD (updated 2007).

Chlordane (CAS No. 57-74-9)

Molecular formula: $C_{10}H_6Cl_8$
Synonyms and trade names: Belt; Chlor Kil; Chlortox; Corodane; Gold Crest C-100; Kilex Lindane; Kypchlor; Niran; Octachlor; Synklor; Termex; Topiclor 20; Toxichlor; Velsicol 1068
IUPAC name: 1,2,4,5,6,7,8,8-Octachloro-2,3,3a,4,7,7a-tetrahydro-4,7-methanoindane
Toxicity class: US EPA: II; WHO: II

Use and exposure

Chlordane is a viscous, amber-colored liquid. Technical grade chlordane is a mixture of many structurally related compounds including trans-chlordane, cis-chlordane, -chlordene, heptachlor, and trans-nonachlor. 14,15 Chlordane was used as a broad-spectrum pesticide in the United States from 1948 to 1988. Its uses included termite control in homes; pest control on agricultural crops such as maize and citrus, on home lawns, gardens, turf, and ornamental plants. Chlordane is a persistent organochlorine insecticide. It kills insects when ingested and on contact. Formulations include dusts, emulsifiable concentrates, granules, oil solutions, and wettable powder.

Toxicity and health effects

Exposures to chlordane cause adverse health effects and poisoning to animals and humans. The acute oral LD50 values of technical grade chlordane for the rat range from 137 to 590 mg/kg and acute dermal LD50 for the rabbit is 1720 mg/kg. Signs of acute chlordane intoxication include ataxia, convulsions, and cyanosis followed by death due to respiratory failure. Rats treated by gavage with 100 mg/kg once a day for 4 days had increased absolute liver weights; fatty infiltration of the liver; and increased serum triglycerides, creatine phosphokinase, and lactic acid dehydrogenase. Sheep treated by stomach tube with 500 mg/kg showed signs of intoxication, but recovered fully within 5–6 days; a dose of 1000 mg/kg resulted in death after 48 h.

Ingestion of chlordane induces vomiting, dry cough, agitation and restlessness, hemorrhagic gastritis, bronchopneumonia, muscle twitching, convulsions, and death among humans. Non-lethal, but accidental poisoning of children has resulted in convulsions, excitability, loss of coordination, dyspnea, and tachycardia. Recovery, however, was complete. Ingestion of chlordane contaminated water (1.2 g/L) caused symptoms of gastrointestinal and neurological disorders. Chronic inhalation of chlordane produced symptoms of poisoning that included, but were not limited to, sinusitis, bronchitis, dermatitis, neuritis, migraine, gastrointestinal distress, fatigue, memory deficits, personality changes, decreased attention span, numbness or paresthesias, blood dyscrasias, disorientation, loss of coordination, dry eyes, and seizures. Chlordane-treated laboratory rats showed blood diseases, including aplastic anemia and acute leukemia.

Chlordane and carcinogenicity

Chlordane can promote cancer and is an endocrine disrupter. Long term and at high doses (30–64 mg/kg/day for 80 weeks), chlordane causes liver cancer in laboratory mice. The US EPA classified chlordane as Group B2, meaning a probable human carcinogen.

Exposure limits

The Acceptable Daily Intake (ADI) for chlordane has been set as 0.0005 mg/kg/day and the PEL as 0.5 mg/m^3 (8 h).

References

Material Safety Data Sheets (MSDS). 1996. Chlordane. Extension Toxicology Network, Pesticide Information Profiles. Oregon State University, USDA/Extension Service, Corvallis, OR.
US Environmental Protection Agency (US EPA). 2000. Chlordane. Technology Transfer Network, Air Toxics web site (updated 2007).

Chlorine (CAS No. 7782-50-5)

Molecular formula: **Cl**

Use and exposure

Chlorine is a yellow-green gas that is heavier than air and has a strong irritating odor. Chlorine is used extensively in the production of paper products, dyestuffs, textiles, petroleum products, medicines, antiseptics, insecticides, food, solvents, paints, plastics, and many other consumer products. Chlorine is mainly used as a bleach in the manufacture of paper and cloth and to make a wide variety of products. Most of the chlorine produced is used in the manufacture of chlorinated compounds for sanitation, pulp bleaching, disinfectants, and textile processing. Further use is in the manufacture of chlorates, chloroform, carbon tetrachloride, and in the extraction of bromine. Organic chemistry demands much from chlorine, both as an oxidizing agent and in substitution. In fact, chlorine was used as a war gas in 1915 as a choking (pulmonary) agent. Chlorine itself is not flammable, but it can react explosively or form explosive compounds with other chemicals such as turpentine and ammonia.

Chlorine is slightly soluble in water. It reacts with water to form hypochlorous acid and hydrochloric acid. Hypochlorous acid breaks down rapidly. Chlorine gas is used to synthesize other chemicals and to make bleaches and disinfectants. Chlorine is a powerful disinfectant, and in small quantities ensures clean drinking water. It is used in swimming pool water to kill harmful bacteria. Chlorine has a huge variety of uses, i.e., as a disinfectant and purifier, in plastics and polymers, solvents, agrochemicals and pharmaceuticals, as well as an intermediate in manufacturing other substances where it is not contained in the final product. Also, a large percentage of pharmaceuticals contain and are manufactured using chlorine. Thus, chlorine is essential in the manufacture of medicines to treat illnesses such as allergies, arthritis, and diabetes.

Toxicity and health effects

Chlorine is a respiratory irritant. It causes irritation to the mucous membranes and the liquid burns the skin. The poisoning caused by chlorine depends on the amount of chlorine a person or an occupational worker is exposed to, and the length of exposure time. Prolonged exposures to high concentrations of chlorine cause poisoning with symptoms that include, but are not limited to, coughing, burning sensation in the nose, throat, and eyes, blurred vision, nausea, vomiting, pain, redness, and blisters on the skin, chest tightness, and pulmonary edema.

Chlorine and carcinogenicity

There is no information indicating that exposures to chlorine cause cancer in animals and humans. The DHHS, the IARC, and the US EPA have not classified chlorine as a human carcinogen.

Exposure limits

The OSHA has set a permissible exposure limit (PEL) of 1 ppm for chlorine for an 8-h workday (TWA), while the ACGIH has set the limit of 0.5 ppm as the TLV for an 8-h day (TWA), and a short-term exposure level (STEL) of 1.0 ppm of chlorine.

References

Agency for Toxic Substances and Disease Registry (ATSDR). 2002. Chlorine. Managing Hazardous Materials Incidents. Volume III – Medical Management Guidelines for Acute Chemical Exposures. US Department of Health and Human Services, Public Health Service, Atlanta, GA (updated 2007).

Department of Health and Human Services (DHHS). 2006. Facts about Chlorine. DHHS, Center for Disease Control and Prevention, Atlanta, GA.

Chlorofluorocarbons

Chlorofluorocarbons (CFCs) are the most important ozone-destroying chemicals. They have been used in many ways since they were first synthesized in 1928. They are stable, non-flammable, low in toxicity, and inexpensive to produce. Over time, CFCs found uses as refrigerants, solvents, foam-blowing agents, aerosols, and in other smaller applications. When released into the air, CFCs rise into the stratosphere. In the stratosphere, CFCs react with other chemicals and reduce the stratospheric ozone layer, which protects Earth's surface from the sun. Reducing CFC emissions and eliminating the production and use of ozone-destroying chemicals is very important to protecting Earth's stratosphere.

Use and exposure

CFCs are a family of organic compounds containing chlorine, fluorine, and carbon, and are also called Freons. CFCs entered the industrial scene in the late 1920s and early 1930s. CFCs were identified as a safer alternative to the sulfur dioxide and ammonia refrigerants used at the time. CFCs are inert and volatile compounds with extensive uses as refrigerants, blowing agents for cleaning agents, in the production of plastic foams, as solvents to clean electronic components, and as propellants in air conditioners and aerosol sprays. These compounds are low in toxicity, non-flammable, non-corrosive, and non-reactive with other chemical species, and have desirable thermal-conductivity and boiling-point characteristics. The primary chlorine-containing products on the market are denoted by the industry nomenclature, such as CFC-11, CFC-12, CFC-113, CFC-114, CFC-115, and the hydrochlorofluorocarbon HCFC-22. CFCs are marketed under many different trade names, i.e., Algcon, Algofrene, Arcton, Eskimon, Flugene, Forane, Freon, Frigen, Genetron, Isceon, and Osotron.

Toxicity and health effects

CFCs (commercial) are persistent in the environment because of their chemical stability. Accumulation of the inert CFCs in the atmosphere leads to depletion of the ozone layer and increased intensity of sunlight. This, in turn, is known to cause health complications, such as skin cancer, eye cataracts, and ecological disasters. At high concentrations, CFCs cause neurological disorders such as tingling sensation, humming in the ears, apprehension, EEG changes, slurred speech, and decreased performance in psychological tests.

Reference

International Programme on Chemical Safety (IPCS). 1990. Fully Halogenated Chlorofluorocarbons. Environmental Health Criteria No. 113. World Health Organization, Geneva, Switzerland.

p-Chloronitrobenzene (CAS No. 100-00-5)

Molecular formula: $p\text{-}NO_2C_6H_4Cl$
IUPAC name: 1-chloro-4-nitrobenzene
Synonyms and trade names: 4-Chloronitrobenzene; p-Chloronitrobenzene; p-Nitrochlorobenzene; 1-Chloro-4-nitrobenzene; 4-Nitrochlorobenzene; PCNB; PNCB; 4-Chloronitrobenzene: $C_6H_4ClNO_2$ (Nitrobenzol, oil of mirbane); Nitrobenzene: $C_6H_5NO_2$ (Nitrobenzol, oil of mirbane); o-Nitrotoluene: $CH_3C_6H_4NO_2$ (m-Nitrotoluene, p-Nitrotoluene)

Use and exposure

p-Chloronitrobenzene is extensively used in different industries as an intermediate in the manufacture of dyes, rubber, and agricultural chemicals. It is incompatible with strong oxidizers and alkalis.

Toxicity and health effects

Repeated exposure to high levels of p-chloronitrobenzene causes adverse health effects. The symptoms of toxicity include, but are not limited to, anoxia, unpleasant taste, anemia, methemoglobinemia, hematuria (blood in the urine), spleen, kidney, bone marrow changes, and reproductive effects. The target organs of p-chloronitrobenzene poisoning have been identified as the blood, liver, kidneys, cardiovascular system, spleen, bone marrow, and reproductive system.

p-Chloronitrobenzene and carcinogenicity

Evaluations on the carcinogenicity of p-chloronitrobenzene are inadequate and the evidence is insufficient. Further testing is required to assess their potential to cause cancer. In view of this, the IARC classified p-chloronitrobenzene as Group 3, meaning not classifiable as a human carcinogen.

Exposure limits

The PEL for p-chloronitro-benzene set by the OSHA is 1 mg/m^3 (skin) (TWA), and the IDLH has been identified as approximately 100 mg/m^3.

References

International Agency for Research on Cancer (IARC). 1996. 2-Chloronitrobenzene, 3-chloronitrobenzene and 4-chloronitrobenzene. In: Printing processes and printing inks, carbon black and some nitro compounds.
———. 1996. IARC Monographs on the evaluation of carcinogenic risks to humans. Vol. 65, 263–96, Lyon, France.

Chlorpyrifos (CAS No. 2921-88-2)

Molecular formula: **C$_9$H$_{11}$Cl$_3$NO$_3$PS**
IUPAC name: 0,0-diethyl 0-(3,5,6-trichloro-2-pyridyl) phospothorothioate
Synonyms and trade name: Dursban; Empire; Eradex; Lorsban; Paqeant; Piridane; Scout; Stipend
Toxicity class: US EPA: II; WHO: II

Use and exposure

Chlorpyrifos belongs to the class of insecticides known as organophosphates. Technical chlorpyrifos is an amber to white crystalline solid with a mild sulfur odor. It is insoluble in water, but soluble in benzene, acetone, chloroform, carbon disulfide, diethyl ether, xylene, methylene chloride, and methanol. Formulations of chlorpyrifos include emulsifiable concentrate, dust, granular wettable powder, microcapsule, pellet, and sprays. Chlorpyrifos is widely used as an active ingredient in many commercial insecticides, such as Dursban and Lorsban, to control household pests, mosquitoes, and pests. Formulations of chlorpyrifos include emulsifiable concentrates, granules, wettable powders, dust, microcapsules, pellets, and sprays. The US EPA has classified chlorpyrifos as a GUP.

Toxicity and health effects

Exposures to chlorpyrifos cause adverse health effects and poisoning. The symptoms include, but are not limited to, headache, dizziness, respiratory problems, muscular and joint pains, numbness, tingling sensations, incoordination, tremor, nausea, abdominal cramps, vomiting, sweating, blurred vision, respiratory depression, slow heart beat, nervousness, weakness, cramps, diarrhea, chest pain, pin-point pupils, tearing, salivation, clear nasal discharge and sputum, muscle twitching, and in severe poisonings convulsions, coma, and death. Exposures to chlorpyrifos cause adverse effects to the nervous system. The effects include phosphorylation of the active site, disturbance in the activity of the acetylcholinesterase (AChE) enzyme (inactivity). AChE enzyme is necessary to stop the transmission of the chemical neurotransmitter.

In occupational workers, high concentrations of chlorpyrifos cause poisoning with symptoms of unconsciousness, convulsions and/or fatal injury. Persons with respiratory ailments and disturbed liver function are known to be at increased health risk. Also, repeated exposures to chlorpyrifos have been reported to cause disturbances in the process of brain development.

Chlorpyrifos and carcinogenicity

The published reports of the NTP, the IARC, and the NIOSH indicate that there are no evidences suggesting that chlorpyrifos is carcinogenic. The US EPA has categorized chlorpyrifos as Group E, meaning no evidence as a human carcinogen.

Exposure limits

The Acceptable Daily Intake (ADI) for chlorpyrifos is set as 0.003 mg/kg/day and the NOEL is set at 0.03 mg/kg/day.

Precautions

Occupational workers should be careful during handling and use of chlorpyrifos. The workplace should have adequate washing facilities at all times and close to the site of handling and use. Eating, drinking, and smoking should be prohibited during handling and before washing after handling. Containers should be kept away from foodstuffs, animal feed and their containers, and out of reach of children.

References

Agency for Toxic Substances and Disease Registry (ATSDR). 1997. Toxicological Profile for Chlorpyrifos. US Department of Health and Human Services, Public Health Service, Atlanta, GA (updated 2010).

Gallo, M.A. and Lawryk, N.J. 1991. Organic phosphorus pesticides. Chapter 16. pp. 917–1123. In: Hayes, W.J. Jr. and Laws, E.R. Jr. (eds.) *Handbook of Pesticide Toxicology*. Academic Press, New York.

Kidd, H. and James, D.R. (eds.). 1991. *The Agrochemicals Handbook*, 3rd ed. Royal Society of Chemistry Information Services, Cambridge, U.K. pp. 5–14.

Material Safety Data Sheets (MSDS). 1996. Chlorpyrifos. Extension Toxicology Network. Pesticide Information Profiles. Oregon State University, USDA/Extension Service, Corvallis, OR.

Slotkin, T.A. 1999. Brain developmental damage occurs from common pesticide Dursban (chlorpyrifos). *Environ. Health Perspect.* 107 (suppl. 1). pp. 71–80.

Tomlin, C.D.S. (ed.). 2007. *The Pesticide Manual: A World Compendium*, 14th ed. British Crop Protection Council, Farnham, Surrey, U.K.

Chlorpyrifos-methyl (CAS No. 5598-13-0)

Molecular formula: $C_7H_7Cl_3NO_3PS$
IUPAC name: O,O-Dimethyl O-3,5,6-trichloro-2-pyridyl phosphorothioate
Synonyms and trade names: DOWCO 214; ENT 27520; OMS-1155; Reldan; Zertell

Use and exposure

Chlorpyrifos-methyl is a general use organophosphate insecticide registered in 1985. It is used for the control of stored grain pests, weevils, moths, borers, beetles and mealworms, red flour beetle, grain moth, for seed treatment. It is effective against rice stem borer, aphids, cutworms, plant and leaf hoppers, mole crickets, moths and stored grain pests. It is poorly soluble in water, moderately soluble in hexane and alcohols, and readily soluble in other organic solvents, such as acetone, benzene, and chloroform.

Toxicity and health effects

Exposures to chlorpyrifos-methyl cause excessive salivation, sweating, rhinorrhea and tearing, muscle twitching, blurred vision/dark vision, slurred speech, weakness, tremor, incoordination, headache, dizziness, nausea, vomiting, abdominal cramps, diarrhea, respiratory depression, tightness in chest, wheezing, productive cough, fluid in lungs, and pin-point pupils. In cases of severe exposure and poisoning, chlorpyrifos-methyl causes seizures, incontinence, respiratory depression, loss of consciousness, respiratory paralysis, and death. Exposures to chlorpyrifos-methyl cause cholinesterase inhibition in workers, and the systemic toxicity includes body weight loss, decreased food consumption, and liver, kidney, and adrenal pathology.

Precautions

Workplaces should have adequate washing facilities at all times during handling and these should be close to the site of handling. Eating, drinking, and smoking should be prohibited during handling and before washing after handling. Containers of chlorpyrifos-methyl should be kept away from foodstuffs, animal feed and their containers, and out of reach of children.

References

Tomlin, C.D.S. (ed.). 2007. *The Pesticide Manual: A World Compendium*, 14th ed. British Crop Protection Council, Farnham, Surrey, U.K.
World Health Organization (WHO). Chlorpyrifos-methyl. Data Sheets on Pesticides No. 33. WHO, Geneva, Switzerland.

Cobalt metal, dust, and fume (as Co) (CAS No. 7440-48-4)

Molecular formula: **Co**

Use and exposure

Cobalt is a silvery, bluish-white, odorless, and magnetic metal. The fume and dust of cobalt metal is odorless and black. The appearance and odor of cobalt compounds and their dusts and fumes vary with the compound. Cobalt metal in powdered form is incompatible with fused ammonium nitrate, hydrozinium nitrate, and strong oxidizing agents and should be avoided. It ignites on contact with bromide pentafluoride. Powdered cobalt ignites spontaneously in air. Exposure to cobalt metal fume and dust can occur through inhalation, ingestion, and eye or skin contact.

Toxicity and health effects

Acute exposure to cobalt metal, dust, and fume is characterized by irritation of the eyes and, to a lesser extent, irritation of the skin. In sensitized individuals, exposure causes an asthma-like attack, with wheezing, bronchospasm, and dyspnea. Ingestion of cobalt may cause nausea, vomiting, diarrhea, and a sensation of heat.

Human exposures to cobalt and cobalt compounds cause cough, tight chest, pain in chest on coughing, dyspnea, malaise, ache, chills, sweating, shivering, and aching pain in back and limbs. After further days of exposures to high concentrations of cadmium, the worker develops more severe pulmonary responses such as severe dyspnea, wheezing, chest pain and precordial constriction, persistent cough, weakness and malaise, anorexia, nausea, diarrhea, nocturia, abdominal pain, diffuse nodular fibrosis, respiratory hypersensitivity, asthma, abdominal pain, cardiomyopathy, lung damage, hemoptysis, prostration, and death.

Cobalt metal, dust, and fume are pulmonary toxins and respiratory and skin sensitizers. Inhalation of cobalt metal fume and dust may cause interstitial fibrosis, interstitial pneumonitis, myocardial and thyroid disorders, and sensitization of the respiratory tract and skin. Chronic cobalt poisoning may also produce polycythemia and hyperplasia of the bone marrow. Myocardial disorders have also been observed in cobalt production workers. Chronic exposure to cobalt metal, dust, or fume may cause respiratory or dermatologic signs and symptoms. Following skin sensitization, contact with cobalt causes eruptions of dermatitis in creases and on frictional surfaces of the arms, legs, and neck. Chronic cobalt poisoning may cause polycythemia, hyperplasia of the bone marrow and thyroid gland, pericardial effusion, and damage to the alpha cells of the pancreas.

Exposure limits

The current OSHA PEL for cobalt metal, dust, and fume (as Co) is 0.1 mg/m^3 of air in an 8-h TWA concentration. The NIOSH has established a REL for cobalt metal, dust, and fume of 0.05 mg/m^3) as a TWA for up to a 10-h workday and a 40-h workweek.

Storage

Cobalt metal dust (powdered metal) should be stored in a cool, dry, well-ventilated area in tightly sealed containers that are labeled in accordance with OSHA standards. Containers of cobalt metal dust should be protected from physical damage and ignition sources, and should be stored separately from strong oxidizers.

References

National Library of Medicine (NLM). 1995. Hazardous substances data bank: Cobalt metal. NLM, Bethesda, MD.

Sittig, M. 1991. *Handbook of Toxic and Hazardous Chemicals*, 3rd ed. Noyes, Park Ridge, NJ.

US Department of Labor. 2008. Occupational Safety & Health Administration (OSHA). Occupational safety and health guideline for cobalt metal, dust, and fume (as co). OSHA, Washington, DC.

Copper (CAS No. 7440-50-8)

Molecular formula: **Cu**

Use and exposure

Copper is a metal that occurs naturally throughout the environment in rocks, soil, water, and air. Copper is an essential element in plants and animals (including humans), which means it is necessary for us to live. Therefore, plants and animals must absorb some copper through eating, drinking, and breathing. Copper in very low levels is essential for good health. Copper is used to make many different kinds of products like wire, plumbing pipes, and sheet metal. Before 1982, US pennies were made of copper, while those made after 1982 are only coated with copper. Copper is also combined with other metals to make brass and bronze pipes and faucets. Copper compounds are commonly used in agriculture to treat plant diseases like mildew, for water treatment and as preservatives for wood, leather, and fabrics.

Toxicity and health effects

Copper is an essential micronutrient for human health. While very low levels of copper are essential for good health, high levels of copper can cause harmful health effects. Occupational breathing of high levels of copper in workplaces is known to cause irritation of the nose and throat. Acute ingestion of excess copper in drinking water can cause gastrointestinal (GI) tract disturbances. Ingestion of high levels of copper cause nausea, vomiting, and diarrhea, and very high doses of copper cause damage to the kidneys and can be fatal.

Copper and carcinogenicity

The US EPA has not classified copper as a human carcinogen.

Reference

Agency for Toxic Substances and Disease Registry (ATSDR). 2004. Toxicological Profile for Copper. US Department of Health and Human Services, Public Health Service, Atlanta, GA (updated 2007).

Copper dust, fume, and mist (as Cu) (CAS No. 7440-50-8)

Molecular formula: **Cu**

Toxicity and health effects

Occupational workers exposed to copper fumes, dust and mists in work areas develop symptoms of poisoning. These include irritation to the mucous membrane, nasal, and pharyngeal irritation; nasal perforation, eye irritation, metallic or sweet taste, dermatitis; prolonged periods of exposure to high concentrations cause anemia, adverse effects to the lung, liver, and kidney. The exposed worker also suffers from metal fume fever; chills, muscle aches, nausea, fever, dry throat, coughing, weakness, lassitude, irritation of the eyes and the upper respiratory tract, discolored skin and hair, and acute lung damage.

Occupational workers exposed to copper dust suffer from gastrointestinal disturbances, headache, vertigo, drowsiness, and hepatomegaly. Vineyard workers chronically exposed to Bordeaux mixture (copper sulfate and lime) exhibit degenerative changes of the lungs and liver. Dermal exposure to copper may cause contact dermatitis in some individuals.

Copper is required for collagen formation. Copper deficiency is associated with athero-sclerosis and other cardiovascular conditions. Any kind of imbalance of copper in the body causes health disorders that include, but are not limited to, arthritis, fatigue, adrenal burnout, insomnia, scoliosis, osteoporosis, heart disease, cancer, migraine headaches, seizures, gum disease, tooth decay, skin and hair problems, and uterine fibroids, endometriosis (in females). Copper deficiency is associated with aneurysms, gout, anemia, and osteoporosis.

Exposures to copper in the form of dusts and mists cause irritation to the eyes, respiratory system, mucous membrane, nasal, pharyngeal irritation cough, dyspnea (breathing difficulty), and wheezing. Prolonged exposures are known to cause nasal perforation. Copper has caused anemia and damage to the lung, liver, and kidney in experimental laboratory animals. Reports have indicated that copper dusts and fumes are potential occupational carcinogens.

Exposure limits

The OSHA, the ACGIH, and the NIOSH have set the PEL for copper in general industry as 1.0 mg/ m^3 (TWA).

References

New Jersey Department of Health and Senior Services. 1986. Copper (Dust, Fume or Mist). Occupational Disease and Injury Services, Trenton, NJ.
US Department of Labor. 2000. Occupational Safety & Health Administration (OSHA). Copper Dusts and Mists (as Cu). OSHA, Washington, DC.

Copper fume (as Cu) (CAS No. 1317-38-0)

Molecular formula: **CuO**

Use and exposure

Copper metal, metal compounds and alloys are often used in "hot" operations in the workplace. The workplace operations include, but are not limited to, welding, brazing, soldering, plating, cutting, and metalizing. At the high temperatures reached in these operations, metals often form metal fumes that have different health effects.

Toxicity and health effects

Exposures to copper fume cause fever, chills, muscle aches, nausea, dry throat, coughing, weakness, lassitude, irritation to the eyes, nose, throat, skin, upper respiratory tract, chest tightness, nose bleed, edema, and lung damage. Symptoms of copper fume poisoning also include metallic or sweet taste, skin itching, skin rash, skin allergy, and a greenish color to the skin, teeth, and hair. Workers have increased risk of Wilson's disease.

Copper fume and carcinogenicity

There is evidence that workers in copper smelters have an increased risk of lung cancer, but this is thought to be due to arsenic trioxide and not copper.

Precautions

Occupational workers should use protective clothing, such as suits, gloves, footwear, and headgear, and promptly change the contaminated clothing/work dress. Workers should not eat, smoke, or drink where copper dust or powder is handled, processed, or stored. Workers should wash hands carefully before eating, drinking, smoking, or using the toilet. The workplace should have a vacuum or a wet method facility to reduce the metal dust during cleanup.

Copper sulfate (CAS No.7758-98-7)

Molecular formula: $CuSO_4 \; 5H_2O$
Synonyms and trade names: Copper sulfate; Cupric sulfate; Bluestone; Blue vitriol; Copper sulfate pentahydrate; Hydrated copper sulfate.

Use and exposure

Copper sulfate (anhydrous form) is green or gray-white powder, whereas pentahydrate, the most commonly encountered salt, is bright blue. The anhydrous form occurs as a rare mineral known as chalcocyanite. Hydrated copper sulfate occurs in nature as chalcanthite. Copper sulfate is made by the action of sulfuric acid with a variety of copper compounds. Copper sulfate is used in hair dyes, coloring glass, processing of leather, textiles, and in pyrotechnics as a green colorant.

Copper sulfate pentahydrate is used as a fungicide and a mixture with lime is called Bordeux mixture and is used to control fungus on grapes, melons, and other berries, as a molluscicide for the destruction of slugs and snails, particularly the snail host of the liver fluke. Copper sulfate is used in Fehling and Benedict's solution to test reducing sugars.

Toxicity and health effects

Workers who accidentally ingest copper sulfate experience abdominal pain and cramps, burning sensation, corrosive effects, nausea, vomiting, loose bowel movement, and a metallic taste. Exposures to copper sulfate by ingestion or skin absorption cause severe irritating effects to the eyes and skin The aerosol is irritating to the respiratory tract, and effects on the blood, kidneys and liver result in hemolytic anemia, kidney impairment, liver impairment, and shock or collapse. At large doses, accidental intake of copper sulfate causes renal failure, comatose, and even death. Long-term exposure to copper sulfate may lead to liver damage, lung diseases, and decreased female fertility.

Storage

Workers should keep copper sulfate stored in a cool, dry area with sufficient ventilation. It should be kept away from alkalis, magnesium, ammonia, acetylene, and sodium hypobromite.

Precautions

During handling and use of copper sulfate, students and occupational workers should wear safety glasses and should not breathe the material in powder form. Copper sulfate is an environmental pollutant and must be carefully incorporated when used in its varied applications. Workers should wear protective clothing, goggles, impermeable gloves, and rubber boots to avoid skin contact.

Reference

Material Safety Data Sheet (MSDS). 2009. Copper sulfate. Chemical safety data: Copper (II) sulfate. Physical Chemistry at Oxford University, Oxford, U.K.

Coumaphos (CAS No. 56-72-4)

Molecular formula: $C_{14}H_{16}ClO5PS$
IUPAC name: O-(3-chloro-4methyl-2-oxo-2H-Chromen-7-ylO,O-diethyl phosphorothioate)
Synonyms and trade names: Agridip; Asunthol; Meldane; Muscatox; Umbethion; Asuntol; Baymix
Toxicity class: US EPA: II; WHO: Ia

Use and exposure

Technical coumaphos is a tan crystalline solid with a slight sulfur odor. It is insoluble in water, slightly soluble in acetone, chloroform, and ethanol, and soluble in organic solvents. It is used for the control of a wide variety of livestock insects, including cattle grubs, screw-worms, lice, scabies, flies, and ticks. It is used against ectoparasites, which are insects that live outside the host animals, such as sheep, goats, horses, pigs, and poultry. Coumaphos is incompatible with pyrethroids and piperonyl butoxide. The US EPA has grouped coumaphos as an RUP. Acute dermal exposures to occupational workers occur during mixing, loading, and application of coumaphos, especially in the livestock dip-vat and hand-held sprayer uses.

Toxicity and health effects

Exposures to coumaphos cause signs of poisoning such as diarrhea, drooling, difficulty in breathing, leg and neck stiffness among occupational workers. Acute inhalation of coumaphos causes headaches, dizziness, and incoordination. Moderate poisoning causes muscle twitching and vomiting while severe poisoning leads to fever, toxic psychosis, lung edema, and high blood pressure. Repeated exposures cause irritability, confusion, headache, speech difficulties, effects on memory concentration, disorientation, severe depressions, sleepwalking, and drowsiness or insomnia among occupational workers. Coumaphos has been classified as non-carcinogenic to humans.

References

EXTOXNET. 1994. Coumaphos. Extension Toxicology Network. USDA, Corvallis, OR.
Gosselin, R.E., et al. 1976. *Clinical Toxicology of Commercial Products*, 4th ed. Williams & Wilkins, Baltimore, MD.
Material Safety Data Sheet (MSDS). 1994. Coumaphos. Extension Toxicology Network (EXTOXNET). USDA. Corvallis, OR.
Occupational Health Services (OHS). 1991. MSDS for Coumaphos. OHS Inc., Secaucus, NJ.
US Environmental Protection Agency (US EPA). 1985. Chemical profile: Coumaphos. Washington, DC.

———. 1989. Coumaphos. Pesticide Fact Sheet no. 207. Office of Pesticides and Toxic Substances, Washington, DC.

———. 2006. Chemical Emergency Preparedness and Prevention: Coumaphos. Emergency First Aid Treatment Guide. US EPA, Atlanta, GA.

Cyanide (CAS No. 57-12-5)

Molecular formula: **CN**

The most common cyanide is hydrogen cyanide (HCN) and its salts—sodium cyanide (NaCN) and potassium cyanide (KCN). Cyanides are ubiquitous in nature, arising from both natural and anthropogenic sources. Cyanide is released into the environment from numerous sources. Metal finishing and organic chemical industries as well as iron and steel production are major sources of cyanide releases to the aquatic environment. More than 90% of emissions to the air are attributed to releases from automobile exhausts. Workers in a wide variety of occupations may be exposed to cyanides. The general population may be exposed to cyanides by inhalation of contaminated air, ingestion of contaminated drinking water, and/or consumption of a variety of foods.

Cyanide most commonly occurs as HCN and its salts—sodium and KCN. Cyanides are both man-made and naturally occurring substances. They are found in several plant species as cyanogenic glycosides and are produced by certain bacteria, fungi, and algae. In very small amounts, cyanide is a necessary requirement in the human diet. Cyanides are released to the environment from industrial sources and car emissions. Cyanides are readily absorbed by inhalation, oral and dermal routes of exposure. HCN and its simple soluble salts, such as sodium and KCN, are among the most rapidly acting poisons. The CNS is the primary target organ for cyanide toxicity. Neurotoxicity has been observed in humans and animals following ingestion and inhalation of cyanides. Neurotoxic effects, including convulsions and coma, preceded death in guinea pigs dermally exposed to HCN. Exposed laboratory animals developed signs of toxicity and death in 3–12 min after the eyes were treated. Cardiac and respiratory effects, possibly CNS-mediated, have also been reported. In tropical regions of Africa, a high incidence of ataxic neuropathy, goiter, amblyopia, and other health disorders have been associated with chronic ingestion of cassava, one of the dietary staples containing cyanogenic glycosides that release HCN when metabolized in vivo.

Cyanide process, also known as cyanidation, is the most widely used process for extracting gold and silver from ores. The ores are powdered grounds and can be concentrated by flotation. It is then mixed with dilute solutions of sodium (or potassium or calcium) cyanide while air is bubbled through it to form the soluble complex ion, $Au(CN)2-1$. The precious metals are precipitated from solution by zinc. The precipitates are smelted to remove the zinc and treated with nitric acid to dissolve the silver.

Exposure limits

The US EPA has set a limit of 0.2 ppm for cyanide in drinking water. The OSHA has set a limit of 10 ppm for HCN and most cyanide salts in the workplace.

Cyanide compounds

Cyanide compounds are a group of chemical substances. They are based on a common structure formed when elemental nitrogen and carbon are combined. Cyanides are produced by certain bacteria, fungi, and algae, and may be found in food and plants. Cyanide itself is an ion, or combining form that carries a positive or negative charge. It is a powerful and rapid-acting poison. Cyanide in combination with metals and organic compounds forms simple and complex salts, such as KCN and NaCN.

Cyanide compounds most commonly occur as hydrogen cyanide and its salts—sodium and KCN. Cyanides are both man-made and naturally occurring substances. They are found in several plant species as cyanogenic glycosides and are produced by certain bacteria, fungi, and algae. In very small amounts, cyanide is a necessary requirement in the human diet. Cyanides are released into the environment from industrial sources and car emissions.

Uses

Cyanide salts are mainly used in electroplating, metallurgy, and the production of organic chemicals; in photographic development; as anticaking agents in road salts; in the extraction of gold and silver from ores; and in the making of plastics. Minor uses of cyanide salts include as insecticides and rodenticides, chelating agents, and in the manufacture of dyes and pigments.

Cyanide is released into the environment from numerous sources. Metal finishing and organic chemical industries as well as iron and steel production are major sources of cyanide releases to the aquatic environment. More than 90% of emissions to the air are attributed to releases from automobile exhausts. Workers in a wide variety of occupations may be exposed to cyanides. The general population may be exposed to cyanides by inhalation of contaminated air, ingestion of contaminated drinking water, and/or consumption of a variety of foods. Cyanogenic glycosides, which produce hydrogen cyanide upon hydrolysis, are found in a number of plant species.

Other cyanide compounds include

Calcium cyanide (CAS No. 592-01-8)

Molecular formula: $Ca(CN)_2$

Copper cyanide (CAS No. 544-92-3)

Molecular formula: $CuCN$

Cyanogen (CAS No. 460-19-5)

Molecular formula: C_2N_2

Cyanogen bromide (CAS No. 506-68-3)

Molecular formula: CBr

Toxicity and health effects

Exposure to cyanogen bromide is dangerous. The chemical substance is poisonous and causes fatal injury if swallowed, inhaled, or absorbed through the skin. It is corrosive and the vapors cause severe irritation to the eyes and respiratory tract, and cause burns to any area of contact. On contact with acids, cyanogen bromide liberates poisonous gas, affecting the blood, cardiovascular system, CNS, and thyroid.

Cyanogen chloride (CAS No. 506-77-4)

Molecular formula: **CNCl**

Hydrogen cyanide (CAS No. 74-90-8)

Molecular formula: **HCN**
Synonyms and trade names: Carbon hydride nitride; Cyanane; Cyclon; Formic anammonide; Formonitrile; Hydrocyanic acid; Prussic acid

Use and exposure

HCN is a colorless to pale blue liquid or gas. It has a distinct odor resembling bitter almonds. HCN reacts with amines, oxidizers, acids, sodium hydroxide, calcium hydroxide, sodium carbonate, caustic substances, and ammonia. HCN was first isolated from a blue dye, Prussian blue, in 1704. HCN is obtainable from fruits that have a pit, such as cherries, apricots, and bitter almonds, from which almond oil and flavoring are made. HCN is used in fumigating, electroplating, mining, and in producing synthetic fibers, plastics, dyes, and pesticides. It is also used as an intermediate in chemical syntheses. Exposures to cyanide occur in workplaces such as the electroplating, metallurgical, firefighting, steel manufacturing, and metal cleaning industries. Human exposures to cyanide also occur from wastewater discharges of industrial organic chemicals, iron and steel works, and wastewater treatment facilities.

Toxicity and health effects

HCN is particularly dangerous because of its toxic and asphyxiating effects on all life requiring oxygen to survive. HCN combines with the enzymes in tissue associated with cellular oxidation. The signs and symptoms of HCN poisoning are non-specific and very rapid. The symptoms include excitement, dizziness, nausea, vomiting, headache, weakness, drowsiness, gasping, thyroid, blood changes, confusion, fainting, tetanic spasm, lockjaw, convulsions, hallucinations, loss of consciousness, coma, and death. When oxygen becomes unavailable to the tissues, it leads to asphyxia and causes death. Children are more vulnerable to HCN exposure. HCN is readily absorbed from the lungs; symptoms of poisoning begin within seconds to minutes. Inhalation of HCN results in the rapid onset of poisoning, producing almost immediate collapse, respiratory arrest, and death within minutes (Table 1).

Hydrogen cyanide and carcinogenicity

Information on the carcinogenicity of HCN in humans or animals for oral exposure is unavailable. Similarly, there are no reports that cyanide causes cancer in animals and humans. The US EPA has classified cyanide as a Group D, meaning not classifiable as to human carcinogenicity.

Exposure limits

The OSHA has set the PEL of HCN as 10 ppm at the workplace and the NIOSH has set the IDLH concentration as 50 ppm.

Precautions

Occupational workers should be very careful in the management of HCN since the gas in air is explosive at concentrations over 5.6%, equivalent to 56,000 ppm and it does not provide adequate warning of hazardous concentrations. HCN at a concentration of 300 mg/m³ in air becomes fatal within about 10 min and HCN at a concentration of 3500 ppm (about 3200 mg/m³) kills a human in about 1 min.

References

Agency for Toxic Substances and Disease Registry (ATSDR). 2006. Toxicological Profile for Cyanide. US Department of Health and Human Services, Public Health Service, Atlanta, GA.
US Environmental Protection Agency (US EPA). 2000. Cyanide compounds hazard summary. Technology Transfer Network Air Toxics web site (updated 2007).

Potassium cyanide (CAS No. 151-50-8)

Molecular formula: **KCN**

Toxicity and health effects

Potassium cyanide is a white solid or colorless water solution with a faint bitter almond odor. As a solution, it is slightly soluble in ethanol. It is a poison that reacts with acid or acid fumes to emit deadly hydrogen cyanide. When heated to decomposition, it emits very toxic fumes. As a solid, potassium cyanide is incompatible with nitrogen trichloride, perchloryl fluoride, sodium nitrite, acids, alkaloids, chloral hydrate, and iodine. A synonym for potassium cyanide is potassium salt of hydrocyanic acid.

Potassium silver cyanide (CAS No. 506-61-6)

Toxicity and health effects

Potassium silver cyanide is a poisonous, white solid made of crystals, which are light sensitive. It is soluble in water and acids, and slightly soluble in ethanol. It emits very toxic fumes when heated to decomposition. Synonyms for potassium silver cyanide are potassium argentocyanide and potassium dicyanoargentate.

Sodium cyanide (CAS No. 143-33-9)

Molecular formula: **NaCN**

Toxicity and health effects

Sodium cyanide is a white crystalline solid that is odorless when dry, but emits a slight odor of hydrogen cyanide in damp air. It is slightly soluble in ethanol and formamide. It is very poisonous. It explodes if melted with nitrite or chlorate at about 450°F. It produces a violent reaction with magnesium, nitrites, nitrates, and nitric acid. On contact with acid, acid fumes, water, or steam, it produces toxic and flammable vapors. Synonyms for sodium cyanide are hydrocyanic acid, sodium salt, and cyanide of sodium.

Reference

Material Safety Data Sheet. 2009. Safety data for hydrogen cyanide. Physical Chemistry at Oxford University, U.K.

Cyanides and carcinogenicity

There are no reports available about the cancer bioassays or epidemiological studies of cyanide. The US EPA categorized cyanides as Group D, meaning not classifiable as a human carcinogen.

References

National Institute for Occupational Safety and Health (NIOSH). 1976. Occupational Exposure to Hydrogen Cyanide and Cyanide Salts. Publ. no. 1-191. US Department of Education and Welfare, Public Health Service, Rockville, MD.

US Environmental Protection Agency (US EPA). 1990. Calcium cyanide. Integrated Risk Information System (IRIS). Environmental Criteria and Assessment Office, Office of Health and Environmental Assessment, Cincinnati, OH.

———. 1990. Cyanogen. Integrated Risk Information System (IRIS). Environmental Criteria and Assessment Office, Office of Health and Environmental Assessment, Cincinnati, OH.

———. 1990. Potassium cyanide. Integrated Risk Information System (IRIS). Environmental Criteria and Assessment Office, Office of Health and Environmental Assessment, Cincinnati, OH.

———. 1990. Potassium silver cyanide. Integrated Risk Information System (IRIS). Environmental Criteria and Assessment Office, Office of Health and Environmental Assessment, Cincinnati, OH.

———. 1990. Silver cyanide. Integrated Risk Information System (IRIS). Environmental Criteria and Assessment Office, Office of Health and Environmental Assessment, Cincinnati, OH.

———. 1991. Sodium cyanide. Integrated Risk Information System (IRIS). Environmental Criteria and Assessment Office, Office of Health and Environmental Assessment, Cincinnati, OH.

(For more information on cyanide compounds refer to the literature)

D

- DDT
- Demeton-S-methyl
- Diazinon
- Dichlorvos
- Dicofol
- Diethylamine
- Dimethoate

- Dimethyl adipate
- Dimethylamine
- 1,4 Dioxane
- Diphenylamine
- Disulfoton

DDT (CAS No. 50-29-3)

Molecular formula: $C_{14}H_9C_{15}$
IUPAC name: 1,1,1'-Trichloro-2,2-bis(4-chlorophenyl) ethane
Synonyms and trade names: Dichlorodiphenyltrichloroethane

Technical p,p'-DDT is a waxy solid, but in its pure form appears as colorless crystals. It is a mixture of three isomers, namely, p,p'-DDT isomer (ca. 85%), o,p'-DDT, and o,o'-DDT (in smaller levels). DDT is very soluble in cyclohexanone, dioxane, benzene, xylene, trichloro-ethylene, dichloromethane, acetone, chloroform, diethyl ether, ethanol, and methanol. The US EPA grouped DDT as an RUP. DDT is often contaminated with DDE (1,1-dichloro-2,2-bis(chlorophenyl) ethylene) and DDD (1,1-dichloro-2,2-bis(p-chlorophenyl) ethane).

Dichloro-diphenyl-trichloroethane (DDT) was first synthesized in 1873 by the German chemist Othmar Zeidler. In 1939, Paul Muller of Geigy Pharmaceutical in Switzerland discovered the effectiveness of DDT as an insecticide. In 1948, Muller was awarded the Nobel Prize in medicine and physiology for his discovery of DDT. The use of DDT increased enormously on a worldwide basis after World War II, primarily because of its effectiveness against the mosquito that spreads malaria and lice that carry typhus. WHO estimates that during the period of its use approximately 25 million lives were saved.

Use and exposure

DDT was extensively used during World War II among the Allied troops and certain civilian populations to control insect typhus and malaria vectors. Application of DDT became extensive because of easy control of a large number of crop pests and vectors of human diseases. Humans are exposed to DDT because of different activities and many factors. These include, but are not limited to, (i) consumption of eating contaminated foods, such as root and leafy vegetable, fatty meat, fish, and poultry, but levels are very low; (ii) eating contaminated imported foods from countries that still allow the use of DDT to control pests; (iii) breathing contaminated air or drinking contaminated water near waste sites and landfills that may contain higher levels of these chemicals; (iv) infants fed on breast milk from mothers who have been exposed; and (v) breathing or swallowing soil particles near waste sites or landfills that contain these chemicals.

By the 1970s, it was obvious that overuse and misuse of DDT was associated with environmental and health effects. Eventually, in June 1972, the US EPA cancelled all use of DDT on crops except in certain cases of disease control where the US EPA allowed a limited use of DDT. However, many tropical countries are still using DDT for the control of malaria.

Use of DDT was banned in Sweden in 1970 and in the United States in 1972. In view of its large-scale use over the decades, many insect pests may have developed resistance to DDT. It is no longer registered for use in the United States barring a public health emergency, e.g., an outbreak of malaria. The latest group meeting by 110 countries on DDT met in Geneva (September 17, 1999) to phase out the production of DDT and impose a total ban on its use even for public health purposes. The conference agree on the conclusion for a global ban on DDT. Absence of a suitable substitute for DDT in the control of malaria and the absence of an antimalaria vaccine necessitates the continuing use of DDT for malaria control.

Toxicity and health effects

Exposures to high concentrations of DDT cause adverse health effects and poisoning among occupational workers. The symptoms of toxicity include, but are not limited to, tremors, diarrhea, dizziness, headache, vomiting, numbness, paresthesias, hyperexcitability, and convulsions. Chronic exposures to DDT caused adverse effects on the nervous system, liver, kidneys, and immune systems in experimental animals. Laboratory rats and mice given DDT (16–32 and 6.5–13 mg/kg for about 26 weeks and 80–140 weeks, respectively) experienced the tremors. Laboratory studies with non-human primates given DDT (10 mg/kg/day over 100 days) showed changes in the cellular chemistry of the CNS and at higher doses (50 mg/kg/day) caused loss of equilibrium in animals. Prolonged exposures to DDT caused adverse effects in species of animals, i.e., rats, mice, hamsters, dogs, and monkeys, showing pathomorphological changes in the liver, kidney, CNS, and adrenal glands.

Humans exposed to DDT have shown many adverse effects, i.e., nausea, diarrhea, increased liver enzyme activity, irritation of the eyes, nose and throat, disturbed gait, malaise and excitability; at higher doses, tremors and convulsions. It has been reported that the health effects of DDT in humans exposed to different doses and for different periods of poisoning include sweating, headache, nausea, and convulsions. It is very important to remember that earlier findings on the toxicological effects of DDT in animals and humans, as reported above, require further confirmatory data since the studies did not observe the GLP regulations. Therefore, the potential hazards of DDT to human health and environmental safety require careful evaluations.

DDT and malaria control

The US government has finally begun to reverse its policy on the insecticide DDT. Let's hope that this policy shift represents the beginning of the end of what can only be called a crime against humanity: the decades-old withholding of the world's most effective antimalarial weapon from billions of adults and children at risk of dying from the disease. WHO has been given the credit for one billion people live free from malaria, and for saving millions of lives. After 30 years of worldwide use of DDT, in 1973, WHO concluded that the benefits of use of DDT were far greater than its possible risks. During its recent discussions, the United Nations Environment Program (UNEP) observed whether or not DDT should have been totally banned along with 11 other persistent organic pollutants. The total ban of DDT was sharply criticized; in South Africa, a temporary total ban on the use of DDT for indoor spraying resulted in a sudden increase in malaria. Eleven countries in Africa, seven in Asia, and five in Latin America still use DDT for vector disease control. In fact, according to the general consensus, limited and strictly controlled proper use of DDT should be allowed for public health purposes, in particular where other effective, safe, and affordable alternative pesticides are not available, as the benefits are clearly far superior to possible risks. In this context, it is important to note the observations that DDT can cause many toxicological effects, but the effects on human beings at likely exposure levels seem to be very slight. However, the perceived rather than the calculated risks from DDT use are an important consideration in maintaining public confidence. Thus, it would seem prudent that if its use was continued for antimalarial campaigns and the benefits of use outweigh the risks, tight control should continue and the effects of spraying DDT should be closely monitored. (See the literature for more information.)

DDT and carcinogenicity

Studies in DDT-exposed workers did not show increases in cancer. Studies in laboratory animals exposed to DDT in food suggested that DDT can cause liver cancer. The DHHS indicated that DDT may reasonably be anticipated to be a human carcinogen. The IARC determined that DDT may possibly cause cancer in humans. The US EPA also observed that DDT, DDE, and DDD are probable human carcinogens.

In most studies in which the relation between exposure to organochlorine compounds and breast cancer was examined, residues were measured in serum, although they are higher in breast adipose tissue, which represents cumulative internal exposure at the target side for breast cancer. In a large study in which DDT and its metabolites DDE and DDD were measured in breast adipose tissue, the concentrations of DDE were higher than those of DDT in both breast cancer patients and controls. After adjustment for age, no relation was found between the concentration of either DDT, DDE, or DDT + DDE + DDD and breast cancer.

Laboratory studies showed that DDT increased incidences of lung tumors and lymphomas in mice, incidences of liver tumors in rats, and incidences of adrenal adenomas in hamsters. Long-term oral administration of DDT to non-human primates caused hepatic toxicity, malignant and benign tumors at various sites and the adverse health effects found in non-human primates were of borderline statistical significance. However, it was reported that no conclusion could be made about the carcinogenicity of DDT in monkeys on the basis of a 130-month study at one dose. The reports on DDT and cancer have become contradictory and inconclusive. However, reports on DDT have become contradictory. For instance, vant't Veer et al. did not support the hypothesis that DDE increases the risk of breast cancer in post-menopausal women in Europe. Thus, several workers in a large and well-designed study found no evidence that exposures to DDT and DDE caused an increased risk of breast cancer. The evidences suggest generation of more adequate and quality data on the toxicity profile of DDT.

Exposure limits

The OSHA has set the limit for DDT as 1.0 mg/m^3 at the workplace for an 8-h day (TWA).

References

Agency for Toxic Substances and Disease Registry (ATSDR). 2002. Toxicological Profile for DDT, DDDE, and DDD. US Department of Health and Human Services, Public Health Service, Atlanta, GA (updated 2007).

Dikshith, T.S.S. (ed.). 1991. *Toxicology of Pesticides in Animals*. CRC Press, Boca Raton, FL.

Material Safety Data Sheet (MSDS). 1996. DDT. Extension Toxicology Network. Oregon State University, USDA/Extension Service, Corvallis, OR.

Material Safety Data Sheets (MSDS). Pesticide Information Profiles (PIP). Oregon State University, USDA/Extension Service, Corvallis, OR.

Smith, A.G. 1991. Chlorinated hydrocarbon insecticides. In: Hayes, W.J. Jr. and Laws, E.R. Jr. (eds.) *Handbook of Pesticide Toxicology*. Academic Press, New York. p. 731–915.

Tomatis, L. and Huff, J. 2000. Evidence of carcinogenicity of DDT in nonhuman primates. *J. Cancer Res. Clin. Oncol.* 126: 246.

van't Veer, P. et al. 1997. DDT and post menopausal breast cancer in Europe – a case control study. *Br. Med. J.* 12: 315.
World Health Organization (WHO). 1997. DDT and its Derivatives. Environmental Health Criteria No. 9. WHO, Geneva, Switzerland.

Demeton-S-methyl (CAS No. 919-86-8)

Molecular formula: $C_6H_{15}O_3PS_2$
IUPAC name: S-2-ethylthioethyl – O, O-dimethyl phosphorothioate
Synonyms and trade names: Metasystox; Metaphor; Meta-isosystox; Azotox; Duratox; Mifatox; Persyst
Toxicity class: US EPA: I; WHO: Ib

Use and exposure

Demeton-S-methyl is pale yellow in color, oily and has a sulfur-like odor. It is sparingly soluble in water, but is very rapidly soluble in common polar organic solvents, such as dichloromethane, 2-propanol, toluene, and n-hexane. It is a highly toxic, systemic, and contact insecticide and acaricide and is classified as Category I. It is used for the control of insects, aphids, saw-flies, and spider mites on cereals, fruits, vegetables, and ornamental plants. On heating, demeton-s-methyl undergoes decomposition and emits very toxic fumes, phosphorus oxides, and sulfur oxides.

Toxicity and health effects

Demeton-S-methyl is highly toxic to animals and humans. Careless occupational exposures to demeton-S-methyl cause severe symptoms of poisoning that include, but are not limited to, headaches, nausea, vomiting, diarrhea, sweating, dizziness, tremors, lack of coordination, hiccough, and memory loss. Prolonged exposures to high concentrations cause pupillary constriction, blurred vision, muscle cramp, excessive salivation, weakness, sweating, abdominal cramps, unconsciousness, respiratory distress, convulsions, respiratory failure, and death.

Demeton-S-methyl and carcinogenicity

There are no reports indicating that demeton-S-methyl has embryotoxic or teratogenic potential, or causes adverse effects on reproduction or development, or carcinogenic action to animals and humans. Demeton-S-methyl is grouped as A4 meaning, not classifiable as a human carcinogen.

Exposure limits

The ADI of demeton-S-methyl has been set as 0.0003 mg/kg, the TLV as 0.5 mg/m³, the STEL should not exceed 1.5 mg/m³ for not more than a total of 30 min, and the NOEL as 1.0 mg/kg.

Storage

Demeton-S-methyl should be kept stored separately and away from food and feedstuffs. Keep in a well-ventilated room.

Precautions

During handling and use of demeton-S-methyl, occupational workers should use a face shield or eye protection in combination with breathing protection, protective gloves, and protective clothing. During use and handling of the chemical substance, workers should not eat, drink, or smoke during work and wash hands before eating. Workers should prevent generation of chemical mists and observe strict hygiene.

References

International Programme on Chemical Safety and the Commission of the European Communities (IPCS-CEC). 1997. Demeton-S-methyl. ICSC card no. 0705. IPCS, Geneva, Switzerland (updated 2005).

Kidd, H. and James, D.R. (eds.). 1991. *The Agrochemicals Handbook*, 3rd ed. Royal Society of Chemistry Information Systems, Unwin Brothers Limited, Surrey, U.K.

Sax, N.I. (ed.). 1984. *Dangerous Properties of Industrial Materials*, 6th ed. Van Nostrand Reinhold, New York.

World Health Organization (WHO). 1997. Dimeton-S-methyl. Environmental Health Criteria No. 197. WHO, Geneva, Switzerland.

Diazinon (CAS No. 333-41-5)

Molecular formula: $C_{12}H_{21}N_2O_3PS$
IUPAC name: O, O-Diethyl 0-2-isopropyl-6-methyl (pyrimidin-4-yl) phosphorothioa
Synonyms and trade names: Basudin; Knox-out; Dazzel; Gardentox; Kayazol; Nucidol
Toxicity class: US EPA: II or III; WHO: II

Use and exposure

Diazinon is available as a colorless or dark brown liquid. It is sparingly soluble in water but very soluble in petroleum ether, alcohol, and benzene. Diazinon is used for the control of a variety of agricultural and household pests. These include pests in soil, on ornamental plants, fruit, vegetable, crops pests, and household pests like flies, fleas, and cockroaches. Diazinon undergoes decomposition on heating above 120°C and produces toxic fumes, such as nitrogen oxides, phosphorous oxides, and sulfur oxides. It reacts with strong acids and alkalis with the possible formation of highly toxic tetra ethyl thiopyrophosphates. Diazinon is classified as an RUP. Depending on the type of formulation, diazinon is classified as toxicity class II, meaning moderately toxic, or toxicity class III, meaning slightly toxic.

Toxicity and health effects

Humans are exposed to diazinon during manufacture and professional applications. Diazinon causes poisoning with symptoms such as headache, dizziness, nausea, weakness, feelings of anxiety, vomiting, pupillary constriction, convulsions, respiratory distress or labored breathing, unconsciousness, muscle cramp, excessive salivation, respiratory failure, and coma.

Diazinon and carcinogenicity

There are no evidences indicating that exposures to diazinon cause mutagenic, teratogenic, and/or carcinogenic effects in laboratory animals or humans. The DHHS, the IARC, the OSHA, and the US EPA have not reported that diazinon causes cancer in humans or in animals and have listed diazinon as Group 3, meaning not classifiable as a human carcinogen.

Precautions

Workers should avoid eye contact with diazinon, wear chemical safety glasses or goggles, protective clothing or equipment, wear waterproof boots, long-sleeved shirts, long pants, and a hat. Workers should avoid contamination of food and feed, wash thoroughly after handling and before eating or smoking. In fact, occupational workers should avoid eating, drinking, or smoking in areas of work with the chemical.

References

Agency for Toxic Substances and Disease Registry (ATSDR). 2008. Toxicological Profile for Diazinon. US Department of Health and Human Services, Public Health Service, Atlanta, GA (updated 2009).

Dikshith, T.S.S. and Diwan, P.V. 2003. *Industrial Guide to Chemicals and Drug Safety*. Wiley, J. Hoboken, NJ.

International Programme on Chemical Safety and the Commission of the European Communities (IPCS-CEC). 1993. Diazinon. ICSC card no. 0137 (updated 2001).

Kidd, H. and James, D.R. (eds.). 1991. *The Agrochemicals Handbook*, 3rd ed. Royal Society of Chemistry Information Services, Cambridge, U.K.

Materials Safety Data Sheets (MSDS). 1996. Diazinon. Extension Toxicology Network Pesticide Information Profiles. Oregon State University, USDA/Extension Service, Corvallis, OR.

Dichlorvos (CAS No. 62-73-7)

Molecular formula: $C_4H_7Cl_2O_4P$
IUPAC name: 2,2-Dichlorovinyl dimethyl phosphate
Synonyms and trade names: DDVP; Vapona; Phosvit; Vantaf; Uniphos; Swing Nuvon
Toxicity class: US EPA: I; WHO: Ib

Use and exposure

Dichlorvos occurs as an oily colorless to amber liquid with slight solubility in water. It has an aromatic chemical odor. Dichlorvos is used as an agricultural insecticide for the control of crop pests, such as flies, aphids, spider mites, caterpillars, and thrips, and also pests in store grains, and parasitic worms in animals, in flea collars for dogs. Occupational workers and the general public can be exposed to dichlorvos while working with the manufacture, formulation, and application on agricultural crops, and when used as a fumigant, and as pest strips. Human exposures also occur through food contamination.

Toxicity and health effects

Exposures to dichlorvos through all routes, namely, oral, dermal, and respiratory, cause adverse effects to species of laboratory animals, such as rats, mice, and rabbits. The symptoms of poisoning include perspiration, nausea, salivation, vomiting, diarrhea, drowsiness, fatigue, headache, and in severe cases, tremors, ataxia, convulsions, and coma.

Humans exposed to dichlorvos show many symptoms of poisoning that include, but are not limited to, irritability, confusion, headache, speech difficulties, sweating, blurred vision, drowsiness or insomnia, numbness, tingling sensations, incoordination, tremor, abdominal cramps, difficulty in breathing or respiratory depression and slow heart beat, impaired memory concentration, disorientation, and severe depressions.

Dichlorvos and carcinogenicity

Laboratory animals exposed to dichlorvos by gavage showed an increased incidence of tumors of the pancreas and leukemia in male rats, tumors of the pancreas and mammary gland in female rats, and tumors of the forestomach in both sexes of mice. However, dichlorvos administered in the diet did not cause tumors in animals. There is no information regarding the carcinogenic effects of dichlorvos in humans. The US EPA classified dichlorvos as Group B2, meaning a probable human carcinogen.

Exposure limits

The OSHA has set a PEL of 0.11 ppm (1 mg/m^3) of dichlorvos in workplace air for 10 h TWA. Similarly, the US EPA has set the limit of 0.02 ppm of dichlorvos in food products. The IDLH has been reported as 100 mg/m^3.

References

Agency for Toxic Substances and Disease Registry (ATSDR). 1997. Toxicological Profile for Dichlorvos. US Department of Health and Human Services, Public Health Service, Atlanta, GA (updated 2006).

Budavari, S. (ed.). 1989. *The Merck Index. An Encyclopedia of Chemicals, Drugs, and Biologicals*, 11th ed. Merck, Rahway, NJ.

International Agency for Research on Cancer (IARC). 1979. IARC Monographs on the Evaluation of the Carcinogenic Risk of Chemicals to Humans: Some Halogenated Hydrocarbons. Volume 20. World Health Organization, Lyon, France.

National Institute of Occupational Safety and Health (NIOSH). 1997. Dichlorvos. IDLH Documentation. NIOSH, Atlanta, GA.

Sittig, M. (ed.). 1985. *Handbook of Toxic and Hazardous Chemicals and Carcinogens*, 2nd ed. Noyes, Park Ridge, NJ.

US Department of Health and Human Services (US DHHS). 1993. Registry of Toxic Effects of Chemical Substances (RTECS). National Toxicology Information Program, National Library of Medicine, Bethesda, MD.

US Environmental Protection Agency (US EPA). 1999. Integrated Risk Information System (IRIS) on Dichlorvos. National Center for Environmental Assessment, Office of Research and Development, Washington, DC.

———. 2000. Dichlorvos hazard summary. Technology Transfer Network Air Toxics web site, US EPA (updated 2007).

Dicofol (CAS No. 115-32-2)

Molecular formula: $C_{14}H_9Cl_5O$
IUPAC name: 2,2,2-Trichloro-1,1-bis(4-chlorophenyl)ethanol
Synonyms and trade names: Acarin; Cekudifol; Decofol; Dicaron; Dicomite; Difol; Hilfol; Kelthane; Mitigan
Toxicity class: US EPA: II or III (depending on the formulation); WHO: III

Use and exposure

Pure dicofol is available as a white or gray powder or colorless solid crystals while the technical dicofol is a red-brown or amber viscous liquid with an odor like fresh-cut hay. Dicofol undergoes decomposition on burning or on contact with acids, acid fumes or bases, producing toxic and corrosive fumes including hydrogen chloride. Dicofol is a persistent OCP used as an acaricide and miticide. It is structurally similar to DDT. Dicofol is combustible and incompatible with strong oxidizing agents. It is soluble in most aliphatic and aromatic solvents and most common organic solvents, but practically insoluble in water and hydrolyzes in basic solution. It is used on a wide variety of fruit, vegetables, ornamental and field crops. Dicofol is manufactured from DDT. Dicofol is corrosive to some metals. It is used for foliar applications, mostly on cotton, apples, and citrus crops. Other crops include strawberries, mint, beans, peppers, tomatoes, pecans, walnuts, stonefruit, cucurbits, and non-residential lawns/ornamentals.

Toxicity and health effects

Exposures to dicofol cause adverse health effects and poisoning. Occupational workers suffer harmful effects on inhalation, ingestion and through skin contact. The symptoms of poisoning include, but are not limited to, nausea, dizziness, weakness, and vomiting from ingestion or respiratory exposure, skin irritation or rash from dermal exposure. Dicofol-poisoned occupational workers show skin sensitization, conjunctivitis of the eyes, and pathomorphological changes in the liver, kidneys, and CNS. After exposures to high concentrations of dicofol, workers show nervousness, hyperactivity, headache, nausea, vomiting, unusual sensations, fatigue, convulsions, coma, respiratory failure, and death. However, published literature is limited and more data is required on occupational workers as well as the general population.

Dicofol and carcinogenicity

There is limited evidence that dicofol may cause cancer in laboratory animals, but there is no evidence that it causes cancer in humans. This observation was based on animal test data that showed an increase in the incidence of liver adenomas (benign tumor) and combined liver adenomas and carcinomas in male mice. Dicofol is an experimental carcinogen and a human mutagen. It has been reported that the results of the experiment in mice provide limited evidence that dicofol is carcinogenic to experimental animals. No data on humans were available. The available data are insufficient and inadequate to evaluate the carcinogenicity of dicofol as a human carcinogen. The US EPA classified dicofol as Group C, meaning a possible human carcinogen.

Exposure limits

The ADI for dicofol has been set as 0.002 mg/kg/day.

References

International Agency for Research on Cancer (IARC). 1983. Dicofol. IARC Monographs on the Evaluation of the Carcinogenic Risk of Chemicals to Humans. Vol. 30, 87–101 (updated 1998).
International Programme on Chemical Safety and the Commission of the European Communities (IPCS-CEC). 2003. Dicofol. ICSC card no. 0752. World Health Organization, Geneva, Switzerland.

Material Safety Data Sheet (MSDS). 1996. Dicofol. Extension Toxicology Network (EXTOXNET), Pesticide Information Profiles (PIP). Oregon State University, USDA/Extension Service, Corvallis, OR.

———. 2003. Safety Data for Dicofol, Physical Chemistry at Oxford University, U.K.

Kidd, H. and James, D.R. (eds.). 1991. *The Agrochemicals Handbook*, 3rd ed. Royal Society of Chemistry Information Services, Cambridge, U.K.

Diethylamine (CAS No. 109-89-7)

Molecular formula: $CH_3CH_2NH_2$
Synonyms and trade names: Aminoethane; DEA; Monoethylamine

Use and exposure

Diethylamine is a colorless, strongly alkaline, highly inflammable, liquid. It is completely soluble in water. On burning, diethylamine releases ammonia, carbon monoxide, carbon dioxide, and nitrogen oxides. It is incompatible with several chemical substances, such as strong oxidizers, acids, cellulose nitrate, some metals, and dicyanofuroxan. N-nitrosamines, many of which are known to be potent carcinogens, may be formed when diethylamine comes in contact with nitrous acid, nitrates, or atmospheres with high nitrous oxide concentrations. It has several applications in industries, such as organic synthesis of resins, as rubber accelerator, pharmaceuticals, pesticides, dyes, electroplating operations, and as a polymerization inhibitor.

Toxicity and health effects

Exposures to diethylamine cause adverse health effects. The symptoms of toxicity include irritation of skin, eyes, and mucous membrane. The acute oral LD_{50} and acute dermal LD_{50} in rats and rabbits are 540 and 580 mg/kg, respectively, and the acute inhalation LC_{50} (4 h) in rats is 4000 ppm. The pathomorphological changes caused by diethylamine include lungs, liver, and kidneys, cellular infiltration, bronchopneumonia, parenchymatous degeneration, and nephritis.

Diethylamine and carcinogenicity

The IARC, the US EPA, and the NTP have not listed diethylamine as a human carcinogen and have classified it as Group A4, meaning not classifiable as a human carcinogen.

Storage

Diethylamine should be protected from physical damage. It should be kept stored in a cool, dry, well-ventilated location, away from incompatible chemical substances and away from fire hazard and smoking areas. The containers should be bonded and grounded for transfer to avoid static sparks. Storage and use areas should be no smoking areas.

Precautions

Occupational workers and users should be very careful during the use and chemical management of diethylamine. Workers should wear impervious protective clothing, including boots, gloves, a laboratory coat, apron or coveralls, as appropriate, to prevent skin contact.

The chemical is very hazardous, corrosive, and harmful, and is a very flammable liquid and vapor. Exposures to vapor may cause flash fire. It causes burns and adverse effects to the cardiovascular system. Workers should use chemical safety goggles and a full-face shield to avoid splashing of the chemical substance. An eye-wash fountain and quick-drench facilities in the work area should be maintained by the chemical management unit.

References

Material Safety Data Sheet (MSDS). 2005. Safety data for diethylamine. Physical Chemistry at Oxford University, U.K.
———. 2008. Diethylamine. MSDS no. D3056. Environmental Health & Safety. Mallinckrodt Baker, Phillipsburg, NJ.

Dimethoate (CAS No. 60-51-5)

Molecular formula: $C_5H_{12}NO_3PS_2$
IUPAC name: O,O-Dimethyl S-methylcarbamoylmethyl phosphorodithioate
Synonyms and trade names: Daphene; Devigon; Dicap; Dimet; Rogodan; Rogodial; Rogor; Sevigor; Trimetion
Toxicity class: US EPA: II; WHO: II

Use and exposure

Dimethoate is a gray-white crystalline solid at room temperature. It is sparingly soluble in water, soluble in methanol and cyclohexane, but very soluble in chloroform, benzene. It has been classified by the US EPA under GUP. Dimethoate is used extensively for the control of crop pests, such as mites, aphids, thrips, plant-hoppers, white-flies, and a wide range of other insects that damage, crops, fruits, vegetables, and ornamental plants. Dimethoate is also used for the control of cattle grubs that infect livestock. Thermal decomposition of dimethoate is highly hazardous owing to the release of fumes of dimethylsulfide, methyl mercaptane, carbon monoxide, carbon dioxide, phosphorus pentoxide, and nitrogenoxides.

Toxicity and health effects

Dimethoate is toxic to animals and humans. Occupational exposures cause poisoning with symptoms that include, but are not limited to, sweating, headache, weakness, giddiness, nausea, vomiting, stomach pains, blurred vision, pupillary constriction, slurred speech, and muscle twitching. Workers repeatedly exposed to dimethoate have shown symptoms of numbness, tingling sensations, incoordination, headache, dizziness, tremor, nausea, abdominal cramps, difficulty breathing or respiratory depression, slow heart beat, and speech difficulties. Prolonged exposures cause severe poisoning with adverse effects on the CNS, leading to incoordination, slurred speech, loss of reflexes, weakness, fatigue, involuntary muscle contractions, twitching, tremors of the tongue or eyelids, and eventually paralysis of the body extremities and the respiratory muscles, psychosis, irregular heart beats, unconsciousness, convulsions, coma, and death caused by respiratory failure or cardiac arrest.

Dimethoate and carcinogenicity

Dimethoate is possibly a human mutagen, teratogen, and carcinogen. The evidences of carcinogenicity, even with high-dose, long-term exposure, are inadequate and inconclusive, suggesting that carcinogenic effects of dimethoate in humans are unlikely.

References

International Programme on Chemical Safety (IPCS). 1989. Dimethoate. Environmental Health Criteria No. 90. World Health Organization, Geneva, Switzerland.
Material Safety Data Sheet (MSDS). 1996. Dimethoate. Extension Toxicology Network (EXTOXNET), Pesticide Information Profiles (PIP). Oregon State University, USDA/Extension Service, Corvallis, OR.
Meister, R.T. (ed.). 1992. *Farm Chemicals Handbook '92*. Meister, Willoughby, OH.
Tomlin, C.D.S. (ed.). 1997. *The Pesticide Manual: A World Compendium*, 11th ed. British Crop Protection Council, Farnham, Surrey, U.K.

Dimethyl adipate (CAS No: 627-93-0)

Molecular formula: $C_8H_{14}O_4$
Synonyms: Dimethyl hexanedioate; Hexanedioic acid dimethyl ester

Use and exposure

Dimethyl adipate (DMA) is a colorless and flammable liquid. It is soluble in alcohol and ether, but sparingly soluble in water. It reacts with acids, alkalis, and strong oxidants. Dimethyl adipate is synthesized by the esterification of adipic acid. Dimethyl adipate is part of a dibasic ester (DBE) blend used as a major ingredient in several paint strippers; the DBE blends used in paint stripping formulations contain a major portion (about 90%) of DMA. Dimethyl adipate is used as a chemical intermediate and as a plasticizer in the production of paper and cellulose resins.

Toxicity and health effects

Exposures to dimethyl adipate cause toxicity and adverse health effects in laboratory animals and humans. Workplace exposures to dimethyl adipate by inhalation, ingestion, or skin absorption cause harmful and irritation effects to users.

Dimethyl adipate and carcinogenicity

There is no published information indicating that dimethyl adipate is a human carcinogen.

Precautions

During handling of dimethyl adipate, occupational workers should be careful and use self-contained breathing apparatus, rubber boots, and heavy rubber gloves and avoid prolonged period of exposures. Workers should avoid contact of dimethyl adipate with skin, eyes and nose.

References

Material Safety Data Sheet (MSDS). 2003. Safety data for DBE-6 dibasic ester. Dept. Physical Chemistry, Oxford University, Oxford, U.K.
Material Safety Data Sheet (MSDS). 2008. Dimethyl adipate pure. Dow Chemicals Co., Ltd., Kingsbury, Middlesex, U.K.

Dimethylamine (CAS No. 124-40-3)

Molecular formula: C_2H_7N
Synonyms and trade names: Anhydrous dimethylamine; Dimethylamine (aqueous solution); DMA; N-methylmethanamine

Use and exposure

Dimethylamine is a colorless flammable gas at room temperature. It has a pungent, fishy, or ammonia-like odor at room temperature and is shipped and marketed in compressed liquid form. The air odor threshold concentration for dimethylamine is 0.34 ppm of air. It is very soluble in water, and soluble in alcohol and ether. It is incompatible with oxidizing materials, acrylaldehyde, fluorine, maleic anhydride, chlorine, or mercury. Dimethylamine is a precursor to several industrially important compounds. For instance, it is used in the manufacture of several products, e.g., for the vulcanization process of rubber, as detergent soaps, in leather tanning, in the manufacture of pharmaceuticals, and also for cellulose acetate rayon treatment.

Toxicity and health effects

Exposures to dimethylamine cause adverse health effects. The symptoms include, but are not limited to, severe pain to the eyes, corneal edema/injury, redness, irritation and burning of the skin, chemical burns, and dermatitis. Severe inhalation exposure causes runny nose, coughing, sneezing, burning of the nose and throat, shortness of breath, and delayed pulmonary effects like tracheitis, bronchitis, pulmonary edema, and pneumonitis.

Dimethylamine and carcinogenicity

There are no reports or evidence for the carcinogenicity of dimethylamine as a human carcinogen.

Exposure limits

The exposure limits of dimethylamine are set as 5 ppm by the OSHA and 10 ppm by the ACGIH and the STEL as 15 ppm.

Storage

Dimethylamine should be stored in a cool, dry, well-ventilated area in tightly sealed containers that are labeled in accordance with OSHA's Hazard Communication Standard [29 CFR 1910.1200]. Containers of dimethylamine should be protected from physical damage and ignition sources, and should be stored separately from oxidizing materials, acrylaldehyde, fluorine, maleic anhydride, chlorine, and mercury. Outside or detached storage is preferred. If stored inside, a standard flammable liquids cabinet or room should be used. Ground and bond metal containers and equipment when transferring liquids. Empty containers of dimethylamine should be handled appropriately.

Precautions

During handling of dimethylamine, workers should use proper fume hoods, personal protective clothing and equipment, avoid skin contact, and use gloves, sleeves, and encapsulating suits. Dimethylamine is extremely flammable and may be ignited by heat, sparks, or open flames. Liquid dimethylamine will attack some forms of plastic, rubber, and coatings and is flammable. The vapors of dimethylamine are an explosion and poison hazard. Containers of dimethylamine may explode in the heat of a fire and require proper disposal. Workers should use dimethylamine with adequate ventilation and containers must be kept properly closed.

References

National Library of Medicine (NLM). 1995. Dimethylamine. Hazardous substances data bank. NLM, Bethesda, MD.

Sittig, M. (ed.). 1991. *Handbook of Toxic and Hazardous Chemicals*, 3rd ed. Noyes, Park Ridge, NJ.

US Department of Labor. 1999. Dimethylamine. Occupational Safety & Health Administration, Washington, DC.

1,4 Dioxane (CAS No. 123-91-1)

Molecular formula: $C_4H_8O_2$

Synonyms and trade names: 1,4-Dioxacyclohexane; Diethylene dioxide; 1,4-Diethylene dioxide; -Dioxacyclohexane; Diethylene; Dioxide; Glycol Ethylene Ether 8; Ethylene Glycol Ethylene Ether; Diox; Dioxane

Use and exposure

Technical 1,4-dioxane is a clear liquid with an ether-like odor. It is highly flammable and forms explosive peroxides in storage (rate of formation is increased by heating, evaporation, or exposure to light). 1,4-Dioxane is incompatible with oxidizing agents, oxygen, halogens, reducing agents, and moisture. 1,4-Dioxane is used as a solvent for cellulose acetate, ethyl cellulose, benzyl cellulose, resins, oils, waxes, some dyes, as a solvent for paper, cotton, and textile processing, and for various organic and inorganic compounds and products. It is also used in automotive coolant liquid, in shampoos and other cosmetics as a degreasing agent, and as a component of paint and varnish. Human exposures to 1,4-dioxane have been traced to multiple occupations, breathing of contaminated workplace air, and drinking polluted water. There are many industrial uses of 1,4-dioxane, i.e., as a solvent for celluloses, resins, lacquers, synthetic rubbers, adhesives, sealants, fats, oils, dyes, and protective coatings; as a stabilizer for chlorinated solvents, printing inks; as a wetting and dispersing agent in textile processing, agrochemicals, and pharmaceuticals, and in the preparation and manufacture of detergents.

Toxicity and health effects

Laboratory studies in experimental animals indicate that repeated exposures to large amounts of 1,4-dioxane in drinking water, in air, or on the skin cause convulsions, collapse, and damage to the liver and kidneys in animals. On inhalation of 1,4-dioxane, occupational workers suffered from severe poisoning. The symptoms of poisoning included coughing, irritation of eyes, drowsiness, vertigo, headache, anorexia, stomach pains, nausea, vomiting, irritation of the upper respiratory passages, coma, and death. 1,4-Dioxane also caused hepatic and renal lesions, and demyelination and edema of the brain among workers. (See the literature for more information.)

1,4-Dioxane and carcinogenicity

Laboratory studies with animals exposed to 1,4-dioxane showed induction of nasal cavity and liver carcinomas in rats, liver carcinomas in mice, and gall bladder carcinomas in guinea pigs. The NIOSH observed dioxane as a potential occupational carcinogen. Further, on the basis of weight-of-evidence characterization, the US EPA classified 1,4-dioxane as Group B2, meaning a probable human carcinogen.

Exposure limits

The NIOSH has set the REL as 1 ppm for 30 min. The OSHA has set the PEL as 100 ppm (skin) (TWA), and the IDLH limit as 500 ppm.

Precautions

Workers Should be careful during handling of 1,4-Dioxane and avoid open flames, sparks and smoking. Workers should wear proper protectives since 1,4-Dioxane in known as hazardous, cause damage to eyes, respiratory tract, liver and kidney.

References

International Programme on Chemical Safety and Commission of the European Communities (IPCS-CEC). 1994. 1,4-Dioxane. ICSC Card No. 0041. IPCS-CEC, Geneva, Switzerland (Updated 2008).

Material Safety Data Sheet (MSDS). 2007. Safety data for 1, 4-Dioxane. Dept. Physical and Theoretical Chemistry, Oxford University, Oxford, U.K.

Diphenylamine (CAS No. 122-39-4)

Molecular formula: $(C_6H_5)_2NH$
Synonyms and trade names: N-phenylaniline; Anilinobenzene; Benzenamine,N-phenyl-

Use and exposure

Diphenylamine is a colorless monoclinic leaflet substance. It is used in the manufacture of a variety of substances, i.e., dyestuffs and their intermediates, pesticides, antihelmintic drugs, and as reagents in analytical chemistry laboratories.

Toxicity and health effects

Diphenylamine is highly toxic and is rapidly absorbed by the skin and through inhalation. It has caused anorexia, hypertension, eczema, and bladder symptoms. Experimental animals exposed to diphenylamine demonstrated cystic lesions but failed to demonstrate cancerous growth. Inhalation of diphenylamine dust may cause systemic poisoning. The symptoms of toxicity include, but are not limited to, anoxia, headache, fatigue, anorexia, cyanosis, vomiting, diarrhea, emaciation, hypothermia, bladder irritation, kidney, heart, and liver damage.

Diphenylamine and carcinogenicity

The ACGIH, the IARC, and the NTP have not grouped diphenylamine as a human carcinogen.

Exposure limits

The ACGIH set the TLV of diphenylamine as 10 mg/m³ (TWA) and the NIOSH set the REL as 10 mg/m³.

Storage

Diphenylamine should be protected from physical damage. Storage of diphenylamine outside or a detached area is preferred. Inside storage should be in a standard flammable liquids storage room or cabinet. Diphenylamine should be kept separately from oxidizing materials and incompatible chemical substances. Storage and work areas should be no smoking areas. Diphenylamine should be kept protected from light.

Precautions

Students and occupational workers should be careful during use and handling of diphenylamine. Workers should wear impervious protective clothing, including boots, gloves, a laboratory coat, apron or coveralls, as appropriate, to prevent skin contact. Finely dispersed particles of diphenylamine form explosive mixtures in air. Diphenylamine is very harmful on exposures by swallowing, inhalation, and/or skin absorption. Diphenylamine causes irritation to the skin, eyes, and respiratory tract, and causes blood vascular changes leading to methemoglobinemia.

References

International Programme on Chemical Safety and the Commission of the European Communities (C) IPCS CEC. 1994. Diphenylamine. IPCS-CEC, Geneva, Switzerland.
Material Safety Data Sheet (MSDS). 2009. Diphenylamine. MSDS no. D7728. Environmental Health & Safety (EHS). Mallinckrodt Baker, Phillipsburg, NJ.
National Institute for Occupational Safety and Health (NIOSH). 2006. Diphenylamine. International Chemical Safety Cards (ICSC). ICSC card no. 0466 (updated 2006).

Disulfoton (CAS No. 298-04-4)

Molecular formula: $C_8H_{19}O_2PS_3$
IUPAC name: O, O-diethyl S-2-ethylthioethyl phosphorodithioate
Synonyms and trade names: Disyston; Disystox; Dithiodemeton; Dithiosystox; Solvigram; Solvirex
Toxicity class: US EPA: I; WHO: Ia

Use and exposure

Disulfoton is a dark yellowish oil with an aromatic, sulfurous odor. It is soluble in most organic solvents and fatty oils. Disulfoton is a selective, systemic insecticide and acaricide. It is used for seed coating and for soil application to protect from insect attacks, for the control of sucking insects, aphids, leaf hoppers, thrips, beet-flies, spider mites, and coffee-leaf miners. Disulfoton has been used extensively in pest control on a variety of crops, such as cotton, tobacco, sugar beets, corn, peanuts, wheat, ornamentals, cereal grains, and potatoes. It is grouped by the US EPA under RUP. Human exposures to disulfoton occur through breathing contaminated air, drinking contaminated water, eating contaminated food, and working in industries that manufacture and formulate the pesticide.

Toxicity and health effects

Disulfoton is highly toxic to animals and humans by all routes of exposures, namely, by dermal absorption, through ingestion, and inhalation by the respiratory route. The symptoms of poisoning include blurred vision, fatigue, headache, dizziness, sweating, tearing, and salivation. It inhibits cholinesterase and affects the nervous system function. It does not cause delayed neurotoxicity. Prolonged period of exposures to high concentrations of disulfoton cause harmful effects to the nervous system with symptoms such as narrowing of the pupils, vomiting, diarrhea, drooling, difficulty in breathing, lung edema, tremors, convulsions, coma, and death. Disulfoton causes no mutagenic or teratogenic effects in laboratory animals. There are no reports indicating that disulfoton causes cancer in animals or humans. The DHHS, the IARC, and the US EPA have not classified disulfoton as to its ability to cause cancer.

References

Agency for Toxic Substances and Disease Registry (ATSDR). 1995. Toxicological Profile for Disulfoton. US Department of Health and Human Services, Public Health Service, Atlanta, GA (updated, 2007)

Tomlin, C.D.S. (ed.). 2006. *The Pesticide Manual: A World Compendium of Pesticides*, 14th ed. British Crop Protection Council, Farnham, Surrey, U.K.

US Environmental Protection Agency (US EPA). 2007. Disulfoton. IRIS, US EPA, Washington DC.

World Health Organization (WHO). 1988. Disulfoton. Data Sheets on Pesticides No. 68. WHO/VBC/DS/88.68. WHO, Geneva, Switzerland.

E

- Endosulfan
- Endrin
- Ethion
- Ethyl alcohol
- Ethyl alcohol completely denatured
- Ethylene dibromide
- Ethylene dichloride
- Ethyleneimine
- Ethylene oxide
- Ethyl silicate
- Ethylene sulfide

Endosulfan (CAS No. 115-29-7)

Molecular formula: $C_9H_6Cl_6O_3S$
Synonyms and trade names: Acmaron; Agrosulfan; Endocel; Endomil; Endol; Endosol; Endotox 555; Fezdion; Hexa-Sulfan ;Kendan; Polydan; Thiodan and many more names

Chemical   name: 6,7,8,9,10,10-Hexachloro-1,5,5a,6,9,9a-hexahydro-6,9-methano-2,4,3-benzodioxathiepine-3-oxide
Toxicity class: US EPA: IB; WHO: II

Use and exposure

Endosulfan is a pesticide. It is a cream- to brown-colored solid that may appear in the form of crystals or flakes. It has a smell like turpentine, but does not burn. It does not occur naturally in the environment. Endosulfan is used to control insects on food and non-food crops and also as a wood preservative. Endosulfan is used for the control of ticks and mites, and the control of rice stem borers. It is an RUP, meaning it can only be used by professional applicators. Endosulfan, commonly known by its trade name Thiodan, is an insecticide and was first introduced in the 1950s. Endosulfan enters the air, water, and soil during its manufacture and use. It is often sprayed onto crops and the spray may travel long distances before it lands on crops, soil, or water. On crops, endosulfan usually breaks down in a few weeks, but it sticks to soil particles and may take years to completely break down. Endosulfan does not dissolve easily in water. In surface water, endosulfan attaches to soil particles floating in water or attaches to soil at the bottom. It can build up in the bodies of animals that live in endosulfan-contaminated water. It is also extremely toxic to fish and other aquatic life. Exposures to endosulfan occur among workers and people working in industries involved in making endosulfan or as pesticide applicators and by skin contact with soil containing endosulfan.

Toxicity and health effects

Endosulfan is readily absorbed by the stomach, the lungs, and through the skin, meaning that all routes of exposure can pose a hazard. Exposures to endosulfan cause adverse health effects and poisoning. The symptoms of toxicity include, but are not limited to, hyperactivity, nausea, dizziness, headache, irritability, restlessness, muscular twitching, and convulsions have been observed in adults exposed to high doses. Severe exposures cause poisoning, disturbances of the CNS, and may result in death. Laboratory studies in experimental animals indicated that long-term exposure to endosulfan can also damage the kidneys, testes, and liver and may possibly affect the body's ability to fight infection. More studies are needed to confirm similar situations in humans. Reports have indicated that the most prominent signs of acute exposure are hyperactivity, tremors, decreased respiration, difficulty in breathing, salivation, and convulsions. Long-term neurotoxic effects have been observed after high acute exposure. The National Poison Control Information Center of the Philippines recorded 278 poisonings, including 85 deaths due to endosulfan in 1990. In Colombia, in 1993, at least 60 people were poisoned and one person died as a result of exposure to thiodan.

Endosulfan and carcinogenicity

There is no confirmatory data available for endosulfan as a human carcinogen. Studies in animals are inconclusive. The NHHS, the NTP, the IARC, and the US EPA indicated that endosulfan is not classifiable as a human carcinogen.

Endosulfan and human poisonings

Many cases of accidental and suicidal poisonings related to endosulfan have been reported. In severe cases, death occurred within a few hours of ingestion of endosulfan. The poisoned worker showed signs and symptoms that include, but are not limited to, vomiting, restlessness, irritability, convulsions, pulmonary edema, cyanosis, and EEG changes.

Exposure limits

The OSHA has set the PEL of endosulfan as 0.1 mg/m^3 in workroom air for 8-h (TWA). The ACGIH has set the TLV as 0.1 mg/m^3, and the NIOSH has set the REL as 0.1 mg/m^3. The ADI for endosulfan is 0.006 mg/kg/day. The US FDA has set the limit of endosulfan on dried tea below 24 ppm. The US EPA recommends that levels of endosulfan in rivers, lakes, and streams should not be more than 74 ppb, and in other raw agricultural products should be below 0.1–2 ppm.

References

Agency for Toxic Substances and Disease Registry (ATSDR). 2001. Toxicological Profile for Endosulfan. US Department of Health and Human Services, Public Health Service, Atlanta, GA (updated 2007).

Dikshith, T.S.S. (ed.). 1991. *Toxicology of Pesticides in Animals*. CRC Press, Boca Baton, FL.

Dikshith, T.S.S. and Diwan, P.V. 2003. *Industrial Guide to Chemical and Drug Safety*. Wiley, J. Hoboken, NJ.

Kidd, H. and James, D.R. (eds.). 1991. *The Agrochemicals Handbook*, 3rd ed. Royal Society of Chemistry Information Services, Cambridge, U.K.

Maier-Bode, H. 1968. Properties, effect, residues, and analytics of the insecticide endosulfan. *Residue Rev.* 22: 10–44.

Material Safety Data Sheets (MSDS). 1996. Endosulfan. Extension Toxicology Network (EXTOXNET), Pesticide Information Profiles (PIP). Oregon State University, USDA/Extension Service, Corvallis, OR.

Meister, R.T. 2004. *Crop Protection Handbook*. Meister Media Worldwide, Willoughby, OH.

National Institute of Occupational Health (NIOH). 2003. Final report of the investigation of unusual illnesses allegedly produced by endosulfan exposure in Padre Village of Kasargod district (North Kerala). National Institute of Occupational Health. Indian Council for Medical Research (ICMR), Ahmedabad, India.

Roberts, D.M., Karunarathna, A., Buckley, N.A., Manuweera, G., Sheriff, M.H. and Eddleston, M. 2003. Influence of pesticide regulation on acute poisoning deaths in Sri Lanka. *Bull. World Health Org.* Vol. 81 (11): 789–98.

Smith, A.G. 1991. Chlorinated hydrocarbon insecticides. In: Hayes, W.J. Jr. and Laws, E.R. Jr. (eds.) *Handbook of Pesticide Toxicology*. Academic Press, New York.

Endrin (CAS No.72-20-8)

Molecular formula: $C_{12}H_8Cl_6O$ (stereoisomer of dieldrin)
Chemical name: 1,2,3,4,10,10-Hexachloro-6,7-epoxy-1,4,4A,5,6,7,8,8A-octahydro-1,4-endo,endo-5,8-dimethanonaphthalene

Synonyms:

Endrex; Endricol; EN 57; Endrine; Endrin isomer; Hexadrin; Mendrin.

Use and exposure

Endrin is an organochlorine compound. Endrin appears as a stable, white or beige crystalline solid. It is incompatible with strong acids and strong oxidizers and corrodes some metals.

Endrin decomposes on heating above 245°C producing hydrogen chloride, phosgene. In fact, the US EPA has sharply restricted the availability and uses of many organochlorine groups of pesticides. These include DDT, aldrin, dieldrin, endrin, heptachlor, mirex, chlordecone, and chlordane, while many other organochlorines, however, remain the active ingredients of various home and garden products and some agricultural, structural, and environmental pest control products.

Toxicity and health effects

Exposures to endrin cause toxicity and adverse health effects. Endrin is highly toxic to humans and species of animal. The symptoms of poisoning include, but are not limited to, headache, dizziness, nausea, vomiting, incoordination, tremor, mental confusion, and hyperexcitable state. Exposures to very high concentrations of endrin and in severe cases of poisoning, symptoms include convulsions, seizures, coma, and respiratory depression. In severe organochlorine compound poisoning, symptoms include myoclonic jerking movements, generalized tonic-clonic convulsions, respiratory depression following the seizures, and coma.

Endrin and carcinogenicity

Endrin is identified as an experimental teratogen and is a reproductive hazard to laboratory animals. The epidemiological study carried out on occupationally exposed workers to endrin did not support cancer risk. The IARC classified endrin as Group 3, meaning not classifiable as a human carcinogen.

Exposure limits

The ACGIH set the TLV of endrin as 0.1 m^3. The NIOSH and the OSHA have established a limit of 0.1 mg endrin per cubic meter of air (0.1 mg/m^3) (8 h).

Precautions

Students and occupational workers should be careful during handling of endrin. Workers should not wash away the chemical waste into the sewer, but sweep spilled chemical substance into sealable containers. Workers should use extra personal protection, face shield and eye protection, a chemical protection suit, and self-contained breathing apparatus. Exposures to endrin cause effects on the CNS leading to convulsions and death. The effects may be delayed. Medical observation is indicated.

References

Agency for Toxic Substances and Disease Registry (ATSDR). 1996. Toxicological Profile for Endrin. US Department of Health and Human Services, Public Health Service, Atlanta, GA (updated 2008).

Echobichon, D.J. 1996. Toxic effects of pesticides. In: C. D. Klaassen (ed.) *Casarett & Doull's Toxicology: The Basic Science of Poisons*, 5th ed. McGraw-Hill, New York. pp. 649–55.

International Programme on Chemical Safety and the Commission of the European Communities (IPCS-CEC). 2001. Endrin. ICSC card no. 1023. IPCS, IARC, Geneva, Switzerland.

Material Safety Data Sheets (MSDS). 2005. Safety data for endrin. Physical Chemistry at Oxford University, U.K.

Ethion (CAS No. 563-12-2)

Molecular formula: $C_9H_{22}O_4P_2S_4$
IUPAC name: O,O,O',O'-Tetraethyl S,S'-methylene bis (phosphorodithioate)
Synonyms and trade names: Cethion; Dhanumit; Ethanox; Ethiol; Hylmox; Nialate; Rhodiacide; Tafethion
Toxicity class: US EPA: II; WHO: II

Use and exposure

Technical ethion is an odorless amber liquid. It is very sparingly soluble in water, but soluble in most organic solvents. Ethion undergoes decomposition on heating or on burning and produces toxic and corrosive fumes, including phosphorus oxides and sulfur oxides. It is used for the control of crop pests and household insects. These include, but are not limited to, aphids, mite, sticks, scales, thrips, leaf hoppers, maggots, leaf-feeding insects, foliar-feeding larvae, and house flies. It may be used on a wide variety of food, fiber, and ornamental crops, including greenhouse crops, lawns, and turf. Ethion is often used on citrus and apples. It is mixed with oil and sprayed on dormant trees to kill eggs and scales. Occupational workers and the general public are exposed to ethion while working in industries that manufacture ethion and when eating raw fruits or vegetables that have been treated with ethion. There are no residential uses for ethion.

Toxicity and health effects

Ethion is highly to moderately toxic to animals and humans by the oral route. It causes toxicity and poisoning with symptoms such as nausea, cramps, diarrhea, excessive salivation, severe depression, irritability, confusion, headache, blurred vision, fatigue, tightness in chest, abnormal heart beat and breathing. On repeated exposures and in high concentrations, ethion causes severe symptoms of poisoning. The symptoms of toxicity and poisoning include, but are not limited to, pupillary constriction, muscle cramp, impaired memory and concentration, disorientation, speech difficulties, delayed reaction times, nightmares, sleepwalking, loss of coordination, convulsions, unconsciousness/coma, and death. Studies on laboratory animals and humans have not indicated any evidence of mutagenic, teratogenic, or carcinogenic effects of ethion.

References

Agency for Toxic Substances and Disease Registry (ATSDR). 2000. Toxicological Profile for Ethion. US Department of Health and Human Services, Public Health Service, Atlanta, GA (updated 2007).

International Programme on Chemical Safety and the European Commission (IPCS-EC). 2004. Ethion. ICSC card no. 0888. IPCS, Geneva, Switzerland.

Material Safety Data Sheets (MSDS). 1996. Ethion. Extension Toxicology Network (EXTOXNET), Pesticide Information Profiles (PIP). Oregon State University, USDA/Extension Service, Corvallis, OR.

National Institute of Occupational Safety and Health (NIOSH). 2005. *Pocket Guide to Chemical Hazards*. Ethion, NIOSH Publication No. 2005-149.

Tomlin, C.D.S. (ed.). 2006. *The Pesticide Manual: A World Compendium*, 14th ed. British Crop Protection Council. Blackwell, Cambridge, U.K.

US Environmental Agency (US EPA). 2007. Re-registration, Ethion Red. US EPA, Atlanta, GA.

Ethyl alcohol (CAS No. 64-17-5)

Molecular formula: C_2H_5OH

Synonyms and trade names: Alcohol; Absolute alcohol; Absolute ethanol; Anhydrous alcohol; Alcohol dehydrated; Cologne spirit, Ethanol; Ethyl hydrate; Ethyl hydroxide; Fermentation alcohol; Grain alcohol; Methylcarbinol; Molasses alcohol; Potato alcohol; Sekundasprit; Spirits of wine

Use and exposure

Ethyl alcohol is a colorless flammable liquid with a typical lower alcohol odor and is miscible in water in all proportions. It is stable and hygroscopic. It is incompatible with strong oxidizing agents, peroxides, acids, acid chlorides, acid anhydrides, alkali metals, ammonia, and moisture. Ethyl alcohol forms explosive mixtures with air. Ethyl alcohol is the most common solvent used in aerosols, cosmetics, pharmaceuticals, alcoholic beverages, vinegar production, and in the chemical synthesis of a large variety of products in different industries. For instance, in the manufacture of plastics, lacquers, polishes, plasticizers, perfumes, adhesives, rubber accelerators, explosives, synthetic resins, nitrocellulose, inks, preservatives, and as a fuel.

Toxicity and health effects

Exposures to ethyl alcohol by ingestion cause dizziness, faintness, drowsiness, decreased awareness and responsiveness, euphoria, abdominal discomfort, nausea, vomiting, staggering gait, lack of coordination, and coma. Ethyl alcohol causes no adverse effects with normal skin, but is potentially harmful when absorbed across markedly abraded skin. Repeated inhalation of ethyl alcohol vapors in high concentrations may cause a burning sensation in the throat and nose, stinging and watering in the eyes with symptoms of irritation, dizziness, faintness, drowsiness, nausea, and vomiting. Direct exposures of the eyes to ethyl alcohol may cause mild to moderate conjunctivitis, seen mainly as redness of the conjunctiva.

Prolonged and repeated oral exposures to ethyl alcohol result in the development of progressive liver injury with fibrosis. Chronic exposures or repeated ingestion of ethyl alcohol by pregnant women are known to adversely affect the CNS of the fetus, producing a collection of effects that together constitute fetal alcohol syndrome. The adverse health effects observed in the fetus include mental and physical retardation, disturbances of learning, motor, and language deficiencies, small size head, and behavioral disorders. The target organs that are damaged by prolonged exposures to ethyl alcohol include the eyes, skin, respiratory system, CNS, liver, blood, and reproductive system.

Ethyl alcohol and carcinogenicity

The ACGIH classified ethyl alcohol as A4, meaning not classifiable as a human carcinogen. The IARC, the NIOSH, the NTP, and the OSHA have not listed ethyl alcohol as a human carcinogen.

Exposure limits

For ethyl alcohol, the OSHA set the PEL as 1000 ppm (TWA), similarly the ACGIH set the TLV as 1000 ppm (TWA). The NIOSH set the IDLH as 3300 ppm.

Storage

Ethyl alcohol should be protected from physical damage. It should be kept stored in a cool, dry, well-ventilated location, away from any area where the fire hazard may be acute. Outside or detached storage is preferred. Separate from incompatibles. Containers should be bonded and grounded for transfer to avoid static sparks. The storage and use areas should be free from smoking areas.

Precautions

During handling of ethyl alcohol, workers should use chemical-resistant shields, monogoggles, proper gloves, laboratory coat/apron, and protective equipment as required. Workers and the workplace should have adequate ventilation vent hoods, class b extinguisher. Workers should avoid sources of heat, sparks, or flames. Waste disposal and spill should be collected in suitable containers or absorbed on a suitable absorbent material for subsequent disposal. Waste material should be disposed of in an approved incinerator or in a designated landfill site, in compliance with all federal, provincial, and local government regulations.

References

International Programme on Chemical Safety and the Commission of the European Communities (IPCS-CEC). 1994. Ethanol (anhydrous). International Chemical Safety Cards. ICSC card no. 0044 (updated 2005).

Material Safety Data Sheet (MSDS). 2008. Absolute alcohol. Physical Chemistry at Oxford University, U.K.

Ethyl alcohol completely denatured (CAS No. not applicable to mixtures)

Molecular formula: Not applicable to mixtures
Synonyms and trade names: Alcohol; Spirits of wine; Potato alcohol; CDA Formula 19
Ingredients: **Ethyl alcohol (CAS No. 64-17-5)**: 95%; **Methyl isobutyl ketone (CAS No. 108-10-1)**: 4%; **Kerosene (CAS No. 8008-20-6)**: less than 1%

Use and exposure

Ethyl alcohol completely denatured is a mixture of many chemicals including methyl isobutyl ketone (MIC). Ethyl alcohol completely denatured is incompatible with strong oxidizing agents, perchlorates, aluminum, alkali metals, acetyl chloride, calcium hypochlorite, chlorine oxides, mercuric nitrate, hydrogen peroxide, nitric acid, bromine pentafluoride, chromyl chloride, permanganic acid, uranium hexafluoride, and acetyl bromide. Ethyl alcohol ignites on contact with phosphorous (III) oxide; platinum; disulfuric acid + nitric acid; potassium tert-butoxide + acids. Ethyl alcohol will ignite and then explode on contact with acetic anhydride + sodium hydrogen sulfate. It forms explosive products in reaction with silver nitrate; ammonia + silver; silver (I) oxide + ammonia or hydrazine. MIC is incompatible with aldehydes, nitric acid, perchloric acid, and strong oxidizers. Violent reaction occurs with potassium-tert-butoxide.

Toxicity and health effects

Exposures of ethyl alcohol completely denatured to skin cause dryness with mild irritation and redness. Workers exposed to vapors of ethyl alcohol completely denatured show symptoms of depression, eye and upper respiratory tract irritation, burning sensation, headache, dizziness, tremors, and nausea. Ingestion of ethyl alcohol completely denatured with significant exposures cause dose-related CNS depression. The symptoms of poisoning include headache, tremor, fatigue, hallucinations, distorted perceptions, narcosis, convulsions, coma, respiratory failure, and death. Chronic exposures to high concentrations cause severe damage to the CNS, liver, blood, and reproductive system. After chronic toxicity of the chemical substance, workers show effects as physical dependence, malnutrition, and neurological disorders like amnesia, dementia, and prolonged sleepiness.

Ethyl alcohol completely denatured and carcinogenicity

Chronic ingestion of ethyl alcohol completely denatured produces cancers of the esophagus, the liver, and the kidneys.

Exposure limits

The ACGIH and the OSHA set the exposure levels of MIC as follows: Threshold limit value (TLV) (ACGIH) and PEL (OSHA) as 50 ppm (TWA), and STEL as 75.

Storage

Ethyl alcohol completely denatured should be kept protected from physical damage. It should be kept stored in a cool, dry, well-ventilated location, away from any area where the fire hazard may be acute. Outside or detached storage is preferred. Separate from incompatibles. Containers should be bonded and grounded for transfer to avoid static sparks. The storage and use areas should be free from smoking areas.

Reference

Material Safety Data Sheet (MSDS). 2009. Ethyl alcohol completely denatured. MSDS no. E2012. Environmental Health & Safety (EHS). Mallinckrodt Baker, Phillipsburg, NJ.

Ethylene dibromide (Dibromoethane) (CAS No. 106-93-4)

Molecular formula: $C_2H_4Br_2$
Synonyms and trade names: Bromofume; Ethylene bromide; EDB; Dowfume; Glycol bromide

Use and exposure

Ethylene dibromide is a heavy, colorless liquid with a mild sweet odor, like chloroform. It is also known as 1,2-dibromomethane. Ethylene dibromide is soluble in alcohols, ethers, acetone, benzene, and most organic solvents, and slightly soluble in water. Ethylene dibromide was once of dominant use, although its use has faded as an additive in leaded gasoline. Ethylene dibromide (1,2-dibromoethane) reacts with lead residues to generate volatile lead bromides. It has been used as a pesticide in soil and various crops. Exposure to ethylene dibromide primarily occurs from its past use as an additive to leaded gasoline and as a fumigant. Most of the uses of ethylene dibromide have been stopped in the United States however, it is still used as a fumigant for the treatment of logs for termites and beetles, for the control of moths and beehives, and as a preparation for dyes and waxes. Ethylene dibromide was used as a fumigant to protect against insects, pests, and nematodes in citrus, vegetable, and grain crops, and as a fumigant for turf, particularly on golf courses. In 1984, the US EPA banned its use as a soil and grain fumigant.

Toxicity and health effects

Exposures to ethylene dibromide cause adverse health effects and poisoning. Ethylene dibromide is extremely toxic to humans. Long-term exposures of ethylene dibromide to laboratory animals cause deleterious effects to the liver, kidney, and the testis, irrespective of the route of exposure. Limited data on men occupationally exposed to ethylene dibromide indicate that long-term exposure to ethylene dibromide can impair reproduction by damaging sperm cells in the testicles. Several animal studies indicate that long-term exposure to ethylene dibromide increases the incidences of a variety of tumors in rats and mice in both sexes by all routes of exposure. The symptoms of toxicity include, but are not limited to, redness, inflammation, skin blisters, and ulcers on accidental swallowing/ingestion. Ethylene dibromide has also been reported to cause birth defects in exposed humans.

Ethylene dibromide and carcinogenicity

There is sufficient evidence in experimental animals for the carcinogenicity of ethylene dibromide while evidences in humans are inadequate. The IARC classified ethylene dibromide as Group 2A, meaning a probable human carcinogen. The US EPA classified ethylene dibromide as Group B2, meaning a probable human carcinogen.

Exposure limits

The Acceptable Daily Intake (ADI) for ethylene dibromide (inorganic bromide) is set as 1.0 mg/kg/day.

References

Agency for Toxic Substances and Disease Registry (ATSDR). 1992. Toxicological Profile for 1,2-Dibromoethane. US Department of Health and Human Services, Public Health Service, Atlanta, GA (updated 2007).

National Institute for Occupational Safety and Health (NIOSH). 1997. *Pocket Guide to Chemical Hazards*. US Department of Health and Human Services, Public Health Service, Centers for Disease Control and Prevention, Cincinnati, OH.

Ethylene dichloride (CAS No. 107-06-2)

Molecular formula: $C_2H_4Cl_2$
Synonyms and trade names: 1,2-Dichloroethane; Dichloroetano; Ethane dichloride; Ethylene chloride

Use and exposure

Ethylene dichloride is one of the highest volume chemicals used in the United States. It is a colorless oily liquid with a chloroform-like odor, detectable over the range of 6–40 ppm, with a sweet taste. Ethylene dichloride (1,2-dichloroethane), which has a carbon-carbon single bond, should be distinguished from 1,2-dichloroethene, which has a carbon-carbon double bond. It is a skin irritant. Ethylene dichloride is also used as an extraction solvent, as a solvent for textile cleaning and metal degreasing, in certain adhesives, and as a component in fumigants for upholstery, carpets, and grain. Other miscellaneous applications include paint, varnish, and finish removers, soaps and scouring compounds, wetting and penetrating agents, organic synthesis, ore flotation, and as a dispersant for nylon, rayon, styrene-butadiene rubber, and other plastics.

Toxicity and health effects

Exposures to ethylene dichloride cause CNS depression (dizziness, drowsiness, trembling, unconsciousness), nausea, vomiting, abdominal pain, skin irritation, dermatitis, eye irritation, corneal opacity, blurred vision, headache, sore throat, cough, bronchitis, pulmonary edema (may be delayed), liver, kidney, cardiovascular system damage, cardiac arrhythmia, acute abdominal cramps, diarrhea, internal bleeding (hemorrhagic gastritis and colitis), and respiratory failure. Ethylene dichloride involves the kidneys, liver, eyes, skin, CNS, and the cardiovascular system as the target organs. Ethylene dichloride is known to cause systemic effects and has been identified as a priority pollutant in many countries.

Prolonged periods of inhalation of the vapors of ethylene dichloride irritate the respiratory tract. Symptoms of severe toxicity are CNS effects, liver damage, kidney damage, adrenal gland damage, cyanosis, weak and rapid pulse, and unconsciousness. Death can occur from respiratory and circulatory failure.

The acute effects of ethylene dichloride are similar for all routes of entry: ingestion, inhalation, and skin absorption. Acute exposures result in nausea, vomiting, dizziness, internal bleeding, bluish-purple discoloration of the mucous membranes and skin (cyanosis), rapid but weak pulse, and unconsciousness. Acute exposures can lead to death from respiratory and circulatory failure. Autopsies in such situations have revealed widespread bleeding and damage in most internal organs. Repeated long-term exposures to ethylene dichloride have resulted in neurologic changes, loss of appetite, and other gastrointestinal problems, irritation of the mucous membranes, liver and kidney impairment, and death.

Ethylene dichloride and carcinogenicity

Ethylene dichloride is a probable human carcinogen and has been reported to cause liver damage and mutagenic effects. It has been identified as an experimental transplacental carcinogen and A4, meaning not classifiable as a human carcinogen.

Storage

Ethylene dichloride should be kept protected against physical damage. Store in a cool, dry, well-ventilated location, away from any area where the fire hazard may be acute. Outside or detached storage is preferred. Separate from incompatibles. Containers should be bonded and grounded for transfer to avoid static sparks.

Precautions

Occupational workers should avoid use of ethylene dichloride along with oxidizing agents, strong alkalis, strong caustics, magnesium, sodium, potassium, active amines, ammonia,

TABLE 1

Statistically Significant Tumors in NCI Bioassay of 1,2-Dicholoroethane (Ethylene Dichloride)

Species/Sex	Adverse Effect	Site
Rats/male	Squamous-cell	Forestomach
	Carcinomas	Circulatory system
	Hemangiosarcomas	Subcutaneous tissue
	Fibromas	
Rats/female	Adenocarcinomas	Mammary gland
Mice/female	Adenocarcinomas	Mammary gland
Mice/female	Stromal polyps	Endometrium
	Stromal sarcomas	Endometrium
Mice/male and female	Adenomas	Alveoli and bronchioli

iron, zinc, nitric acid, and aluminium. On decomposition, ethylene dichloride emits toxic fumes of phosgene, hydrogen chloride, acetylene, and vinyl chloride.

Exposure limits

The OSHA set the PEL as 50 ppm (TWA) with a ceiling limit of 100 ppm and a maximum limit of 200 ppm/5 min, the ACGIH set the TLV as 10 ppm (TWA), and the NIOSH set the IDLH as 50 ppm.

References

International Chemical Safety Cards (WHO/IPCS/ILO). 2004. 1,2-Dichloroethane. Occupational Safety & Health Administration (OSHA), Ethylene Dichloride Washington, DC.

Material Safety Data Sheet (MSDS). 2006. Ethylene dichloride. MSDS no. E4700. Mallinckrodt Baker, Phillipsburg, NJ.

National Institute for Occupational Safety and Health (NIOSH). 1978. Ethylene dichloride. NIOSH *Pocket Guide to Chemical Hazards*. NIOSH, Atlanta, GA.

National Institute for Occupational Safety and Health (NIOSH). 1978. Ethylene dichloride (1,2-dichloroethane). Cincinnati, OH (updated 1997).

Ethyleneimine (CAS No. 151-56-4)

Molecular formula: C_2H_5N

Synonyms and trade names: Azacyclopropane; Aziridine; Aziran; Azirane; Dihydro-1H-azirine; Dihydroazirene; Dimethyleneimine; Ethylenimine

Use and exposure

Ethyleneimine is a colorless liquid with an ammonia-like smell or pungent odor. It is highly flammable and reacts with a wide variety of materials. Ethyleneimine is used in polymerization products, as a monomer for polyethyleneimin, and as a comonomer for polymers, e.g., with ethylenediamine. Polymerized ethylenimine is used in paper, textile chemicals, adhesive binders, petroleum, refining chemicals, fuels, lubricants, coating resins, varnishes, lacquers, agricultural chemicals, cosmetics, ion-exchange resins, photographic chemicals, colloid flocculants, and surfactants.

Ethyleneimine readily polymerizes, and it behaves like a secondary amine. Ethyleneimine is highly caustic, attacking materials such as cork, rubber, many plastics, metals, and glass except those without carbonate or borax. It polymerizes explosively on contact with silver, aluminum, or acid. The activity of ethyleneimine is similar to that of nitrogen and sulfur mustards. Ethyleneimine is used as an intermediate in the production of triethylenemelamine. Polymerized ethyleneimine is used in paper, textile chemicals, adhesive binders, petroleum, refining chemicals, fuels, lubricants, coating resins, varnishes, lacquers, agricultural chemicals, cosmetics, ion-exchange resins, photographic chemicals, colloid flocculants, and surfactants.

Toxicity and health effects

Exposures to ethyleneimine cause adverse health effects and poisoning. Ingestion/swallowing, inhalation, or absorption through exposures to skin cause severe irritation, blisters, severe deep burns, and effects of sensitization. Ethyleninime is corrosive to the eye tissue and may cause permanent corneal opacity and conjunctival scarring, severe respiratory tract irritation, and effects of inflammation in workers. Ethyleneimine is a severe blistering agent, causing third degree chemical burns of the skin. The symptoms of toxicity include, but are not limited to, cough, dizziness, headache, labored breathing, nausea, vomiting, tearing and burning of the eyes, sore throat, nasal secretion, bronchitis, shortness of breath, laryngeal edema, pronounced changes of the trachea and bronchi of lungs. Ethyleneimine with its corrosive effects cause injury to the mucous membranes and acute oral exposure may cause scarring of the esophagus in humans. The onset of symptoms and health effects caused by ethyleneimine depends on exposure concentration.

Ethyleneimine and carcinogenicity

There are no reported data on the potential carcinogenicity of ethylenimine in laboratory animals exposed by inhalation. The NIOSH reported ethyleneimine as a potential carcinogen and the IARC classified it as Group 3, meaning not classifiable as a human carcinogen. However, ethyleneimine has not undergone a complete evaluation and determination under the US EPA's IRIS program for evidence of human carcinogenic potential. The US EPA has not classified ethyleneimine as a carcinogen.

Exposure limits

The ACGIH set the TLV of ethyleninime as 0.5 ppm on skin exposure in workplace area. Also there are no reports regarding the measurement of personal exposure to ethyleneimine.

Precautions

During use of ethyleninime, students and occupational workers should wear protective equipment, such as gloves, safety glasses, and should have good ventilation. Ethyleninime should be handled as a carcinogen. Ethyleninime vapor/air mixtures are explosive and pose a risk of fire and explosion on contact with acid(s), oxidants.

References

Material Safety Data Sheet (MSDS). 2005. Safety data for ethyleneimine. Dept. Physical Chemistry, Oxford, Oxford University, U.K.

National Institute for Occupational Safety and Health (NIOSH). 1997. *Pocket Guide to Chemical Hazards*. US Department of Health and Human Services, Public Health Service, Centers for Disease Control and Prevention, Cincinnati, OH.

US Environmental Protection Agency (US EPA). 2002. Ethyleneimine (Aziridine) hazard summary. Technology Transfer Network Air Toxics web site (updated 2007).

Ethylene oxide (CAS No. 75-21-8)

Molecular formula: C_2H_4O
Synonyms and trade names: EO; EtO; Alkene Oxide; Ethylene Oxide 1,2-Epoxyethane; Dihydrooxirene Oxacyclopropane; Dimethylene Oxide; Oxane; Oxirane; Oxidoethane; Epoxyethane

Use and exposure

Ethylene oxide is the simplest cyclic ether. It is a colorless gas or liquid and has a sweet, etheric odor. Ethylene oxide is a flammable, very reactive and explosive chemical substance. On decomposition, vapors of pure ethylene oxide mix with air or inert gases and become highly explosive. Ethylene oxide, is used in large scale as an intermediate in the production of monoethylene glycol, diethylene glycol, triethylene glycol, poly(ethylene) glycols, ethylene glycol ethers, ethanolamine, ethoxylation products of fatty alcohols, fatty amines, alkyl phenols, cellulose, and poly(propylene glycol). It is also used as a fumigant for food and cosmetics, and in hospital sterilization of surgical equipment and heat sensitive materials.

Toxicity and health effects

Ethylene oxide is toxic by inhalation. Symptoms of overexposure include headache, dizziness, lethargy, behavioral disturbances, weakness, cyanosis, loss of sensation in the extremities, reduction in the sense of smell and/or taste, progressing with increasing exposure to convulsions, seizure and coma. Ethylene oxide is also an irritant to the skin and the respiratory tract, and inhaling the vapors may cause the lungs to fill with fluid several hours after exposure. Inhalation may cause dizziness or drowsiness. Liquid contact may cause frostbite, and an allergic skin reaction. After oral exposure (ingestion) to ethylene oxide, laboratory animals show adverse effects in the blood, damage to the liver, kidneys, reproductive effects, miscarriages/spontaneous abortion, and cancer.

Ethylene oxide and carcinogenicity

Prolonged exposures of ethylene oxide to laboratory animals produced incidences of liver cancer. The OSHA classified ethylene oxide as a carcinogenic agent, the ACGIH classified ethylene oxide as A2, meaning a suspected human carcinogen, while the NTP classified ethylene oxide as a known human carcinogen. Similarly, the NIOSH classified ethylene oxide as a potential human carcinogen and the IARC classified ethylene oxide as Group 1, meaning a human carcinogen.

Exposure limits

For ethylene oxide, the OSHA set a limit of 1 ppm over an 8-h workday, 40-h workweek with a STEL (not to exceed 15 min) of 5 ppm; the NIOSH recommends that average workplace air should contain less than 0.1 ppm ethylene oxide averaged over a 10-h workday, 40-h workweek.

Precautions

Ethylene oxide is dangerously explosive under fire condition; it is flammable over an extremely large range of concentrations in air and burns in the absence of oxygen.

References

Agency for Toxic Substances and Disease Registry (ATSDR). 1999. Managing Hazardous Materials Incidents. Volume III – Medical Management Guidelines for Acute Chemical Exposures: Ethylene oxide. US Department of Health and Human Services, Public Health Service, Atlanta, GA (updated 2008).

National Institute for Occupational Safety and Health (NIOSH). 2005. Ethylene oxide. NIOSH *Pocket Guide to Chemical Hazards*. DHHS, NIOSH Publication no. 2005-149 (updated 2008).

Lynch, D.W. and Lewis, T.T. et al. 1997. Carcinogenic and toxicologic effects of inhaled ethylene oxide and propylene oxide in F 344 rats. *Toxicol. Appl. Pharmacol.* 76: 69–84.

McClellan, P.P. 1950. Manufacture and uses of ethylene oxide and ethylene glycol. *Ind. Eng. Chem.* 42: 2402–7.

Ethyl silicate (CAS No. 78-10-4)

Molecular formula: $(CH_2H_5O)_4$, Si

Synonyms and trade names: Ethyl silicate; Condensed silicon ethoxide; Tetraethoxysilane; Tetraethyl orthosilicate; Tetraethyl ester orthosilicic acid; Tetraethyl silicate; Ethyl ortho-silicat; Tetraethoxysilane

Use and exposure

Ethyl silicate is a flammable, colorless liquid with a mild, sweet, alcohol-like odor. Exposure to ethyl silicate can occur through inhalation, ingestion, and eye or skin contact. It is practically insoluble in water, soluble in alcohol, and slightly soluble in benzene. Occupational workers are exposed to ethyl silicate at workplaces associated with the manufacture and transportation of ethyl silicate, during use as a bonding agent for industrial buildings and investment castings, ceramic shells, crucibles, refractory bricks, and other molded objects, as a protective coating for heat- and chemical-resistant paints, lacquers, and films, in the manufacture of protective and preservative coatings for protection from corrosion (primarily as a binder for zinc dust paints), chemicals, heat, scratches, and fire. Workers are also exposed to the chemical substance in the production of silicones; as a chemical intermediate in the preparation of soluble silica; as a gelling agent in organic liquids, as a coating agent inside electric lamp bulbs, in the synthesis of fused quartz, and during industrial use in the textile industry in aqueous emulsions, deluster, and fireproofing; as a component of lubricants; as a mold-release agent; and as a heat-resistant adhesive.

Toxicity and health effects

Exposures to ethyl silicate cause adverse health effects. The symptoms of poisoning include, but are not limited to, irritation of the eye, mucous membrane, respiratory tract, respiratory difficulty, tremor, fatigue, narcosis, nausea, and vomiting. Prolonged periods of skin contact may produce drying, cracking, inflammation, and dermatitis. As observed in laboratory animals, occupational workers exposed to the chemical substance may suffer from liver and kidney damage, CNS depression, and anemia. At concentrations of 3000 ppm, ethyl silicate causes extreme and intolerable irritation of the eyes and mucous membranes; at 1200 ppm, it produces tearing of the eyes; at 700 ppm, it causes mild stinging of the eyes and nose; and at 250 ppm, it produces slight irritation of the eyes and nose.

Exposure limits

The current OSHA PEL for ethyl silicate is 100 ppm as an 8-h TWA concentration. The NIOSH has established an REL for ethyl silicate of 10 ppm as a TWA for up to a 10-h workday and a 40-h workweek. The ACGIH has assigned ethyl silicate a TLV of 10 ppm as a TWA for a normal 8-h workday and a 40-h workweek.

Storage

Ethyl silicate should be kept stored in a cool, dry, well-ventilated area in tightly sealed containers that are labeled in accordance with OSHA's Hazard Communication Standard. Containers of ethyl silicate should be protected from physical damage and should be stored separately from strong oxidizers, water, mineral acids, and alkalis.

Precautions

Occupational workers should avoid contact between ethyl silicate and strong oxidizers, water, mineral acids, and alkalis. Workers should use appropriate personal protective clothing and equipment that must be carefully selected, used, and maintained to be effective in preventing skin contact with ethyl silicate. The selection of the appropriate personal protective equipment (PPE) (e.g., gloves, sleeves, encapsulating suits) should be based on the extent of the worker's potential exposure to ethyl silicate. There are no published reports on the resistance of various materials to permeation by ethyl silicate.

References

Occupational Safety & Health Administration (OSHA). *Occupational Safety and Health Guideline for Ethyl Silicate*. OSHA, Washington, DC.

Hathaway, G.J., Proctor, N.H., Hughes, J.P. and Fischman, M.L. 1991. *Proctor and Hughes' Chemical Hazards of the Workplace*, 3rd ed. Van Nostrand Reinhold, New York.

National Library of Medicine (NLM). 1992. Hazardous substances data bank: Ethyl Silicate. National Library of Medicine, Bethesda, MD.

Sittig, M. 1991. *Handbook of Toxic and Hazardous Chemicals*, 3rd ed. Noyes, Park Ridge, NJ.

F

- Fenamiphos
- Fenitrothion
- Fenoxycarb

- Fenthion
- Formaldehyde
- Fluvalinate
- Fonophos
- Fumaric acid (furfuran)

Fenamiphos (CAS No. 22224-92-6)

Molecular formula: $C_{13}H_{22}NO_3PS$
IUPAC name: Ethyl-4-methylthio-m-tolyl isopropylphosphoramidate
Synonyms and trade names: Nemacur; Namiphos; Phenamiphos; Methaphenamiphos
Toxicity class: US EPA: I; WHO: Ia

Use and exposure

Fenamiphos is a colorless crystal or a tan, waxy solid. It is non-corrosive to metals and breaks down readily in strong acids and bases. Fenamiphos is used as a nematicide and an insecticide. It is sparingly soluble in water, but readily soluble in dichloromethane, isopropanol, and toluene. It is used primarily for the control of nematodes and thrips on citrus, grapes, peanuts, pineapples, tobacco, turf, and ornamentals. There are no residential uses for fenamiphos. The US EPA has grouped fenamiphos as an RUP owing to its high acute toxicity and toxicity to wildlife. Reports have indicated that the manufacture and sale of fenamiphos is to be phased out by 2007–2008.

Toxicity and health effects

Occupational exposures to fenamiphos cause severe toxicity and adverse health effects. It inhibits the activity of the cholinesterase enzyme in humans, leading to over stimulation of the nervous system. The symptoms of poisoning include, but are not restricted to, nausea, dizziness, confusion, impaired memory, disorientation, severe depression, irritability, headache, speech difficulties, delayed reaction times, nightmares, sleepwalking, and drowsiness or insomnia. Very high exposures, such as accidental ingestion and/or major spillage, cause respiratory paralysis and death. Laboratory studies have indicated that the teratogenic effects of fenamiphos occurred only at levels that caused overt maternal toxicity and are likely indirect consequences of this toxicity. Other studies have indicated that fenamiphos is non-mutagenic and also have no evidence suggesting that fenamiphos is carcinogenic to animals and humans.

References

Material Safety Data Sheet (MSDS). 1994. Fenamiphos. MSDS. Extension Toxicology Network (EXTOXNET), Pesticide Information Profiles (PIP). Oregon State University, USDA/Extension Service, Corvallis, OR.

International Programme on Chemical Safety and the Commission of the European Communities (IPCS-CEC). 1999. Fenamiphos. ICSC card no. 0483. IPCS-CEC, Geneva, Switzerland.

Institute of Food and Agricultural Sciences (IFAS). 2006. Fenamiphos Use Facts and Phaseout. University of Florida, Gainesville, FL.

US Environmental Protection Agency (US EPA). 1988. Health Advisories for 50 Pesticides. Fenamiphos. Office of Drinking Water. US EPA, Atlanta, GA.
———. 2002. Pesticides: Re-registration, Fenamiphos Facts. (EPA738-F-02-003) (updated 2007).

Fenitrothion (CAS No. 122-14-5)

Molecular formula: $C_9H_{12}NO_5PS$
IUPAC name: O,O-Dimethyl O-4-nitro-m-tolyl phosphorothioate
Synonyms and trade names: Accothion; Agrothion; Cytel; Dicofen; Fenstan; Folithion; Metathion; Novathion; Nuvano; Pestroy; Sumanone; Sumithion
Toxicity class: US EPA: II; WHO: II

Use and exposure

Pure fenitrothion is a yellowish brown liquid with an unpleasant odor. It is insoluble in water, but readily soluble in common organic solvents, such as acetone, alcohol, benzene, chlorinated hydrocarbons, dichloromethane, 2-propanol, toluene, in ethers, methanol, and xylene. It decomposes explosively. Fenitrothion is a contact insecticide and a selective acaricide of low ovicidal properties. Fenitrothion is effective against a wide range of pests, namely, penetrating, chewing, and sucking insect pests (coffee leaf-miners, locusts, rice stem borers, wheat bugs, flour beetles, grain beetles, grain weevils) on cereals, cotton, orchard fruits, rice, vegetables, and forests. It may also be used as a fly, mosquito, and cockroach residual contact spray for farms and public health programs. Fenitrothion is also effective against household insects and all nuisance insects. WHO confirmed its effectiveness as a vector control agent for malaria. It is extensively used in other countries, including Japan, where parathion has been banned. Occupational workers are exposed to fenitrothion during mixing, loading/transportation, and field applications.

Toxicity and health effects

Fenitrothion is toxic to animals and humans. After prolonged periods of exposures to high concentrations of fenitrothion, occupational workers show poisoning. The symptoms include, but are not limited to, general malaise, fatigue, headache, loss of memory and ability to concentrate, anorexia, nausea, thirst, loss of weight, cramps, muscular weakness, and tremors, and at sufficiently high dosage produce typical cholinergic poisoning. The formulation product, sumithion 50EC, causes delayed neurotoxicity in adult rats, as well as humans.

Fenitrothion and carcinogenicity

Male and female rats fed with fenitrothion for a period of 2 years did not show any dose-related increase in tumor incidence. There are no published reports indicating that fenitrothion is a human carcinogen.

Exposure limits

The ADI of fenitrothion has been set as 0.003 mg/kg and the TLV has not been established.

References

International Programme on Chemical Safety and the Commission of the European Communities (IPCS-CEC). 2001. Fenitrothion. ICSC no. 0622. IPCS-CEC, Geneva, Switzerland.

Gallo, M.A. and Lawryk, N.J. 1991. Organic phosphorus pesticides Chapter 16, pp. 917–1123. In: W. J. Hayes Jr. and E. R. Laws Jr. (eds.) *Handbook of Pesticide Toxicology*. Academic Press, New York.

Hayes, W. Jr. (ed.). 1992. *Pesticides Studied in Man*. Williams & Wilkins, Baltimore, MD.

Kidd, H. and James, D.R. (eds.). 1991. *The Agrochemicals Handbook*, 3rd ed. Royal Society of Chemistry Information Services, Cambridge, U.K.

Material Safety Data Sheet (MSDS). 1995. Fenitrothion. Extension Toxicology Network (EXTOXNET), Pesticide Information Profiles (PIP). Oregon State University, USDA/Extension Service, Corvallis, OR.

Meister, R.T. 1995. *Farm Chemicals Handbook '95*. Meister, Willoughby, OH.

Spencer, E.Y. 1981. *Guide to the Chemicals Used in Crop Protection*, 7th ed. Publication 1093. Research Branch. Agriculture, Canada.

World Health Organization (WHO) International Programme on Chemical Safety (IPCS). 1992. Fenitrothion. Environmental Health Criteria No. 133. WHO, Geneva, Switzerland.

Fenoxycarb (CAS No. 79127-80-3)

Molecular formula: $C_{17}H_{19}NO_4$
IUPAC name: Ethyl 2-(4-phenoxy-phenoxy)-ethyl carbamate
Synonyms and trade names: Comply; Insegar; Logic; Pictyl; Torus; Varikill
Toxicity class: US EPA: IV

Use and exposure

Fenoxycarb is a yellow granular solid, broad-spectrum insect growth regulator and non-neurotoxic carbamate. Fenoxycarb is almost insoluble in water, but soluble and very soluble in hexane, acetone, chloroform, diethyl ether, and methanol. Fenoxycarb is a GUP, meaning the user or the pesticide applicator does not need a license. It is used to control a wide variety of insect pests, as a fire ant bait and for flea and mosquito control. It is also used for the control of cockroaches, butterflies, moths, beetles, scale, and sucking insects, pests of stored products. As a growth regulator, fenoxycarb blocks the ability of an insect to change into the adult stage from the juvenile stage (metamorphosis). Fenoxycarb interferes with larval molting, the periodic shedding or molting of the old exoskeleton and production of a new exoskeleton. Although fenoxycarb is a carbamate insecticide, it exhibits no anticholinesterase activity and is thus considered non-neurotoxic. It is useful for the control of fire ants, fleas, mosquitoes, cockroaches, moths, scale insects, and insects attacking vines, olives, cotton, and fruit. It is also used to control these pests on stored products, and is often formulated as a grit or corncob bait. It mimics the action of the juvenile hormones (JH) on a number of physiological processes, such as molting and reproduction in insects.

Toxicity and health effects

Fenoxycarb is practically non-toxic to mammals after oral ingestion. The oral LD_{50} for rats is greater than 10,000 mg/kg and the dermal LD_{50} for rats is greater than 2,000 mg/kg. Direct application of fenoxycarb on the skin of laboratory rats caused labored breathing and diarrhea in animals. Although fenoxycarb does not irritate the skin, it is an eye irritant. The liver is the primary organ affected by fenoxycarb in long-term animal studies. Prolonged period of oral exposures to rats and dogs with very low doses of fenoxycarb caused no health effects in the animals, while high concentrations caused adverse effects to the liver of rats, mice, and dogs. Reports on the teratogenicity and mutagenicity of fenoxy-carb are not available in the literature.

Fenoxycarb and carcinogenicity

The observations of US EPA peer review and the observations of the Ciba-Geigy Corporation's cancer risk calculations have indicated that fenoxycarb is not a human carcinogen. However, results of animal toxicology lifetime feeding studies indicated that fenoxycarb was carcinogenic at high dose levels to male mice with regard to an increase in lung and Hardarian gland tumors. No significant increase in carcinogenic effects was observed in either female mice or the rat lifetime feeding study. There are no other reports on the carcinogenicity of fenoxycarb to animals and humans.

References

Material Safety Data Sheet (MSDS). 1986. Fenoxycarb. Chemical Data Sheet No. 2/86. PMEP. Cornel University Ithaca, N.Y.
US Environmental Protection Agency (USEPA). 1986. Fenoxycarb. Pesticide Fact Sheet No. 78. USEPA, Washington D.C., p. 3–37.

Fenthion (CAS No. 55-38-9)

Molecular formula: $C_{10}H_{15}O_3PS_2$
IUPAC name: O,O-Dimethyl O-4-methylthio-m-tolyl phosphorothiaote
Synonyms and trade names: Lebaycid; Mercaptophos; Prentox; Queletox; Spotton; Talodex
Toxicity class: US EPA: II; WHO: II

Use and exposure

Pure fenthion is a colorless liquid. Technical fenthion is a yellow or brown oily liquid with a weak garlic odor. It is insoluble or very sparingly soluble in water, but soluble in all organic solvents, alcohols, ethers, esters, halogenated aromatics, and petroleum ethers. It is grouped by the US EPA under RUP and hence requires handling by qualified, certified, and trained workers. Fenthion is used for the control of sucking and biting pests, i.e., fruit flies, stem borers, mosquitoes, and cereal bugs. In mosquitoes, it is toxic to both the adult and immature forms (larvae). The formulations of fenthion include dust, emulsifiable concentrate, granular, liquid concentrate, spray concentrate, ULV, and wettable powder.

Toxicity and health effects

Fenthion is moderately toxic to mammals, and highly toxic to birds. Exposures to fenthion cause poisoning with symptoms among occupational workers as observed with organophosphate pesticide-induced toxicity. These include, but are not limited to, numbness, tingling sensations, incoordination, headache, dizziness, tremor, nausea, abdominal cramps, sweating, blurred vision, difficulty breathing or respiratory depression, and slow heart beat. Very high doses may result in unconsciousness, incontinence, and convulsions or fatality. Reports have indicated that exposures to fenthion cause adverse effects on the central and peripheral nervous systems, and the heart of exposed workers.

Fenthion and carcinogenicity

Reports have indicated that evidences are neither sufficient nor adequate to draw conclusions on the mutagenicity, teratogenicity, and carcinogenicity of fenthion on animals and humans or to classify fenthion as a human carcinogen.

References

Francis, J.I. and Branes, J.M. 1963. Studies on the mammalian toxicity of fenthion. *Bull. World Health. Org.* 29: 205.

Gallo, M.A. and Lawryk, N.J. 1991. Organic phosphorus pesticides. Chapter 16, p. 917–1123. In: Hayes, W.J. Jr. and Laws, E.R. Jr. (eds.) *Handbook of Pesticide Toxicology*. Academic Press, New York.

Kidd, H. and James, D.R. (eds.). 1991. *The Agrochemicals Handbook*, 3rd ed. Royal Society of Chemistry Information Services, Cambridge, U.K.

Material Safety Data Sheet (MSDS). 1996. Fenthion. Extension Toxicology Network (EXTOXNET), Pesticide Information Profiles (PIP). Oregon State University, USDA/Extension Service, Corvallis, OR.

Tomlin, C.D.S. (ed.). 2006. *The Pesticide Manual: A World Compendium*, 14th ed. British Crop Protection Council, Farnham, Surrey, U.K.

US Public Health Service. 1995. Hazardous Substance Data Bank. Washington, DC.

Fluvalinate (CAS No. 102851-06-9)

Chemical name: (RS)-Alpha-cyano-3-phenoxybenzyl N-(2-chloro-a,a,a-trifluoro-p-tolyl)-D-valinate

Synonyms and trade names: Apistan; Klartan; Mavrik; Mavrik Aqua Flow; Spur; Taufluvalinate; Yardex

Use and exposure

Fluvalinate is a viscous, yellow oil in appearance. It is very soluble in organic solvents and aromatic hydrocarbons and insoluble in water. Fluvalinate is a synthetic pyrethroid. It is used as a broad-spectrum insecticide against moths, beetles, and other insect pests on cotton, cereal, grape, potato, fruit tree, vegetable and plantation crops, fleas, and turf and ornamental insects. It is available in emulsifiable concentrates, suspensions, and flowable formulations.

Fluvalinate is a moderately toxic compound in the US EPA toxicity class II. Some formulations may have the capacity to cause corrosion of the eyes. Pesticides containing fluvalinate must bear the signal word DANGER on the product label. Fluvalinate is classified as an RUP because of its high toxicity to fish and aquatic invertebrates.

Toxicity and health effects

Exposures to fluvalinate cause coughing, sneezing, throat irritation, itching, or burning sensations on the arms or face with or without a rash, headache, and nausea. Prolonged period of exposures to pyrethroids cause adverse effects on the CNS, liver, and kidneys. Fluvalinate is slightly toxic to birds.

Fluvalinate and carcinogenicity

Laboratory animals exposed to fluvalinate showed no tumors. No tumors were observed in mice given doses up to 20 mg/kg/day or in rats given doses as high as 2.5 mg/kg/day for over 2 years.

Exposure limits

For fluvalinate, no exposure limits such as ADI, PEL, or TLV have been reported in the literature.

Storage

Fluvalinate should be kept stored only in its original container at a temperature not exceeding 40°C. It should be kept away from food, drink, and animal feedstuffs.

Precautions

Occupational workers should avoid contact of fluvalinate with the skin and eyes and avoid breathing contaminated fumes. Wear suitable protective clothing, nitrile rubber gloves, and safety glasses or a face shield. Fluvalinate is an RUP and should be used, handled, and/or purchased only by certified applicators.

References

Elliot, M. (ed.). 1977. *Pyrethroids*. American Chemical Society, Philadelphia, PA.
US Environmental Protection Agency (US EPA). 1986. Fluvalinate. Extension Toxicology Network (EXTOXNET), Pesticide Information Profiles (PIP). Oregon State University, USDA/Extension Service, Corvallis, OR.
———. 1989. Cypermethrin. Pesticide Fact Sheet No. 199. Office of Pesticides and Toxic Substances, Washington, DC.
World Health Organization (WHO). 1989. Allethrins: Allethrin, D-allethrin, Bioallethrin, S-bioallethrin. Environmental Health Criteria No. 87. International Programme on Chemical Safety. WHO, Geneva, Switzerland.
———. 1990. Permethrin. Environmental Health Criteria No. 94. International Programme on Chemical Safety. WHO, Geneva, Switzerland.

Fonophos (CAS No. 944-22-9)

Molecular formula: $C_{10}H_{15}OPS_2$
IUPAC name: O-ethyl S-phenyl (RS)-ethylphosphonodithioate
Synonyms and trade names: Capfos; Cudgel; Difonate; Dyfonate; Dyphonate; Stauffer
Toxicity class: US EPA: I or II; WHO: Ia

Use and exposure

Fonofos is a highly toxic compound. It is sparingly soluble in water, but soluble in acetone, ethanol, xylene, and kerosene. It has been grouped by the US EPA under RUP and hence requires special handling by qualified, certified, and trained workers. Fonofos was used as a soil insecticide, which resulted in its direct release to the environment. It was primarily used on corn crops, sugar cane, peanuts, tobacco, turf, and some vegetable crops. It controls aphids, corn borer, corn root-worm, corn wire-worm, cutworms, white grubs, and some maggots. Formulations of fonofos include granular, microgranular, emulsifiable concentrate, suspension concentrate, microcapsule suspension, and for seed treatment.

Toxicity and health effects

Fonofos is highly toxic like many other organophosphate pesticides to humans and animals. Exposure to fonofos induces clinical signs of toxicity with typical symptoms of poisoning and cholinesterase inhibition. Accidental ingestion of fonofos by occupational workers results in signs and symptoms of acute intoxication, including muscarinic, nicotinic, and CNS manifestations. Symptoms of fonofos poisoning is a delayed process and occur within a few minutes to 12 h after exposure. Early symptoms of poisoning include, but are not limited to, blurred vision, pinpoint pupils, headache, dizziness, depression, tremors, salivation, diarrhea, and labored breathing. Skin absorption of fonofos causes sweating and muscle twitching, while eye contact leads to severe tearing, pain, and blurred vision. Prolonged exposures to high concentrations of fonofos lead to respiratory failure and death. There are no reports indicating that fonophos is mutagenic, teratogenic, or carcinogenic in animals and humans. Fonofos registration was cancelled in 1999, therefore, it is not considered an environmental contaminant of concern at the present time.

References

Hazardous Substances Data Bank (HSDB). 2004. MTBE. Division of Specialized Information Services, Hazardous National Library of Medicine.

Material Safety Data Sheet (MSDS). 1996. Fonofos. Extension Toxicology Network (EXTOXNET), Pesticide Information Profiles (PIP). Oregon State University, USDA/Extension Service, Corvallis, OR.

Tomlin, C.D.S. (ed.). 1997. *The Pesticide Manual: A World Compendium*, 11th ed. British Crop Protection Council, Farnham, Surrey, U.K.

US Environmental Protection Agency (US EPA). 1984. Fonofos. Pesticide Fact Sheet no. 36. Office of Pesticides and Toxic Substances, Washington, DC.

———. 1987. Draft Health Advisory Summary: Fonofos. Office of Drinking Water, Washington, DC. pp. 5–70.

———. 2006. Fonofos. Health Effects Support (US EPA Document no. EPA-822-R-06-009 August 2006). US EPA, Atlanta, GA.

Formaldehyde (CAS No. 50-0-0)

Molecular formula: CH_2O
Synonyms and trade names: FA; Fannoform; Formalith; Formalin; Formalin 40; Formic aldehyde; Formol; Fyde; Hoch; Karsan; Lysoform; Methyl aldehyde; Methylene glycol; Methylene oxide; Methanal; Morbicid; Oxomethane; Oxymethylene; Paraform; Polyoxymethylene glycols; Superlysoform

Use and exposure

Formaldehyde is an important chemical widely used by industry to manufacture building materials and numerous household products. It is also a by-product of combustion and certain other natural processes. It is present in substantial concentrations both indoors and outdoors.

Formaldehyde is well known as a preservative in medical laboratories, as an embalming fluid, and as a sterilizer. Its primary use is in the production of resins and as a chemical intermediate. Urea formaldehyde (uf) and phenol formaldehyde (pf) resins are used in foam insulations, as adhesives in the production of particle board and plywood, and in the treating of textiles.

Sources of formaldehyde in the home include building materials, smoking, household products, and the use of unvented, fuel-burning appliances, like gas stoves or kerosene space heaters. Formaldehyde, by itself or in combination with other chemicals, serves a number of purposes in manufactured products. It has been reported that the use and production of formaldehyde in 1998 was about 11.3 billion pounds and the international production crossed over 46 billion pounds in 2004.

Toxicity and health effects

Formaldehyde is a colorless, pungent-smelling gas. Exposures to low levels of formaldehyde cause irritation of the eyes, nose, throat, skin, nausea, and difficulty in breathing. Short-term exposure to formaldehyde can be fatal. Long-term exposure to low levels of formaldehyde may cause respiratory difficulty, eczema, and sensitization. Occupational workers with asthma have been found to be more sensitive to the effects of inhaled formaldehyde; in high concentrations, formaldehyde triggers attacks in people with asthma. Also, intake/drinking large amounts of formaldehyde causes severe pain, vomiting, and coma leading to death. Acute and chronic health effects of formaldehyde vary depending on the individual. The typical threshold for development of acute symptoms due to inhaled formaldehyde is 800 ppb; however, sensitive individuals have reported symptoms at formaldehyde levels around 100 ppb.

Formaldehyde and carcinogenicity

Formaldehyde is common to the chemical industry. The 11th report on carcinogens classifies it as "reasonably anticipated to be a human carcinogen." According to the IARC, formaldehyde is classified as a human carcinogen and has been linked to nasal and lung cancer, with possible links to brain cancer and leukemia. The DHSS has determined that formaldehyde may reasonably be anticipated to be a carcinogen.

Exposure limits

The US EPA recommends that an adult should not drink water containing more than 1 mg/L of formaldehyde for a lifetime exposure, and a child should not drink water containing more than 10 mg/L for 1 day or 5 mg/L for 10 days. The OSHA has set a PEL for formaldehyde as 0.75 ppm for an 8-h workday, 40-h workweek. The NIOSH recommends an exposure limit of 0.016 ppm.

Reference

Agency for Toxic Substances and Disease Registry (ATSDR). 1999. Toxicological Profile for Formaldehyde. Department of Health and Human Services, Public Health Service, Atlanta, GA (updated 2008).

G

- Glyphosate
- Guthion

Glyphosate (CAS No. 1071-83-6)

Molecular formula: $C_3H_8NO_5P$
Chemical name: N-(phosphonomethyl) glycine
Synonyms and trade names: Roundup

Use and exposure

Glyphosate is a broad-spectrum, non-selective systemic herbicide. It is a colorless crystal at room temperature and is soluble in acetone, ethanol, xylene, and water. Glyphosate is used for the control of annual and perennial plants, including grasses, sedges, broad-leaved weeds, and woody plants. It can be used on non-cropland as well as on many varieties of crops. Glyphosate itself is an acid, but it is commonly used in salt form, most commonly isopropylamine salt. It may also be available in acidic or trimethylsulfonium salt forms. It is generally distributed as water-soluble concentrates and powders. Glyphosate is a GUP.

Toxicity and health effects

Glyphosate is practically non-toxic if ingested, with a reported acute oral LD_{50} of 5600 mg/kg in the rat. The toxicities of the technical acid (glyphosate) and the formulated product (Roundup) are nearly the same. Laboratory animals, such as rats, dogs, mice, and rabbits, exposed to glyphosate for 2 years did not indicate any kind of adverse health effects.

Glyphosate and carcinogenicity

Studies with rats indicated that glyphosate is not carcinogenic. Rats given oral doses of up to 400 mg/kg/day did not show any signs of cancer, nor did dogs given oral doses of up to 500 mg/kg/day, or mice fed glyphosate at doses of up to 4500 mg/kg/day. It appears that glyphosate is not carcinogenic.

Exposure limits

The exposure limit (ADI) of glyphosate is set as 0.3 mg/kg/day.

References

Kidd, H. and James, D.R. (eds.). 1991. *The Agrochemicals Handbook*, 3rd ed. Royal Society of Chemistry Information Services, Cambridge, U.K.

Material Safety Data Sheet (MSDS). 1996. Glyphosate. Extension Toxicology Network (EXTOXNET), Pesticide Information Profiles. Oregon State University, USDA/Extension Service, Corvallis, OR.

Monsanto Company. 1985. Toxicology of Glyphosate and Roundup Herbicide. St. Louis, MO.

Guthion (CAS No. 86-50-0)

Molecular formula: $C_{10}H_{12}N_3O_3PS_2$ (For details, refer: Azinphos methyl)

Reference

Agency for Toxic Substances and Disease Registry (ATSDR). 2008. Toxicological Profile for Guthion. US Department of Health and Human Services, Public Health Service, Atlanta, GA.

H

- Heptachlor and heptachlor epoxide
- n-Hexane
- Hexanone
- Holmium
- Hydrazines
- Hydrochloric acid
- Hydrofluoric acid
- Hydroquinone

Heptachlor and heptachlor epoxide

Heptachlor (CAS No. 76-44-8)

Molecular formula: $C_{10}H_5Cl_7$
IUPAC name: 1,4,5,6,7,8,8-Heptachloro-3a 4,7,7a-tetrahydro-4,7-methano-l *H*-indene
Synonyms and trade names: Aahepta; Agroceres; Basaklor; Drinox; Heptachlorane; Heptagran; Heptagranox; Heptamak; Heptamul; Heptasol

Heptachlor epoxide (CAS No. 1024-57-3)

Molecular formula: $C_{10}H_5Cl_70$
IUPAC name: 1,4,5,6,7,8,8-Heptachloro-3a,4,7,7a-tetrahydro-4,7-methanoindene
Synonyms and trade names: Biarbinex; Cupincida; Drinox; Fennotox; Heptamul; Heptox; Termide; Epoxyheptachlor; Hepox; Heptepoxide
Toxicity class: US EPA: II; WHO: II

Use and exposure

Heptachlor is an organochlorine cyclodiene insecticide isolated from technical chlordane. It is available in the form of white crystals or a tan-colored waxy solid with a characteristic camphor-like or cedar-like odor. It is sparingly soluble or insoluble in water, but fairly soluble in acetone, benzene, ethanol, xylene, and other organic solvents. It is used for the control of termites, ants, and soil insects in cultivated and non-cultivated soils. Heptachlor epoxide is formed in nature when heptachlor is released into the environment and mixes with oxygen. Heptachlor epoxide remains in the soil for long periods of time. Heptachlor and heptachlor epoxide may also be present at numerous hazardous waste sites. Although the use of heptachlor is restricted, exposure to the general population does occur through the ingestion of contaminated food.

Toxicity and health effects

Exposures to heptachlor epoxide cause adverse health effects to animals and humans. Exposure to heptachlor is toxic by mouth, by skin contact, as well as by inhalation of dust from powder concentrates. Heptachlor acts as a CNS stimulant. Prolonged period of exposures to high concentrations of heptachlor cause headache, dizziness, nausea, vomiting, weakness, irritability, salivation, lethargy, respiration distress, muscle tremors, convulsions, and paralysis. Severe cases of poisoning lead to respiratory failure and death. In fact, seizures and cortical excitability are the prime CNS symptoms following acute heptachlor exposure. The photoisomer of heptachlor (photoheptachlor) and the major metabolite of heptachlor, namely, heptachlor epoxide are more toxic than the parent compound. Heptachlor induces tremors, convulsions, paralysis, and hypothermia in rats and young calves. The poisoned animals showed muscle spasms in the head and neck region, convulsive seizures, elevated body temperatures, and engorged brain blood vessels.

Humans exposed to heptachlor in the home during termite control operations showed signs of neurotoxicity, i.e., irritability, salivation, lethargy, dizziness, labored respiration, muscle tremors, convulsions, and death due to respiratory failure. Heptachlor interfered with nerve transmission, caused hyperexcitation of the CNS, lethargy, incoordination, tremors, convulsions, stomach cramps or pain, leading to coma and death.

Heptachlor and carcinogenicity

Lifetime exposure to heptachlor resulted in liver tumors in animals. The IARC and the US EPA classified heptachlor as a possible human carcinogen. The US EPA reported that heptachlor epoxide is a possible human carcinogen. The IARC reported that while sufficient evidences are found in experimental animals for the carcinogenicity of heptachlor, the evidences are inadequate to consider it as a human carcinogen. Studies on heptachlor carcinogenicity and human exposure are inconclusive. In the overall evaluation, heptachlor has been classified as Group 2B, meaning possibly carcinogenic to humans. However, it is a confirmed animal carcinogen with unknown relevance to humans and the US EPA has grouped it as 3B, meaning not a human carcinogen.

Precautions

Heptachlor decomposes on heating above 160°C, producing toxic fumes including hydrogen chloride. It reacts with strong oxidants and attacks metal. During use and handling of liquid and powder formulations of heptachlor, occupational workers should wear protective neoprene or PVC gloves, cotton overalls, rubber boots, and a face shield or dust mask. Heptachlor should be kept stored in locked buildings.

Exposure limits

For heptachlor, the OSHA set the PEL as 0.5 mg/m^3 (skin) and the ACGIH set the TLV as 0.05 mg/m^3 TWA (skin). The ADI for heptachlor has been set as 0.0001 $mg/kg/day$.

References

Agency for Toxic Substances and Disease Registry (ATSDR). 2005. Toxicological Profile for Heptachlor and Heptachlor Epoxide. US Department of Health and Human Services, Public Health Service, Atlanta, GA (Draft for Public Comment) (updated 2007).

International Programme on Chemical Safety (IPCS/CEC). 1999. Heptachlor (Technical Product). International Chemical Safety Card No. 0743.

Lu, F.C.A. 1995. A review of the acceptable daily intakes of pesticides assessed by the World Health Organization. *Regul. Toxicol. Pharmacol.* 21: 351–64.

Material Safety Data Sheet (MSDS). 1996. Heptachlor. Extension Toxicology Network (EXTOXNET), Pesticide Information Profiles (PIP). Oregon State University, USDA/Extension Service, Corvallis, OR.

Tomlin, C.D.S. (ed.). 1994. *The Pesticide Manual: A World Compendium*, 10th ed. British Crop Protection Council (BCPC), Thornton Weath, U.K.

US Environmental Protection Agency (US EPA). 2000. Heptachlor. Technology Transfer Network Air Toxics web site (updated 2007).

World Health Organization (WHO). 2006. International Chemical Assessment Document 70, Heptachlor. WHO, Geneva, Switzerland.

n-Heptane (CAS No. 142-82-5)

Molecular formula: $CH_3(CH_2)_5CH_3$

Use and exposure

n-Heptane is a flammable liquid, present in crude oil and widely used in the automobile industry. For example, as a solvent, as a gasoline knock testing standard, as automotive starter fluid, and paraffinic naphtha. n-Heptane causes adverse health effects in occupational workers, such as CNS depression, skin irritation, and pain. Other compounds such as n-octane ($CH_3(CH_2)_6CH_3$), n-nonane ($CH_3(CH_2)_7CH_3$), and n-decane ($CH_3(CH_2)_8CH_3$) have different industrial applications. Occupational workers exposed to these compounds also show adverse health effects. In principle, management of these aliphatic compounds requires proper handling and disposal to avoid health problems and to maintain chemical safety standards for safety to workers and the living environment.

Reference

Cheremisinoff, N.A. 2003, *Industrial solvents handbook*, 2nd ed., New York. Marcel Dekker. Inc.

n-Hexane (CAS No. 110-54-3)

Molecular formula: $CH_3(CH_2)_4CH_3$

Use and exposure

n-Hexane is a highly flammable liquid, usually isolated from crude oil, and has extensive industrial applications as a solvent in adhesive bandage factories and other industries. It is highly toxic, triggering several adverse health effects, i.e., nausea, skin irritation, dizziness, numbness of limbs, CNS depression, vertigo, and respiratory tract irritation to animals and humans. Occupational exposure of industrial workers has demonstrated motor polyneuropathy. Workers associated with long-term glue sniffing showed adverse effects in the form of degeneration of axons and nerve terminals.

Reference

Cheremisinoff, N.A. 2003, *Industrial solvents handbook*, 2nd ed., New York. Marcel Dekker. Inc.

2-Hexanone (CAS No. 591-78-6)

Molecular formula: $C_6H_{12}O$
Synonym and trade names: Methyl n-butyl ketone; MBK

Use and exposure

2-Hexanone is a colorless to pale yellow liquid with a sharp odor. It dissolves very easily in water and is miscible in ethanol, methanol, and benzene. 2-Hexanone evaporates easily into the air as a vapor. It is stable, flammable, and incompatible with oxidizing agents, strong bases, and reducing agents. 2-Hexanone is a waste product of wood pulping, coal gasification, and oil shale operations. Formerly, 2-hexanone was in use as a paint and paint thinner with other chemical substances, and to dissolve oils and waxes.and is used as a solvent and organic synthesis intermediates. However, the industrial uses of 2-hexanone is now very much restricted.

Toxicity and health effects

Exposures to 2-hexanone in industrial workplaces cause adverse health effects to occupational workers. The symptoms of toxicity and poisoning include, but are not restricted to, weakness, numbness, and tingling of the hands and feet. Similar effects were seen in animals that ate or breathed high levels of 2-hexanone; these effects included weakness, clumsiness, and paralysis.

2-Hexanone and carcinogenicity

There is no published information about the carcinogenicity of 2-hexanone.

Exposure limits

The OSHA and the ACGIH have set a limit of 5 ppm of 2-hexanone in workplace air (TWA), while the NIOSH set 1 ppm—a much lower limit in workplace air.

Storage

2-Hexanone should be kept stored in a tightly closed container, in a cool, dry place, away from sources of ignition.

Precautions

Occupational workers should be careful when using and handling 2-hexanone. Workers should wash thoroughly after handling the chemical substance. Workers should avoid contact of 2-hexanone with the eyes, skin, and clothing and also avoid contact with heat, sparks, and flame.

Reference

Agency for Toxic Substances and Disease Registry (ATSDR). 1992. Toxicological Profile for 2-Hexanone. US Department of Health and Human Services, Public Health Service, Atlanta, GA (updated 2008).

Hydrazines

The term "hydrazines" is a generic name that includes a group of three structurally related chemicals: (i) hydrazine, (ii) l,l-dimethylhydrazine, and (iii) 1,2-dimethylhydrazine. Exposure to relatively high concentrations of hydrazines in air can be lethal and suggest that hydrazine may be more toxic than l,l-dimethylhydrazine.

1,1-Dimethylhydrazine (CAS No. 57-14-7)

Use and exposure

1,1-Dimethylhydrazine is a colorless, flammable, hygroscopic liquid that gradually turns yellow on contact with air and is miscible with water. 1,1-Dimethylhydrazine is primarily used as a high-energy fuel in military applications, as a rocket propellant and fuel for thrusters, and small electrical power-generating units. 1,1-Dimethylhydrazine is also used in the manufacture of a plant growth regulator, in chemical synthesis, in photographic chemicals, as a stabilizer for fuel additives, and as an absorbent for acid gases. Exposure to 1,1-dimethylhydrazine usually occurs at the workplace during use and handling of the chemical substance.

No information is available on the carcinogenic effects of 1,1-dimethylhydrazine in humans. Carcinogenic effects were observed in mice and rats exposed to 1,1-dimethylhydrazine by inhalation, but the carcinogenicity could not be definitively attributed to 1,1-dimethylhydrazine because of the presence of contaminants in the study. The US EPA has not classified 1,1-dimethylhydrazine for potential carcinogenicity, while the IARC has classified 1,1-dimethylhydrazine as Group 2B, meaning a possible human carcinogen.

Reference

US Environmental Protection Agency (US EPA). 1999. Integrated Risk Information System (IRIS) on 1,1-Dimethylhydrazine. National Center for Environmental Assessment, Office of Research and Development, Washington, DC.

1,2-Dimethylhydrazine (CAS No. 540-73-8)

Molecular formula: $C_2H_8N_2$
Synonyms and trade names: Dimethylhydrazine; N,N'-Dimethylhydrazine

Use and exposure

1,2-Dimethylhydrazine is a colorless or yellow hygroscopic liquid and releases fumes in air. 1,2-Dimethylhydrazine is stable and incompatible with oxidizing agents, water, and moisture.

Hydrazines include 1,1-dimethylhydrazine and 1,2-dimethylhydrazine, collectively. Hydrazines in pure form are clear, colorless liquids that quickly evaporate in air. Hydrazines smell somewhat like ammonia. Hydrazines are highly reactive and easily catch fire. 1,1-Dimethylhydrazine, and 1,2-dimethylhydrazine are somewhat similar in chemical structure and reactivity. However, there are some clear differences in their production, uses, and adverse health effects. There are many other hydrazine compounds. Hydrazines are manufactured from chemicals such as ammonia, dimethylamine, hydrogen peroxide, and sodium hypochlorite. A small amount of hydrazine occurs naturally in some plants. Large amounts of hydrazines are in different countries. 1,2-Dimethylhydrazine is a research chemical and the quantities produced are likely to be much less.

Hydrazine has been used as a fuel source for rocket propellant and spacecraft, including the space shuttle. Hydrazine is also used to treat boiler water to reduce corrosion and to reduce other chemical substances. Hydrazine is used as a medicine and in the manufacture of other medicines, farm chemicals, and plastic foams. 1,1-Dimethylhydrazine has been used as a rocket propellant and in the manufacture of other chemical substances, while the literature indicate that 1,2-dimethylhydrazine has no commercial use but is used in researches to study colon cancer in experimental animals.

Toxicity and health effects

Inhalation exposure to hydrazines, cause adverse health effects and poisoning to animals and humans. The symptoms of toxicity and poisoning include, but are not limited to, injury to the lungs, liver, kidney, vomiting, uncontrolled shaking, lethargy (sluggishness), coma, and neuritis (an inflammation of the nerves), convulsions, tremors, seizures, and the CNS. Effects on the nervous system have also been seen in animals exposed to hydrazine and 1,1-dimethylhydrazine, but not to 1,2-dimethylhydrazine.

Hydrazines and carcinogenicity

Studies in laboratory animals indicate that exposures to 1,2-dimethyl-hydrazine cause colon cancer. The observations of different regulatory agencies suggesting that hydrazine is a carcinogenic chemical substance are comparable; for instance, (i) the DHHS report that hydrazine and 1,1-dimethylhydrazine may reasonably be anticipated as a carcinogen; (ii) the IARC classify hydrazine, 1,1-dimethylhydrazine, and 1,2-dimethylhydrazine as possibly carcinogenic to humans; (iii) the US EPA report that hydrazine, 1,1-dimethylhydrazine, and 1,2-dimethylhydrazine are probable human carcinogens; and (iv) the ACGIH lists hydrazine and 1,1-dimethylhydrazine as suspected human carcinogens.

Exposure limits

The US EPA report that hydrazine and 1,1-dimethylhydrazine are hazardous air pollutants. (i) The OSHA limits the amount of hydrazine and 1,1-dimethylhydrazine to 0.1 and 0.5 ppm, respectively, in workplace air (8-h workday); (ii) the NIOSH recommends that the levels of hydrazine and 1,1-dimethylhydrazine in workplace air not exceed 0.03 and 0.06 ppm, respectively (2-h); and (iii) the US FDA has ruled that hydrazine and should not be added to water.

Reference

Agency for Toxic Substances and Disease Registry (ATSDR). 1997. Toxicological Profile for Hydrazines. US Department of Health and Human Services, Public Health Service, Atlanta, GA (updated 2008).

Hydrochloric acid (CAS No. 7647-01-0)

Molecular formula: **HCl** in **H$_2$O** (in water)

Use and exposure

Hydrochloric acid, or hydrogen chloride, is either a colorless liquid with a pungent odor, or a colorless to slightly yellow gas that can be shipped as a liquefied compressed gas. The acid is used in the production of fertilizers, dyes, dyestuffs, artificial silk, and paint pigments, and in refining edible oils and fats. Hydrochloric acid is also used in electroplating, leather tanning, ore refining, soap refining, petroleum extraction, and pickling of metals, and is used in the photographic, textile, and rubber industries. In addition, hydrochloric acid is used as an antiseptic in toilet bowls against animal pathogenic bacteria, and in food processing as a starch modifier.

Toxicity and health effects

Exposures to hydrochloric acid cause severe health effects and corrosive reactions. Concentrated hydrochloric acid (fuming hydrochloric acid) forms acidic mists. Both the mist and the solution have a corrosive effect on human tissue, with the potential to damage the respiratory organs, eyes, skin, and intestines. Inhalation of vapors can cause coughing, choking, inflammation of the nose, throat, and upper respiratory tract, and in severe cases, pulmonary edema, circulatory failure, and death. Accidental ingestion and/or swallowing of hydrochloric acid at workplaces causes immediate pain and burns of the mouth, throat, esophagus, and gastrointestinal tract. It also causes nausea, vomiting, and diarrhea, and in severe cases, death. Any kind of contact of the skin surfaces to hydrochloric acid causes redness, pain, and severe skin burns. Concentrated solutions of hydrochloric acid cause deep ulcers and discolor the skin. Vapors of hydrochloric acid cause irritating effects to the eyes and eye damage, leading to severe burns and permanent eye damage. Long-term exposures to concentrated vapors of hydrochloric acid cause erosion of the teeth. Occupational workers and persons with pre-existing skin disorders or eye disease are more susceptible to the effects of hydrochloric acid.

Precautions

Users and occupational workers should use personal protective equipments (PPE) such as rubber or PVC gloves, protective eye goggles, and chemical-resistant clothing and shoes to minimize risks when handling hydrochloric acid. The hazards of solutions of hydrochloric acid depend on the concentration. Users/students/occupational workers should note: Hydrochloric acid—Caution/POISON! DANGER! CORROSIVE. LIQUID AND MIST CAUSE SEVERE BURNS TO ALL BODY TISSUE. MAY BE FATAL IF SWALLOWED OR INHALED.

References

Material Safety Data Sheet (MSDS). 2004. Safety data for Hydrochloric acid (concentrated). Department of Physical Chemistry, Oxford University, Oxford, U.K.

US Environmental Protection Agency (USEPA). 2000. Hydrochloric acid (Hydrogen chloride). Technology Transfer Network. USEPA, Washington D.C., (Updated 2007).

Hydrocyanic acid (CAS no. 74-90-8)

Molecular formula: **HCN**

Synonyms and trade names: Formonitrile; Hydrogen cyanide; Hydrocyanic acid (liquefied); Prussic acid

Use and exposure

Hydrocyanic acid (hydrogen cyanide) is a clear colorless liquid with a faint odor of bitter almonds. It evaporates easily (or boils) at room temperature and the vapors are slightly lighter than air. It is soluble in water. It is reactive and incompatible with amines, oxidizers, acids, sodium hydroxide, calcium hydroxide, sodium carbonate, caustics, and ammonia.

Hydrogen cyanide is manufactured by the oxidation of ammonia–methane mixtures under controlled conditions and by the catalytic decomposition of formamide. It may be generated by treating cyanide salts with acid, and it is a combustion by-product of nitrogen-containing materials such as wool, silk, and plastics. It is also produced by enzymatic hydrolysis of nitriles and related chemicals. Hydrogen cyanide gas is a by-product of coke-oven and blast furnace operations. Industrial applications of hydrogen cyanide are many. For instance, in fumigation, electroplating, mining, metallurgical, firefighting, steel manufacturing, and metal cleaning industries, to producing synthetic fibers, plastics, dyes, pesticides, and also as an intermediate in chemical syntheses.

Toxicity and health effects

Exposures to hydrogen cyanide cause adverse health effects to animals and humans. Hydrogen cyanide is readily absorbed from the lungs and the symptoms of poisoning begin within seconds to minutes. The symptoms of toxicity and poisoning include, but are not restricted to, asphyxia, lassitude or weakness, exhaustion, headache, confusion, nausea, vomiting, increased rate and depth of respiration, or respiration slow and gasping, thyroid and blood changes. Inhalation of hydrogen cyanide causes headache, dizziness, confusion, nausea, shortness of breath, convulsions, vomiting, weakness, anxiety, irregular heart beat, tightness in the chest, and unconsciousness, and these effects may be delayed. The target organs of induced toxicity and poisoning include the CNS, cardiovascular system, thyroid, and blood.

Hydrogen cyanide and carcinogenicity

Hydrogen cyanide has not been classified for carcinogenic effects or as a human carcinogen. The US EPA classified cyanide as a Group D chemical substance, meaning not classifiable as a human carcinogen.

Exposure limits

The OSHA set the PEL for hydrocyanic acid as 10 ppm (skin) and the ACGIH set the TLV as 0.2 ppm (TWA).

Storage

Hydrogen cyanide should be stored in a cool, dry, well-ventilated area in tightly sealed containers and with the correct label. Containers of hydrogen cyanide should be protected from physical damage and should be stored separately from amines and oxidizers, such as perchlorates, peroxides, permanganates, chlorates, and nitrates. It should be kept separated from strong acids, such as hydrochloric, sulfuric, and nitric acids, away from sodium hydroxide, calcium hydroxide, sodium carbonate, water, ammonia, acetaldehyde, and caustics.

Precautions

Hydrocyanic acid (hydrogen cyanide) is a deadly poison by all routes of exposure (absorption of liquid through skin, inhalation of vapors). Breathing in a small amount of the gas or swallowing a very small amount may be fatal. Average fatal dose is 50–60 mg. A few minutes of exposure to 300 ppm may result in death. Exposure to 150 ppm half an hour leads to fatal injury. Prolonged exposure of closed containers to heat may cause violent rupture and rocketing. Occupational workers and users should wear appropriate personal protective clothing to prevent skin contact and appropriate eye protection to prevent eye contact (PPE). On contamination with the chemical substance, workers should immediately wash the skin.

References

Sax, N.I. and Lewis, R.J. (eds). 1989. *Dangerous Properties of Industrial Materials*, 7th ed. Van Nostrand Reinhold, New York.
Sittig, M. (ed.) 1991. *Handbook of Toxic and Hazardous Chemicals*, 3rd ed. Noyes, Park Ridge, NJ.
National Institute for Occupational Safety and Health (NiOSH). 2005. Hydrogen cyanide. NiOSH pub. no. 2005-149. NiOSH, Atlanta, GA (updated 2009).

Hydroquinone (CAS No. 123-31-9)

Molecular formula: $C_6H_4(OH)_2$
Synonyms: (Quinol, hydroquinol; p-diphenol; dihydroxybenzene; 1,4-benzenediol)

Use and exposure

Hydroquinone, a colorless, hexagonal prism, has been reported to be a good antimitotic and tumor-inhibiting agent. It is a reducing agent used in a photographic developer, which polymerizes in the presence of oxidizing agents. In the manufacturing industry it may occur include bacteriostatic agent, drug, fur processing, motor fuel, paint, organic chemicals, plastics, stone coating, and styrene monomers.

Toxicity and health effects

Exposures to hydroquinone in large quantities by accidental oral ingestion produce toxicity and poisoning. The symptoms of poisoning include, but are not limited to, blurred speech, tinnitus, tremors, sense of suffocation, vomiting, muscular twitching, headache, convulsions, dyspnea and cyanosis from methemoglobinemia, coma, and collapse from respiratory failure. Occupational workers should be allowed to work with protective clothing and dust masks with full-face or goggles to protect the eyes, and under proper management.

I

- Iodine
- Isocyanates
- Isofenphos
- Isopropyle alcohol

Iodine (CAS No. 7553-56-2)

Molecular formula: I_2
Chemical name: Iodine

Use and exposure

Iodine is available as bluish-black crystals with a metallic luster and a pungent odor. It is slightly soluble in water (0.03 g/100 g). It is stable under ordinary conditions of use and storage.

Iodine is incompatible with ammonia, powdered metals, alkali metals, or strong reducing agents. It reacts violently or explosively with acetaldehyde and acetylene, and reacts with ammonium hydroxide to form shock-sensitive iodides on drying.

Iodine is a naturally occurring element that is essential for the good health of people and animals. Iodine is found in small amounts in seawater and in certain rocks and sediments. Iodine occurs in many different forms that can be blue, brown, yellow, red, white, or colorless. Most forms of iodine easily dissolve in water or alcohol. Iodine has many uses. Its most important use is as a disinfectant for cleaning surfaces and storage containers. It is also used in skin soaps and bandages, and for purifying water. Iodine is used in medicines and is added to food, such as table salt, to ensure that people have enough iodine in their bodies to form essential thyroid hormones. Iodine is put into animal feeds for the same reason. Iodine is used in the chemical industry for making inks and coloring agents, chemicals used in photography, and in making batteries, fuels, and lubricants. Radioactive iodine also occurs naturally. Radioactive iodine is used in medical tests and to treat certain diseases, such as over-activity or cancer of the thyroid gland. Iodine is important for the thyroid gland to produce thyroid hormones.

Toxicity and health effects

Exposures to iodine cause very serious health effects and poisoning. The symptoms of toxicity and poisoning include, but are not limited to, salivation, excessive tearing, swelling of eyelids and salivary glands, metallic taste, irritation of the eyes, skin, nose, and throat, lacrimation, headache, cough, sore throat, tight chest, an allergic skin or respiratory reaction, and pulmonary edema. It also causes pain, burns, blurred vision, rash, skin hypersensitivity, joint swelling and pain, deficiency of thyroid hormones -iodism, fever and enlarged lymph glands. Exposures cause a burning sensation in the throat, abdominal cramps, vomiting, diarrhea, shock (in larger doses), and affects the cardiovascular system and the CNS.

Babies and children need iodine to form thyroid hormones, which are important for growth and health. Excessive iodine from the mother causes adverse effects to the baby's thyroid gland; the gland becomes so large that it makes breathing difficult or impossible. Also low levels of iodine from the mother can cause a baby to produce insufficient thyroid hormone, disturbing the growth and mental development of the growing baby.

Exposure limits

For iodine, the OSHA set the PEL as 0.1 ppm, the ACGIH set the TLV as 0.01 ppm (TWA) inhalable fraction and vapor for iodine and iodides, and the NIOSH set the REL as 0.1 ppm.

Precautions

Students, users, and occupational workers should specially note iodine as: Poison, Danger, and Corrosive. Exposures cause severe irritation or burns to every area of contact. It may be fatal if ingested/swallowed/inhaled. The vapors cause severe irritation to the skin, eyes, and respiratory tract. Iodine is a strong oxidizer and contact with other material may cause fire.

Occupational workers should wear impervious protective clothing, boots, gloves, a laboratory coat, apron or coveralls, as appropriate, to prevent skin contact of iodine. Also, workers should use chemical safety goggles and/or a full-face shield where splashing is possible. Maintain an eye-wash fountain and quick-drench facilities in the work area.

References

Agency for Toxic Substances and Disease Registry (ATSDR). 2004. Toxicological Profile for Iodine. US Department of Health and Human Services, Public Health Service, Atlanta, GA (updated 2010).

Material Safety Data Sheet (MSDS). 2009. Iodine. MSDS no. I2680. Environmental Health & Safety (EHS). Mallinckrodt Baker, Phillipsburg, NJ.

Occupational Safety & Health Administration (OSHA). 2004. Safety and health of iodine. OSHA, Washington, DC.

Isocyanates

Isocyanates are compounds containing the isocyanate group (-NCO). Isocyanates are a family of highly reactive, low molecular weight chemicals. Isocyanates have extensive industrial use in the manufacture of flexible and rigid foams, fibers, coatings such as paints and varnishes, and elastomers, and are increasingly used in the automobile industry, autobody repair, and building insulation materials. Spray-on polyurethane products containing isocyanates have been developed for a wide range of retail, commercial, and industrial uses to protect cement, wood, fiberglass, steel and aluminum, including protective coatings for truck beds, trailers, boats, foundations, and decks. Isocyanates are the raw materials that make up all polyurethane products. Isocyanate-related occupations include painting, foam-blowing, and the manufacture of many polyurethane products, such as chemicals, polyurethane foam, insulation materials, surface coatings, car seats, furniture, foam mattresses, under-carpet padding, packaging materials, shoes, laminated fabrics, polyurethane rubber, and adhesives, and during the thermal degradation of polyurethane products.

Isocyanate compounds

The most widely used compounds are diisocyanates, which contain two isocyanate groups, and polyisocyanates, which are usually derived from diisocyanates and may contain several isocyanate groups. The most commonly used diisocyanates include methylenebis(phenyl isocyanate) (MDI), toluene diisocyanate (TDI), and hexamethylene diisocyanate (HDI). Other common diisocyanates include naphthalene diisocyanate (NDI), methylene bis-cyclohexylisocyanate (HMDI) (hydrogenated MDI), and isophorone diisocyanate (IPDI). Examples of widely used polyisocyanates include HDI biuret and HDI isocyanurate.

Use and exposure

The world production of isocyanates is estimated to be 3 billion pounds annually. The major route of occupational exposure to isocyanates is inhalation of the vapor or aerosol; exposure may also occur through skin contact during the handling of liquid isocyanates. Occupational exposure normally occurs during the production and use of isocyanates—particularly during the mixing and foaming processes in the polyurethane foam industry. Exposures to airborne isocyanates may also occur from the melting or burning of polyurethane foams during firefighting.

Toxicity and health effects

Exposures to isocyanates cause severe irritation to the mucous membranes of the eyes and gastrointestinal and respiratory tracts. The flu-like symptoms of toxicity include, but are not limited to, fever, muscle aches, and headaches, dry cough, chest tightness, and breathing difficulty. Studies have also indicated that any kind of direct skin contact with TDI causes marked effects of inflammation. With repeated exposures, occupational workers with symptoms of respiratory irritation develop chemical bronchitis and severe bronchospasm. With repeated exposures to isocyanates, occupational workers develop sensitization, attacks of asthma, and death.

Isocyanates and carcinogenicity

The IARC and WHO concluded that data were found to be sufficient to show that toluene diisocyanate (TDI) causes cancer in laboratory animals and should be treated as a potential human carcinogen.

Precautions

- Occupational workers should be very careful during use, handling, and disposal of isocyanates.
- Workers should observe strict regulatory methods to protect themselves from diisocyanate exposure.

- Occupational workers should be aware that the highest diisocyanate concentrations may occur inside containment structures.

- Use appropriate respiratory protection when working with diisocyanates.

- Workers should wash hands and face before eating, drinking, or smoking outside the work area.

- Workers should shower and change into clean clothes before leaving the worksite.

- Workplace management should use and implement appropriate engineering controls and work practices to reduce the exposure limits of the workers to the lowest feasible concentration.

References

Lewis, R.J. Sr. 1993. *Hazardous Chemicals Desk Reference*, 3rd ed. Van Nostrand Reinhold, New York.

National Institute for Occupational Safety and Health (NIOSH). 1973. Criteria for a recommended standard: occupational exposure to toluene diisocyanate. DHEW (NIOH) Publication No. HSM 73-11022. US Department of Health, Education, and Welfare, Public Health Service, Center for Disease Control, Cincinnati, OH.

World Health Organization (WHO). International Agency for Research on Cancer (IARC). 1986. IARC Monographs on the Evaluation of the Carcinogenic Risk of Chemicals to Humans, Some Chemicals Used in Plastics and Elastomers, Vol. 39. World Health Organization, Lyon, France.

World Health Organization (WHO). 1987. Toluene diisocyanate. Environmental Health Criteria No. 75. IPCS, Geneva, Switzerland.

US Environmental Protection Agency (USEPA). 2000. 2,4-Toluene diisocyanate hazard summary. Technology Transfer Network Air Toxics Web site. US EPA. Washington D.C. (updated 2007).

Isofenphos (CAS No. 25311-71-1)

Molecular formula: $C_{15}H_{24}NO_4PS$
IUPA name: O-Ethyl O-2-isopropoxycarbonylphenyl N- isopropyl-phosphoramidothioate
Synonyms and trade names: Amaze; Oftanol; Pryfon
Toxicity class: US EPA: I; WHO: Ib

Use and exposure

Isofenphos is a colorless oil at room temperature. It is sparingly soluble in water, but soluble in cyclohexone, toluene, acetone, and diethyl ether. The US EPA has grouped isofenphos under RUP, which indicates that qualified, certified, and trained workers are required in the safety management of isofenphos. It is used on turf and ornamental trees and shrubs to control white grubs, mole crickets, and other insects, such as soil-dwelling insects, cabbage root flies, corn roundworms, and wire worms.

Toxicity and health effects

Exposures to isofenphos are highly toxic to animals and humans. Acute and prolonged exposures to high concentrations of isofenphos has caused poisoning with symptoms such as increased secretions, difficulty in breathing, diarrhea, urination, pupil contraction, slowness of the heart, convulsions, and coma. Isofenphos in combination with malathion cause severe poisoning in humans.

In 1990, an outbreak in the north of England, involving a large number of pigs, consumption of contaminated feed, and development of ataxia, has suggested that isofenphos causes delayed neuropathy. In view of the hazards and based on the requests of the manufacturers, the US EPA indicated the proposals to phase out isofenphos. Studies have indicated that isofenphos is not mutagenic, teratogenic, or carcinogenic to animals and humans.

References

Broadberg, R.K. 1990. Estimation of Exposure of Persons in California to Pesticide Products Containing Isofenphos. California Department of Food and Agriculture. Division of Pest Management, Sacramento, CA.

Hazardous Substance Data Bank. 1995. US Public Health Service (PHS). Washington, DC.

Kidd, H. and James, D.R. (eds). 1991. *The Agrochemicals Handbook*, 3rd ed. Royal Society of Chemistry Information Services, Cambridge, U.K.

Material Safety Data Sheet (MSDS). 1996. Isofenphos. Extension Toxicology Network (EXTOXNET), Pesticide Information Profiles (PIP). Oregon State University, USDA/Extension Service, Corvallis, OR.

Shaw, I.C., Parker, R.M., Porter, S., Quick, M.P., Lamont, M.H., Patel, R.K., Norman, I.M. and Johnson, M.K. 1995. Delayed neuropathy in pigs induced by isofenphos. *Vet. Record* 136 (4): 95–97.

US Environmental Protection Agency (US EPA). 1990. Pesticide Environmental Fate One-Line Summary: Isofenphos. Environmental Fate and Effects Division, Washington, DC.

Isopropyl alcohol (CAS No. 67-63-0)

Synonyms and trade names: Isopropanol; 2-Propanol; Propan-2-ol; sec-Propyl alcohol; Dimethylcarbinol; Isohol; Petrohol

Use and exposure

Isopropyl alcohol is a flammable, colorless liquid with an odor resembling alcohol. It is miscible with water, ethyl ether, and ethyl alcohol. Isopropyl alcohol is incompatible with strong oxidizers, acetaldehyde, chlorine, ethylene oxide, acids, and isocyanates.

Toxicity and health effects

Exposures to isopropyl alcohol cause irritation to the eyes and mucous membranes. Exposures to isopropyl alcohol for 3–5 min (400 ppm) caused mild irritation of the eyes, nose, and throat, and at 800 ppm these symptoms became severe. Ingestion or an oral dose of 25 mL in 100 mL of water produced hypotension, facial flushing, bradycardia, and dizziness. Ingestion in large quantities caused extensive hemorrhagic tracheobronchitis, bronchopneumonia, and hemorrhagic pulmonary edema. Prolonged skin contact with isopropyl alcohol caused eczema and sensitivity. Delayed dermal absorption is attributed to a number of pediatric poisonings that have occurred following repeated or prolonged sponge bathing with isopropyl alcohol to reduce fever. In several cases, symptoms included respiratory distress, stupor, and coma. Laboratory animals exposed to isopropyl alcohol develop poisoning with symptoms of hind leg paralysis, unsteadiness, lack of muscular coordination, respiratory depression, and stupor. Isopropyl alcohol is a potent CNS depressant, and in large doses causes cardiovascular depression.

Epidemiological studies suggested an association between isopropyl alcohol and para-nasal sinus cancer; however, subsequent analysis suggests that the "strong-acid" process used to manufacture isopropyl alcohol may be responsible for these cancers.

Isopropyl alcohol and carcinogenicity

Isopropyl alcohol is not listed by the NTP or the IARC as an anticipated carcinogen or human carcinogen. The IARC has concluded that the evidence for the carcinogenicity of isopropyl alcohol is inadequate.

Storage and transport

Isopropyl alcohol should be stored in a cool, dry, well-ventilated area in tightly sealed containers with a proper label. Outside or detached storage is preferable. Inside storage should be a flammable liquids storage room or cabinet. Workers should not store isopropyl alcohol above 37°C (100°F). Containers of isopropyl alcohol should be protected from physical damage and contact with air, and should be stored separately from strong oxidizers, acetaldehyde, chlorine, ethylene oxide, acids, and isocyanates. Isopropyl alcohol should be transported to the nearest laboratory as quickly as possible in cool containers.

Precautions

Workers should wash hands and face thoroughly after handling isopropyl alcohol. Workers should wear gloves, safety glasses and a face shield, boots, apron, and a full impermeable suit is recommended if exposure is possible to a large portion of the body.

References

Lewis, R.J. (ed.). 1993. *Lewis Condensed Chemical Dictionary*, 12th ed. Van Nostrand Reinhold, New York.
Material Safety Data Sheet (MSDS). 2009. 2-Propanol. MSDS no. P6401. Environmental Health and Safety (EHS). Mallinckrodt Baker, Phillipsburg, NJ.

K

- Kepone
- Kerosene

Kepone (CAS No. 143-50-0)

Synonyms and trade names: Chlordecone; Decachlorooctahydro-1,3,4-metheno-2H-cyclobuta(cd)-pentalen-2-one; Decachlorooctahydro-kepone-2-one; Decachloro-tetrahydro-4,7-methanoindeneone; Chlordecone

Kerosene (Stove oil)

Use and exposure

Kerosene is a white to pale yellow, flammable liquid that has wide use in household and industrial activities. For instance, in heating, as cooking fuel, in cleaning, degreasing, as a solvent, for paints, enamels, polishes, varnishes, and in asphalt coating.

Toxicity and health effects

Kerosene toxicity is variable and is based on the composition. It is rapidly absorbed by the skin and accidental ingestion results in mucous membrane irritation, gastrointestinal irritation, vomiting, diarrhea, pneumonitis, CNS depression, drowsiness, coma, and may lead to death. Prolonged contact is also known to cause skin blisters and dermatitis. Studies with non-human primates have demonstrated that aerosols and kerosene aspiration into the lungs cause cellular damage. Household activities and possible long-term exposures to kerosine require proper care to avoid skin contact and possible damage. Kerosene should never be sucked by mouth.

M

- Malathion
- Methane
- Methamidophos
- Methidathion
- Methylamine
- Methyl isocyanate
- Methyl bromide (Bromomethane)
- Methylene chloride
- Methyl parathion
- Mevinphos
- Mirex and chlordecone
- Molybdenum
- Mustard gas

Malathion (CAS No. 121-75-5)

Molecular formula: $C_{10}H_{19}O_6PS_2$
IUPAC name: Diethyl(dimethoxythiophosphorylthio) succinate
Synonyms and trade names: Celthion; Cythion; Dielathion; Emmaton; Exathios; Fyfanon; Hilthion; Karbofos; Maltox
Toxicity class: US EPA: III; WHO: III

Use and exposure

Malathion is a clear amber liquid. It is sparingly soluble in water, but soluble in a majority of organic solvents. The US EPA grouped malathion under GUP. It is used as an insecticide as well as an acaricide for the control of pests. Malathion is used for the control of sucking insects and chewing insects on fruits and vegetables, and is an effective insecticide for the control of several household pests, such as houseflies, cockroaches, mosquitos, aphids, animal ectoparasites, and human head and body lice. It is also found in formulations with many other pesticides.

Toxicity and health effects

Acute and prolonged period of exposures to high concentrations of malathion cause poisoning in animals and humans. The symptoms of poisoning include, but are not limited to, numbness, tingling sensations, incoordination, headache, dizziness, tremor, nausea, abdominal cramps, sweating, blurred vision, difficulty breathing or respiratory depression, and slow heart beat. Very high doses may result in unconsciousness, incontinence, and convulsions, or fatality. Malathion did not indicate any kind of delayed neurotoxicity in experimental studies with hens. Reports have indicated that because of and accidental exposures through severe skin absorption, malathion caused poisoning and fatalities among workers associated with the malaria control operations in Pakistan. In certain cases, development of pulmonary fibrosis following the poisoning has also been observed.

Reports have indicated that malathion is neither mutagenic nor teratogenic to animals and humans. In animals, malathion induced liver carcinogenicity at doses that were considered excessive. However, available information is not adequate to confirm the carcinogenicity of malathion to animals and humans. The IARC has determined that malathion is unclassifiable as to its carcinogenicity to humans. The IARC observed that the available data do not provide evidence that malathion or its metabolite malaoxon is carcinogenic to experimental animals and there is no data on humans and classified as Group 3 meaning, not classifiable as to carcinogenicity for humans.

References

Agency for Toxic Substances and Disease Registry (ATSDR). 2003. Toxicological Profile for Malathion. US Department of Health and Human Services, Public Health Service, Atlanta, GA (updated 2006).

Baker, E.L. 1956. Epidemic malathion poisoning in Pakistan malaria workers. *Lancet* 1: 31.

Dive, A., Mahieu, P., Van Binst, R. et al. 1994. Unusual manifestations after malathion poisoning. *Hum. Exp. Toxicol.* 13: 271–74.

International Programme on Chemical Safety and the Commission of the European Communities (IPCS-CEC). 2005. Malathion. ICSC card no. 0172. IPCS, CEC, Geneva, Switzerland.

Material Safety Data Sheet (MSDS). 1996. Malathion. Extension Toxicology Network (EXTOXNET), Pesticide Information Profiles (PIP). Oregon State University, USDA/Extension Service, Corvallis, OR.

World Health Organization (WHO). 1987. Malathion. Data Sheet on Pesticides No. 29. WHO, Geneva, Switzerland.

Methane (CAS No. 74-82-8)

Molecular formula: CH_4
Chemical family: Hydrocarbons, gas
Trade name: Methyl hydride; Marsh gas; Fire damp; Biogas; Natural gas.

Use and exposure

Methane is a natural, colorless, odorless, and tasteless gas. It is used primarily as fuel to make heat and light. It is also used to manufacture organic chemicals. Methane can be formed by the decay of natural materials and is common in landfills, marshes, septic systems, and sewers. It is soluble in alcohol, ether, benzene, and organic solvents. Methane is incompatible with halogens, oxidizing materials, and combustible materials. Methane evaporates quickly. Methane gas is present in coal mines, marsh gas, and in sludge degradations. Methane can also be found in coal gas. Pockets of methane exist naturally underground. In homes, methane may be used to fuel a water heater, stove, and clothes dryer. Incomplete combustion of gas also produces carbon monoxide. Methane gas is flammable and may cause flash fire. Methane forms an explosive mixture in air at levels as low as 5%. Electrostatic charges may be generated by flow and agitation.

Toxicity and health effects

Methane is a relatively potent gas. It is the simplest alkane and the principal component of natural gas. Exposures to methane gas cause toxicity and adverse health effects. The signs and symptoms of toxicity include, but are not limited to, nausea, vomiting, difficulty breathing, irregular heart beat, headache, drowsiness, fatigue, dizziness, disorientation, mood swings, tingling sensation, loss of coordination, suffocation, convulsions, unconsciousness, and coma. While at low concentrations methane causes no toxicity, high doses lead to asphyxiation in animals and humans. Displacement of air by methane gas is known to cause shortness of breath, unconsciousness, and death from hypoxemia. Methane gas does not pass readily through intact skin. However, in its extremely cold liquefied form, methane can cause burns to the skin and eyes. No long-term health effects are currently associated with exposure to methane.

Methane and carcinogenicity

There are no published reports indicating that methane is a cancer-causing agent or a human carcinogen. Methane is not listed as a carcinogen or potential carcinogen by NTP, IARC, or OSHA.

Exposure limits

No occupational exposure limits have been established for methane gas. However, the ACGIH has set 1000 ppm (TWA).

Storage

Occupational workers should store methane gas containers away from incompatible substances and handle in accordance with standard set regulations and grounding and bonding if required.

Precautions

Occupational workers should be careful during handling and management of methane gas because of its severe fire and explosion hazard, particularly with pressurized containers. The containers may rupture or explode if exposed to sufficient heat. Workers should avoid heat, flames, sparks, and other sources of ignition, and stop any leak if possible without personal risk. Workers should wear appropriate chemical-resistant gloves. Also, vapors should be reduced with water spray and keep unnecessary workers/people away from the place of chemical hazard. The closed spaces should be well ventilated before the workers enter. Methane is not toxic; however, it is highly flammable and may form explosive mixtures with air. Methane is violently reactive with oxidizers, halogens, and some halogen-containing compounds. Methane is also an asphyxiant and in enclosed areas displaces oxygen. Septic tanks, cesspools, and drywells present serious hazards, including septic cave-in or collapse, methane gas explosion hazards, and asphyxiation hazards. Occupational workers/work area supervisor should note the indications of methane gas poisoning: Soon after exposure to oxygen levels of less than 15% in air, if the workers feel symptoms of dizziness, headache, and tiredness, medical advice should be provided.

References

Material Safety Data Sheet (MSDS). 2006. Safety data for Methane. Department of Physical Chemistry, Oxford University, Oxford, U.K.
US Environmental Protection Agency (US EPA). 2007. Methane. US EPA, Atlanta, GA.

Methamidophos (CAS No. 10265-92-6)

Molecular formula: $C_2H_8NO_2PS$
IUPAC name: O,S-Dimethylphosramidothioate
Synonyms and trade names: Monitor; Tamaron; Nitofol; Swipe; Nuratron; Vetaron; Filitox; Patrole; Tamanox
Toxicity class: US EPA: I; WHO: Ib

Use and exposure

Methamidophos is a colorless crystalline solid with a pungent odor. It is readily soluble in water, alcohols, ketones, and aliphatic chlorinated hydrocarbons, but sparingly soluble in ether and practically insoluble in petroleum ether. It is highly toxic, systemic with properties of an insecticide, acaricide, and avicide. The US EPA has grouped methamidophos as an RUP, meaning use and handling of this chemical substance requires qualified, certified, and trained workers. Methamidophos is effective against chewing and sucking insects and is used to control aphids, flea beetles, worms, white-flies, thrips, cabbage loopers, Colorado potato beetles, potato tube-worms, army-worms, mites, leafhoppers, and many others. Crop uses include broccoli, brussels sprouts, cauliflower, grapes, celery, sugar beets, cotton, tobacco, and potatoes. It is used on many vegetables, hops, corn, peaches, and other crops. Methamidophos is also a breakdown product of another OP, namely, acephate. Methamidophos is slightly corrosive to mild steel and copper alloys.

Toxicity and health effects

Methamidophos is highly toxic to mammals. Inhalation of methamidophos causing weakness, tightness in the chest, wheezing, headache, blurred vision, pinpoint pupils, tearing, and runny nose are common early symptoms. On accidental ingestion and with severe poisoning, methamidiophos causes nausea, vomiting, diarrhea, and cramps, sweating and twitching, weakness, shakiness, blurred vision, pinpoint pupils, dyspnea (shortness of breath), tightness in the chest, sweating, confusion, changes in heart rate, convulsions, coma, respiratory failure, and death. People with health disorders, such as high blood pressure, problems of the gastrointestinal, heart, liver, lung, or nervous system, have been reported to be more sensitive to methamidophos-induced toxicity. Reports have indicated that occupational workers exposed to methamidophos who developed poisoning with symptoms such as pain (needle type) in the feet, legs, and hands, high blood pressure, gastrointestinal disorders, heart, liver, lung, or nervous system problems, may be more sensitive to methamidophos.

Methamidophos and carcinogenicity

There is no evidence/published reports about the carcinogenicity of methamidiophos in experimental rats and/or mice. Methamidiophos has not been reported as a human carcinogen.

References

International Programme on Chemical Safety and the Commission of the European Communities (IPCS-CEC). 1994. Methamidophos. ICSC card no. 0176. IPCS, CEC, Geneva, Switzerland.
Material Safety Data Sheet (MSDS). 1995. Methamidophos. Extension Toxicology Network (EXTOXNET), Pesticide Information Profiles (PIP). Oregon State University, USDA/Extension Service, Corvallis, OR.
———. 2008. Safety data for methamidophos. Department of Physical Chemistry, Oxford University, U.K.
US Environmental Protection Agency (US EPA). 2006. Methamidophos. Emergency First Aid Treatment Guide, US EPA, Atlanta, GA.
World Health Organization. (WHO), International Programme on Chemical Safety (IPCS). 1993. Methamidophos Health and Safety Guide No. 79. WHO, Geneva, Switzerland.

Methidathion (CAS No. 950-37-8)

Molecular formula: $C_6H_{11}N_2O_4PS_3$
IUPAC name: O, O-Dimethyl S-(2,3-dihydro-5-methoxy-2-oxo-1,3,4-thiadiazol-3-methyl) Phosphorodithioate
Synonyms and trade names: Somonic; Somonil; Supracide; Suprathion; Ultracide
Toxicity class: US EPA: I; WHO: Ib

Use and exposure

Methidathion is a colorless crystalline pesticide at room temperature. It is sparingly soluble in water, but very soluble in octanol, ethanol, xylene, acetone, and cyclohexane. Methidathion is a non-systemic organophosphorous insecticide and acaricide with stomach and contact action. It is used to control a variety of insects and mites on crops and fruit plants. Methidathion is highly toxic to animals and humans. The US EPA grouped methidathion as a class I toxic substance and as an RUP.

Toxicity and health effects

Acute and prolonged exposures to methidathion cause poisoning in animals and humans. The symptoms include, but are not limited to, nausea, vomiting, cramps, diarrhea, salivation, headache, dizziness, muscle twitching, difficulty breathing, blurred vision, and tightness in the chest, pulmonary edema, respiratory depression, and respiratory paralysis. Acute exposures to high concentrations of methidathion cause intense breathing problems, including paralysis of the respiratory muscles.

Methidathion and carcinogenicity

Methidathion did not induce any genetic changes in experimental animals. Information on the teratogenicity and carcinogenicity of methidathion is inadequate in animals and humans.

Exposure limits

The recommended ADI for methidathion has been set as ADI 0.001 mg/kg/day.

Precautions

Occupational workers should be very careful during use and chemical management of methidathion. As this chemical substance is a highly toxic pesticide, it has been grouped by the US EPA as toxicity class I. The containers and labels of the products should bear the signal word DANGER. Methidathion is an RUP, except for use in nurseries, and on safflower and sunflowers.

References

Material Safety Data Sheet (MSDS). 1996. Methidathion. Extension Toxicology Network (EXTOXNET), Pesticide Information Profiles (PIP). Oregon State University, USDA/Extension Service, Corvallis, OR.

Lu, F.C. 1995. A review of the acceptable daily intakes of pesticides assessed by the World Health Organization. *Regul. Toxicol. Pharmacol.* 21: 351–64.

Meister, R.T. (ed.). 1989. *Farm Chemical Handbook '89*, 75th ed. Meister, Willoughby, OH.

National Library of Medicine's Hazardous Substance Database, 2006.

US Environmental Protection Agency (US EPA). 2006. Emergency First Aid Treatment Guide Methidathion.

Methylamine (CAS No. 74-89-5)

Molecular formula: CH_3NH_2
Synonyms and trade names: Aminomethane; Methanamine; Monomethylamine

Use and exposure

Methylamine is a colorless, fish-like smelling gas at room temperature. It is used in a variety of industries, such as the manufacture of dyestuffs, treatment of cellulose, acetate rayon, as a fuel additive, rocket propellant, and in leather tanning processes.

Toxicity and health effects

Exposures to methylamine are known to cause adverse health effects among occupational workers. The workers demonstrate symptoms of toxicity that include, but are not limited to, irritation to the eyes, nose, and throat. Studies have indicated that the compound causes injury to the eyes through corneal opacities and edema hemorrhages in the conjunctiva, and injury to the liver. Studies of Guest and Varma indicated no significant deleterious effects on the internal organs or skeletal deformities in experimental mice.

Methyl isocyanate (CAS No. 624-83-9)

Molecular formula: C_2H_3NO
Synonyms and trade names: Isocyanatomethane; Methyl carbylamine; MIC.

Use and exposure

Methyl isocyanate is used as a chemical intermediate for the production of carbamate insecticides and herbicides. Occupational exposure to methyl isocyanate may occur in those workers who use insecticides and herbicides produced from methyl isocyanate.

Toxicity and health effects

Exposures to methyl isocyanate are extremely toxic to humans. In Bhopal, India, in 1984, accidental exposures to methyl isocyanate caused poisoning and fatal injuries to large numbers of occupational workers as well as the general public, because of a Union Carbide gas leak. The accidental gas leakage resulted in the deaths of more than 2,000 people and adverse health effects in greater than 170,000 survivors. Pulmonary edema was the cause of death in most cases, with many deaths resulting from secondary respiratory infections such as bronchitis and bronchial pneumonia. The acute inhalation exposure to methyl isocyanate in humans caused poisoning with symptoms that include, but are not limited to, respiratory tract irritation, difficulty breathing, blindness, nausea, gastritis, sweating, fever, chills, and liver and kidney damage. Survivors continued to exhibit damage to the lungs with bronchoalveolar lesions and decreased lung function. The damage to the eyes included loss of vision, loss of visual acuity, and cataracts. There is no report on the chronic/long-term effects of methyl isocyanate in humans or animals.

Studies on animals have indicated increased incidence of fetal deaths and decreased fertility, live litter size, fetal body weight, and neonatal survival following inhalation exposure to methyl isocyanate during pregnancy.

Methyl isocyanate and carcinogenicity

There is no information available on the carcinogenicity of methyl isocyanate in humans. The US EPA classified methyl isocyanate as a Group D chemical substance, meaning not classifiable as a human carcinogen. Laboratory mice and rats that were exposed to methyl isocyanate by inhalation showed no tumors.

References

Agency for Toxic Substances and Disease Registry (ATSDR). 2002. Managing Hazardous Materials Incidents. Volume III – Medical Management Guidelines for Acute Chemical exposures. Methyl

isocyanate, Atlanta, Georgia. US Department of Health and Human Services (USDHHS), Public Health Service (PHS).
US Environmental Protection Agency (USEPA). 2000. Methyl isocyanate hazard summary. Technology Transfer Network Air Toxics web site (updated 2007).

Methyl bromide (CAS No. 74-83-9)

Molecular formula: $CH_3 Br$
Synonyms and trade names: Bromomethane; Monobromomethane

Use and exposure

Methyl bromide (Bromomethane) is an odorless, sweetish, colorless gas that has been used as a soil fumigant and structural fumigant to control pests across a wide range of agricultural sectors. Methyl bromide is soluble in ethanol, benzene, carbon disulfide, and sparingly in water. During the 1920s, methyl bromide was used as an industrial fire extinguishing agent. The current uses of methyl bromide include the fumigation of homes and other structures for the control of termites and other pests.

Because methyl bromide depletes the stratospheric ozone layer, the amount produced and imported in the United States was reduced incrementally until it was phased out on January 1, 2005, pursuant to our obligations under the Montreal Protocol on Substances that Deplete the Ozone Layer (Protocol) and the Clean Air Act (CAA).

Toxicity and health effects

Exposures to methyl bromide by inhalation cause injury to the brain, nerves, lungs, and throat. At high doses, breathing methyl bromide causes injury to the kidneys and liver. The symptoms of methyl bromide toxicity and poisoning include, but are not limited to, dizziness, headache, abdominal pain, vomiting, weakness, hallucinations, loss of speech, incoordination, labored breathing, and convulsions. Contact with the skin and eyes can lead to irritation and burns. After serious exposures to methyl bromide, occupational workers suffer with lung and/or nervous system-related problems and permanent brain/ nerve damage. Laboratory study with species of animals indicated that bromomethane does not cause birth defects and does not interfere with normal reproduction except at high exposure levels.

Methyl bromide and carcinogenicity

The IARC and the US EPA found inadequate evidence in humans and limited evidence in experimental animals for the carcinogenicity of methyl bromide and hence reported methyl bromide (Bromomethane) as a Group 3 chemical substance, meaning not classifiable as a human carcinogen.

Exposure limits

The OSHA has set the limits of methyl bromide (Bromomethane) in workplace air as 5.0 ppm and recommends that the exposures to this chemical substance be reduced to the lowest level feasible. The TLV of methyl bromide is set as 1.0 ppm TWA (skin).

Storage

Methyl bromide (Bromomethane) should be kept stored in sealed containers to keep it from evaporating.

Precautions

Occupational workers should be careful during use and management of methyl bromide, which is very reactive and produces toxic and corrosive fumes, including hydrogen bromide, bromine, and carbon oxybromide. Methyl bromide reacts with strong oxidants, attacks many metals in the presence of water, attacks aluminium, zinc, and magnesium with the formation of pyrophoric compounds, and causes fire and explosion hazard.

References

Agency for Toxic Substances and Disease Registry (ATSDR). 1992. Toxicological Profile for Bromomethane. US Department of Health and Human Services, Public Health Service, Atlanta, GA (updated 2008).

Alexeeff, G.V. and Kilgore, W.W. 1983. Methyl bromide. *Residue Rev.* 88: 102–53.

Anger, W.K., Setzer, J.V., Russo, J.M., Brightwell, W.S., Wait, R.G. and Johnson, B.L. 1981. Neurobehavioral effects of methyl bromide inhalation exposures. *Scand. J. Work Environ. Health* 7 (Suppl. 4): 40–47.

Hazardous Substance Data Bank (HSDB). 1999. National Library of Medicine, Bethesda, MD.

US Environmental Protection Agency (US EPA). 1999. Integrated Risk Information System (IRIS) database. Reference concentration (RfC) for methyl bromide. US EPA, Atlanta, GA.

World Health Organization (WHO). 1995. International Programme on Chemical Safety (IPCS). Methyl bromide. Environmental Health Criteria No. 166. WHO, IPCS, Geneva, Switzerland.

Methylene chloride (CAS No. 75-09-2)

Molecular formula: CH_2Cl_2
Synonyms and trade names: Dichloromethane

Use and exposure

Methylene chloride is a colorless liquid with a mild, sweet odor. It does not occur naturally in the environment. It is made from methane gas or wood alcohol. Industrial uses of methylene chloride are extensive, as a solvent in paint strippers, as a propellant in aerosols, and as a process solvent in the manufacturing of drugs. Methylene chloride is also used as a metal cleaning and finishing solvent, and it is approved as an extraction solvent for spices and hops. Exposure to methylene chloride occurs in workplaces by breathing fumes from paint strippers that contain it (check the label), breathing fumes from aerosol cans that use it (check the label), and breathing contaminated air near waste sites.

Toxicity and health effects

Exposures to methylene chloride cause adverse health effects and poisoning to users. Methylene chloride harms the human CNS. The symptoms of poisoning include, but are not limited to, dizziness, nausea, tingling, and numbness in the fingers and toes. Laboratory animals exposed to very high levels of methylene chloride suffer unconsciousness and fatal injury/death. Occupational workers who are exposed to direct skin contact with methylene chloride indicate symptoms of intense burning and mild redness of the skin, damage to the eyes and cornea.

Methylene chloride and carcinogenicity

The DHHS has reported that methylene chloride may reasonably be anticipated to be a carcinogen. Breathing high concentrations of it for long periods did increase the incidence of cancer in mice. However, methylene chloride has not been shown to cause cancer in humans exposed to vapors in the workplace.

Exposure limits

The US EPA recommends that water containing more than 13.3 ppm of methylene chloride for longer than 1 day or with more than 1.5 ppm for longer than 10 days is not suitable and the NIOSH recommends the permissible limit of 75 ppm of methylene chloride in air over a 10-h workday.

References

Agency for Toxic Substances and Disease Registry (ATSDR). 1990. Case studies in environmental medicine: Methylene chloride toxicity. US Department of Health and Human Services, Public Health Service, Atlanta, GA.
————. 1993. Toxicological Profile for Methylene Chloride. US Department of Health and Human Services, Public Health Service, Atlanta, GA.

Methyl parathion (CAS No. 298-00-0)

Molecular formula: $C_8H_{10}NO_5PS$
IUPAC name: O,O-dimethyl-O-4-nitrophenylphosphorothioate
Synonyms and trade names: Folidol-M; Metacide; Dhanuman; Sweeper; Metaphos; Metron
Toxicity class: US EPA: I; WHO: I or Ia

Use and exposure

Pure methyl parathion exists as white crystals, while the technical product is light to dark tan in color. Impure methyl parathion is a brownish liquid that smells like rotten eggs. It is sparingly soluble in water, but soluble in dichloromethane, 2-propanol, toluene, and in many organic solvents. Methyl parathion is used to kill insects on farm crops. Methyl parathion is a contact insecticides and acaricide used for the control of boll weevils and many biting and sucking insect pests of agricultural crops. It kills insects by contact, stomach and respiratory action. The US EPA now restricts how methyl parathion can be used and applied; only trained people are allowed to spray it. Methyl parathion can no longer be used on food crops commonly consumed by children.

Farm workers, chemical sprayers, and people who work in factories that make methyl parathion are most likely to be exposed. People who live near farms where methyl parathion is used or near landfills where methyl parathion has been dumped may be exposed. Individuals may also be exposed by going into fields too soon after spraying.

Toxicity and health effects

Methyl parathion is highly toxic by inhalation and ingestion, and moderately toxic by dermal adsorption. Exposures to very high concentrations of methyl parathion for a short period cause poisoning and death. Repeated exposures cause dullness, loss of consciousness, dizziness, confusion, headaches, respiratory distress, chest tightness, wheezing, vomiting, diarrhea, cramps, tremors, and blurred vision. Very high doses are known to cause unconsciousness, incontinence, convulsions, and/or fatal injury.

Methyl parathion and carcinogenicity

There are no reports indicating that methyl parathion is teratogenic, mutagenic, or carcinogenic to animals and humans. The US EPA and the IARC have observed that methyl parathion is not classifiable as a human carcinogen.

Exposure limits

The recommended ADI for methyl parathion has been set as 0.02 mg/kg/day.

References

Agency for Toxic Substances and Disease Registry (ATSDR).2001. Toxicological Profile for Methyl Parathion. US Department of Health and Human Services, Public Health Service, Atlanta, GA (updated 2007).

Material Safety Data Sheet (MSDS). 1996. Methyl parathion. Extension Toxicology Network (EXTOXNET), Pesticide Information Profiles (PIP). Oregon State University, USDA/Extension Service, Corvallis, OR.

National Cancer Institute (NCI). 1979. Bioassay of Methyl Parathion for Possible Carcinogenicity. DHEW Pub. No. (NIH) 79-1713. National Institute of Health, Bethesda, MD.

Mevinphos (CAS No. 7786-34-7)

Molecular formula: $C_7H_{13}O_6P$
IUPAC name: 2-Methoxycarbonyl-1-methylvinyl dimethyl phosphate
Synonyms and trade names: Apavinphos; CMDP; ENT 22374; Fosdrin; Gesfid; Meniphos; Menite; Mevinox; Mevinphos; OS-2046; PD5; Phosdrin; Phosfene
Toxicity class: US EPA: I; WHO: Ia

Use and exposure

Pure mevinphos is a colorless liquid while technical grade mevinphos is a pale yellow liquid with a very mild odor. It is soluble in water, but very soluble in alcohols, ketones, chlorinated hydrocarbons, aromatic hydrocarbons, and many organic solvents. It is both an insecticide and an acaricide. It has been grouped by the US EPA under RUP, therefore it should be purchased and used only by certified and trained applicators. It is not registered for use in many countries including the United States. Mevinphos is used for the control of a broad spectrum of insects, including aphids, grasshoppers, leafhoppers, cutworms, caterpillars, and many other insects on a wide range of field, forage, vegetable, and fruit crops. It is also an acaricide that kills or controls mites and ticks. It acts quickly both as a contact insecticide, acting through direct contact with target pests, and as a systemic insecticide, which becomes absorbed by plants on which insects feed.

Toxicity and health effects

Mevinphos is highly toxic and causes poisoning in animals and humans. The symptoms of poisoning include, but are not restricted to, nausea, dizziness, blurred vision, confusion, and at very high exposures (like accidental spills), numbness, tingling sensations, incoordination, headache, dizziness, tremor, abdominal cramps, respiratory paralysis, respiratory depression, slow heart beat, and death. Very high doses may result in unconsciousness, incontinence, convulsions, and death. Prolonged exposures to methyl parathion also cause acute pulmonary edema.

Mevinphos and carcinogenicity

There are no reports indicating that mevinphos produces mutagenic, teratogenic, or carcinogenic effects to animals and humans. There are no evidences to indicate that mevinphos is a human carcinogen.

Precautions

Mevinphos is a highly toxic OP pesticide due to its acute oral and dermal toxicity and residual effects on species of animals. The US EPA has indicated it as a toxicity class I chemical substance. All emulsifiable and liquid concentrates of mevinphos are classified as RUPs by the US EPA. Therefore, mevinphos and other similar RUPs should be purchased and used only by certified and trained applicators. Products containing mevinphos must bear the signal words DANGER—POISON.

Exposure limits

The ADI for mevinphos has been set as 0.0015 mg/kg/day. The TLV of mevinphos is set as 0.092 mg/m^3 (8-h).

References

Gallo, M.A. and Lawryk, N.J. 1991. Organic phosphorus pesticides. Chapter 16, pp. 917–1123. In: W. J. Hayes Jr. and E. R. Laws Jr. (eds.) *Handbook of Pesticide Toxicology*. Academic Press, New York.

Lu, F.C. 1995. A review of the acceptable daily intakes of pesticides assessed by the World Health Organization. *Regul. Toxicol. Pharmacol.* 21: 351–64.

Material Safety Data Sheet (MSDS). 1996. Metvinphos. Extension Toxicology Network (EXTOXNET), Pesticide Information Profiles (PIP). Oregon State University, USDA/Extension Service, Corvallis, OR.

Tomlin, C.D.S (ed.). 2006. *The Pesticide Manual: A World Compendium*, 14th ed. British Crop Protection Council, Farnham, Surrey, U.K.

US Environmental Protection Agency (US EPA). 2000. Mevinphos Facts United States Prevention, Pesticides EPA 738-F-00-012.

Mirex (CAS No. 2385-85-5); Chlordecone (CAS No. 143-50-5)

Molecular formula: $C_{10}Cl_{12}$
Synonyms and trade names: 1,3,4-Metheno-1H-cyclobuta[cd]pentalene, 1,1a,2,2,3,3a,4,5,5,5a,5b,6, -dodecachlorooctahydro-; Dechlorane

Use and exposure

Mirex and chlordecone are two separate, but chemically similar, manufactured insecticides that do not occur naturally in the environment. Mirex is a white crystalline solid, and chlordecone is a tan-white crystalline solid. Both chemicals are odorless.

Mirex and chlordecone have not been manufactured or used in the United States since 1978. Mirex was used to control fire ants, and as a flame retardant in plastics, rubber, paint, paper, and electrical goods from 1959 to 1972. Chlordecone was used as an insecticide on tobacco, ornamental shrubs, bananas, and citrus trees, and in ant and roach traps. Mirex was sold as a flame retardant under the trade name Dechlorane, and chlordecone was also known as Kepone. Mirex and chlordecone break down slowly in the environment, and may stay for years in soil and water. Mirex and chlordecone do not dissolve easily in water, but they easily stick to soil and sediment particles.

Human exposures to mirex and chlordecone occur by touching or ingesting contaminated soil near hazardous waste sites; by ingesting contaminated fish or other animals living near hazardous waste sites; and nursing infants of mothers living near hazardous waste sites may be exposed to mirex through their mothers' milk. Drinking water or breathing air is not likely to cause exposure because these compounds do not easily dissolve in water or evaporate.

Toxicity and health effects

Exposures to high levels of mirex and chlordecone over a long period cause harmful health effects. The symptoms of poisoning involve the nervous system, skin, liver, and male reproductive system. Laboratory animals exposed to high concentrations of mirex and chlordecone demonstrated tissue damage in the stomach, intestine, liver, kidneys, eyes, thyroid, and the nervous and reproductive systems.

Mirex and chlordecone and carcinogenicity

There are no studies available on whether mirex and chlordecone are carcinogenic in people. However, studies in mice and rats have shown that ingesting mirex and chlordecone can cause liver, adrenal gland, and kidney tumors. The DHHS has determined that mirex and chlordecone may be reasonably anticipated to be carcinogens.

Exposure limits

The US EPA set a limit of 1.0 ppt of mirex in surface water (1 ppt) to protect fish and other aquatic life from harmful effects. The NIOSH recommends that average workplace air levels should not exceed 1 mg/m^3 of chlordecone over a 10-h period. The US FDA observed that eating fish and other foods with concentrations below 100 ppt of mirex or concentrations of chlordecone below 400 ppt has not caused any adverse health effects in people.

Reference

Agency for Toxic Substances and Disease Registry (ATSDR). 1995. Toxicological Profile for Mirex and Chlordecone. US Department of Health and Human Services, Public Health Service, Atlanta, GA (updated 2007).

Molybdenum (as Mo), Insoluble Compounds (Total Dust)

Use and exposure

Molybdenum is an essential trace mineral associated with several enzyme systems required for normal body functions. However, a prolonged period of exposure to high concentrations of molybdenum dust is known to cause poisoning among occupational workers. Mine workers have developed symptoms of molybdenosis.

Toxicity and health effects

Chronic exposures to molybdenum in high concentrations cause health disorders that include, but are not limited to, irritation to eyes, nose, and throat, weight loss, dizziness, listlessness, anorexia, diarrhea, weakness, fatigue, headache, lack of appetite, epigastric pain, and joint and muscle pain. It also causes redness and moist skin, tremors of the hands, sweating, liver and kidney damage, blood disorders, dyspnea, anemia, gout-like symptoms, such as arthralgia, articular deformities, erythema, and edema. Available reports show that high concentrations of molybdenum cause knock-knees (*genu valgum*), gastrointestinal irritation with diarrhea, coma, and death from cardiac failure in exposed individuals.

Mustard gas (CAS No. 505-60-2)

Mustard gas (H) (sulfur mustard)
Molecular formula: $C_4H_8Cl_2S$
Synonyms and trade names: Sulfur mustard; 1,1′ thiobis (2 chloroethane); bis-(2-chloroethyl) sulfide; beta, beta′ dichloroethyl sulfide; 2, 2′ dichloroethyl sulfide; Agent HD; H; HS; Mustard gas; bis (beta-chloroethyl) sulfide; Bis(2-chloroethyl)sulfide; 1-chloro-2 (beta-chlorodiethylthio) ethane; Yellow cross liquid; Kampfstoff "Lost"; Yperite
Chemical class: Alkylating agent

Use and exposure

Mustard gas is the common name given to 1,1-thiobis(2-chloroethane), a chemical warfare agent that is believed to have first been used near Ypres in Flanders on July 12, 1917. Mustard gas is a thick liquid at ambient temperature. It is heavier than water as a liquid and heavier than air as a vapor. It does not occur naturally in the environment. Pure liquid mustard gas is colorless and odorless. It is stable, combustible, and incompatible with strong oxidizing agents. On mixing with other chemical substances, mustard gas appears brown in color and gives of a garlic-like smell. When heated, it decomposes and emits highly toxic, corrosive fumes and fumes of oxides of sulfur and chlorine-containing compounds. It is soluble in fats and oils, gasoline, kerosene, acetone, carbon tetrachloride, alcohol, tetrachloroethane, ethylbenzoate, and ether; solubility in water is negligible. It is miscible with organophosphorus nerve agents. During earlier years, mustard gas was used as an important chemical warfare agent. In fact, it was used in large amounts during World War I and II. Mustard gas was first used by the German army in 1917. It was one of the most lethal poisonous chemicals used during the war. It was reportedly used in the Iran-Iraq war in 1980–1988. Currently, it is not used in the United States, except for research purposes.

Toxicity and health effects

Mustard gas is a vesicant or blister-causing, alkylating chemical substance. Exposures to mustard gas cause adverse health effects. The symptoms of poisoning include, but are not limited to, dyspnea, cough, fever, headache, severe irritation, vesicant, skin burns, skin blisters, swelling of eyelids, coughing, bronchitis, photophobia (sensitivity to light), lacrimation (tearing), and blindness, irritation or ulceration of the respiratory tract, respiratory disorders eventually leading to death. The skin disorders become very severe in hot and humid climatic conditions. It causes necrosis of the skin, eyes, and respiratory tract. Exposures to mustard gas cause vomiting, internal and external bleeding, damage to the mucous membrane, damage to the bronchial tubes, and severe respiratory impairment. It produces cytotoxic action on the hematopoietic (blood-forming) tissues that are especially sensitive. The target organs of mustard gas include the lung, larynx, pharynx, oral cavity, bone marrow, and sexual organs. Users may suffer with hemorrhage and anemia several days after exposure. (Table. Mustard gas.)

Mustard gas and carcinogenicity

The DHHS and the IARC have classified mustard gas as a human carcinogen. Reports indicate that workers exposed to mustard gas during the production process as well as during war developed respiratory cancer.

Mustard gas and global wars

The Germans first used mustard gas at Ypres, Belgium, on July 12, 1917. Mustard gas was particularly dangerous to troops because several hours must elapse before the individual detects a disabling or lethal dose. After the introduction of mustard gas, the warfare environment changed drastically. Since World War I, a number of isolated incidents have occurred in which mustard agent was reportedly used. In 1935, Italy probably used mustard gas against Ethiopia. Japan allegedly used mustard gas against the Chinese from 1937 to 1944. Egypt apparently used mustard gas in the 1960s against Yemen. More recently, the Iraqis used it against the Iranians and the Kurdish population in northern Iraq. Mustard agent remains an important chemical warfare agent today.

Exposures to mustard gas in the war fields caused rapid visual damage and incapacitation of large numbers of people. The great majority of people with eye injuries were visually disabled for approximately 10 days with conjunctivitis, photophobia, and minimal corneal swelling.

The use of mustard gas in the past two decades and the continuation of the manufacture of chemical weapons have increased the risk that soldiers in particular and civilian populations around the world may be exposed to these highly toxic chemical substances.

Exposure limits

The federal government recommends a maximum concentration for long-term exposure to sulfur mustard by the general population of 0.00002 mg/m^3 in air.

Storage

Mustard gas should be kept stored in containers made of glass for research, development, test, and evaluation (RDTE) quantities or one-ton steel containers for large quantities. Agent containers should be stored in a single-containment system within a laboratory hood or in a double-containment system. Mustard gas is forbidden for transport other than via military (Technical Escort Unit) transport according to 49 CFR 172.

Precautions

Mustard gas is an extremely hazardous chemical substance. On inhalation, mustard gas causes fatal injury. It reacts violently with oxidizing materials, with water or steam to produce toxic and corrosive fumes. Workers should avoid high heat, contact with acid and or acid fumes. Users should have protective equipment, wear a positive pressure, pressure-demand, full-face, self-contained breathing apparatus (SCBA) or pressure-demand supplied-air respirator with escape SCBA and a fully-encapsulating, chemical-resistant suit, safety glasses, gloves, and good ventilation. It should be handled as a carcinogen. This material must not be used unless a full risk assessment has been prepared in advance. It should not be used if safer alternatives are available. Disposal of waste must be carried out in a safe fashion—anything other than trivial amounts must not be vented through a fume cupboard. Use of this chemical for research purposes may be legally prohibited or restricted.

References

Agency for Toxic Substances and Disease Registry (ATSDR). 2003. Toxicological Profile for Sulfur Mustard. US Department of Health and Human Services, Public Health Service, Atlanta, GA (updated 2007).

Material Safety Data Sheet (MSDS). 2005. Mustard gas. Physical Chemistry at Oxford University, U.K.

NOAA's Ocean Service. Mustard gas. National Oceanic and Atmospheric Administration, USA.

TABLE

Exposures to Mustard Gas and Symptoms of Poisoning

Ocular toxicity:
- Appearance of mild ocular toxicity and health effects after 4–12 h. The symptoms include tearing, itching, gritty feeling, and burning sensation.
- Appearance of moderate ocular toxicity and health effects after 3–6 h. The symptoms include tearing, itching, gritty feeling, burning sensation, redness, eyelid edema, and moderate pain.
- Appearance of severe ocular toxicity and health effects after 1–2 h. The symptoms include marked lid edema, corneal damage, severe pain, rhinorrhea, and sneezing.

Respiratory toxicity:
- The symptoms of mild respiratory effects include hacking cough, hoarseness, and epistaxis.

Dermal toxicity:
- The symptoms of mild dermal toxicity include erythema after 2–24 h.
- The symptoms of severe dermal toxicity include vesication (skin blisters) after 2–24 h.

N

- Naphthalene
- Nitrogen mustards

Naphthalene (CAS No. 91-20-3)

Molecular formula: $C_{10}H_8$
Synonyms and trade names: Moth flakes; Mothballs; Camphor tar; White tar; NCI–C52904.

Use and exposure

Naphthalene is a commercially important aromatic hydrocarbon. Naphthalene occurs as a white solid or powder. It is soluble in alcohol and ether and insoluble in water. Naphthalene occurs in coal tar in large quantities and is easily isolated from this source in pure condition. Transparent prismatic plates are also available as white scales, powder balls, or cakes with a characteristic odor and aromatic taste. It volatilizes at room temperature with a characteristic mothball or strong coal tar odor. The primary use for naphthalene is in the production of phthalic anhydride, carbamate insecticides, surface active agents, and resins, as a dye intermediate, as a synthetic tanning agent, as a moth repellent, and in miscellaneous organic chemicals. Naphthalene is used in the production of phthalic anhydride; it is also used in mothballs. Naphthalene is also used in the manufacture of phthalic and anthranilic acids to make indigo, indanthrene, and triphenyl methane dyes, for synthetic resins, lubricants, celluloid, lampblack, smokeless powder, and hydronaphthalenes. Naphthalene is used in dusting powders, lavatory deodorant discs, wood preservatives, fungicides, mothballs and as an insecticide. It has been used as an intestinal antiseptic, a vermicide, and in the treatment of pediculosis and scabies. Occupational exposure to naphthalene occurs in the dye industry and other chemical synthetic industries.

Toxicity and health effects

Exposures cause adverse health effects and toxicity. Acute exposure of humans to naphthalene by inhalation, ingestion, and dermal contact is associated with hemolytic anemia, damage to the liver, and, in infants, neurological damage. The symptoms of acute exposure include headache, nausea, vomiting, diarrhea, malaise, confusion, anemia, jaundice, hematuria and an acute hemolytic reaction, convulsions, and coma. Cataracts and optic neuritis have been reported in humans acutely exposed to naphthalene by inhalation and ingestion.

Cataracts have also been reported in animals following acute oral exposure. Chronic exposure of workers to naphthalene has been reported to cause cataracts and retinal hemorrhage. Chronic inflammation of the lung, chronic nasal inflammation, hyperplasia of the respiratory epithelium in the nose, and metaplasia of the olfactory epithelium were reported in mice chronically exposed to naphthalene via inhalation.

Naphthalene and carcinogenicity

The US EPA has classified naphthalene as Group C, meaning possible a human carcinogen. In addition, reports indicate that naphthalene and coal tar exposure have been associated with laryngeal and intestinal carcinoma. The IARC reported that evidences of naphthalene as a human carcinogen is inadequate while evidences in experimental animals are sufficient. Therefore, naphthalene is classified as Group 2B, meaning possibly carcinogenic to humans.

Precautions

Naphthalene causes hemolysis with subsequent blocking of the renal tubules by precipitated hemoglobin. Hepatic necrosis can occur. Hemolysis is more likely to occur in individuals with a hereditary deficiency of glucose-6-phosphate dehydrogenase, sickle cell anemia, and sickle cell trait.

References

Agency for Toxic Substances and Disease Registry (ATSDR). 1995. Toxicological Profile for Naphthalene (update). US Department of Health and Human Services, Public Health Service, Atlanta, GA (updated 2007).

Nitrogen Mustards

Blister Agents: HN-1, HN-2, HN-3 (CAS No. HN-1 538-07-8); (CAS No. HN-2 51-75-2); (CAS No. HN-3 555-77-1)

Nitrogen mustards, produced during the 1920s and 1930s as chemical warfare weapons, were potential chemical substances of yesteryear. They are vesicants (or blister agents) similar to sulfur mustards. They smell fishy, musty, soapy, or fruity and are either in the form of an oily-textured liquid or a vapor (the gaseous form of a liquid) or a solid. Nitrogen mustard is liquid at normal room temperature (70°F) with a clear, pale amber, or yellow color. HN-1, HN-2, and HN-3 are the military designations of nitrogen mustard. (Also refer Mustard gas for more data).

Use and exposure

Nitrogen mustards (HN-1, HN-2, HN-3) are colorless to yellow, oily liquids that evaporate very slowly. HN-1 has a faint, fishy or musty odor. HN-2 has a soapy odor at low concentrations and a fruity odor at higher concentrations. HN-3 may smell like butter almond. Use of nitrogen mustards is very much restricted other than for chemical warfare. Currently, there are no records of its use. Previously, HN-1 was used to remove warts; HN-2 has been used sparingly in chemotherapy.

Toxicity and health effects

Exposures to nitrogen mustards cause adverse health effects and poisoning to humans. The symptoms of poisoning caused after accidental exposures to nitrogen mustard vapor by breathing include, but are not limited to, nasal and sinus pain or discomfort, pharyngitis, laryngitis, cough, abdominal pain, diarrhea, nausea, vomiting, and shortness of breath. Cells lining the nasal airways suffer immediate damage. Exposure to high levels of nitrogen mustards leads to fatal injury. On skin contact, nitrogen mustard vapors or liquid cause swelling and rash, followed by blistering, and in high concentrations, the chemical substance causes second- and third-degree skin burns. Contact of the vapor of nitrogen mustard to the eyes causes inflammation, pain, swelling, corneal damage, burns, and loss of vision. Accidental ingestion of nitrogen mustards causes burning of the mouth, esophagus, and stomach, and may further lead to damage to the immune system and bone marrow. On exposures to very high concentrations of nitrogen mustard vapor, occupational workers suffer tremors, incoordination, and seizures. Prolonged or repeated period of exposures to nitrogen mustard has caused cancer in animals and leukemia in humans.

Nitrogen mustard and carcinogenicity

The IARC has classified nitrogen mustard HN-2 as probably carcinogenic to humans, based on evidence that it causes leukemia in humans and cancers of the lung, liver, uterus, and large intestine in animals.

Reference

Agency for Toxic Substances and Disease Registry (ATSDR). 2002. Managing Hazardous Materials Incidents. Volume III – Medical Management Guidelines for Acute Chemical Exposures: Blister Agents: HN-1, HN-2, HN-3 (Nitrogen Mustards). US Department of Health and Human Services, Public Health Service, Atlanta, GA.

Nitroglycerin (CAS No. 55-63-0)

Molecular formula: $C_3H_5(ONO_2)_3$
Synonyms and trade names: Anginine; Basting gelatin; Blasting oil; Glyceryl nitrate; GTN; Nitroglyn; Nitrol; NG; NTG; Nitroglycerine; Nitroglycerol; Trinitroglycerol; Trinitrin

Use and exposure

Nitroglycerin or glyceryl trinitrate is a pale yellow oily liquid and is also available in the form of rhombic crystals. It is highly explosive. It is used in combination with ethylene glycol dinitrite in the manufacture of dynamite. It is slightly soluble in water, and miscible with acetone, ether, benzene, and other organic solvents. Nitroglycerin is incompatible with strong acids, such as hydrochloric acid, sulfuric acid, nitric acid, and causes violent reactions with ozone. Nitroglycerin is a powerful explosive in its pure form and very sensitive to mechanical shock, heat, or UV radiation. It is a severe explosion hazard when shocked or exposed to O_3. This product is hygroscopic. Exposures to nitroglycerin occur among workers associated with occupations/operations, such as the manufacture and transportation of nitroglycerin, the manufacture of gun powder, dynamite, smokeless powders, guncotton, and other explosives and the use of the chemical substance in rocket propellants and in medicines.

Toxicity and health effects

Exposures to nitroglycerin cause severe health effects to users and occupational workers. Nitroglycerin is highly toxic to humans and causes rapid vasodilatation and severe throbbing. Nitroglycerin is rapidly absorbed through the exposed skin surface and the respiratory route. Repeated exposures at high concentrations to nitroglycerin include, but are not limited to, vomiting, nausea, cyanosis, coma, and death. Exposures to nitroglycerin leads to an increase in the heart rate, myocardial contraction, and angina pectoris. Occupational exposures to nitroglycerine have also been traced to work adaptation to the chemical and the consequences of withdrawal syndrome. Nitroglycerin on skin contact causes violent headache. Therefore, whenever the spirit glyceryl trinitrate is spilled, a solution of sodium hydroxide should be added immediately to check the chemical reaction. The worker must remember that on evaporation of alcohol, glyceryl trinitrate results in severe explosion. This needs care and caution in the proper management of work places and the health of workers.

Exposure limits

The OSHA set the PEL for nitroglycerin as 0.2 ppm of air; at no time should occupational workers exceed this exposure limit. The NIOSH set a STEL of 0.1 mg/m^3 for periods not to exceed 15 min in the workplace; the REL for nitroglycerin is not established. For nitroglycerin, the ACGIH set a TLV of 0.05 ppm (TWA) for a normal 8-h workday and a 40-h workweek.

Storage

Nitroglycerin should be kept stored in a cool, dry, well-ventilated area in tightly sealed containers with proper labels. Containers of nitroglycerin should be protected from physical damage, ignition sources, shocks (or jolts), and ultraviolet radiation, and should be stored away from strong acids.

Precautions

During handling of nitroglycerin, occupational workers should use appropriate personal protective equipment (PPE) and clothing and avoid skin contact with nitroglycerin. The selection of the appropriate PPE includes gloves, sleeves, encapsulating suits and should be based on the extent of the worker's potential exposure to nitroglycerin. On skin contact with nitroglycerin, occupational workers should IMMEDIATELY flood the affected skin with water and remove and isolate all contaminated clothing. Gently wash all affected skin areas thoroughly with soap and water. Workers should seek immediate medical attention.

References

Clayton, G., and Clayton, F. (eds.). 1993. *Patty's Industrial Hygiene and Toxicology*, 4th ed. Volume I, Part A and Part B. *General Principles*. Wiley, J. New York.

National Library of Medicine (NLM). 1992. Hazardous substances data bank: Nitroglycerin. NLM, Bethesda, MD.

Patnaik, P. (ed.) 1992. *A Comprehensive Guide to the Hazardous Properties of Chemical Substances*. Van Nostrand Reinhold, New York.

Sittig, M. (ed.). 1991. Handbook of Toxic and Hazardous Chemicals, 3rd ed. Noyes, Park Ridge, NJ.

O-Nitrotoluene (CAS No. 88-72-2)

Molecular formula: $CH_3C_6H_4NO_2$
Synonyms and trade names: Nitrotolul, methylnitrobenzene, 2-nitrotoluene, 2-NT

Use and exposure

O-Nitrotoluene is a yellow-colored liquid. The compound is used for the synthesis of a variety of industrial products. These include azo dyes, agricultural chemicals, explosives, sulfur dyes, and rubber chemicals.

Toxicity and health effects

Exposures to O-nitrotoluene cause adverse effects to animals, users, and occupational workers. The symptoms of acute and chronic toxicity and poisoning include, but are not limited to, irritation of the skin and mucous membrane, hypoxia, anemia, and depression in occupational workers. More reports are available regarding the genotoxicity, carcinogenicity, and reproductive effects of O-nitrotoluene in experimental animals in literature.

O

- Osmium tetroxide

Osmium tetroxide (CAS No. 20816-12-0)

Molecular formula: OsO_4
Synonyms and trade names: Osmic acid anhydride; Osmium oxide

Osmium and compounds

Osmium is an extremely dense, blue-gray, hard but brittle metal that remains lustrous at high temperatures. Osmium possesses quite remarkable chemical and physical properties. It has the highest melting point and the lowest vapor pressure in the olatinum family. Osmium is highly volatile, extremely toxic, and is rarely used in its pure state, often alloyed with other metals. Those alloys are utilized in high-wear applications. Osmium alloys such as osmiridium are very hard and, along with other platinum group metals, are used in the tips of fountain pens, instrument pivots, and electrical contacts, as they can resist wear from frequent operation.

Use and exposure

Osmium tetroxide is a colorless to pale yellow crystalline solid with an odor that has been described as pungent or chlorine-like. Osmium tetroxide is used as a tissue fixative for electron microscopy. It has been used in fingerprint detection and for staining fatty tissues for optical and electron microscopy studies. Students and workers must take precautions in the use of osmium tetroxide. Human poisoning and symptoms of osmium tetroxide are modulated by the route of exposure. Exposures to osmium tetroxide trigger and cause adverse health effects very rapidly at very low concentrations.

Toxicity and health effects

Occupational exposures to osmium tetroxide cause toxicity and adverse health effect to users and workers. The symptoms of toxicity and poisoning include, but are not limited to, the development of corrosive effects to the area of contact, severe chemical burns to the skin, blisters, discoloration, pain, eyes, burning sensation, tearing, cough, headache, wheezing, shortness of breath, pulmonary edema, and fatal injury/death. Exposures to osmium tetroxide cause skin redness or rash, visual disturbances, severe conjunctivitis, and possible permanent loss of vision. Accidental ingestion of osmium tetroxide causes abdominal cramps, burning sensation, vomiting, and collapse. Prolonged period of inhalation exposure to osmium tetroxide causes insomnia, digestive disturbance, and distress to the pharynx and larynx.

P

- Paraffin
- Parathaion
- Phenol and phenolic compounds
- Phosphamidon
- Picric acid
- Propetamphos

Paraffin (Paraffin wax, Hard paraffin)

Molecular formula: $C_nH_{2n}^{+2}$

Use and exposure

Paraffin wax is colorless or white with an odorless mass. It consists of a mixture of solid aliphatic hydrocarbons. Paraffin is used in the manufacture of paraffin papers, candles, food packaging materials, varnishes, floor polishes, to extract perfumes from flowers, in lubricants, and cosmetics. It is also used in waterproofing wood and cork.

Toxicity and health effects

Exposures to paraffin for a prolonged period cause several types of skin disorders, The adverse health effects to skin include chronic dermatitis, wax boils, folliculitis, comedones, papules, melanoderma, and hyperkeratoses. Studies of Hendricks et al. indicated the development of carcinoma of the scrotum in workers exposed to crude petroleum wax. Carcinoma of the scrotum in occupational workers began with a normal hyperkeratotic nevus-like lesion, which subsequently resulted in a squamous cell carcinoma.

Parathion (CAS No. 56-38-2)

Molecular formula: $C_{10}H_{14}NO_5PS$
IUPAC name: O, O-diethyl O-4-nitrophenyl phosphorothioate
Synonyms and trade names: Folidol; Fostox; Ethyl parathion; Etilon; Niran; Orthophos; Panthion; Paramar; Paraphos; Parathene
Toxicity class: US EPA: I; WHO: Ia

Uses and exposures

Pure parathion is a pale yellow liquid with a faint odor of garlic, while technical parathion is a deep brown to yellow liquid. It is sparingly soluble in water, but soluble in alcohols, aromatic hydrocarbons, esters, ethers, n-hexane, dichloromethane, 2-propanol, toluene, and ketones. Parathion is one of the most acutely toxic pesticides and the US EPA has classified parathion as an RUP, meaning it should only be handled by qualified, trained, and certified workers. In January 1992, the US EPA announced the cancellation of parathion for all uses on fruit, nut, and vegetable crops.

Parathion was used for the control of pests of fruits, nuts, and vegetable crops. The only uses retained are those on alfalfa, barley, corn, cotton, sorghum, soybeans, sunflowers, and wheat. Further, to reduce exposure of agricultural workers, parathion may only be applied to these crops by commercially certified aerial applicators and treated crops may not be harvested by hand. Parathion is a broad spectrum, organophosphate pesticide used to control many insects and mites.

Toxicity and health effects

Parathion is highly toxic by all routes of exposure. Parathion, like all organophosphate pesticides, inhibits acetylcholinesterase and alters cholinergic synaptic transmission at neuroeffector junctions (muscarinic effects), at skeletal myoneural junctions, in autonomic ganglia (nicotinic effects), and in the CNS. Exposures to parathion cause symptoms of poisoning that include, but are not limited to, abdominal cramps, vomiting, diarrhea, pinpoint pupils, blurred vision, excessive sweating, salivation and lacrimation, wheezing, excessive tracheobronchial secretions, agitation, seizures, bradycardia or tachycardia, muscle twitching and weakness, and urinary bladder and fecal incontinence. Seizures are much more common in children than in adults. Severe exposures cause loss of consciousness, coma, excessive bronchial secretions, respiratory depression, cardiac irregularity, eventually leading to death. Occupational workers and the general public with health disorders and abnormalities, such as cardiovascular, liver or kidney diseases, glaucoma, or CNS, are at an increased risk of parathion poisoning. Further, high environmental temperatures enhance the severity of parathion poisoning.

Parathion and human fatalities

Human fatalities associated with parathion due to accidental ingestion, spillage of sprays, suicide, and improper use are many. Reports have shown that many school-going children suffered fatalities when exposed to parathion-contaminated air and soil (earlier sprayed with parathion). Also, in Kerala, India, during a shipment, food was contaminated with parathion and the consumption of the same food eventually resulted in large human fatalities. Similarly, consumption of contaminated wheat flour in Jamaica caused poisoning and fatalities. There are more reported cases of poisoning with parathion than with any other pesticide currently in use. There have been a number of cases where intoxication and death have resulted from ingestion of foodstuffs that have been grossly contaminated with parathion. In one Asian country, there were 828 cases of poisoning with 106 deaths caused by flour, sugar, and other foodstuffs becoming contaminated because parathion was transported in the same ship's hold as the food. In another Asian country, barley became contaminated with parathion. There were 38 cases of poisoning with 9 deaths. Similarly, in a country in the Americas, there were 559 cases of poisoning and 16 deaths when sacks of sugar, and possibly flour, absorbed parathion from the floor of a truck. In another country in the Americas, there were 165 known and 445 more suspected cases of poisoning with 63 deaths, when parathion from broken containers contaminated sacks of flour during transportation in a truck. In a European country, parathion caused 26 cases of poisoning owing to transportation of food material in a

parathion-contaminated wagon. There is no evidence suggesting that parathion is mutagenic. Parathion produced embryocidal effects and fetal growth retardation, but no malformations, in mice and rats at doses that were generally below the level that was toxic
for the mother. Evidence is inadequate to evaluate the carcinogenicity of parathion in
experimental animals, and the available data are insufficient to evaluate the carcinogenicity of parathion for humans. The recommended ADI for parathion has been reported
as 0.004 mg/kg body weight.

References

Agency for Toxic Substances and Disease Registry (ATSDR). 2006. Medical Management Guidelines
    for Parathion. Division of Toxicology and Environmental Medicine, Atlanta, GA.
Diggory, H.J.P., Landrigan, P.J., Latimer, K.P., Ellington, A.C., Kimbrough, R.D. et al. 1977. Fatal para-
    thion poisoning caused by contamination of flour in international commerce. *Am. J. Epidemiol.*
    106: 145–53.
International Programme on Chemical Safety (IPCS). 1992. Parathion. Health and Safety Guide No.
    74. World Health Organization, Geneva, Switzerland.
International Programme on Chemical Safety and the Commission of the European Communities
    (IPCS, CEC). 1990. Parathion. ICSC card no. 0006. Commission of the European Communities,
    Luxembourg (updated 1999).
Material Safety Data Sheet (MSDS). 1993. Parathion. Extension Toxicology Network (EXTOXNET),
    Pesticide Information Profiles (PIP). Oregon State University, USDA/Extension Service,
    Corvallis, OR.
US Environmental Protection Agency (US EPA). 1992. Ethyl Parathion – Correction to the Amended
    Cancellation Order, OPP. US EPA, Washington DC (February 4, 1992).

n-Pentane (CAS No. 109-66-0)

Molecular formula: $CH_3(CH_2)_3CH_3$

Use and exposure

n-Pentane is a flammable liquid. It has applications in industry as an aerosol propellant
and as an important component of engine fuel. N-propane is a CNS depressant. Studies
with dogs have indicated that it induces cardiac sensitization. In high concentrations it
causes incoordination and inhibition of the righting reflexes.

Petroleum ether (Ligroin petroleum benzene)

Use and exposure

Petroleum ether is a flammable liquid used as a universal solvent and extractant during
the processing of different chemicals, like fats, waxes, paints, varnishes, furniture polish
thinning, as detergent, and as fuel. The major components include paraffins, olefins, naph-
thenes, aromatics, and about 10%–40% ethyl alcohol.

Toxicity and health effects

After acute and prolonged exposures to petroleum ether in work places, occupational
workers suffer and demonstrate a variety of health disorder. The symptoms of toxicity
include, but are not limited to, erythema, edema, skin peeling, loss of appetite, muscle
weakness, paresthesia, CNS depression, peripheral nerve disorders, skin and respiratory

irritation, and chemical pneumonia in children. Occupational workers exposed to rubber solvent, varnish, thinner, and petroleum spirits develop skin irritation, respiratory problems, and hematologic effects.

Phenol (CAS No. 108-95-2)

Molecular formula: C_6H_5OH
Synonyms and trade names: Carbolic acid; Phenic acid; Hydroxybenzene; Phenyl hydrate

Use and exposure

Phenol is a stable chemical substance of colorless/white crystals with a characteristically distinct aromatic/acrid odor. It is reactive and incompatible with strong oxidizing agents, strong bases, strong acids, alkalis, and calcium hypochlorite. It is flammable and discolors in light. Phenol is used in the manufacture or production of explosives, fertilizer, coke, illuminating gas, lampblack, paints, paint removers, rubber, perfumes, asbestos goods, wood preservatives, synthetic resins, textiles, drugs, and pharmaceutical preparations. It is also used extensively as a disinfectant in the petroleum, leather, paper, soap, toy, tanning, dye, and agricultural industries. Phenol is a systemic poison and constitutes a serious health hazard. The risks of using it in the laboratory must be fully assessed before work begins. Typical MEL 2 ppm; typical OEL 1 ppm.

Toxicity and health effects

Exposures to phenol cause adverse health effects and poisoning. Phenol is absorbed very rapidly through surfaces of the skin, lungs, and stomach. The symptoms of prolonged exposures and poisoning include, but are not limited to, vomiting, difficulty in swallowing, diarrhea, lack of appetite, headache, fainting, dizziness, mental disturbances, and skin rash. Direct contact with phenol causes burning of the mouth, irritation to the eyes, nose, and dermatitis, discoloration of the skin, and damage to the liver and kidneys. Exposure to phenol in different concentrations is known to cause mental disturbances, depression of the CNS, and coma.

Precautions

Acute poisoning of phenol by ingestion, inhalation or skin contact may lead to death. Phenol is readily absorbed through the skin. It is highly toxic by inhalation. It is corrosive and causes burns and severe irritation effects. During use and handling of phenol, occupational workers should be very careful. Workers should use protective clothing, rubber boots, and goggles to protect the eyes from vapors and spillage.

Phenol and Phenolic Compounds

Phenols are the simplest group of compounds. Phenols have a wide application in industries in the manufacture of pharmaceuticals. Phenols, in principle, are a group of chemicals that cause severe irritation to the body system. For example, they cause severe irritation to the eyes, skin, nose, respiratory tract, and mucous membranes. They are highly corrosive to the skin and tissues. Creosate, a mixture of phenolic and aromatic substances, is well known as a carcinogenic agent to the skin. Phenols also cause adverse effects to the CNS, cardiovascular, renal, and hepatic systems of animals and humans. In view of this, proper use, disposal, and management are a must for all occupational workers.

Phosphamidon (CAS No. 13171-21-6; mixture); (CAS No. 297-99-4; trans-isomer); (CAS No. 23783-98-4; cis-isomer)

Molecular formula: $C_{10}H_{19}Cl\ NO_5P$
IUPAC name: 2-Chloro-2-diethyl-carbamoyl-1-methylvinyl-dimethylphosphate
Synonyms and trade names: Aimphon; Dimecron; Kinadon; Phosron; Rilan; Rimdon
Toxicity class: US EPA: I; WHO: Ia

Use and exposure

Phosphamidon is a pale yellow to colorless oily liquid with a faint odor. It is miscible with water and is soluble in aromatic hydrocarbons. Phosphamidon decomposes on heating and releases highly toxic fumes, such as phosphorus oxides, hydrogen chloride, and nitrogen oxides. It reacts with bases (hydrolysis) and attacks metals such as iron, tin, and aluminium. Phosphamidon should be handled by trained personnel wearing protective clothing. Phosphamidon is used as a broad-spectrum insecticide and acaricide for the control of pests and vectors on crops like sugar cane, rice, citrus orchards, and cotton. Occupational exposures to phosphamidon occur among factory workers involved in synthesizing formulation and dispensing spray operations. Human exposures also occur among crop harvesters and in vector control operations.

Toxicity and health effects

Phosphamidon is readily absorbed from the gastrointestinal tract, through the intact skin, and by inhalation of spray mists and dusts. Prolonged exposures to phosphamidon cause adverse effects and impairment on the respiratory, myocardial, and neuromuscular transmission in animals and humans. Phosphamidon does not cause delayed peripheral neuropathy in hens. The symptoms of poisoning include, but are not limited to, nausea, vomiting, diarrhea, abdominal cramps, headache, dizziness, eye pain, blurred vision, constriction or dilation of the pupils, tears, salivation, sweating, and confusion. Prolonged period of exposures to phosphamidon cause incoordination, slurred speech, loss of reflexes, weakness, fatigue, involuntary muscle contractions, twitching, tremors of the tongue or eyelids, and eventually paralysis of the body extremities and the respiratory muscles, involuntary defecation or urination, psychosis, irregular heart beat, unconsciousness, convulsions, coma, respiratory failure or cardiac arrest leading to death. Phosphamidon has caused clastogenic effects in the bone marrow cells of rats and mice. However, the studies are found to be inadequate to arrive at meaningful conclusions about phosphamidon as a human carcinogen and no data are available. The ADI for phosphamidon has been reported as 0.0005 mg/kg body weight. (ADI: acceptable daily intake is an estimate of the amount of a pesticide, expressed on a body weight basis, which can be ingested daily over a lifetime without appreciable health risk.)

References

International Programme on Chemical Safety (IPCS). 2001. Phosphamidon. Poisons Information Monograph 454.

International Programme on Chemical Safety and the Commission of the European Communities (IPCS, CEC). 2005. Phosphamidon. ICSC card no. 0189. Commission of the European Communities, Luxembourg.

Jokanovic, M., Maksimovic, M. and Stepanovic, R.M. 1995. Interaction of phosphamidon with neuropathy target esterase and acetylcholinesterase of hen brain. *Arch. Toxicol.* 69 (6): 425–28.
World Health Organization (WHO). 2001. Phosphamidon, Data Sheets on Pesticides No. 74 (WHO/VBC/DS/87.74).

Picric acid (CAS No. 88-89-1)

Molecular formula: $HOC_6H_2(NO_2)_3$
Synonyms and trade names: Picronitric acid, 2,4,6-Trinitrophenol, Trinitrophenol

Use and exposure

Picric acid is a highly flammable, white to yellowish crystalline substance. It used in the manufacture of fireworks, matches, electric batteries, colored glass, explosives, and disinfectants. Pharmaceutical, textile, and leather industries also make use of picric acid.

Toxicity and health effects

Exposures to picric acid cause different adverse effects on the skin of animals and humans, such as allergies, dermatitis, irritation, and sensitization. Absorption of picric acid by the system causes headache, fever, nausea, diarrhea, and coma. In high concentrations, picric acid is known to cause damage to the erythrocytes, kidneys, and liver.

Precautions

Workers should use protective clothing, avoid skin contact, and use goggles and face masks to avoid dust.

References

Chicago National Safety Council. 1969. Picric acid. Data Sheet No. 351. Chicago National Safety Council, Chicago, Ellinois, USA.
Dikshith (ed.). 2009. *Safe use of Chemicals – A practical guide*. CRC Press, Boco Raton, FL, pp. 66–67.

Propane (CAS No. 74–98–6)

Molecular formula: $CH_3CH_2CH_3$

Use and exposure

Propane is released to the living environment from automobile exhausts, burning furnaces, natural gas sources, and during combustion of polyethylene and phenolic resins. Propane is both highly inflammable and explosive and needs proper care and management of workplaces. Its use in industry includes as a source for fuel and propellant for aerosols. Occupational workers exposed to liquefied propane have demonstrated skin burns and frostbite. Propane also causes depression effects on the CNS.

Propetamphos (CAS No. 31218-83-4)

Molecular formula: $C_{10}H_{20}NO_4PS$

IUPAC name: (E)-O-2-Isopropoxycarbonyl-1-methylvinyl O-methyl ethyl-phosphoramidothioate
Synonyms and trade names: Blotic; Safrotin; Seraphos
Toxicity class: US EPA: II; WHO: Ib

Use and exposure

Propetamphos technical is a yellowish, oily liquid at room temperature and a moderately toxic acaricide and insecticide. It is sparingly soluble in water, but very soluble in acetone, chloroform, ethanol, and hexane. The US EPA has classified propetamphos as a GUP and RUP, indicating that its handling should be done by certified, qualified applicators. Propetamphos is used for the control of cockroaches, flies, ants, ticks, moths, fleas, and mosquitoes in households and where vector eradication is necessary to protect public health. It is also used in veterinary applications to combat parasites, such as ticks, lice, and mites in livestock. The formulations of propetamphos include aerosols, emulsified concentrates, liquids, and powders. Commercial products include aerosols, emulsified concentrates, liquids, and powders.

Toxicity and health effects

Propetamphos is a moderately toxic organophosphate insecticide. It inhibits the cholinesterase enzyme in animals and humans leading to overstimulation of the CNS. Prolonged period of exposures to high concentrations of propetamphos cause poisoning with symptoms that include, but are not limited to, nausea, headache, dizziness, confusion, numbness, tingling sensations, incoordination, tremor, abdominal cramps, sweating, blurred vision, breathing difficulty, unconsciousness, and convulsions, slow heart beat, respiratory paralysis, and death. There is no evidence to suggest that propetamphos causes mutagenic, teratogenic, or carcinogenic effects in animals or humans.

References

Material Safety Data Sheet (MSDS). 1996. Propetamphos. Extension Toxicology Network (EXTOXNET), Pesticide Information Profiles (PIP). Oregon State University, USDA/Extension Service, Corvallis, OR.
Tomlin, C.D.S. (ed.). 1997. *The Pesticide Manual: A World Compendium*, 11th ed. British Crop Protection Council, Farnham, Surrey, U.K.
US Environmental Protection Agency (US EPA). 2000. Propetamphos Facts, United States Prevention, Pesticides, Environmental Protection and Toxic Substances (EPA 738-F-00-16 October 2000).
US Public Health Service. 1996. Hazardous Substance Data Bank. Washington, DC.

Q

- Quinalphos
- Quinone

Quinalphos (CAS No. 82-68-8)

Molecular formula: $C_6Cl_5NO_2$
Synonyms and trade names: Nitropentachlorobenzene; PCNB; Avicol; Batrilex; Botrilex; Brassicol; Chinozan; Folosan; Fomac 2; Fungichlor; Kobu; Kobutol; Marisan forte; Olpisan; Pentagen; Quintocene; Quintozene; Terrachlor; Tilcarex; Tri-pcnb; Tritisan

Use and exposure

Quinalphos is an off-white powder. It is stable and incompatible with strong bases and strong oxidizing agents. Quinalphos is uses as a fungicide and herbicide.

Toxicity and health effects

Exposures to quinalphos cause toxicity and adverse health effects. On contact with the skin, quinalphos causes the effects of sensitization.

Quinalphos and carcinogenicity

Quinalphos is an experimental carcinogen and neoplastigen.

References

Material Safety Data Sheet (MSDS). 2005. Quinalphos. Physical Chemistry at Oxford University, U.K.
Dikshith, T.S.S. and Diwan, P.V. 2003. *Industrial Guide to Chemical and Drug Safety*. Wiley, J. Hoboken, NJ.

Quinone (CAS No. 78919-13-8)

Molecular formula: $C_6H_4O_2$
Synonyms and trade names: Benzoquinone; Chinone; p-Benzoquinone; 1,4-Benzoquinone

Use and exposure

Quinone exists as a large, yellow, monoclinic prism with an irritating odor. Quinone is extensively used in the dye, textile, chemical, tanning, and cosmetic industries. In chemical synthesis for hydroquinone and other chemicals, quinone is used as an intermediate. It is also used in the manufacturing industries and chemical laboratories associated with protein fiber, photographic film, hydrogen peroxide, and gelatin making.

Toxicity and health effects

Exposures to quinine vapor are highly irritating to the eyes and may be followed by corneal opacities, structural changes in the cornea, and loss of vision. Solid quinone may produce discoloration, severe irritation, swelling, and form papules and vesicles.

References

Anderson, B. and Oglesby, F. 1958. Corneal changes from quinone-hydroquinone exposure. *Arch. Ophthalmol.* 59: 495.

Seutter, E. and Sutorius, A.H.M. 1972. Quantitative analysis of hydroquinone in urine. *Clin. Chim. Acta.*, **38**, 231.

S

- Selenium
- Styrene
- Sulfur dioxide

Selenium (CAS No.7782-49-2)

Molecular formula: **Se**

Use and exposure

Jewelers most frequently encounter selenium in the form of brass-black and gun-bluing compounds. Selenium print toner used by photographers is sometimes used by jewelers as a metal-coloring solution. These coloring mixtures usually contain selenic acid. Selenic acid can release hydrogen selenide gas that can cause illness, and used daily, it might enlarge the liver and spleen. Tellurium is sometimes used in association with selenium.

Toxicity and health effect

Exposures to selenium cause adverse health effects. selenium dioxide is a by-product of copper and nickel melting, or of heating alloys containing selenium. The dusts are very irritating to the mucous membranes and the lungs. Contact can cause dermatitis.

Selenium and carcinogenicity

There are no published reports available regarding the carcinogenicity, mutagenic effects and teratogenic effects of sodium. International Agency for Research on Cancer observed that selenium and selenium compounds are not classifiable as human carcinogens.

Precautions

During use and handling of selenium, occupational workers should be careful to avoid contact with the skin. Selenium compounds are considered very damaging to the liver, and hazardous.

Reference

Agency for Toxic Substances and Disease Registry (ATSDR). 2003. Toxicological profile for selenium. US Department of Health and Human Services, Public Health Service. Atlanta, GA.

Glover, J.R. 1970. Selenium and its Industrial Toxicology. Ind.Med.Sug. 30, 50.

International Agency for Research on Cancer (IARC). 1998. Selenium and Selenium Compound. IARC. Vol. 9. p. 245.

International Programme on Chemical Safety and the Commission of the European Communities (IPCS CEC). 1999. Selenium. ICSC card no. 0072. IPCS, CEC ICSC, Geneva, Switzerland.

Styrene (CAS No. 100-4205)

Molecular formula: C_8H_8
Synonyms and trade names: Cinnamene; Cinnamol; Phenyl ethyl benzene; Phenylethane; Styrol; Styrolene

Use and exposure

Styrene is a colorless or yellow, sweet odor liquid. It is produced during alkylation of benzene with ethylene. It is highly reactive and polymerizes rapidly with a violent explosive reaction. This demands proper handling, transportation, and storage by adding polymerization inhibitors in adequate quantities during these operations. Styrene monomer has been extensively used in the manufacture of chemical intermediates, filling components, plastics, resins, and stabilizing agents.

Toxicity and health effects

Exposures to styrene induce adverse health effects, which include irritation to the eyes, mucous membrane, loss of appetite, vomiting, and nausea. Prolonged exposure results in skin damage in the form of dermatitis, rough, and fissured skin.

Styrene and carcinogenicity

Laboratory studies indicated styrene has carcinogenic potential to rodents and, accordingly, the IARC has classified the chemical under Group 2B, meaning a possible human carcinogen.

References

Agency for Toxic Substances and Disease Registry (ATSDR). 2007. Toxicological Profile for Styrene. US Department of Health and Human Services, Public Health Service, Atlanta, GA (updated 2010).
Material Safety Data Sheets (MSDS). 2008. Styrene. Physical Chemistry at Oxford University, U.K.

Sulfur dioxide (CAS No.7446-09-5)

Molecular formula: SO_2
Synonyms and trade names: Sulfur superoxide; Sulfur dioxide; Sulfurous acid anhydride; Fermenicide liquid; Fermenicide powder

Use and exposure

Sulfur dioxide is a colorless gas. It is stable, and non-corrosive when dry to common materials except zinc. Sulfur dioxide is corrosive when wet and incompatible with strong reducing or oxidizing agents, moisture, zinc, and its alloys. Sulfur dioxide has a large number of industrial applications. For instance, sulfur dioxide is used in the manufacture of sodium sulfite, sulfuric acid, sulfuryl chloride, thionyl chloride, organic sulfonate, disinfectants, fumigants, industrial and edible proteins, etc. Sulfur dioxide is also used extensively as a bleaching agent, particularly in the bleaching of beet sugar, flour, straw, textiles, and wood pulp. Sulfur dioxide has industrial utility in the tanning of leather, in brewing and preserving. Sulfur dioxide is a colorless gas with a characteristic and strong suffocating odor.

Sulfur dioxide gas is released primarily from the combustion of fossil fuels (75%–85% of the industrial sources), the smelting of sulfide ores, volcanic emissions, and several other natural sources. It is a US EPA priority air pollutant, but has many industrial and agricultural uses. It is sometimes added as a warning marker and fire retardant to liquid grain fumigants.

Toxicity and health effects

Exposures to sulfur dioxide cause adverse health effects to users and occupational workers. The gaseous sulfur dioxide is particularly irritating to the mucous membranes of the upper respiratory tract. Chronic exposure to sulfur dioxide produces dryness of the throat, cough, rhinitis, conjunctivitis, corneal burns, and corneal opacity. Acute exposure to high concentrations of sulfur dioxide may also result in death due to asphyxia. By contrast, chronic exposures to sulfur dioxide lead to nasopharyngitis, fatigue, and disturbances of the pulmonary function. Animals exposed to chronic doses of sulfur dioxide have shown thickening of the mucous layer in the trachea and also hypertrophy of goblet cells and mucous glands resembling the pathology of chronic bronchitis. It has been found that penetration of sulfur dioxide into the lungs is greater during mouth breathing than during nose breathing. In fact, an increase in the flow rate of the gas would markedly increase the penetration. Human subjects exposed for very brief periods to sulfur dioxide also showed alterations in pulmonary mechanics. More information on the adverse effects of sulfur dioxide and the manner of its potentiation in association with other chemicals may be found in literature.

Sulfur dioxide and carcinogenicity

The IARC classified sulfur dioxide as Group 3, meaning not classifiable as a human carcinogen.

Exposure limits

The OSHA set the PEL as 5 ppm (TWA 8-h workshift).

Precautions

Occupational workers should be careful at workplaces because exposure to sulfur dioxide occurs from breathing it in the air. It affects the lungs and at high levels may result in burning of the nose and throat, breathing difficulties, and severe airway obstructions.

References

Agency for Toxic Substances and Disease Registry (ATSDR). 1998. Toxicological Profile for Sulfur Dioxide. US Department of Health and Human Services, Public Health Service, Atlanta, GA (updated 2010).

Material Safety Data Sheet (MSDS). 2006. Safety data for sulfur dioxide. Department of Physical Chemistry, Oxford University, Oxford, U.K.

Skalpe, I.O. 1964. Long-term effects of sulfur dioxide exposure in pulp mill. *Br. J. Ind. Med.* 21: 69.

T

- Temephos
- Tin
- Toluene
- Toluene diisocyanate
- O-Toluidine
- Triethanolamine
- Turpentine

Temephos (CAS No. 3383-96-8)

Molecular formula: $C_{16}H_{20}O_6P_2S_3$
IUPAC name: O,O,O,O'-Tetramethyl-O,O'-thiodi-p-phenylene-bis(phosphorothioate)
Synonyms and trade names: Abate; Bathion; Bithion; Difennthos; Ecopro; Nimitox; Temeguard
Toxicity class: US EPA: III; WHO: III

Use and exposure

Temephos is a solid at room temperature and is composed of colorless crystals. As a liquid, it is brown and viscous. Temephos is a non-systemic insecticide. It is insoluble in water, hexane, and methyl cyclohexane, but soluble in common organic solvents. It is combustible and the liquid formulations contain organic solvents that may be flammable. Temophos decomposes on heating or burning, producing toxic fumes such as phosphorus oxides and sulfur oxides. Temephos reacts with strong acids and strong bases. The US EPA has grouped temephos as a GUP. Temephos is used for the control of mosquito, midge, and black fly larvae. It is used in lakes, ponds, and wetlands. It may also be used to control fleas on dogs and cats and to control lice on humans.

Toxicity and health effects

Exposures to high concentrations of temephos for a prolonged period of time cause poisoning to occupational workers. The symptoms include, but are not limited to, nausea, salivation, headache, loss of muscle coordination, and difficulty breathing. Workers exposed to a combination of temephos and malathion suffer potentiation of toxicity of temephos. Exposures to granules of temephos formulation may cause irritation. Prolonged and repeated skin contact may cause irritation. Inhalation may cause respiratory irritation and chemical pneumonities. There are no reports on the prolonged effects of temephos.

Temephos and carcinogenicity

Reports have indicated that temephos is not mutagenic or carcinogenic to animals and humans.

Tin (CAS No. 25583-20-4)

Use and exposure

Tin is a soft, white, silvery metal that is insoluble in water. Tin metal is used to line cans for food, beverages, and aerosols. It is present in brass, bronze, pewter, and some soldering materials. Tin can combine with other chemicals to form various compounds. When tin is combined with chlorine, sulfur, or oxygen, it is called an inorganic tin compound. Inorganic tin compounds are found in small amounts in the earth's crust. They are also present in toothpaste, perfumes, soaps, coloring agents, food additives, and dyes. Tin when combined with carbon forms organotin compounds. These compounds are used in making plastics, food packages, plastic pipes, pesticides, paints, wood preservatives, and rodent (rats and mice) repellants. Organic tin compounds stick to soil, sediment, and particles in water. Humans are usually exposed to tin at far less than 1 ppm in air and water. The amounts in air and water near hazardous waste sites could be higher. The most common use for tin coating is for edge retention and corrosion resistance on machine tooling, such as drill bits and milling cutters, often improving their lifetime by a factor of three or more.

Toxicity and health effects

Inhalation breathing in, oral eating or drinking, or dermal exposure skin contact to some organotin compounds has been shown to cause harmful effects in humans, but the main effect will depend on the particular organotin compound. There have been reports of skin and eye irritation, respiratory irritation, gastrointestinal effects, and neurological problems in humans exposed for a short period of time to high amounts of certain organotin compounds. Some neurological problems have persisted for years after the poisoning occurred. Lethal cases have been reported following ingestion of very high amounts. Studies in animals have shown that certain organotins mainly affect the immune system, but a different type primarily affects the nervous system. Yet, there are some organotins that exhibit very low toxicity. Exposure of pregnant rats and mice to some organotin compounds has reduced fertility and caused stillbirth, but scientists are still not sure whether this occurs only with doses that are also toxic to the mother. Some animal studies also suggest that the reproductive organs of males may be affected.

Tin and carcinogenicity

Inorganic tin compounds are not known to cause cancer. There are no studies and/or published reports indicating that exposures to organotin compounds cause cancer in humans. The US EPA reported that organotin and tributyl-tin oxide are not classifiable as human carcinogens.

Reference

Agency for Toxic Substances and Disease Registry (ATSDR). 2005. Toxicological Profile for Tin. US Department of Health and Human Services, Public Health Service, Atlanta, GA (updated 2007).

Toluene (CAS No. 108-88-3)

Molecular formula: $C_6H_5CH_3$
Synonyms and trade names: Methacide; Methylbenzene; Phenylmethane; Toluol; Toluene (dissolved)

Use and exposure

Toluene is a clear, colorless, flammable liquid with a sweet/pungent odor. It is extensively used as a solvent in different industries, i.e., rubber chemical manufacturing, drugs and pharmaceuticals, thinner for inks, paints dyes, and perfume manufacturing. It is a natural constituent of crude oil and is produced from petroleum refining and coke-oven operations. Toluene occurs naturally as a component of crude oil and occurs in petroleum refining and coke oven operations.

Occupational workers associated with several kinds of activities, such as manufacturing of dyes, printing inks, painting automobile mechanics, gasoline manufacturers, shippers, and retailers, adhesives and coatings manufacturers and applicators, audio-equipment product workers, chemical industry workers, coke-oven workers, fabric manufacturers (fabric coating), sites of hazardous wastes, linoleum manufacturers, in pharmaceutical manufacturing, printing works, shoe manufacturing industry, become exposed to toluene.

Toxicity and health effects

Exposures to toluene cause adverse health effects to animals and humans. The symptoms of toxicity and poisoning include, but are not limited to, mild irritation to the skin, headache, nausea, and effects on the CNS. Prolonged exposure to high concentrations of toluene causes disturbances in vision, dizziness, nausea, CNS depression, paresthesia, and sudden collapse. The acute oral LD_{50} value of toluene in laboratory rats has been reported as 636–7300 mg/kg. Exposure to toluene has been reported to cause rapid and severe corneal damage and conjunctiva inflammation. The acute dermal LD_{50} in rabbits was found to be between 1200 and 1400 mg/kg.

Toluene and carcinogenicity

Studies in humans and animals generally indicate that toluene does not cause cancer. The US EPA indicated that the carcinogenicity of toluene cannot be classified. In view of the inadequate evidence for the carcinogenicity of toluene in humans and lack of evidence on carcinogenicity in experimental animals, the IARC classified toluene as Group 3, meaning not classifiable as a human carcinogen.

Exposure limits

The US EPA has set a limit of 1.0 mg/L in drinking water. Regulatory agencies further state that any kind of discharges, releases, or spills of more than 1000 pounds of toluene must be reported to the National Response Center. The OSHA has set a limit of 200 ppm of toluene at the workplace air.

References

Agency for Toxic Substances and Disease Registry (ATSDR). 2000. Toxicological Profile for Toluene. US Department of Health and Human Services, Public Health Service, Atlanta, GA.

Juntunen, J.E., et al. 1985. Nervous system effects of long-term occupational exposure to toluene. *Acta Neurol. Scand.* 72: 512–17 (cited in NTP, 1990).

National Toxicology Program. (NTP). 1990. Toxicology and Carcinogenesis Studies of Toluene in F344/N Rats and B6C3F1 Mice (Inhalation Studies). Technical Report Series No. 371. US Department of Health and Human Services, Public Health Service, National Institutes of Health, Research Triangle Park, NC.

US Environmental Protection Agency (US EPA). 1992. Integrated Risk Information System (IRIS) on Toluene. Environmental Criteria and Assessment Office. Office of Health and Environmental Assessment, Cincinnati, OH.

World Health Organization (WHO). 1985. Toluene. Environmental Health Criteria 52. World Health Organization, Geneva (cited in US Air Force, 1989).

2,4-Toluene diisocyanate (CAS No. 584-84-9)

Molecular formula $C_9H_6N_2O_2$

Synonyms and trade names: Toluene-2,4-diisocyanate; 2,4-Diisocyanato-1-methylbenzene; Di-iso-cyanatotoluene; TDI, creorcinol diisocyanate; m-Toluene diisocyanate; 4-Methyl-1,3-phenylene diisocyanate; 4-Methyl-phenylene diisocyanate; Hylene T; Hylene TCPA; Mondur TD; Niax TDI

Use and exposure

Toluene diisocyanate exists in two isomeric forms (2,4-toluene diisocyanate and 2,6-toluene diisocyanate), which have similar properties and effects. Toluene diisocyanate is produced commercially as an 80:20 (2,4-toluene diisocyanate:2,6-toluene diisocyanate) mixture of the two isomers. At room temperature, the mixture is a clear, pale yellow liquid with a sharp, pungent odor. It should be stored under refrigeration, away from light and moisture in a tightly closed container in an inert atmosphere. Toluene diisocyanate is insoluble in water and miscible with most common organic solvents.

Toluene diisocyanate is made by reacting toluene diamine with carbonyl chloride (phosgene). Toluene diisocyanate is commonly used as a chemical intermediate in the production of polyurethane foams, elastomers, and coatings, paints, varnishes, wire enamels, sealants, adhesives, and binders. It is also used as a cross-linking agent in the manufacture of nylon polymers. 2,4-Toluene diisocyanate is used as a chemical inter-mediate in the production of polyurethane products, such as foams, coatings, and elastomers.

Toxicity and health effects

Exposures to toluene diisocyanate cause severely irritating effects especially to the mucous membranes and tissues. Inhalation of toluene diisocyanate produces euphoria, ataxia, mental aberrations, vomiting, abdominal pain, respiratory sensitization, bronchitis, emphysema, and asthma. The mechanism by which toluene diisocyanate produces toxic symptoms is not known, but the compound is highly reactive and may inactivate tissue biomolecules by covalent binding. Acute and chronic exposures to toluene diisocyanate produce health disorders of the skin, the respiratory system, the CNS, and the gastroin-testinal tract. A number of occupational studies have reported that chronic exposures to toluene diisocyanate reduce lung function among workers associated with the production of polyurethane foam.

Exposure to toluene diisocyanate produces severe respiratory problems and individuals with pre-existing breathing difficulties may be more susceptible to its effects. It causes irritation of the respiratory tract. Concentration-dependent effects occur, often after a delay of 4–8 h and may persist for 3–7 days. Exposures to high-concentration inhalation of toluene diisocyanate cause symptoms of toxicity, such as chest tightness, cough, breathlessness, inflammation of the bronchi with sputum production and wheezing with possible accumulation of fluid in the lungs. Previously exposed occupational workers and users have been reported to develop symptoms of inflammation of the lungs on re-exposure to even extremely low levels of toluene diisocyanate. Flu-like symptoms, such as fever, malaise, shortness of breath, and cough, can develop 4–6 h after exposure and persist for 12 h or longer. In sensitized workers/individuals, asthmatic attacks occur after exposure to extremely low toluene diisocyanate air concentrations (0.0001 ppm). The asthmatic reactions occur immediately and/or delayed (4–8 h). Exposure to toluene diisocyanate leads to reactive airway dysfunction syndrome (RADS), a chemically or irritant-induced type of asthma. Children may be more vulnerable because of relatively increased minute ventilation per kilogram and failure to evacuate an area promptly when exposed.

2,4 Toluene diisocyanate and carcinogenicity

The DHSS has determined that toluene diisocyanate may reasonably be anticipated to be a carcinogen. The US EPA has not classified 2,4-toluene diisocyanate for carcinogenicity while the IARC classified 2,4-toluene diisocyanate as Group 2B, meaning a possible human carcinogen. However, toluene diisocyanate has not undergone a complete evaluation and determination under the US EPA's IRIS program for evidence of human carcinogenic potential.

Exposure limits

The OSHA has set a PEL of 0.02 ppm for toluene diisocyanate, and the NIOSH has set the IDLH as 2.5 ppm.

Precautions

Occupational workers should be careful during use and waste disposal of toluene diisocyanate. Use and storage of toluene diisocyanate requires precautions. It polymerizes under the influence of bases, tertiary amines, and acyl chlorides with fire or explosion hazard. On combustion, it forms toxic vapors and gases, including nitrogen oxides and isocyanates. Toluene diisocyanate reacts readily with water, acids, and alcohols, and causes explosion hazard.

The development of any unusual signs or symptoms, such as headache, increased pain or a discharge from the eyes, increased redness or pain or a pus-like discharge in the area of a skin burn within 24 h after exposure to toluene diisocyanate, requires immediate medical support to the exposed worker.

References

Adams, W.G.F. 1975. Long-term effects on the health of men engaged in the manufacture of tolulene di-isocyanate. *Br. J. Ind. Med.* 32: 72–78.

International Agency for Research on Cancer (IARC). 1986. IARC Monographs on the Evaluation of the Carcinogenic Risk of Chemicals to Humans, Vol. 39. World Health Organization, Lyon, France.

International Programme on Chemical Safety and the Commission of the European Communities (IPCS, CEC). 1995. ICSC 2,4-Toluene diisocyanate. ICSC card no. 0339. International Labor Organization (ILO), Geneva, Switzerland (updated 2004).

US Environmental Protection Agency (US EPA). 1999. Toluene diisocyanate mixture. Integrated Risk Information System (IRIS). National Center for Environmental Assessment, Office of Research and Development, Washington, DC.

———. 2000. 2,4-Toluene diisocyanate hazard summary. Technology Transfer Network Air Toxics web site (updated 2007).

O-Toluidine (CAS No. 95-53-4)

Molecular formula: C_7H_9N
Synonyms and trade names: O-Methylamine; O-Aminotoluene; 2-Methylaniline

Use and exposure

O-Toluidine is a light yellow to reddish brown liquid. On exposure to light and atmospheric air, the compound quickly turns a dark color. The compound has extensive use in a large number of industries around the world. For instance, as an intermediate in the manufacture of azo and indigo dyes, pigments, sulfur dyes, pesticides, pharmaceutical products, rubber and vulcanizing chemicals, and similar products.

Toxicity and health effects

Exposures to O-toluidine cause toxicity and poisoning to animals and occupational workers. It is highly toxic to animals and humans and is rapidly absorbed by oral, dermal, and inhalation by mammals. The acute oral LD_{50} to rats ranges from 900 to 940 mg/kg. The compound is known to cause adverse effects in workers, which include headache, irritation of skin, eye, kidneys, bladder, and hematuria. O-Toluidine has caused hepatocellular adenoma and carcinoma in experimental laboratory mice and rats.

Occupational workers exposed to O-toluidine have also demonstrated bladder cancer although the role of aniline cannot be ruled out. However, the IARC working group, because of insufficient data, classify O-toluidine as a Group 2B agent, meaning possibly carcinogenic to humans, while the NIOSH classify this compound as an occupational carcinogen, and the ACGIH label it as a suspected human carcinogen under A2 class.

Triethanolamine (CAS No. 102-71-6)

Molecular formula: $(HOCH_2CH_2)_2N$
Synonyms and trade names: Triethyloamine; Trolamine; Trihydroxytriethylamine

Use and exposure

Triethanolamine is a pale yellow and viscous liquid. It is hygroscopic with an irritant and ammoniacal odor. There are multiple industrial and domestic applications for this compound, i.e., in the manufacture of toilet products, cosmetics formulations, solvents for waxes, resins, dyes, paraffins and polishes, herbicides, and lubricants for textile products. In the pharmaceutical industry, triethanolamine is used as a non-steroidal, anti-inflammatory agent, an emulsifier, and an alkylating agent.

Toxicity and health effects

Exposures to triethanolamine, in contrast with other chemical compounds, is known to cause low toxicity to animals and the acute oral LD_{50} to rats and guinea pigs ranges from 8000 to 9000 mg/kg. Triethanolamine was found to be a moderate eye irritant. A 5%–10% solution of triethanolamine did not induce skin irritation or skin sensitization. Studies of Inoue et al. and many other workers have indicated the absence of the mutagenic potential of triethanolamine as evidenced by both in vivo and in vitro studies (*Salmonella typhimurium* tests, Chinese hamster ovary cells, and rat liver chromosome analysis). Further, extensive studies have demonstrated the absence of potential carcinogenicity of triethanolamine in rats and mice, suggesting a low or lack of acute or chronic toxicity of the chemical to mammals.

Turpentine (CAS No. 8006-64-2)

Molecular formula: $C_{10}H_{16}$
Synonyms and trade names: Gum turpentine; Spirit of turpentine; Gum spirit; Wood turpentine

Use and exposure

Turpentine is oleorosin extracted from trees of **pinus** (pinacae). It is a yellowish, opaque, sticky mass with a characteristic odor and taste. It is used extensively in different industries associated with the manufacturing of polishes, grinding fluids, paint thinners, resins, degreasing solutions, clearing materials, and ink making. The two primary uses of turpentine in industry are as a solvent and as a source of materials for organic synthesis. As a solvent, turpentine is used for thinning oil-based paints for producing varnishes and as a raw material in the chemical industry.

Toxicity and health effects

Occupational exposures to turpentine cause adverse health effects on absorption through the skin, lungs, and intestine. The vapor of turpentine causes severe irritation to the nose, eyes, and respiratory system has a whole. Aspiration of liquid turpentine causes direct irritation to the lungs and results in pulmonary edema and hemorrhage. It also causes dermatitis, eczema, and hypersensitivity among occupational workers. Splashing of liquid turpentine in the eyes causes corneal burns. Turpentine is also known to cause skin eruption, irritation to the gastrointestinal tract, kidney and bladder damage, delirium, ataxia, and benign skin tumor.

Turpentine and carcinogenicity

The ACGIH classified turpentine as A4, meaning not classifiable as a human carcinogen.

Precautions

During handling of turpentine, occupational workers should always use protective clothing, rubber gloves, and face masks to avoid adverse health effects to the skin and respiratory tract.

Reference

Dikshith, T.S.S. and Diwan, P.V. 2003. *Industrial Guide to Chemical and Drug Safety*. Wiley, J. Hoboken, NJ.

V

- Vanadium

Vanadium (CAS No. 7440-62-2)

Molecular formula: **V**

Use and exposure

Vanadium is a soft, ductile, silver-gray metal. It has good resistance to corrosion by alkalis, sulfuric and hydrochloric acid, and salt water. Vanadium metal, sheet, strip, foil, bar, wire, and tubing are used in industries. It is used in high-temperature service, in the production of rust-resistant, high-speed tools, and is an important carbide stabilizer in making steels. In fact, most vanadium is used as an additive to improve steels. Vanadium steel is especially strong and hard, with improved resistance to shock. Vanadium pentoxide (V_2O_5) is perhaps vanadium's most useful compound. It is used as a mordant—a material that permanently fixes dyes to fabrics. Vanadium pentoxide is used as a catalyst in chemical reactions and in the manufacture of ceramics. Vanadium pentoxide can also be mixed with gallium to form superconductive magnets.

Toxicity and health effects

Exposures to high levels of vanadium cause harmful health effects. The major effects from breathing high levels of vanadium are on the lungs, throat, and eyes. Workers who breathe vanadium for short and long periods show lung irritation, coughing, wheezing, chest pain, runny nose, and sore throat. Prolonged period of exposures to respirable dusts and vanadium fume have caused potential symptoms of toxicity among occupational workers. The symptoms of poisoning include, but are not limited to, irritation of the eyes and throat, green tongue, metallic taste, sore throat, cough, drowsiness, wheezing, bronchitis, abdominal cramps, nausea, vomiting, diarrhea, respiratory distress, pulmonary edema, bronchial damage, epistaxis (bloody nose), eczema, conjunctivitis, headache, dry mouth, dizziness, nervousness, insomnia, and tremor. It is not classifiable as a human carcinogen. Vanadium is a natural component of fuel oil, and workers have developed vanadium poisoning during cleaning operations on oil-fired furnaces.

Vanadium and carcinogenicity

The DHHS, the IARC, and the US EPA have not classified vanadium as to its human carcinogenicity.

References

Agency for Toxic Substances and Disease Registry (ATSDR). 1992. Toxicological Profile for Vanadium. US Department of Health and Human Services, Public Health Service, Atlanta, GA.

Dikshith, T.S.S. (ed.). 2008. *Safe Use of Chemicals: A Practical Guide*. CRC Press, Boca Raton, FL.

X

- Xylene

Xylene (CAS No. 1330-20-7)

Molecular formula: C_8H_{10}

Use and exposure

Xylene occurs in the manufacture of different petroleum products and as an impurity in benzene and toluene. It is a colorless and flammable liquid. Commercial xylene is a mixture of three isomers, namely, ortho, meta, and para isomer. It is extensively used in different industries associated with paints, rubber, inks, resins, adhesives, paper coating, solvents, and emulsifiers. Also, xylene is used as an important raw material in the manufacture of plasticizers, glass-reinforced polyesters, and alkyd resins.

Toxicity and health effects

Exposures to xylene cause toxicity and adverse health effects to animals and humans. Acute and chronic exposure to xylene induces adverse effects on the skin and respiratory system of animals and humans. Prolonged exposure to xylene demonstrated burning effect, drying, defatting of skin, eye irritation, lung congestion, CNS excitation, depression, mucosal hemorrhage, and mild liver damage.

References

Browning, E. (ed.). 1965. *Toxicity and Metabolism of Industrial Solvents*. Elsevier, New York. pp. 285–88.

Cheremisinoff, N. P. (ed.). 2003. *Industrial Solvents Handbook*, New York, 2nd ed. Marcel Dekker.

Dikshith, T. S. S. (ed.) 2009. *Safe Use of Chemicals: A Practical Guide*. CRC Press, Boca Raton, FL.

Dikshith, T. S. S. and P. V. Diwan (eds.). 2003. *Industrial Guide to Chemical and Drug Safety*. Wiley, J. Hoboken, NJ.

Flick, E. W. (ed.). 1998. *Industrial Solvents Handbook*, 5th ed. Noyes, William Andrew, Norwich, NY.

Sittig, M. (ed.). 1985. *Handbook of Toxic and Hazardous Chemicals and Carcinogens*, 2nd ed. Noyes, Park Ridge, NJ.

Z

- Zinc
- Zinc chloride
- Zinc oxide

Zinc (CAS No. 7440-66-6)

Molecular formula: Zn

Use and exposure

Zinc is one of the most common elements in the earth's crust. Metal zinc was first produced in India and China during the middle ages. Industrially important compounds of zinc are zinc chloride ($ZnCl_2$), zinc oxide (ZnO), zinc stearate ($Zn(C_{16}H_{35}O_2)_2$), and zinc sulfide (Sphalerite, ZnS) found in hazardous waste sites. It is found in air, soil, and water, and is present in all foods. Pure zinc is a bluish-white shiny metal. Zinc has many commercial uses as coatings to prevent rust, in dry-cell batteries, and mixed with other metals to make alloys like brass and bronze. Zinc combines with other elements to form zinc compounds. Zinc compounds are widely used in industry to make paint, rubber, dye, wood preservatives, and ointments.

Toxicity and health effects

Zinc and its compounds are relatively non-toxic, but very large doses can produce an acute gastroenteritis characterized by nausea, vomiting, and diarrhea. The recommended dietary allowance (RDA) for zinc is 15 mg/day for men, 12 mg/day for women, 10 mg/day for children, and 5 mg/day for infants. Insufficient zinc in the diet can result in a loss of appetite, a decreased sense of taste and smell, slow wound healing and skin sores, or a damaged immune system.

Pregnant women with low zinc intake have babies with growth retardation. Exposure to zinc in excess, however, can also be damaging to health. Harmful health effects generally begin at levels from 10–15 times the RDA (in the 100–250 mg/day range). Eating large amounts of zinc, even for a short time, can cause stomach cramps, nausea, and vomiting. Chronic exposures to zinc chloride fumes cause irritation, pulmonary edema, bronchopneumonia, pulmonary fibrosis, and cyanosis. It also causes anemia, pancreas damage, and lower levels of high-density lipoprotein cholesterol. Breathing large amounts of zinc (as dust or fumes) can cause a specific short-term disease, called metal fume fever, including disturbances in the adrenal secretion. Information on the possible toxicological effects following prolonged period of exposures to high concentrations of zinc is not known.

Zinc compounds

Zinc chloride and Zinc oxide fumes

Use and exposure

Exposures to high concentrations of fumes of zinc chloride and zinc oxide for a chronic period have caused health effects to occupational workers. The symptoms of toxicity and poisoning include headache, blurred vision, low back pain, vomiting, fever, chills, muscle ache, dry throat, cough, weakness, exhaustion, metallic taste in the mouth, chest tightness, respiratory distress, and decreased pulmonary function. High concentrations of zinc chloride fumes cause irritation and redness in the eyes, irritation to the skin, nose, and throat, conjunctivitis, burning effect, cough with sputum, breathing problems, chest pain, pulmonary edema, pneumonitis, and pulmonary fibrosis.

Precautions

Occupational workers should be careful during use and handling of zinc and zinc compounds. It is harmful if swallowed or inhaled and may cause irritation to the skin, eyes, and respiratory tract. It may form combustible dust concentrations in air. Zinc powder can react violently with water, sulfur, and halogens. Zinc powder is dangerous or potentially dangerous with strong oxidizing agents, lower molecular weight chlorinated hydrocarbons, strong acids, and alkalis.

Zinc and carcinogenicity

The DHHS and the IARC have not classified zinc for carcinogenicity. Based on incomplete information from human and animal studies, the US EPA grouped zinc as a nonclassifiable human carcinogen.

References

Agency for Toxic Substances and Disease Registry (ATSDR). 2005. Toxicological Profile for Zinc. US Department of Health and Human Services, Public Health Service, Atlanta, GA (updated 2010).
Clayton, G.D. and Clayton, F.E. (eds.). *Patty's Industrial Hygiene and Toxicology: Volume 2A, 2B, 2C: Toxicology*, 3rd ed. Wiley, J. New York. 1981–82.
Dikshith, T.S.S. and Diwan, P.V. (eds.). 2003. *Industrial Guide to Chemical and Drug Safety*. Wiley, J. Hoboken, NJ.

Zinc chloride (CAS No. 7646-85-7)

Molecular formula: $ZnCl_2$
Synonyms and trade names: Zinc chloride; Zinc dichloride; Zinc butter

Use and exposure

Zinc chloride is white/colorless crystalline granules.

Toxicity and health effects

Exposures to zinc chloride cause adverse health effects and poisoning. On contact with the skin, zinc chloride causes skin burns and ulcerations, redness, eyes develop pain and blurred vision, and any splashes from solutions may cause eye damage. It is extremely destructive to the tissues of the mucous membranes and upper respiratory tract. The symptoms of toxicity include, but are not limited to, burning sensation, coughing, wheezing, laryngitis, shortness of breath, headache, nausea and vomiting, and irritation or corrosion to the gastrointestinal tract with abdominal pain. After repeated exposures of zinc chloride through skin contact, occupational workers develop varying degrees of skin problems, such as dermatitis and skin ulcerations. Repeated inhalation of zinc chloride causes occupational asthma among workers.

Zinc chloride and carcinogenicity

There are no reports indicating that zinc chloride is a human carcinogen. The NTP, the IARC, and the OSHA have not listed zinc oxide as a human carcinogen.

Exposure limits

For zinc chloride (as fume) the OSHA set the PEL as 1 mg/m^3 (TWA) 8 h and the ACGIH set the TLV as 1 mg/m^3 (TWA).

Precautions

Exposures to zinc chloride are dangerous, corrosive, and cause burns to any area of contact. Harmful if swallowed or inhaled. Affects the cardiovascular system.

References

Material Safety Data Sheet (MSDS). 2009. Zinc chloride. MSDS no. Z2280. Mallinckrodt Baker, Phillipsburg, NJ.

Pohanish, R.P. (ed.). 2002. Zinc chloride. In: *Sittig's Handbook of Toxic and Hazardous Chemicals and Carcinogens*, 4th ed., Vol. 2. Noyes, Andrew, W. Norwich, NY. pp. 2354–56.

Zinc oxide (CAS No. 1314-13-2)

Molecular formula: **ZnO**

Use and exposure

Zinc oxide is yellowish and powder like.

Toxicity and health effects

Exposures to zinc oxide metal fume cause several health disorders. The symptoms of toxicity include, but are not limited to, fever, chills, muscle ache, nausea, fever, dry throat, cough; lassitude (weakness, exhaustion), metallic taste, headache, blurred vision, low back pain, vomiting, malaise (vague feeling of discomfort), chest tightness, dyspnea (breathing difficulty) and decreased pulmonary function. The overexposure to zinc oxide fumes in workplaces produce symptoms known as metal fume fever or "zinc shakes"; an acute, self-limiting condition. Chronic exposure to zinc oxide may cause respiratory tract irritation with nasopharyngitis and laryngitis.

Zinc oxide and carcinogenicity

There are no reports indicating that zinc oxide is a human carcinogen. The NTP, the IARC, and the OSHA have not listed zinc oxide as a human carcinogen.

References

Dikshith, T.S.S. (ed.). 2009. *Safe Use of Chemicals: A Practical Guide*. CRC Press, Boca Raton, FL.

Fine, J.M., et al. 2000. Characterization of clinical tolerance to inhaled zinc oxide in naïve subjects and sheet metal workers. *J. Occup. Environ. Med.* 42 (11): 1085–91.

International Programme on Chemical Safety and the Commission of the European Communities (IPCS CEC). 1994. Zinc oxide. International Chemical Safety Cards. ICSC No. 0208 (updated 2004).

Pohanish, R.P. (ed.). 2002. Zinc oxide. In: *Sittig's Handbook of Toxic and Hazardous Chemicals and Carcinogens*, 4th ed., Vol. 2. Noyes, Andrew, W. Norwich, NY. pp. 2358–60.

5

Chemical Substances: Future Perspectives

To date, the presence of a variety of chemical substances has been found in drinking water, food, fruits, vegetables, ambient air, and household products. Chemical substances have also been introduced into our food, fabrics, onto our fields, and into our medicines. Identification of newer molecules and compounds, and preparations of newer formulations, such as organochlorines; PCBs, dioxins, epoxy resins and polycarbonate plastics; drugs and pharmaceutical products; food additives; insecticides and pesticides; pigments and colors; phthalates and paints; metals like lead, cadmium, and mercury; metal finishing, plastics, lubricating oils, inks, hairspray and insect repellents; breakdown products of industrial solvents, detergents, common industrial and laboratory solvents, synthetic leather products, and many more, is probably an unending process. Proper application of all these and many other chemical substances has caused rapid changes in global scientific technology and will make life very different in the future.

The identification of chemical substances as "safe" is largely based on specific scientific studies. Given the vast number of chemicals and their many purposes, screening chemical substances for their ability to interfere with the health of animals, humans and the environment is a significant task.

Global regulatory agencies, governments, and international organizations have been active with appropriate legislation and recommendations in an attempt to reduce the vast numbers of toxic chemical substances, thereby reducing the ill effects of their misuse. Additional research studies are required to understand fully the mechanisms and scientific certainties of chemical substances that have not been completely evaluated. Yet, different types of chemical substances have entered the common market. The consequences of negligence in the handling and management of chemical substances, the possible health disorders, and overall human safety need to be understood by each user.

To wait for conclusive scientific proof that these chemicals are adversely affecting human and wildlife populations before taking action could have devastating consequences for future generations. Phasing out of chemical substances has been in practice over decades. However, to achieve this requires an orderly way to eliminate each chemical substance from further use, and only after identifying an alternative and safer substance. Industries and governments require ready information to make correct and proper judgments about chemical substances.

The meaning of the word "safety" clearly suggests the proper methods of use, handling, waste disposal, and management of chemical substances. However, in present society, an important question is how to deal with the vast wealth of chemicals in existence and forthcoming, newer chemical molecules safely. The answer is very simple. Users, occupational workers, and chemical management teams must know the toxic elements of chemical substances, their judicious use, proper management, and the ethics of good laboratory practice (GLP) and good manufacturing practice (GMP). This is possible with a ready, easily understandable resource in all workplaces. This handbook, with updated information on as large a number of chemical substances as possible, is an attempt to help in this direction.

Glossary

Abatement: Reducing the degree or intensity of, or eliminating, pollution.

Abdomen: The area of the body that contains the pancreas, stomach, intestines, liver, gallbladder, and other organs.

Abdominal: Relating to the abdomen, the belly, that part of the body that contains all of the structures between the chest and the pelvis. The abdomen is separated anatomically from the chest by the diaphragm, the powerful muscle spanning the body cavity below the lungs.

Abdominal pain (pain in the belly—the abdomen): Abdominal pain can come from conditions affecting a variety of organs. The abdomen is an anatomical area bounded by the lower margin of the ribs above, the pelvic bone (pubic ramus) below, and the flanks on each side. Although abdominal pain can arise from the tissues of the abdominal wall that surround the abdominal cavity (the skin and abdominal wall muscles), the term abdominal pain is generally used to describe pain originating from organs within the abdominal cavity (from beneath the skin and muscles). These organs include the stomach, small intestine, colon, liver, gallbladder, and pancreas.

Abiotic transformation: Abiotic transformation is any process in which a chemical in the environment is modified by non-biological mechanisms.

Abiotic: Non-biological. The term suggests describing anything that is characterized by the absence of life or incompatible with life. In toxicology and ecotoxicology, the term indicates physical (heat, sunlight) or chemical processes (hydrolysis) that modify chemical structures. Thus, biotic transformation is a process in which a chemical substance in the living environment is modified by non-biological mechanisms.

Abnormal: Not normal; different from the usual structure, position, condition, or behavior. With reference to a tissue growth, abnormal may mean that it is cancerous or premalignant (likely to become cancer).

Abortifacient: Chemical substance that induces an abortion.

ABS (acrylonitrile-butadienne-styrene): ABS is a terpolymer and an amorphous resin. It is manufactured by combining three different compounds (acrylonitrile, butadiene, and styrene). ABS has the unique position of being the "bridge" between utility and engineering thermoplastics.

Absence seizure: The term absence seizure describes a kind of brief seizure with an accompanying loss of awareness or alertness; also often called a petit mal seizure.

Absolute lethal concentration (LC100): The lowest concentration of a chemical substance in an environmental medium that kills 100% of test organisms or species under defined conditions.

Absolute lethal dose (LD100): The lowest amount of a chemical substance that kills 100% of test animals under defined conditions. (It is important to note that the value of absolute lethal dose is dependent on the number of organisms used in its assessment.)

Absolute risk: The excess risk due to exposure to a hazard.

Absorbance: The logarithm to the base of 10 of the reciprocal of transmittance.

Absorbate: Substance/chemical that has been retained by the process of absorption.

Absorbed dose (of a chemical substance): The amount of a chemical substance taken up by an organism or into organs or tissues of interest (internal dose).

Absorbent: Material/chemical substance in which absorption occurs.

Absorption: (i) Penetration of a chemical substance into the body of another; (ii) process of soaking up or taking up hazardous chemical substances to prevent enlargement of the contaminated area; (iii) movement of chemical substances into the blood vascular system or into the tissues of the organism/animal/human.

Abstracts Service: A division of the American Chemical Society.

Abuse (of drugs, chemical substances, solvents, etc.): Improper use of drugs, industrial chemical substances, toxic materials, and many other chemical substances.

Accelerator: A chemical that accelerates a chemical reaction in the production of rubber or plastics.

Acceptable daily intake (ADI): (i) ADI is an estimate of the amount of a chemical substance in food or drinking water, expressed on a body weight basis, which can be ingested daily over a lifetime without appreciable risk (standard human = 60 kg). The ADI is listed in units of milligrams per kilogram of body weight. (ii) The concept of the ADI has been developed principally by WHO and FAO and is relevant to chemical substances such as food additives, residues of pesticides and veterinary drugs in foods. The ADI of chemical substances is derived from data of laboratory toxicity studies and from human experiences of such chemical substances when these are available, and incorporate a safety factor. (for contaminants in food and drinking water, Tolerable Intakes—daily or weekly—are used). (Refer: Tolerable daily intake.) The ADI is considered a safe intake level for a healthy adult of normal weight who consumes an average daily amount of the substance in question. Persons with health problems should seek medical advice.

Acceptable risk: (i) The term denotes to the probability of suffering disease or injury that will be tolerated by an individual, group, or society. Acceptability of risk depends on scientific data, social, economic, and political factors, and on the perceived benefits arising from a chemical or process. (ii) Probability of suffering disease or injury that is considered sufficiently small to be "negligible." The calculated risk of an increase of one case in a million people per year for cancer is usually considered to be negligible.

Accession number: An identification number assigned (for cataloging purposes) to volumes of studies submitted to regulatory authorities.

Acclimatization: The physiological and behavioral adjustments of an organism to changes in its environment.

Accretion: Phenomenon consisting of the increase in size of particles by the process of external additions.

Accumulation: Repeated doses of a chemical substance may result in its concentration in an organism, organ or tissue increasing progressively and the toxic effects may become more marked with successive doses. Factors involved in accumulation include selective binding of the chemical to tissue molecules, concentration of fat-soluble chemicals in body fat, absent or slow metabolism, and slow excretion. Accumulation is a mass balance effect where input exceeds output.

Acetals (polyoxymethylene): Acetals were introduced in the 1960s. They are crystalline thermoplastics made by the polymerization of formaldehyde. Acetals have a unique balance of physical properties not available with metals or most other plastics.

Acetaminophen: Acetaminophen is a pain reliever and fever reducing pharmaceutical drug. (Brand name: Tylenol.) Acetaminophen relieves pain by elevating the pain threshold (i.e., by requiring a greater amount of pain to develop before it is felt by a person). Acetaminophen reduces fever through its action on the heat-regulating center (the "thermostat") of the brain.

Acetylating: This is the introduction of an acetyl group using a reactant such as acetyl chloride or acetic anhydride. An acetyl group is an acyl group having the formula -C-CH3.

ACGIH (American Conference of Governmental Industrial Hygiene): A professional organization composed of personnel in governmental agencies or educational institutions engaged in occupational safety and health/industrial hygiene programs; develops and publishes recommended occupational exposure limits (TLVs) for hundreds of chemical substances and physical agents.

ACGIH carcinogen: 1 = Confirmed Human; 2 = Suspected Human; 3 = Confirmed Animal; 4 = Not Classifiable; 5 = Not Suspected.

Acid rain: Acidified particulate matter in the atmosphere that is deposited by precipitation onto a surface, often eroding the surface.

Acid: (i) A chemical substance that dissolves in water and releases hydrogen ions (H+); acids cause irritation, burns, or more serious damage to tissue, depending on the strength of the acid, which is measured by pH. (ii) Any kind of typically water-soluble and sour compounds that in solution are capable of reacting with a base to form a salt, that redden litmus, that have a pH less than 7, and that are hydrogen-containing molecules or ions able to give up a proton to a base or are substances able to accept an unshared pair of electrons from a base.

Acidification: In the gas phase, this process happens when compounds like nitrogen oxides and sulfur oxides are converted in a chemical reaction in the gas phase or in clouds into acidic substances. These acids are rained-out or dry deposited.

Acidosis: An abnormal increase in the acidity of the body's fluids, caused either by accumulation of acids or by depletion of bicarbonates.

Acrylics: Polymethyl methacrylate offers excellent clarity and weatherability, making it a very good candidate for exterior applications. Acrylic can be offered in many colors, transparent, translucent, opaque, frosted, and special effects.

Action level: (i) A level of a chemical similar to a tolerance level except it is not established through formal regulatory proceedings. It is an informal judgment by a regulatory agency on the amount of a chemical that should be allowed in food products. (ii) A concentration designated by OSHA for a specific chemical substance, and calculated as an 8-h TWA, which initiates certain required activities such as exposure monitoring and medical surveillance.

Activated charcoal: Charcoal that has been heated to increase its absorptive capacity. Activated charcoal is sold as an over-the-counter (OTC) product to help relieve intestinal gas. It is also used to absorb poisons (as in gas mask filters), neutralize poisons that have been swallowed, and filter and purify liquids.

Active ingredient: (i) In any pesticide product, the component that kills, or otherwise controls, target pests. Pesticides are regulated primarily on the basis of active ingredients. (ii) Chemical component of a pesticide product that can kill, repel, attract, mitigate, or control a pest or that acts as a plant growth regulator, desiccant, or nitrogen stabilizer. The remainder of a formulated pesticide product consists of one or more inert ingredients, such as water, solvents, emulsifiers, surfactants, clay, and propellants, which are also present for reasons other than pesticidal activity.

Acuity: Acuteness of vision or perception.

Acute: (i) Effects occurring over a short time, of abrupt onset, in reference to a disease. Acute also connotes an illness of short duration, rapidly progressive, and in need of urgent care. (ii) In animal testing, it pertains to administration of an agent in a single dose; not to be confused with the clinical term for a disease having a short and relatively severe course. (iii) In clinical medicine, it means sudden and severe, having a rapid onset.

Acute bronchitis: Inflammation of the tubes that carry air into the lungs.

Acute dermal toxicity: Adverse effects occurring within a short time of dermal application of a singular dose of a test chemical.

Acute exposure: A single exposure to a toxic chemical substance that may result in severe biological harm or death. Acute exposures are usually characterized as lasting no longer than a day, compared to continued exposures over a period of time.

Acute hepatitis: A newly acquired symptomatic hepatitis virus infection, usually less than 6 months' duration.

Acute inhalation toxicity: Adverse effects produced by a test chemical following a single uninterrupted exposure through/respiratory route over a short period of time (24 h or less) and the chemical capable of being inhaled.

Acute-moderate syndrome: The onset is acute, but not life threatening. For chemicals, the length of exposure is less than 24 h. The patient is likely to seek medical attention, but not be admitted to hospital.

Acute oral toxicity: Adverse effects produced within a short time of oral administration of either a single dose of a test chemical substance or multiple doses given within 24 h.

Acute-severe syndrome: The onset is acute, severe, and life threatening. For chemical substances, the length of exposure is less than 24 h. The patient is likely to be admitted to hospital.

Acute test: A test lasting for a short period of time—14 days.

Acute toxicity: (i) The capacity of a chemical substance to cause adverse health effects/poisonous effects or death as a result of a single or short-term exposure; (ii) any poisonous effect produced within a short period of time following exposure, usually up to 24–96 h, resulting in biological harm and often death; (iii) adverse effects occurring within a short time of administration of a single dose of a chemical, or immediately following short or continuous exposure and/or multiple doses over 24 h or less.

Additive effect: (i) A biologic response to exposure to multiple chemical substances that equals the sum of responses of all the individual chemical substances; (ii) an additive effect is the overall consequence, which is the result of two chemical substances acting together and is the simple sum of the effects of the chemical substances acting independently (also refer: Antagonistic effect, Synergistic effect).

Adduct: A covalent compound formed between a carcinogen or its metabolites and a protein or nucleic acid (either deoxyribonucleic acid (DNA) or ribonucleic acid (RNA)).

Adenocanthoma, ovarian: (i) The term denotes an adenocarcinoma in which some or the majority of the cells exhibit squamous (scaly or plate-like) differentiation; (ii) endometrial adenocarcinomas commonly contain foci of squamous epithelium, in addition to the glandular elements. In the past, these mixed tumors were known as adenocanthomas if they were well differentiated and as adeno-squamous carcinomas if they were poorly differentiated.

Adenocarcinoma: (i) A malignant neoplasm of epithelial cells with a glandular or gland-like pattern (synonym: glandular cancer/glandular carcinoma); (ii) a term applied to a malignant tumor originating in glandular tissue; (iii) tumors of the linings of organs; (iv) carcinoma derived from glandular tissue in which the tumor cells form recognizable glandular structures.

Adenofibrosis: A fibroid change in a gland.

Adenoma: This term refers to a tumor usually benign, in glandular tissue. (i) Adenoma can be pre-cancerous in cases such as polyps in the colon; (ii) a benign epithelial tumor of glandular origin; (iii) a benign epithelial neoplasm in which the tumor cells form glands or gland-like structures; usually well circumscribed, tending to compress rather than infiltrate or invade adjacent tissue.

Adenomatoid: A tissue change resembling an adenoma.

Adenomatous: Relating to an adenoma, and to some types of glandular hyperplasia.

Adenosis: Any disease of the glands (adenosis of the breast = fibrocystic disease).

ADH: Antidiuretic hormone.

ADI: Refer: Acceptable daily intake.

Adipose tissue: Fatty tissue.

Adjuvant therapy: A therapy involving both a primary therapeutic agent and an additional material that enhances the action of the primary agent.

Adjuvant: In immunology, this is a substance injected with antigens (usually mixed with them but sometimes given before or after the antigens) that non-specifically enhances or modifies the immune response to the antigens.

ADME: Absorption, distribution, metabolism, and excretion.

ADR: Adverse drug reaction.

Adrenal cortical steroids: Steroid hormones produced in the cortex of the adrenal gland.

Adrenal gland: A hormone-secreting organ located above each kidney.

Adrenergic: Secreting adrenaline (epinephrine) and (or) related substances; in particular referring to sympathetic nerve fibers.

Adsorbate: Chemical that has been retained by the process of adsorption.

Adsorbent: A solid material on the surface of which adsorption takes place.

Adsorption: A physical process in which molecules of gas, of dissolved chemicals, or of liquids adhere in an extremely thin layer to the surfaces of solid bodies with which they are in contact.

Adulterated food: The term refers to food and food products that are generally, impure, unsafe, or unwholesome; in terms of the Federal Food, Drug, and Cosmetic Act (FFDCA), the Federal Meat Inspection Act, the Poultry Products Inspection Act, and the Egg Products Inspection Act, "adulterated food" in separate language define in very specific (and lengthy) terms how the term "adulterated" will be applied to the foods each of these laws regulate. Products found to be adulterated under these laws cannot enter into commerce for human food use.

Adverse effect: An abnormal, undesirable, or harmful effect to an organism. The symptoms normally include mortality, altered food consumption, altered body and organ weights, altered enzyme levels, or visible pathological change. An effect may be classed as adverse if it causes functional or anatomical damage, causes irreversible change in the homeostasis of the organism, or increases the susceptibility of the organism to other chemical or biological stress. Adverse effects occur within a short time of administration of a single dose of a chemical substance, or immediately following short or multiple doses over 24 h or less.

Adverse health effect: A change in body functions or cell structure that might lead to disease or health problems.

AE: Adverse event.

AEGLs: Acute exposure guideline levels.

Aerobic conditions: In the presence of oxygen.

Aerobic: (i) Requiring oxygen; (ii) oxygen-dependent.

Aerosol particles: One of the components of an atmospheric air mixture comprised of minute solids particles, part of which is almost certainly water.

Aerosol: This is a very broad term applied to any suspension of solid or liquid particles in a gas. The small particles, usually in the range of 0.01 to 100 μm, are dispersed in air; includes liquid (mist) and solid particles (dust).

Aetiology: In medicine, the term refers to the science of the investigation of the cause or origin of disease.

Aflotoxins: Toxins of biological origin produced by common molds, such as *Aspergillus flavus*, and species in different types of foods.

Agent Orange: Refer: 2,4-dichlorophenoxyacetic acid (2, 4-D) and 2,4,5-trichlorophenoxy-acetic acid (2,4,5-T), Dioxin.

Agglomeration: A process of contact and adhesion whereby the particles of dispersion form clusters of increasing size.

AIHA: American Industrial Hygiene Association.

AIN: Acute interstitial nephritis.

Air pollution: (i) Contamination of atmospheric air with substances/chemicals; (ii) contamination of the atmosphere by any toxic or radioactive gases and particulate matter as a result of human activity; (iii) presence of substances in the atmosphere resulting either from human activity or natural processes, in sufficient concentration, for a sufficient time, and under circumstances such as to interfere with the comfort, health, or welfare of persons or to harm the environment.

Air quality standard (AQS): Refer: Environmental quality standard.

Alanine transaminase: A liver protein (ALT).

Albino: An organism exhibiting deficient pigmentation in skin, eyes, and/or hair.

Albuminuria: Presence of albumin, derived from plasma, in the urine.

Alcohol(s): Alcohol is an organic chemical in which one or more hydroxyl (OH) groups are attached to carbon (C) atoms in place of hydrogen (H) atoms. Alcohols are grouped into different classes depending on how the -OH group is positioned on the chain of carbon atoms: (i) in a *primary* (1°) alcohol, the carbon that carries the -OH group is only attached to *one alkyl group*; (ii) in a *secondary* (2°) alcohol, the carbon with the -OH group attached is joined directly to *two alkyl groups*; (iii) in a *tertiary* (3°) alcohol, the carbon atom holding the -OH group is attached directly to *three alkyl groups*. Common alcohols include ethyl alcohol or ethanol (found in alcoholic beverages), methyl alcohol or methanol (can cause blindness), and propyl alcohol or propanol (used as a solvent and antiseptic). Rubbing alcohol is a mixture of acetone, methyl isobutyl ketone, and ethyl alcohol. In everyday talk, alcohol usually refers to ethanol as, for example, in wine, beer, and liquor. It can cause changes in behavior and be addictive.

Algaecide: A pesticide that controls algae.

Alkaline: (Basic, as opposed to acidic); a material whose pH is higher than 7.0 (8 to 14pH).

Alkylating agent: (i) The term describes any chemical substance that introduces an alkyl radical into a compound in place of a hydrogen atom; (ii) a substance that causes the incorporation of single bonded carbon atoms into another molecule.

Allergen: (i) A descriptor term for a chemical substance that produces an allergy; (ii) an immunostimulant antigenic chemical substance that may or may not cause a clinically significant effect, but which is capable of producing immediate hypersensitivity.

Allergy (hypersensitivity): (i) A broad term applied to disease symptoms following exposure to a previously encountered chemical substance (allergen), often one that would otherwise be classified as harmless; essentially a malfunction of the immune system. (ii) Symptoms or signs occurring in sensitized individuals following exposure to a previously encountered chemical substance (allergen) which would otherwise not cause such symptoms or signs in non-sensitized individuals. The most common forms of allergy are rhinitis, urticaria, asthma, and contact dermatitis. (iii) An altered immune response to a specific substance on re-exposure. (iv) An exaggerated immune response to a foreign substance causing tissue inflammation and organ dysfunction.

Allometry: A term in biology to measure the rate of growth of a part or parts of an organism relative to the growth of the whole organism.

Alloy: A metallic material, homogeneous to the naked eye, consisting of two or more elements so combined that they cannot be readily separated by mechanical means. Alloys are considered mixtures for the purpose of classification under the globally harmonized system (GHS).

Alopecia: Baldness; absence or thinning of hair from areas of skin where it is usually present.

Alpha particle: Nucleus of a helium atom emitted by certain radioisotopes on disintegration.

Alveoli: Tiny sacs at the ends of bronchioles in the lungs where carbon dioxide and oxygen are exchanged with red blood cells (RBCs) in adjacent capillaries. (i) The term alveoli is the plural of alveolus.

Ambient: Surrounding or encompassing, for instance, the ambient environment.

Ambient air: The surrounding air.

Ambient standard: Refer: Environmental quality standard.

Ames test: (i) A method performed using bacteria as a test system to determine the mutagenic potential of a substance/chemical; (ii) *in vitro* test for mutagenicity using mutant strains of the bacterium *Salmonella typhimurium*, which cannot grow in a given histidine-deficient medium: mutagens can cause reverse mutations that enable the bacterium to grow on the medium. The test can be carried out in the presence of a given microsomal fraction (S-9) from rat liver (see microsome) to allow metabolic transformation of mutagen precursors to active derivatives.

Amino acids: Amino acids are critical to life, and have many functions in metabolism. Amino acids play central roles both as building blocks of proteins by cells and as intermediates in metabolism; there are about 20 amino acids.

Amnesic shellfish poisoning (ASP): A serious illness caused by the consumption of bivalve shellfish (mollusks) such as mussels, oysters, and clams that have ingested, by filter feeding, large quantities of micro-algae containing poisonous acid; acute symptoms include vomiting, diarrhea, and in some cases, confusion, loss of memory, disorientation, and even coma.

Anabolism: Biochemical processes by which smaller molecules are joined to make larger molecules.

Anaerobic respiration: Living or functioning in the absence of oxygen; cellular respiration in the absence of oxygen.

Anaerobic: Requiring the absence of oxygen.

Analgesic: A drug that alleviates pain without causing loss of consciousness; a pain-reliever; with an effective analgesic, there is an inability to feel pain while still conscious. From the Greek an-, without + algesis, sense of pain and structural support.

Analyte: Any chemical substance measured in the laboratory.

Analytic epidemiologic study: A study that evaluates the association between exposure to hazardous chemical substance(s) and disease by testing scientific hypotheses.

Anaphylaxis: Life-threatening type 1 hypersensitivity allergic reaction occurring in a person or animal exposed to an antigen or hapten to which they have previously been sensitized. The consequences of the reaction may include angio-edema, vascular collapse, shock, and respiratory distress.

Androgen suppression: Treatment designed to suppress or block the production of male hormones.

Anemia: (i) An abnormal deficiency in the oxygen-carrying component of the blood—a condition suggesting lack of RBCs; (ii) decreased hemoglobin or number of RBCs; (iii) a qualitative or quantitative deficiency of hemoglobin, a protein found in RBCs.

Anesthesia: Temporary loss of consciousness induced by high concentrations of organic solvents.

Anesthetic: An agent that causes loss of sensation with or without loss of consciousness.

Aneuploidy: The occurrence of one or more extra or missing chromosomes leading to an unbalanced chromosome complement or any chromosome number that is not an exact multiple of the haploid number.

Angioedema: A reaction in the skin and underlying tissue showing swelling and red blotches.

Antagonism: (i) Combined effect of two or more factors that is smaller than the solitary effect of any one of those factors; (ii) adverse effect or risk from two or more chemicals interacting with each other is less than what it would be if each chemical was acting separately.

Antagonist: (i) A chemical substance that acts within the body to reduce the physiological activity of another chemical substance (as an opiate), especially one that opposes the action on the nervous system of a drug or a substance occurring naturally in the body by combining with and blocking its nervous receptor; (ii) any chemical substance that binds to a cell receptor normally responding to a naturally occurring chemical substance, preventing a response to the natural substance.

Antagonistic: Reduction of the effect of one chemical substance by the presence of another.

Antagonistic effect: (i) A biologic response to exposure to multiple substances that is less than would be expected if the known effects of the individual substances were added together; (ii) antagonistic effect is the consequence of one chemical substance or group of chemical substances counteracting the effects of another: in other words, the situation where exposure to two or a group of chemical substances together has less effect than the simple sum of their independent effects; such chemical substances are said to show antagonism.

Anthelmint (antihelminth): (i) A chemical substance intended to kill or cause the expulsion of parasitic intestinal worms, such as helminths; (ii) a chemical substance acting to expel or kill parasitic intestinal worms.

Anthracosis (coal miners' pneumoconiosis): A form of pneumoconiosis caused by the accumulation of anthracite carbon deposits in the lungs due to inhalation of smoke or coal dust.

Anthropogenic: (i) Effects produced as a result of human activities; (ii) effects caused by or influenced by human activities and the impact on nature.

Anthrax: A disease of mammals and humans caused by a spore-forming bacterium called *Bacillus anthracis*. Anthrax has an almost worldwide distribution and is a zoonotic disease, meaning it may spread from animals to humans. All mammals appear to be susceptible to anthrax to some degree, but ruminants such as cattle, sheep, and goats are the most susceptible and commonly affected, followed by horses, and then swine.

Antibiotics: (i) A chemical substance that can destroy or inhibit the growth of microorganisms. Antibiotics are widely used in the prevention and treatment of infectious diseases. (ii) A chemical substance produced by and obtained from certain living cells (especially bacteria, yeasts, and molds), or an equivalent synthetic substance, which is biostatic or biocidal at low concentrations to some other form of life, especially pathogenic or noxious organisms. (iii) Chemical substances produced by microorganisms or synthetically that inhibit the growth of, or destroy, bacteria. Rules guiding the use of veterinary drugs and medicated animal feeds, including tolerance levels for drug residues in meats for human consumption, are set by the Center for Veterinary Medicine of the Food and Drug Administration (FDA). The Food Safety and Inspection Service (FSIS) enforces the FDA rules through a sampling and testing program that is part of its overall meat and poultry inspection program.

Antibodies: (i) Specific proteins produced by the body's immune system that bind with foreign proteins (antigens); (ii) protein (immunoglobulin) produced by the immune system in response to exposure to an antigenic molecule and characterized by its specific binding to a site on that molecule.

Anticoagulant: Chemical substance that prevents blood clotting, e.g., warfarin.

Antidote: (i) A chemical substance capable of specifically counteracting or reducing the effect of a potentially toxic substance in an organism by a relatively specific chemical or pharmacological action; (ii) an agent that counteracts a poison and neutralizes its effects.

Antiepileptics (antiepileptic drugs): Antiepileptic drugs are medicines that reduce the frequency of epileptic seizures. Antiepileptic drugs have a large number of side effects and possible adverse effects.

Antigen: A foreign substance that provokes an immune response when introduced into the body and the body reacts by making antibodies. The descriptor term is applied to any chemical substance that produces a specific immune response when it enters the tissues of an animal.

Antihistamine: Chemical substance that blocks or counteracts the action of histamine.

Antimicrobic: A drug used to inhibit or kill microorganisms.

Antimuscarinic: (i) inhibiting or preventing the actions of muscarine and muscarine-like agents on the muscarinic acetylcholine receptors.

Antimycotic: A chemical substance used to kill a fungus or to inhibit its growth—a fungicide.

Antinicotinic: (i) inhibiting or preventing the actions of nicotine and nicotine-like agents on the nicotinic acetylcholine receptors.

Antioxidants: Chemical substances added to food to prevent the oxygen present in the air from causing undesirable changes in flavor or color. BHA, BHT, and tocopherols are examples of antioxidants.

Antipyretic: A chemical substance that relieves or reduces fever.

Anxiolytic: An anti-anxiety drug.

Aphasia: Loss or impairment of the power of speech or writing, or of the ability to understand written or spoken language or signs, due to a brain injury or disease.

Aphicide: A chemical substance intended to kill aphids.

Aphid: Common name for a harmful plant parasite in the family Aphididae, some species of which are vectors of plant virus diseases.

Aphotic zone: The deeper part of lakes/sea/ocean where light does not penetrate.

Aphytic zone: Part(s) of the lake floor where vegetation is not available.

Aplasia: Lack of development of an organ or tissue, or of the cellular products from an organ or tissue.

Aplastic anemia: (i) Bone marrow failure with markedly decreased production of white blood cells, RBCs, and platelets leading to increased risk of infection and bleeding; (ii) one type of anemia caused by injury to blood-forming tissues and associated with occupational exposure to chemical substances such as TNT, benzene, and ionizing radiation.

Apnea: Temporary absence or cessation of breathing.

Application factor: Refers to number used to estimate concentration of a substance/chemical that will not produce significant adverse effects/harm to a population during chronic exposure. The factor is based on the formula: application factor = MATP.

Aquaculture: Breeding and rearing of fish in captivity; also termed pisciculture.

Aquatic organism(s): Organism(s)/animals related to/living water bodies.

Aqueous: Related to watery solution.

Arboreal: The term relating to, resembling, or consisting of trees/plants.

ARF: Acute renal failure.

Argyria and argyrosis: Pathological condition characterized by grey-bluish or black pigmentation of tissues (such as skin, retina, mucous membranes, internal organs) caused by the accumulation of metallic silver, owing to the reduction of a silver compound that has entered the organism during (prolonged) administration or exposure and/or workplace exposure.

Aromatic amines: Petrochemical compounds with a pungent odor (known to produce cancer).

Aromatic hydrocarbons: Hydrocarbon compounds in which the carbon atoms are connected by a ring structure that is planar and joined by sigma and pie bonds between the carbon atoms. A class of synthetic compounds used as solvents and grease cutters, these are members of the carcinogenic benzene family of chemicals.

Aromatic: (i) Technical term for a chemical compound that contains one or more benzene rings; (ii) an organic chemical (hydrocarbon) characterized by the presence of a benzene ring.

Arrhythmia: Any variation from the normal rhythm of the heart beat.

Arteriosclerosis: Hardening and thickening of the walls of the arteries.

Arthralgia: Pain in a joint or joints.

Arthritis: Chronic inflammation of a joint, usually accompanied by pain and often by changes in structure.

Artificial coloring: A coloring chemical substance containing any dye or pigment manufactured by a process of synthesis or other similar artifice, or a coloring that is manufactured by extracting naturally produced dyes or pigments from a plant or other material.

Artificial flavoring: Artificial flavors are restricted to an ingredient that is manufactured by a process of synthesis or similar process. The principal components of artificial flavors are usually esters, ketones, and aldehyde groups. These ingredients are declared in the ingredients statement as "artificial flavors" without naming the individual components.

Asbestos: A naturally occurring fibrous mineral found in certain types of rock formations.

Asbestosis: Form of pneumoconiosis caused by inhalation of asbestos fibers.

Ascaricide: Chemical substance intended to kill roundworms.

Ash: Mineral content of a product that remains after complete combustion.

Asphyxiant: A substance capable of inducing asphyxia, which is a lack of oxygen or an excess of carbon dioxide in the body, usually caused by the interruption of breathing and resulting in unconsciousness.

Aspiration: Entry of a liquid or solid chemical product into the trachea and lower respiratory system directly through the oral or nasal cavity, or indirectly from vomiting.

Aspirin: A good example of a trade name that entered into the language, Aspirin was once the Bayer trademark for acetylsalicylic acid.

AST: Aspartate aminotransferase (also known as serum glutamic oxaloacetic transaminase (SGOT)) is an enzyme normally present in the liver and heart cells. AST is released in the blood when the liver or heart is damaged. The blood AST levels are thus elevated with liver damage, e.g., from viral hepatitis, or with an insult to the heart, e.g., from a heart attack. Some medications can also raise AST levels.

Asthma: (i) Respiratory condition caused by narrowing of the airways; symptoms include recurrent attacks of wheezing, coughing, shortness of breath, and labored breathing; (ii) reversible bronchoconstriction (narrowing of bronchioles) initiated by the inhalation of irritating or allergenic agents.

Astringent: (i) Chemical substance causing contraction, usually locally after topical application; (ii) a chemical substance causing cells to shrink, thus causing tissue contraction or stoppage of secretions and discharges; such substances may be applied to skin to harden and protect it.

Ataxia: Unsteady or irregular manner of walking or movement caused by loss or failure of muscular coordination.

Atelectasis: Atelectasis is the collapse of part or all of a lung. It is caused by a blockage of the air passages (bronchus or bronchioles) or by pressure on the lung.

Atherosclerosis: Pathological condition in which there is thickening, hardening, and loss of elasticity of the walls of blood vessels, characterized by a variable combination of changes of the innermost layer consisting of local accumulation of lipids, complex carbohydrates, blood and blood components, fibrous tissue, and calcium deposits. In addition, the outer layer becomes thickened and there is fatty degeneration of the middle layer.

Atmosphere, an: A unit of pressure equal to the pressure exerted by a vertical column of mercury 760 mm high at a temperature of 0° and under standard gravity.

Atmosphere, the: The gaseous envelope surrounding a planet; the earth's atmosphere is surrounded by a whole mass of air largely composed of oxygen (20.9%) and nitrogen (79.1%) by volume, carbon dioxide (0.03%), and traces of noble gases, water vapor, organic matter, suspended solid particles, etc.

Atmosphere: The sum total of all the gases surrounding the earth, extending several hundred kilometers above the surface in a mechanical mixture of various gases in fluid-like motion.

Atmospheric dispersion: The mechanism of dilution of gaseous or smoke pollution leading to the progressive decrease of pollutants.

Atrophy: A wasting or decrease in size of a body organ, tissue, or part, owing to disease, injury, or lack of use. The process is observed during the wasting of a tissue.

ATSDR: Agency for Toxic Substances and Diseases Registry.

Attention: The ability to focus selectively on a selected stimulus, sustaining that focus and shifting it at will; the ability to concentrate.

Autoimmune disease: Pathological condition resulting when an organism produces antibodies or specific cells that bind to constituents of its own tissues (autoantigens) and cause tissue injury: examples of such disease may include rheumatoid arthritis, myasthenia gravis, systemic lupus erythematosus, and scleroderma.

Autoimmunity: A health condition in which the immune responses of an animal are directed against its own tissues.

Autooxidation: Self-catalyzed oxidation reaction that occurs spontaneously in an aerobic environment.

Autophagosome: Name of the membrane-bound body occurring inside a cell and containing decomposing cell organelles.

Autopsy (necropsy): Postmortem examination of the organs and body tissue to determine cause of death or pathological condition.

Autotrophic: Related to organisms that produce their own organic constituents from inorganic compounds utilizing sunlight for energy or by oxidation process.

B

Backbone: The bones, muscles, tendons, and other tissues that reach from the base of the skull to the tailbone. The backbone encloses the spinal cord and the fluid surrounding the spinal cord. Also called spinal column, spine, and vertebral column.

Bacteria: (i) A large group of single-cell microorganisms. Some cause infections and disease in animals and humans. The singular of bacteria is bacterium. (ii) Living single-cell organisms. Bacteria can be carried by water, wind, insects, plants, animals, and people, and survive well on skin, clothes, and in human hair. They also thrive in scabs, scars, the mouth, nose, throat, intestines, and room-temperature foods. Often, bacteria are maligned as the causes of human and animal disease, but there are certain types that are beneficial for all types of living matter.

Bactericide: (i) A chemical substance intended to kill bacteria; (ii) a pesticide used to control or destroy bacteria, typically in the home, schools, and on hospital equipment.

Bagassosis: Lung disease caused by the inhalation of dust from sugar-cane.

Barbiturates: A drug used to treat insomnia, seizures, and convulsions, and to relieve anxiety and tension before surgery. It belongs to the family of drugs called central nervous system (CNS) depressants.

Base-pairing: A process involving the linking of the complementary pair of polynucleotide chains of DNA. The linking process is by means of hydrogen bonds between the opposite purine and pyrimidine pairs. Stable base pairs only form between adenine and thymine (A-T) and guanine and cytosine (G-C). In RNA, uracil replaces thymine and can form a base pair with adenine.

BCF: Bioconcentration factor.

BEI: Biological Exposure Index.

BEN: Balkan endemic nephropathy.

Benign: (i) An adjective applied to any growth that does not invade surrounding tissue (refer: Malignant, Tumor); (ii) not cancerous; cannot invade neighboring tissues or spread to other parts of the body; a condition of tissue growth that is harmless.

Benign tumor: (i) A slow-growing set of cells with abnormal appearance (this term refers to any tissue growth that does not invade surrounding tissues in the body); (ii) a growth that is not cancer, it does not invade nearby tissue or spread to other parts of the body.

Benzene: Benzene is an aromatic hydrocarbon chemical. It is a widely used chemical formed from both natural processes and human activities. Breathing benzene can cause drowsiness, dizziness, and unconsciousness. Breathing very high levels of benzene can result in death.

Beryllium disease (berylliosis): Serious and usually permanent lung damage resulting from chronic inhalation of beryllium.

Best management practices (BMP): Procedures or controls other than effluent limitations to prevent or reduce pollution of surface water, including runoff control, spill prevention, and operating procedures.

Beta burns: Beta-emitting isotopes from smoke and fallout can cause desquamation from high-dose local radiation delivered to exposed skin surfaces, but only if these isotopes are in contact with the skin for longer than 1 h. Since beta radiation is not as penetrating as gamma radiation, dry desquamating skin lesions secondary to beta burns may not be as serious as web desquamating lesions.

Bilateral: Affecting both sides of the body.

Bile: A fluid made by the liver and stored in the gallbladder. Bile is excreted into the small intestine, where it helps digest fat.

Biliary tract: The organs and ducts that make and store bile (a fluid made by the liver that helps digest fat), and release it into the small intestine. The biliary tract includes the gallbladder and bile ducts inside and outside the liver. The biliary tract is also called the biliary system.

Bilirubin: (i) Bilirubin is the yellow breakdown product of normal heme; (ii) orange-yellow pigment, a breakdown product of heme-containing proteins (hemoglobin, myoglobin, cytochromes), which circulates in the blood plasma bound to albumin or as water soluble glucuronide conjugates, and is excreted in the bile by the liver.

Bilirubinuria (urinary bilirubin): (i) Excretion of bilirubin in the urine; (ii) the presence of bile pigments in the urine.

Bioaccumulants: Substances that increase in concentration in living organisms as they take in contaminated air, water, or food because of the very slow metabolic conversion or excretion of the substances.

Bioaccumulation factor: The ratio of concentration of a chemical in an organism to its concentration in food. (i) Bioaccumulation is the process by which chemicals concentrate in an organism. Bioaccumulation refers to the uptake of chemicals both from water (bioconcentration) and from ingested food and sediment. (ii) Progressive increase in the amount of a substance in an organism or part of an organism that occurs because the rate of intake exceeds the organism's ability to remove the substance from the body. (iii) A process where chemical substances are retained in fatty body tissue and increase in concentration over time. Biomagnification is the increase of tissue accumulation in species higher in the natural food chain as contaminated food species are eaten.

Bio-activation: Metabolic conversion of a xenobiotic to a more toxic derivative or one that has a greater effect on living organisms.

Biopsy: The removal of cells or tissues for examination by a pathologist. The pathologist may study the tissue under a microscope or perform other tests on the cells or tissue. There are many different types of biopsy procedures. The most common types include: (i) incisional biopsy, in which only a sample of tissue is removed; (ii) excisional biopsy, in which an entire lump or suspicious area is removed; (iii) needle biopsy, in which a sample of tissue or fluid is removed with a needle. When a wide needle is used, the procedure is called a core biopsy. When a thin needle is used, the procedure is called a fine-needle aspiration biopsy.

Bio-assay: (i) The quantitative measurement of the effects of a chemical substance on the organism under standard conditions; (ii) procedure for estimating the concentration or biological activity of a substance such as a vitamin, hormone, plant growth factor, antibiotic, or enzyme by measuring its effect on a living system compared to a standard system.

Bioavailability: Bioavailability is a measurement to the extent of a therapeutically active drug reaches the blood and is available at the site of action (in the body).

Biochemical mechanism: (i) This is the general term for any chemical reaction or series of reactions, usually enzyme catalyzed, which produces a given physiological effect in a living organism; (ii) reaction or series of reactions, usually enzyme catalyzed, associated with a specific physiological event in a living organism.

Biochemicals: Chemical substances that are either naturally occurring or identical to naturally occurring substances. Examples include hormones, pheromones, and enzymes. Biochemicals function as pesticides through non-toxic, non-lethal modes of action, such as disrupting the mating pattern of insects, regulating growth, or acting as repellants. Biochemicals tend to be environmentally compatible and are thus important to integrated pest management programs (IPM).

Biochemical oxygen demand (BOD): The amount of oxygen used for biochemical oxidation by a unit volume of water at a given temperature and for a given period of time. BOD is used for measuring the degree of water pollution.

Biocide: A general term for any substance that kills or inhibits the growth of microorganisms (mold, slime, bacterium, fungus).

Bioconcentration: (i) A process whereby living organisms acquire chemical substances from water through gills/integument and store it in their bodies at concentration(s) higher than in the environment; (ii) a process leading to a higher concentration of a substance in an organism than in the environmental media to which it is exposed.

Biodegradable: The capability of an organism/biological system to breakdown organic chemical substances.

Biodegradation: (i) Biodegradation is the process by which organic chemical substance breaks down through the action of micro oraganisms; (ii) The molecular degradation of an organic substance resulting from the complex action of living organisms. A substance is said to be biodegraded to an environmentally acceptable extent when environmentally undesirable properties are lost.

Bioelimination: Removal, usually from the aqueous phase, of a test substance in the presence of living organisms by biological processes supplemented by physicochemical reactions.

Bioequivalence: Relationship between two preparations of the same drug in the same dosage form that have a similar bioavailability.

Biologic monitoring: The estimation of hazardous chemical substances in biologic materials (such as blood, hair, urine, or breath) to determine whether exposure has occurred. A blood test for lead is an example of biologic monitoring.

Biological pesticide: A chemical derived from plants, fungi, bacteria, or other non-man-made synthesis, which can be used for pest control. These agents usually do not have toxic effects on animals and people and do not leave toxic or persistent chemical residues in the environment.

Biological agent: A living organism that can cause disease, sickness, and mortality in humans.

Biological cycle: Complete circulatory process through which a substance passes in the biosphere. It may involve transport through the various media (air, water, soil), followed by environmental transformation, and carriage through various ecosystems.

Biological effect monitoring (BEM): Continuous or repeated measurement of early biological effects of exposure to a substance to evaluate ambient exposure and health risk in comparison with appropriate reference values based on knowledge of the probable relationship between ambient exposure and biological effects.

Biological half-life (t1/2): The term used for the time required for the amount of a particular substance in a biological system to be reduced to one-half its value by biological processes when the rate of removal is approximately exponential. Substances with a long biological half-life will tend to accumulate in the body and are, therefore, to be avoided. Substances with a short biological half-life may accumulate if some become tightly bound, even if most are rapidly cleared from the body. There is also the possibility of cumulative effects of chemical substances that have a short residence time in the body.

Biological indicator: Species or a group of species that is representative and typical for a specific status of an ecosystem, which appears frequently to monitoring and whose population shows a sensitive response to changes, e.g., the appearance of a toxicant in an ecosystem.

Biological monitoring: (i) This is a procedure of periodic examination of the biological specimens for the purposes of monitoring. It is usually applied to exposure monitoring but can also apply to effect monitoring. It is an/the analysis of the amounts of potentially toxic substances or their metabolites present in body tissues and fluids, as a means of assessing exposure to these substances and aiding timely action to prevent adverse effects. The term is also used to mean assessment of the biological status of populations and communities of organisms at risk, in order to protect them and to have early warning of possible hazards to human health. (ii) Continuous or repeated measurement of any naturally occurring or synthetic chemical, including potentially toxic substances or their metabolites or biochemical effects in tissues, secreta, excreta, expired air, or any combination of these, in order to evaluate occupational or environmental exposure and health risk in comparison with appropriate reference values based on knowledge of the probable relationship between ambient exposure and resultant adverse health effects.

Biological oxygen demand (BOD): An indirect measure of the concentration of biologically degradable material present in organic wastes. It usually reflects the amount of oxygen consumed in 5 days by biological processes breaking down organic waste.

Biological pesticide: A chemical substance derived from plants, fungi, bacteria, or other non-man-made synthesis, which can be used for pest control.

Biologic uptake: The transfer of substances from the environment to plants, animals, and humans.

Biomagnification: (i) A phenomenon where bioaccumulated chemical substances increase in concentration as they pass upward through two or more trophic levels. (ii) A general term applied to the sequence of processes in an ecosystem by which higher concentrations are attained in organisms of higher trophic level, meaning of higher levels in the food chain. This is a process by which xenobiotics increase in body concentration in organisms through a series of prey-predator relationships from primary producers to ultimate predators, often human beings. The tissue concentration increases as the trophic level increases. (iii) A parameter used to identify a toxic effect in an individual organism and can be used in extrapolation between species. (iv) A chemical, biochemical, or functional indicator to environmental chemical, physical, or biological agent(s). (v) An indicator signaling an event or condition in a biological system or sample and giving a measure of exposure, effect, or susceptibility.

Biomass: The term suggests (i) total amount of biotic material, usually expressed per unit surface area or volume, in a medium such as water, (ii) material produced by the growth of microorganisms, plants, or animals.

Biomedical testing: The testing of persons/workers to find out whether or not a change in the body function might have occurred because of exposure to hazardous chemical substance(s).

Biomolecule: A chemical substance that is synthesized by and occurs naturally in living organisms.

Biopesticide: Biological agent with pesticidal activity, e.g., the bacterium *Bacillus thuringiensis* when used to kill insects.

Biopolymers: Macromolecules (including proteins, nucleic acids, and polysaccharides) formed by living organisms.

Biopsy: Excision of a small piece of living tissue for microscopic or biochemical examination; usually performed to establish a diagnosis.

Biota: The plants and animals in an environment. Some of these plants and animals might be sources of food, clothing, or medicines for people.

Biotechnology: (i) The application of living organisms to produce new products/substances; (ii) agricultural biotechnology is a collection of scientific techniques, including genetic engineering, which are used to create, improve, or modify plants, animals, and microorganisms. Using conventional techniques, such as selective breeding, scientists have been working to improve plants and animals for human benefit for hundreds of years. Modern techniques now enable scientists to move genes (and therefore desirable traits) in ways not previously possible and with greater ease and precision.

Biotransformation: (i) In this process, a chemical substance is modified by a living organism in contrast to abiotic processes referred to earlier. The enzyme-mediated transformation of xenobiotics, frequently involving phase 1 and phase 2 reactions. (ii) A chemical conversion of a substance that is mediated by living organisms or enzyme preparations.

Black carbon: Emitted during the burning of coal, diesel fuel, natural gas, and biomass.

Bladder: The organ that stores urine.

Blights: Diseases that hurt and sometimes destroy plants. Blights cause a plant to wither, stop growing, or cause all or parts of it to die.

BLL (blood lead level): Lead in the human body can be measured in the blood, urine, bones, teeth, and hair. The most frequent test is to measure the BLL. Measuring an individual's BLL can detect lead poisoning in adults or children. RBCs increase erythrocyte protoporphyrin (EP) when blood lead is high. Children with an EP of 35 μg/dL should have their BLL measured. Children with a BLL of 20 μg/dL or higher should be screened for lead poisoning. Medical treatment is necessary if the BLL is higher than 45 μg/dL.

Blood: (i) The tissue with RBCs, white blood cells, platelets, and other substances suspended in a fluid called plasma. Blood takes oxygen and nutrients to the tissues, and carries away wastes. (ii) The blood is transported throughout the body by the circulatory system. Blood functions in two directions: arterial and venous. Arterial blood is the means by which oxygen and nutrients are transported to tissues while venous blood is the means by which carbon dioxide and metabolic by-products are transported to the lungs and kidneys, respectively, for removal from the body, bodily organs, and tissues.

Blood pressure: The force of circulating blood on the walls of the arteries. Blood pressure is taken using two measurements: systolic (measured when the heart beats, when blood pressure is at its highest) and diastolic (measured between heart beats, when blood pressure is at its lowest). Blood pressure is written with the systolic blood pressure first, followed by the diastolic blood pressure (e.g., 120/80).

Blood urea nitrogen (BUN): Nitrogen in the blood that comes from urea (a substance formed by the breakdown of protein in the liver). The kidneys filter urea out of the blood and into the urine. A high level of urea nitrogen in the blood may be a sign of a kidney problem.

Blood-brain barrier: Physiological interface between brain tissues and circulating blood created by a mechanism that alters the permeability of brain capillaries, so that some substances are prevented from entering brain tissue, while other substances are allowed to enter freely.

Blood-placenta barrier: Physiological interface between maternal and fetal blood circulations that filters out some substances that could harm the fetus while favoring the passage of others, such as nutrients; many fat-soluble substances such as alcohol are not filtered out and several types of viruses can also cross this barrier. It is important to remember that the effectiveness of the interface as a barrier varies with species and different forms of placentation.

Body burden: Body burden, also known as chemical load, is the amount of harmful chemicals present in the body of a person. Biomonitoring studies estimate the exposure by measuring the chemicals or their metabolites in human specimens such as the blood or urine. Results are usually expressed in mass units, such as grams and milligrams. These chemical residues—termed the "chemical body burden"—are detected in the blood, urine, and breast milk. It is important that we are aware of the chemical compounds in our bodies, more so occupational workers in chemical industries.

Bolus: The term suggests (i) a single dose of a chemical substance, originally a large pill; (ii) a dose of a chemical substance administered by a single rapid intravenous injection; (iii) a concentrated mass of food ready to be swallowed.

Bone marrow: Bone marrow is the flexible tissue found in the hollow interior of bones, such as the legs, arms, and hips. Marrow produces platelets, RBCs, and white blood cells, the primary agents of the body's immune system.

Botanical pesticide: A chemical substance with toxic activity against pests, which is produced naturally within a plant and may act as a defense against predators.

Botulism: Acute food poisoning caused by botulinum toxin produced in food by the bacterium *Clostridium botulinum* and characterized by muscle weakness and paralysis, disturbances of vision, swallowing, and speech, and a high mortality rate. A rare and serious paralytic illness caused by a nerve toxin that is produced by the bacterium *Cl. botulinum*. There are three main kinds of botulism, one of which is foodborne botulism caused by eating foods that contain the botulism toxin. Foodborne botulism can be especially dangerous because many people can be poisoned by eating a contaminated food. All forms of botulism can be fatal and are considered medical emergencies. Good supportive care in a hospital is the mainstay of therapy for all forms of botulism.

Bowman capsule (glomerular capsule): Bowman's capsule is a cup-like sac at the beginning of the tubular component of a nephron in the mammalian kidney. A glomerulus is enclosed in the sac. It is the outer cortex of the kidney and the main unit for blood filtering.

Bradycardia: Slow heart rate, usually fewer than 60 beats per minute in an adult human.

Bradypnoea: Abnormally slow breathing.

Breathing zone: (i) Space within a radius of 0.5 m from a person's face; (ii) the zone/location in the atmosphere at which individuals (animals/humans) breath.

Bremsstrahlung radiation: Secondary photon radiation (x-ray) produced by the deceleration of charged particles through matter.

Brine: To treat with or steep in brine. A strong solution of water and salt; a sweetener such as sugar, molasses, honey, or corn syrup may be added to the solution for flavor and to improve browning.

Bronchiole: (i) One of the small airways leading to the lungs; (ii) bronchioles—the narrowest airways that branch from the bronchi of the trachea.

Bronchiolitis: Inflammation of the bronchioles, usually caused by a viral infection.

Bronchitis: Chronic or acute inflammation of the large airways. Chronic bronchitis is persistent coughing and production of phlegm for at least 3 months per year for at least two successive years (American Thoracic Society).

Bronchoconstriction: Narrowing of the air passages through the bronchi of the lungs.

Bronchodilator: A drug that relaxes the smooth muscles of the airways and relieves constriction of the bronchi.

Bronchopneumonia: Inflammation of small areas of the lung.

Bronchopulmonary: Pertaining to the lungs and air passages.

Bronchospasm: (i) Spasmodic narrowing of the large airways; (ii) intermittent violent contraction of the air passages of the lungs.

Buffer: A solution that maintains constant pH by resisting changes in pH by dilution or addition of small amounts of acids and bases.

Burn: A burn is a type of injury that may be caused by heat, chemical substances, cold, electricity, radiation, or friction. Burns are characterized as: first degree—redness; second degree—blisters; and third degree—ulcers that heal by scarring. The severity of a burn is variable and depends on the damaged with subsequent pain due to profound injury to tissue.

Byssinosis: Pneumoconiosis caused by inhalation of dust and associated microbial contaminants and observed in cotton, flax, and hemp workers.

Bystander exposure: Liability of members of the general public to come in contact with chemical substances arising from operations or processes carried out by other individuals in their vicinity. The most common examples are (a) exposures to cigarette smoke, (b) industrial fumes, and (c) automobile exhausts.

C

Cachexia: Loss of body weight and muscle mass, and weakness that may occur in patients with cancer, AIDS, or other chronic diseases.

CAD (coronary artery disease): A disease in which there is a narrowing or blockage of the coronary arteries (blood vessels that carry blood and oxygen to the heart). CAD is usually caused by atherosclerosis (a buildup of fatty material and plaque inside the coronary arteries). The disease may cause chest pain, shortness of breath during exercise, and heart attacks. The risk of CAD is increased by having a family history of CAD before age 50, older age, smoking tobacco, high blood pressure, high cholesterol, diabetes, lack of exercise, and obesity.

Calcification: Deposits of calcium in the body tissues. Calcification in the breast can be seen on a mammogram, but cannot be detected by touch. There are two types of breast calcification: macrocalcification and microcalcification. Macrocalcifications are large deposits and are usually not related to cancer. Microcalcifications are specks of calcium that may be found in an area of rapidly dividing cells. Many microcalcifications clustered together may be a sign of cancer.

Calcitonin: A hormone formed by the C cells of the thyroid gland. It helps in maintaining a healthy level of calcium in the blood. When the calcium level is too high, calcitonin lowers it.

Calcium gluconate: The mineral calcium combined with a form of the sugar glucose. It is used to prevent and treat bone loss. It is also being studied in the treatment of bone loss and nerve damage caused by chemotherapy.

Calcium: The mineral needed for healthy teeth, bones, and other body tissues. It is the most common mineral in the body. A deposit of calcium in body tissues, such as breast tissue, may be a sign of disease.

Calorie: The measurement of the energy content of food. The body needs calories to perform its functions, such as breathing, circulating the blood, and physical activity. When a person is sick, their body may need extra calories to fight fever or other problems.

Cancer: (i) The term for diseases where abnormal cells divide without control and can invade nearby tissues. Cancer cells can also spread to other parts of the body through the blood and lymph systems. There are several main types of cancer. Carcinoma is a cancer that begins in the skin or in tissues that line or cover internal organs. Sarcoma is a cancer that begins in bone, cartilage, fat, muscle, blood vessels, or other connective or supportive tissue. Leukemia is a cancer that starts in blood-forming tissue such as the bone marrow, and causes large numbers of abnormal blood cells to be produced and enter the blood. Lymphoma and multiple myeloma are cancers that begin in the cells of the immune system. CNS cancers are cancers that begin in the tissues of the brain and spinal cord. Also called malignancy. (ii) The injurious malignant growth of potentially unlimited size of cells and tissue invading local tissues and spreading to distant areas of the body. (iii) Cancer is a disease that results from the development of a malignant tumor and spreads into surrounding tissues. (iv) Cancer disease occurs when a cell, or group of cells, grows in an unchecked, uncontrolled, or unregulated manner. Cancer disease can involve any tissue of the body and can have many different forms in each body area. Most cancers are named for the type of cell or the organ in which they begin, such as leukemia or lung cancer.

Cancer effect level (CEL): The lowest dose of chemical in a study, or group of studies, that produces significant increases in the incidence of cancer/tumors between the exposed population and its appropriate control.

Cancer risk: A theoretical risk for getting cancer if exposed to a chemical substance every day for 70 years (a lifetime exposure). The true risk might be lower.

Capillaries: The tiniest blood vessels; capillary networks connect the arterioles (the smallest arteries) and the venules (the smallest veins).

Capillary: The smallest type of blood vessel. A capillary connects an arteriole (small artery) to a venule (small vein) to form a network of blood vessels in almost all parts of the body. The wall of a capillary is thin and leaky, and capillaries are involved in the exchange of fluids and gases between tissues and the blood.

Capsule: A sac of tissue and blood vessels that surrounds an organ, joint, or tumor. A capsule is also a form of medicine that is taken by mouth. It usually has a shell made of gelatin with the medicine inside.

Carbamates: Insecticides belonging to class carbamate interrupt nerve conduction and cause the accumulation of acetylcholine at nerve endings by reversibly binding with the acetylcholinesterase enzyme.

Carbamide: A substance formed by the breakdown of protein in the liver. The kidneys filter carbamide out of the blood and into the urine. Carbamide can also be made in the laboratory. A topical form of carbamide is being studied in the treatment of hand-foot syndrome (pain, swelling, numbness, tingling, or redness of the hands or feet that may occur as a side effect of certain anticancer drugs).

Carbohydrate: A sugar molecule. Carbohydrates can be small and simple (e.g., glucose) or they can be large and complex (e.g., polysaccharides such as starch, chitin, or cellulose).

Carboxylhemoglobin: A compound that is formed between carbon monoxide and hemoglobin in the blood of animals and humans, which is incapable of transporting oxygen.

Carcinogen: (i) A carcinogen is any agent, chemical, physical or biological, that can act on living tissue in such a way as to cause a malignant neoplasm. More simply, a carcinogen is any substance that causes cancer. (ii) A chemical that can increase the incidence of cancer in exposed populations. Chemicals are classified by the International Agency for Research on Cancer (IARC) as known, probable, or possible human carcinogens based on available epidemiologic and toxicological evidence.

Carcinogenesis: (i) A biological process involving the transformation of a normal cell into cancer cell. Carcinogenesis is the process that leads to the development of cancer. Carcinogenesis may be a matter of induction by chemical, physical, or biological agents, of neoplasms that are usually not observed, an earlier induction of neoplasms that are usually observed, and/or the induction of more neoplasms than are usually found. (ii) Induction, by chemical, physical, or biological agents, of malignant neoplasms and thus cancer.

Carcinogenic: A term applied to any chemical substance or physical agent that can cause cancer, or produce cancer.

Carcinogenicity test: Long-term (chronic) test designed to detect any possible carcinogenic effect of a test substance.

Carcinogenicity: (i) The process of induction of malignant neoplasms, and thus cancer, by chemical, physical, or biological agents; (ii) cancer-causing potential (of a chemical substance or agent).

Carcinoma: (i) A general term applied to a malignant epithelial tumor; (ii) a malignant tumor of the cells involving the lung, gut, skin, and also epithelial tissues; carcinoma accounts for about 90% of all types of cancer.

Cardiopulmonary: The term is related to or involves both the heart and the lungs.

Cardiotoxic: Chemically harmful to the cells of the heart.

CAS Number: (i) Chemical Abstracts Service registry number, a unique number for each chemical in the format CAS xxx-xx-x; (ii) a unique number assigned to a substance or mixture by the American Society Abstract Service.

Case control study: A study that starts with the identification of persons with the disease (or other outcome variable) of interest, and a suitable control (comparison, reference) group of persons without the disease. The relationship of an attribute to the disease is examined by comparing the diseased and non-diseased with regard to how frequently the attribute is present or, if quantitative, the levels of the attribute, in the two groups.

Case study: A medical or epidemiologic evaluation of one person or a small group of people to gather information about specific health conditions and past exposures.

Catabolism: The term indicates reactions involving the oxidation of organic substrates to provide chemically available energy (e.g., ATP) and to generate metabolic intermediates. Generally, the process involves the breakdown of complex molecules into simpler ones, often providing biologically available energy.

Catalase: Catalase is a heme-based enzyme that catalyzes the breakdown of hydrogen peroxide into oxygen and water. It is found in all living cells, especially the peroxisomes.

Catatonia: Schizophrenia marked by excessive, sometimes violent, motor activity and excitement, or by generalized inhibition.

CBD: Chronic beryllium disease.

CDC (Centers for Disease Control and Prevention): An agency within the US Department of Health and Human Services that monitors and investigates foodborne disease outbreaks and compiles baseline data against which to measure the success of changes in food safety programs.

Ceiling: "The concentration that should not be exceeded during any part of the working exposure." (ACGIH) (Refer: Threshold limit value.)

CCPR: The Codex Committee on Pesticide Residues. The Codex Alimentarius Commission was created in 1963 by FAO and WHO to develop food standards, guidelines, and related texts such as codes of practice under the joint FAO/WHO Food Standards Program.

CCR: Coal combustion residue.

CCRIS: Chemical carcinogenesis research information system.

CDDs: Chlorinated dibenzo-p-dioxins (also refer: CFCs).

CDFs: Chlorinated dibenzofurans (also refer: CFCs).

Ceiling value (CV): (i) The ceiling value is the maximum permissible airborne concentration of a potentially toxic substance and is a concentration that should never be exceeded in the breathing zone; (ii) a concentration of a chemical substance that should not be exceeded, even instantaneously; (iii) the maximum permissible concentration of a material in the working environment that should never be exceeded for any duration.

Cell: The smallest structural unit of all living organisms.

Cell cycle: Regulated biochemical steps that cells go through involving DNA replication.

Cell line: This is a general term applied to a defined population of cells that have been maintained in a culture for an extended period and have usually undergone a spontaneous process of transformation conferring an unlimited culture life span on the cells.

Cell proliferation: Rapid increase in cell number.

Cell-mediated hypersensitivity: State in which an individual reacts with allergic effects caused by the reaction of antigen-specific T-lymphocytes following exposure to a certain substance (allergen) after being exposed previously to the same substance or chemical group.

Cell-mediated immunity: Immune response mediated by antigen-specific T-lymphocytes.

CERCLA: Comprehensive Environmental Response, Compensation, and Liability Act of 1980. CERCLA, also known as Superfund, is the federal law that concerns the removal or cleanup of hazardous substances in the environment and at hazardous waste sites. ATSDR, which was created by CERCLA, is responsible for assessing health issues and supporting public health activities related to hazardous waste sites or other environmental releases of hazardous substances. This law was later amended by the Superfund Amendments and Reauthorization Act (SARA).

Cerebral edema: Accumulation of fluid in and resultant swelling of the brain.

Certified applicator: A person who is authorized to apply "restricted use pesticides" (RUP) as a result of meeting the requirements for certification under FIFRA-mandated programs. Applicator certification programs are conducted by states, territories, and tribes in accordance with national standards set by the EPA. RUPs may only be used by or under the direct supervision of specially trained and certified applicators.

Certified pesticide applicator: Any individual who is certified under Section 4 of the Federal Insecticide, Fungicide, and Rodenticide Act (FIFRA) is authorized to use or supervise the use of any pesticide that is classified for restricted use. Any applicator who applies registered pesticides, only to provide a service of controlling pests without delivering any additional pesticide supplies, is not deemed to be a seller or distributor of pesticides under FIFRA.

CFCs (chlorofluorocarbons): A family of inert, non-toxic, and easily liquefied chemicals used in refrigeration, air conditioning, packaging, insulation, or as solvents and aerosol propellants. Because CFCs are not destroyed in the lower atmosphere, they drift into the upper atmosphere where their chlorine components destroy ozone.

CFR: The Code of Federal Regulations.

Chelation therapy: Treatment with a chelating agent to enhance the elimination or reduce the toxicity of a metal ion.

Chelation: Describes the combination of a metallic ion with heterocyclic ring structures in such a way that the ion is held by bonds from each of the rings.

Chelators: An organic chemical that bonds with and removes free metal ions from solutions.

Chemical asphyxiant: A poison that blocks either the transport or use of oxygen by living organisms.

Chemical hygiene officer: A designated person who provides technical guidance in the development and implementation of the chemical hygiene plan.

Chemical hygiene plan: A written program that outlines procedures, equipment, and work practices that protect employees from the health hazards present in the workplace.

Chemical incident: An unforeseen event involving any non-radioactive substance, resulting in a potential toxic risk to public health or leading to the exposure of two or more individuals, resulting in illness or potential illness.

Chemical interaction: The process by which two or more chemical substances interact with each other, resulting in either antagonistic or synergistic effects.

Chemical oxygen demand (COD): A measure of the oxygen required to oxidize all compounds, organic and inorganic matter in a sample of water. COD is expressed as parts per million of oxygen taken from a solution of boiling potassium dichromate for 2 h. The COD test is used to assess the strength of sewage and waste.

Chemical Safety and Surveillance Committee (CSSC): The CSSC is advisory to the chancellor on all matters relating to the safe use of hazardous chemicals. The primary charge to the committee is to reduce risks associated with hazardous chemicals; establish policies and procedures that meet or exceed applicable norms; monitor new regulations; and implement adopted policies and procedures for hazardous chemicals. Should there be a willful or negligent violation of the University of California at San Diego (UCSD) established chemical safety practices and procedures, the committee has the authority to impose disciplinary measures that are subject to review and/or modification by the chancellor or his/her designated representative.

Chemical safety: Practical certainty that there will be no exposure of organisms to toxic amounts of any substance or group of substances; this implies attaining an acceptably low risk of exposure to potentially toxic substances.

Chemosis: The term in medicine given to any swelling around the eye caused by exposure to a chemical substance as a result of edema of the conjunctiva.

Chloracne: A skin disease characterized by acne that is caused by exposure to dioxin, pentachlorophenol, PCBs, and other chlorinated hydrocarbon compounds.

Chlorinated hydrocarbons: (i) Chemical substances containing only chlorine, carbon, and hydrogen. These include a class of persistent, broad-spectrum insecticides that linger in the environment and accumulate in the food chain. Among them are DDT, aldrin, dieldrin, heptachlor, chlordane, lindane, endrin, mirex, hexachloride, and toxaphene. Other examples include TCE, used as an industrial solvent. (ii) Any chlorinated organic compounds including chlorinated solvents such as dichloromethane, trichloromethylene, and chloroform.

Chlorination: Adding chlorine to water or wastewater, generally for the purpose of disinfection, but frequently for accomplishing other biological or chemical results. Chlorine is also used almost universally in manufacturing processes, particularly for the plastics industry.

Choline acetylase: An enzyme system that helps to generate acetylcholine, an important neurotransmitter. Acetylcholine is synthesized from choline and acetyl CoA by the enzyme choline acetylase.

Chlorofluorocarbons (CFCs): A family of chemicals commonly used in air conditioners and refrigerators as coolants and also as solvents and aerosol propellants. CFCs drift into the upper atmosphere where their chlorine components destroy ozone. CFCs are thought to be a major cause of the ozone hole over Antarctica.

Choline acetyltransferase: An enzyme that controls the production of acetylcholine; appears to be depleted in the brains of individuals with Alzheimer's disease.

Cholinesterase and pseudocholinesterase inhibitors: These are chemical substances that inhibit the enzyme cholinesterase, thereby enhancing and subsequently preventing transmission of nerve impulses from one nerve cell to another or to a muscle.

Cholinesterase: The enzyme that is found primarily at nerve endings that catalyzes the breakdown of acetylcholine in animals. It regulates nerve impulses by the inhibition of acetylcholine; cholinesterase inhibition in animal indicate a variety of acute symptoms such as nausea, vomiting, blurred vision, stomach cramps, and rapid heart rate.

CHRIS: Chemical hazards response information system.

Chromatid: A chromatid is one of the two identical copies of DNA making up a replicated chromosome, which are joined at their centromere for the process of cell division.

Chromosomal aberration: Any abnormality of chromosome number or structure could be described as an aberration.

Chromosomes: The structure found in the nucleus of a cell and these structures are made up of DNA. Chromosomes form the basis of heredity and carry genetic information in DNA in the form of a sequence of nitrogenous bases.

Chromosome-type aberration: The damage expressed in both sister chromatids at the same locus.

Chronic effect: An adverse effect on any living organism in which symptoms develop slowly over a long period of time or recur frequently.

Chronic exposure: (i) The exposures to a chemical substance over a long period of time for more than 1 year; (ii) the long-term exposure or continued exposure or exposures occurring over an extended period of time, or a significant fraction of a test species' or a group of individuals', or a population's lifetime.

Chronic interstitial nephritis (CIN): Chronic interstitial nephritis is a form of nephritis affecting the interstitium of the surrounding tubules of the kidneys. This disease can be either acute or chronic. CIN eventually leads to kidney failure.

Chronic syndrome: The onset of symptoms is gradual over a period longer than 2 months. A chronic syndrome induced by chemical substances may represent: (i) a cumulative exposure with a long latency; (ii) adverse effects that persist for 2 months or longer after a brief high exposure; (iii) the onset of symptoms is gradual over a period longer than 2 months. A chronic syndrome induced by chemicals may represent (a) a cumulative exposure with a long latency or (b) adverse effects that persist for 2 months or longer after a brief high exposure.

Chronic toxicity: (i) The capacity of a substance to cause adverse effects/harmful effects in an organism after long-term exposure; (ii) adverse health effects from repeated doses of a toxic chemical or other toxic substance over a relatively prolonged period of time, generally greater than 1 year. With experimental animals, this usually means a period of exposure of more than 3 months. Chronic exposure studies over 2 years using rats or mice are used to assess the carcinogenic potential of chemical substances.

Chronic: (i) An event or occurrence that persists over a long period of time; lasting a long time; this term in medicine comes from the Greek chronos, time and means lasting a long time; (ii) in experimental toxicology, the term chronic refers to mammalian studies lasting considerably more than 90 days or to studies ranging a large part of the lifetime of an organism; (iii) adverse effects resulting from repeated doses of, or exposures to, a chemical substance by any route for more than 3 months.

CIN: Chronic interstitial nephritis.

Cirrhosis: An abnormal liver condition characterized by irreversible scarring of the liver. Alcohol and viral hepatitis B and C are among the many causes of cirrhosis.

Cirrhosis can cause yellowing of the skin (jaundice), itching, and fatigue. Diagnosis of cirrhosis can be suggested by physical examination and blood tests, and can be confirmed by liver biopsy in some patients. Complications of cirrhosis include mental confusion, coma, fluid accumulation (ascites), internal bleeding, and kidney failure. Treatment of cirrhosis is designed to limit any further damage to the liver as well as complications. Liver transplantation is becoming an important option for patients with advanced cirrhosis.

Clastogen: A clastogen is any chemical substance that causes chromosomal breaks and the consequent gain, loss, or rearrangement of pieces of chromosomes.

Clastogenesis: A process resulting in chromosomal breaks and the gain, loss, or rearrangement of pieces of chromosomes.

Clastogenic: A term applied to a chemical substance or process that causes chromosomal breaks.

Clinical toxicology: Scientific study involving research, education, prevention, and treatment of diseases caused by chemical substances such as drugs and toxins. Clinical toxicology often refers specifically to the application of toxicological principles to the treatment of human poisoning.

Clinical: Medical process to do with the examination and treatment of patients; applicable to patients; a laboratory test may be of clinical value (of use to patients).

Clostridium botulinum: The name of a group of bacteria commonly found in soil. These rod-shaped organisms grow best in low oxygen conditions. The bacteria form spores that allow them to survive in a dormant state until exposed to conditions that can support their growth. *Clostridium botulinum* is the bacterium that produces the nerve toxin that causes botulism.

CNS depression: Reduced level of consciousness.

CNS solvent syndrome: Organic solvents can affect the CNS both acutely (increased reaction time and anesthesia) and chronically (permanent brain damage).

CNS: The central nervous system, the part of the nervous system that consists of the brain and the spinal cord.

Co-carcinogen: (i) A chemical substance/agent that assists carcinogens to cause cancer; (ii) chemical, physical, or biological factor that intensifies the effect of a carcinogen.

Code of Federal Regulations (CFR): (i) A document that codifies all rules of the executive departments and agencies of the federal government. It is divided into 50 volumes, known as titles. Title 40 of the CFR (referenced as 40 CFR) lists all environmental regulations. (ii) The codification of the general and permanent rules published in the federal register by the executive departments and agencies of the federal government. The code is divided into 50 titles that represent broad areas subject to regulation. Most regulations directly related to agriculture are in Title 7. Each title is divided into chapters that usually bear the name of the issuing agency, followed by subdivisions into parts covering specific regulatory areas. Title 9, Chapter III covers the FSIS.

Codex Alimentarius Commission (CAC): (i) The Codex Alimentarius Commission is the highest international body on food standards; (ii) a joint commission of the Food and Agriculture Organization (FAO) and WHO, comprised of some 146 member countries, created in 1962 to ensure consumer food safety, establish fair practices in food trade, and promote the development of international food standards. The commission drafts non-binding standards for food additives, veterinary drugs, pesticide residues, and other substances that affect consumer food safety. It publishes these standards in a listing called the "Codex Alimentarius."

Cohort study: A study in which a group of people with a past exposure to chemical substances or other risk factors are followed over time and their disease experience is compared to that of a group of people without the exposure.

Cohort: A cohort is a group of individuals, identified by a common characteristic, who are studied over a period of time as part of an epidemiological investigation.

Colic: Acute abdominal pain, especially in infants.

Colloid: Material in the nanometer to micrometer size range whose characteristics and reactions are largely controlled by surface properties.

Coma: A state of deep unarousable unconsciousness; a state of profound loss of consciousness.

Combustible liquid: A liquid with a flashpoint at a temperature lower than boiling point; according to the National Fire Protection Association and the US Department of Transportation, it is a liquid with a flash point of 100°F (37.8°C) or higher.

Combustible: A chemical substance that ignites, burns, or supports combustion.

Commercial applicator: A person applying pesticides as part of a business applying pesticides for hire or a person applying pesticides as part of his/her job with another (not for hire) type of business, organization, or agency. Commercial applicators are often certified, but only need to be if they use RUPs.

Common mechanism of toxicity: Two or more chemicals or other substances that cause a common toxic effect(s) by the same, or essentially the same, sequence of major biochemical events (i.e., interpreted as mode of action).

Comparative effect level (CEL): Dose by which potency of chemicals may be compared, e.g., the dose causing a maximum of 15% cholinesterase inhibition.

Compatible materials: Chemical substances that do not react together to cause a fire, explosion, or violent reaction, or lead to the evolution of flammable gases or otherwise lead to injury to people or danger to property.

Complete carcinogens: Chemical substances that both initiate and promote cancer.

Compressed gas: A substance in a container with an absolute pressure greater than 276 kPa or 40 psi at 21°C, or an absolute pressure greater than 717 kPa (40 psi) at 54°C.

Concentration: The amount of a substance present in a certain amount of soil, water, air, food, blood, hair, urine, breath, or any other media.

Concentration-response curve: A graph produced to show the relation between the exposure concentration of a drug or xenobiotic and the degree of response it produces, as measured by the percentage of the exposed population showing a defined, often quantal, effect. If the effect determined is death, the curve may be used to estimate an LC50 value.

Condensation: The process of converting a chemical in the gaseous phase to a liquid/solid state by decreasing temperature, by increasing pressure or both.

Confined space: A confined space has limited or restricted means for entry or exit and is not designed for continuous occupancy. Confined spaces include storage tanks, bins, boilers, ventilation and exhaust ducts, pits, manholes, vats, and reactor vessels.

Congenital: A trait, condition, or disorder that exists in the organism/animal from birth.

Conjugate: In chemistry, this is a water-soluble derivative of a chemical formed by its combination with glucuronic acid, glutathione, sulfate, acetate, glycine, etc. Usually, conjugation takes place in the liver and facilitates excretion of a chemical substance that would otherwise tend to accumulate in the body because of its solubility in body fat.

Conjunctiva: The term applied to the mucous membrane that covers the eyeball and the undersurface of the eyelids.

Conjunctivitis: Inflammation of the membranes of the eye (conjuctiva).

Consumer Product Safety Commission (CPSC): An independent US federal regulatory agency that protects the public against unreasonable risk of injury and death associated with consumer products.

Contact dermatitis: Dermatitis caused by contact with irritating or allergenic chemical substances.

Contact sensitizer: A substance that induces an allergic response following skin contact. The definition for "contact sensitizer" is equivalent to "skin sensitizer."

Contaminant: A chemical substance that is present in an environment at levels sufficient to cause harmful or adverse health effects to organisms, animals, and humans.

Contaminated water: Water rendered unwholesome by contaminants and pollution.

Contamination: Introduction into water, air, and soil of microorganisms, chemicals, toxic substances, wastes, or wastewater in a concentration that makes the medium unfit for its next intended use. Also applies to surfaces of objects, buildings, and various household and agricultural use products.

Control limit: A regulatory value applied to the airborne concentration in the workplace of a potentially toxic substance that is judged to be "reasonably practicable" for the whole spectrum of work activities and which must not normally be exceeded.

Convulsion: Abnormal and involuntary jerks and quick movements of the body.

COPD: Chronic obstructive pulmonary disease.

Corrosive of tissue: The descriptor applied to any chemical substance that destroys tissues on direct contact.

Corrosive: (i) Any liquid or solid chemical substance that causes visible destruction/irreversible alteration of skin tissue at the place of contact; (ii) a chemical substance capable of causing visible destruction of, and/or irreversible changes to living tissue by chemical action at the site of contact (i.e., strong acids, strong bases, dehydrating agents, and oxidizing agents).

Corrosive to metal: A substance or a mixture that, by chemical action, will materially damage, or even destroy, metals.

Cost-benefit analysis: A quantitative evaluation and decision-making technique where comparisons are made between the costs of a proposed regulatory action on the use of a substance/chemical and the overall benefits to society of the proposed action; often converting both the estimated costs and benefits into health and monetary units.

Counts per minutes (cpm): The number of counts or nuclear events detected by a radiation survey device such as a Geiger counter. Since not all events that occur are detected, cpm are always less than actual disintegrations per minute (dpm) emanating from a radioactive material.

CPSC: Consumer Product and Safety Commission.

Crepitations: Abnormal respiratory sounds heard on auscultation of the chest, produced by passage of air through passages that contain secretion or exudate or that are constricted by spasm or a thickening of their walls.

Critical end-point: Toxic effect used by the US Environmental Protection Agency (US EPA) as the basis for a reference dose.

Critical control point: An operation (practice, procedure, process, or location) at or by which preventive or control measures can be exercised that will eliminate, prevent, or minimize one or more hazards. Critical control points are fundamental to hazard analysis and critical control point (HACCP) systems.

Critical organ: That part of the body that is most susceptible to the action of chemical substances and or radiation damage under specific conditions.

Cross-contamination: The transfer of harmful substances or disease-causing microorganisms to food by hands, food-contact surfaces, sponges, cloth towels, and utensils that touch raw food, and then touch ready-to-eat foods. Cross-contamination can also occur when raw food touches or drips onto cooked or ready-to-eat foods.

CSF: Cerebrospinal fluid.

Cumulative effect: Overall change that occurs in an organism/animal after exposures to repeated doses of a chemical substance or radiation.

CWP: Coal workers' pneumoconiosis.

Cyanosis: Bluish discoloration of the skin due to deficient oxygenation of the blood.

Cytochrome P450: (i) This is a heme-containing protein that takes part in phase I reactions of xenobiotics during biotransformation processes; (ii) the iron-containing proteins are important in cell respiration as catalysts of oxidation-reduction reactions.

Cytogenetics: Cytogenetics part of the science of genetics that correlates the structure and number of chromosomes with heredity and genetic variability.

Cytokine: Any of a group of soluble proteins that are released by a cell, causing a change in function or development of the same cell (autocrine), an adjacent cell (paracrine), or a distant cell (endocrine); cytokines are involved in reproduction, growth and development, normal homeostatic regulation, response to injury and repair, blood clotting, and host resistance (immunity and tolerance).

Cytoplasm: In cell biology, this term is applied to the ground substance of the cell which contains the cell organelles, such as the nucleus, mitochondria, endoplasmic reticulum, ribosomes, etc.

Cytotoxic: Chemical substance causing harmful effects/damage to cell structure and function and ultimately causing cell death.

D

Dechlorination: Removal of chlorine and chemical replacement with hydrogen or hydroxide ions to detoxify a substance.

Decontamination: To make safe by eliminating poisonous or otherwise harmful substances, such as noxious chemicals or radioactive material, from people, buildings, equipment, and the landscape.

Defoliant: A chemical substance used for removal of leaves by its toxic action on living plants.

Dehydrogenase: An enzyme that catalyzes oxidation of compounds by removing hydrogen.

Delaney Clause: The Delaney Clause in the Federal Food, Drug, and Cosmetic Act (FFDCA) states that no additive shall be deemed to be safe for human food if it is found to induce cancer in man or animals. It is an example of the zero tolerance concept in food safety policy. The Delaney prohibition appears in three separate parts of the FFDCA: Section 409 on food additives; Section 512, relating to animal drugs in meat and poultry; and Section 721 on color additives. The Section 409 prohibition applied to many pesticide residues until enactment of the Food Quality

Protection Act of 1996 (P.L. 104–170, August 3, 1996). This legislation removed pesticide residue tolerances from Delaney Clause constraints.

Delayed effects (latent effect): Consequences of effects occurring after a latent period following the end of exposures to toxic chemical substance or other harmful environmental factors.

Dementia: Marked decline in mental function.

Demyelination: Destruction of the myelin sheath of a nerve.

Denaturation: Addition of methanol, acetone, or other suitable chemical(s) to alcohol to make it unfit for drinking.

Dendrite: Any of the branched extensions, or processes, of the neuron along which nerve impulses travel toward the cell body.

Denitrification: Reduction of nitrates to nitrites, nitrogen oxides, or dinitrogen (N2) catalyzed by facultative aerobic soil bacteria under anaerobic conditions.

Deoxyribonucleic acid (DNA): Deoxyribonucleic acid is the constituent of chromosomes that stores hereditary information in the form of a sequence of nitrogenous bases, purine and pyrimidine. Much of the information is related with the synthesis of proteins and acts as a determinant of all physical and functional activities of the cell, and consequently of the whole organism.

Department of Transportation (DOT): A US federal agency that regulates the labeling and transportation of hazardous materials. The US EPA, a US federal agency that develops and enforces regulations to protect human health and the natural environment.

Depilatory: Chemical substances that cause loss of hair.

Dermal irritation: A localized skin reaction resulting from either single or multiple exposures to a physical or chemical agent at the same site. It is characterized by redness and swelling and may be accompanied by local cell death.

Dermatitis: Inflammation of the skin. Occurrence of contact dermatitis is due to local exposure to chemical substances and may be caused by irritation, allergy, or infection.

Decibel (dB): A unit used to measure sound intensity and other physical quantities.

Descriptive epidemiology: Study of the occurrence of disease or other health-related characteristics in populations, including general observations concerning the relationship of disease to basic characteristics such as age, sex, race, occupation, and social class; it may also be concerned with geographic location. The major characteristics in descriptive epidemiology can be classified under the headings: individuals, time, and place.

Detergent: A detergent is a cleaning or wetting agent, classed as anionic if it has a negative charge and cationic if it has a positive charge.

Detoxification (detoxication): (i) The process, or processes, of chemical modification that make a toxic molecule less toxic; (ii) treatment of patients suffering from poisoning in such a way as to promote physiological processes that reduce the probability or severity of adverse effects.

Detoxify: To reduce the toxicity of a chemical substance either (i) by making it less harmful; or (ii) by treating patients suffering from poisoning in such a way as to reduce the probability and/or severity of harmful effects; (iii) to reduce or eliminate the toxicity of a chemical substance or poison; to promote the recovery of a person from an addictive drug such as alcohol or heroin.

Developmental toxicity: Adverse effects on the developing organism (including structural abnormality, altered growth, functional deficiency, or death) resulting from exposure through conception, gestation (including organogenesis), and postnatally up to the time of sexual maturation.

Diaphoretic: Chemical substances that cause sweating.

Dioxins: A group of chemical compounds that share certain similar chemical structures and biological characteristics. Dioxins are present in the environment all over the world. Within animals, dioxins tend to accumulate in fat. About 95% of the average person's exposure to dioxins occurs through consumption of food, especially food containing animal fat. Scientists and health experts are concerned about dioxins because studies have shown that exposure may cause a number of adverse health effects.

Diploid: The state in which the chromosomes are present in homologous pairs. It is important to remember that the normal human somatic (non-reproductive) cells are diploid (they have 46 chromosomes), whereas reproductive cells, with 23 chromosomes, are haploid.

Diplopia: Appearance of temporary double vision.

Disease: (i) Literally, dis-ease, lack of ease; pathological condition that presents a group of symptoms peculiar to it and which establishes the condition as an abnormal entity different from other normal or pathological body states; (ii) sickness often characterized by typical patient problems (symptoms) and physical findings (signs).

Disinfectant: A chemical that destroys vegetative forms of harmful microorganisms but does not ordinarily kill bacterial spores.

Distribution: A general term for the dispersal of a xenobiotic and its derivatives throughout an organism or environmental system.

Diuresis: Excretion of urine, especially in excess.

Diuretic: A chemical substance or agent that increases urine production.

DNA (deoxynucleic acid): A nucleic acid that carries the genetic information in the cell and is capable of self-replication and synthesis of RNA.

Dominant: Allele that expresses its phenotypic effect when present in either the homozygous or heterozygous state.

Dosage (of a chemical substance): Dose divided by product of mass of organism and time of dose.

Dose: (i) The total quantity of a chemical substance administered to, taken up, or absorbed by an organism, organ, or tissue; (ii) the amount of a toxic chemical substance taken into the body over a given period of time.

Dose-effect curve: (i) A graph showing the relation between the dose of a drug or xenobiotic and the magnitude of the graded effect that it produces; (ii) a graph showing the relation between the dose of a drug or xenobiotic and the degree of response it produces, as measured by the percentage of the exposed population showing a defined, often quantal, effect. If the effect determined is death, such a curve may be used to estimate an LD50 value.

Draize test: Evaluation of materials for their potential to cause dermal or ocular irritation and corrosion following local exposure; generally using the rabbit model (almost exclusively the New Zealand White), although other animal species have been used.

Drug: Any chemical substance that, when absorbed into a living organism, may modify one or more of its functions. A common term generally accepted for a chemical substance taken for a therapeutic purpose, but is also commonly used for substances of abuse.

DSL: Domestic substances list.

Dysarthria: Imperfect articulation of speech due to neuromuscular damage.

Dysfunction: Abnormal, impaired, or incomplete functioning of an organism, organ, tissue, or cell.

Dysphagia: Difficulty in swallowing.

Dysplasia: Abnormal development of an organ or tissue identified by morphological examination.

Dyspnea: Difficulty in breathing, or shortness of breath.

E

ECD: Electron capture detector.

ECn: A commonly used abbreviation for the exposure concentration of a toxicant causing a defined effect on n% of a test population.

Ecology: Branch of biology that studies the interactions between living organisms and all factors (including other organisms) in their environment; such interactions encompass environmental factors that determine the distributions of living organisms.

Ecosystem: (i) Grouping of organisms (microorganisms, plants, animals) interacting together, with and through their physical and chemical environments, to form a functional entity within a defined environment; (ii) the interacting synergism of all living organisms in a particular environment; every plant, insect, aquatic animal, bird, or land species that forms a complex web of interdependency. An action taken at any level in the food chain, use of a pesticide for example, has a potential domino effect on every other occupant of that system.

Ecotoxicology: (i) Ecotoxicology is the science devoted to the study of the production of harmful effects by substances entering the natural environment, especially effects on populations, communities, and ecosystems; an essential part of ecotoxicology is the assessment of movement of potentially toxic substances through environmental compartments and through food webs. (ii) Ecotoxicology is a research field that explores how exposure to a toxicant negatively affects single organisms, populations, communities, and ecosystems.

Eczema: Acute or chronic skin inflammation with erythema, papules, vesicles, pustules, scales, crusts, or scabs, alone or in combination, of varied etiology.

Edema (oedema): Presence of abnormally large amounts of fluid in the intercellular spaces of body tissues.

EDn: A commonly used abbreviation for the dose of a toxicant causing a defined effect on n% of a test population.

Effective concentration (EC): Concentration of a substance that causes a defined magnitude of response in a given system. Note that the EC50 is the median concentration that causes 50% of maximal response.

Effective dose (ED): Dose of a chemical substance that causes a defined magnitude of response in a given system. Note that the ED50 is the median dose that causes 50% of maximal response.

Effluent: Fluid, solid, or gas discharged from a given source into the external environment.

Electroneurography (ENG): Recording and measuring the electrical signals generated by nerves by means of an electromyograph. ENG is used in testing the effects of neurotoxic substances on humans.

Electrophoresis: Electrophoresis is based on the migration of charged molecules in solution in response to an electric field.

Electrophysiology: Measuring and recording the electrical activity of the brain or nerve cells by means of electrodes.

Encephalopathy: Degeneration of the brain.

Element: One of the 103 known chemical substances that cannot be broken down further.

Elimination: The process whereby a substance or other material is expelled from the body (or a defined part thereof), usually by a process of extrusion or exclusion, but sometimes through metabolic transformation. The combination of chemical degradation of a xenobiotic in the body and excretion by the intestine, kidneys, lungs, skin, in sweat, expired air, milk, semen, menstrual fluid, or secreted fluids.

Embryo: The term is applied to the earliest stages of development of a plant or animal. The embryo is generally contained in another structure, the seed, egg, or uterus.

Embryonic period: Period from fertilization to the end of major organogenesis.

Embryotoxic: The adjective term applied to any chemical substance that is harmful, in any sense, to an embryo.

Embryotoxicity: (i) Production of toxic effects by a chemical substance in progeny in the first period of pregnancy between conception and the fetal stage; (ii) any kind of toxic effect on the conceptus as a result of prenatal exposure to toxic substances/agents during the embryonic stages of development; these effects may include malformations and variations, malfunctions, altered growth, prenatal death, and altered postnatal function.

Emesis: Vomiting.

Emission: The release or discharge of a substance into the environment. Generally refers to the release of gases or particulates into the air.

Emission standard: This regulatory value is a quantitative limit on the emission or discharge of a potentially toxic substance from a source. The simplest form for regulatory purposes is a uniform emission standard (UES), where the same limit is placed on all emissions of a particular contaminant (refer: Limit values).

Endemic: Present in a community or among a group of people; said of a disease prevailing continually in a region.

Endocrine: Pertaining to hormones or to the glands that secrete hormones directly into the bloodstream.

Endocrine system: The endocrine system consists of a series of ductless glands that produce chemical messages or extra-cellular signaling molecules called hormones. The endocrine system is important or responsible for regulating metabolism, growth, development and puberty, and tissue function, and also plays a part in determining mood.

Endocrine toxicity: Any adverse structural and/or functional changes to the endocrine system (the system that controls hormones in the body) that may result from exposure to chemical substances. Endocrine toxicity can harm human and animal reproduction and development.

Endocrine disruptor chemicals: (i) Different groups of exogenous chemical substances that act like hormones in the endocrine system and disrupt the physiologic function of endogenous hormones; (ii) exogenous chemical substances that alter the function(s) of the endocrine system and consequently cause adverse health effects in an intact organism, its progeny or (sub)populations.

Endogenous: Produced within or caused by factors within an organism/animal system.

Endoplasmic reticulum: (i) In cell biology, this is a complex pipe-like system of membranes that occupies much of the cytoplasm in cells and contains many of the enzymes that mediate biodegradation of xenobiotics; (ii) intracellular complex of membranes in which proteins and lipids, as well as molecules for export, are synthesized and where the biotransformation reactions of the monooxygenase enzyme systems occur. It may also be isolated as microsomes following cell fractionation procedures.

Endothelial: Pertaining to the layer of flat cells lining the inner surface of blood and lymphatic vessels, and the surface lining of serous and synovial membranes.

Endothelium: Layer of flattened epithelial cells lining the heart, blood vessels, and lymphatic vessels.

Endotoxin: Toxin that forms an integral part of the cell wall of certain bacteria and is released only on breakdown of the bacterial cell; endotoxins do not form toxoids.

Enteritis: Intestinal inflammation.

Environment: (i) The term environment has several meanings and depends on the specific context. In general, the term environment refers to the natural environment, which includes all living and non-living things that occur naturally on Earth. Environment consists of all, or any, of the following: the air; the water—oceans, rivers, lakes, and ponds; the land—mountains, forests, fields, and gardens. (ii) Aggregate, at a given moment, of all external conditions and influences to which a system under study is subjected.

Environmental damage: Adverse effects to the natural environment.

Environmental exposure level (EEL): The level (concentration or amount or a time integral of either) of a chemical substance to which an organism or animal or other component of the environment is exposed in its natural surroundings.

Environmental fate: Destiny of a chemical substance or biological pollutant after release into the natural environment.

Environmental hazard: Chemical or physical agent capable of causing harm to the environment.

Environmental health criteria documents: Critical publications of the IPCS containing reviews of methodologies and existing knowledge—expressed, if possible, in quantitative terms—of selected chemical substances (or groups of chemical substances) on identifiable, immediate, and long-term effects on human health and welfare.

Environmental health impact assessment (EHIA): Estimate of the adverse effects to health or risks likely to follow from a proposed or expected environmental change or development.

Environmental health: Human welfare and its influence by the environment, including technical and administrative measures for improving the human environment from a health point of view.

Environmental hygiene (environmental sanitation): Practical control measures used to improve the basic environmental conditions affecting human health, e.g., clean water supply, human and animal waste disposal, protection of food from biological contamination, and housing conditions, all of which are concerned with the quality of the human environment.

Environmental impact assessment (EIA): Appraisal of the possible environmental consequences of a past, ongoing, or planned action, resulting in the production of an environmental impact statement or "finding of no significant impact (FONSI)."

Environmental monitoring: Continuous or repeated measurement of agents in the environment to evaluate environmental exposure and possible damage by comparison with appropriate reference values based on knowledge of the probable relationship between ambient exposure and resultant adverse effects.

Environmental protection: The term environmental protection indicates: (i) actions taken to prevent or minimize adverse effects to the natural environment; (ii) complex of measures including monitoring of environmental pollution, development and practice of environmental protection principles (legal, technical, and hygienic), including risk assessment, risk management, and risk communication.

Environmental Protection Agency (EPA): The Environmental Protection Agency is an organization/agency of the federal government charged with a variety of responsibilities relating to protection of the quality of the natural environment. It includes research and monitoring, promulgation of standards for air and water quality, and control of the introduction of pesticides and other hazardous materials into the environment. The US EPA leads the nation's environmental science, research, education, and assessment efforts. The mission of the US EPA is to protect human health and the environment. Since 1970, the US EPA has been working for a cleaner, healthier environment for the American people. The US EPA works closely with other federal agencies, state and local governments, and Indian tribes to develop and enforce regulations under existing environmental laws. The US EPA is responsible for researching and setting national standards for a variety of environmental programs and delegates responsibility to states and tribes for issuing permits, and monitoring and enforcing compliance.

Environmental quality objective (EQO): A regulatory value defining the quality to be aimed for in a particular aspect of the environment, e.g., "the quality of water in a river such that coarse fish can maintain healthy populations." Unlike an environmental quality standard, an EQO is not usually expressed in quantitative terms and is not legally enforceable.

Environmental quality standard (EQS): This regulatory value defines the maximum concentration of a potentially toxic substance that can be allowed in an environmental compartment, usually air (AQS) or water, over a defined period.

Enzyme induction: A process whereby an enzyme is synthesized in response to a specific chemical substance or to other agents such as heat or a metal species.

Enzyme: (i) An enzyme is a protein that acts as a selective catalyst permitting reactions to take place rapidly in living cells under physiological conditions; (ii) a macromolecule that functions as a biocatalyst by increasing the specific reaction.

Enzymic (enzymatic) process: Any chemical reaction or series of reactions catalyzed by an enzyme or enzymes.

Epidemiologist: A medical scientist who studies the various factors involved in the incidence, distribution, and control of disease in a population.

Epidemiology: (i) Epidemiology is the science devoted to the statistical study of categories of persons and the patterns of diseases from which they suffer, with the aim of determining the events or circumstances causing these diseases; (ii) study of the distribution of disease, or other health-related conditions and events in human or animal populations, in order to identify health problems and possible causes.

Epigastric: Pertaining to the upper-middle region of the abdomen.

Epigenetic changes: Any changes in an organism brought about by alterations in the action of genes are called epigenetic changes. Epigenetic transformation refers to those processes that cause normal cells to become tumor cells without any mutations having occurred (also refer: Mutation, Transformation, Tumor).

Epilepsy: Epilepsy is a collective term for a variety of different types of seizures, all forms of epilepsy start with a random discharge of nerve impulses into the brain. Antiepileptic drugs act by either raising the seizure threshold or by limiting the spread of impulses from one nerve to another inside the brain.

Epileptiform: Reactions that occur in severe or sudden spasms, as in convulsion or epilepsy.

Epithelioma: Any tumor derived from epithelium.

Epithelium: Sheet of one or more layers of cells covering the internal and external surfaces of the body and hollow organs.

Erosion: The wearing away of soil by wind or water, intensified by land-clearing practices related to farming, residential or industrial development, road building, or logging.

Erythema: In medicine, this term is applied to redness of the skin due to blood vessel distension and by congestion of the capillaries.

Eschar: In medicine, this term describes a slough or dry scab that forms, for example, on an area of skin that has been burnt or exposed to corrosive agents.

ESIS: European Chemical Substances Information System.

Estuary: A complex ecosystem between a river and near-shore ocean waters where fresh and salt water mix. These brackish areas include bays, mouths of rivers, salt marshes, wetlands, and lagoons and are influenced by tides and currents. Estuaries provide valuable habitat for marine animals, birds, and other wildlife.

ET50: The exposure time required to produce a defined effect when a test population is exposed to a fixed concentration or specified dose of a toxicant.

Etiology (aetiology): The science dealing with the cause or origin of disease.

Excipient: Any largely inert substance added to a drug to give suitable consistency or form to the drug.

Excitotoxicity: Pathological process by which neurons are damaged and killed by the overactivation of receptors for the excitatory neurotransmitter glutamate, such as the NMDA receptor and AMPA receptor.

Excretion: A general term for the removal of substances from the body.

Explosive: A substance that causes a sudden, almost instantaneous release of pressure, gas, and heat when subjected to sudden shock, pressure, or high temperature.

Explosive limits (chemical substance): The amounts of vapor in air that form explosive mixtures. These limits are expressed as lower and upper values and give the range of vapor concentrations in air that will explode if an ignition source is present.

Exposure: (i) Any kind of contact with a chemical substance by swallowing, breathing, or touching; (ii) radiation or pollutants that come in contact with the body and present a potential health threat. The most common routes of exposure are through the skin, mouth, or by inhalation.

Exposure limits: The concentration of a substance in the workplace to which most workers can be exposed during a normal daily and weekly work schedule without adverse effects.

Eukaryote: A eukaryote is an organism (microorganism, plant, or animal) whose cells contain a membrane-bound nucleus and other membrane-bound organelles (compare: prokaryote).

Eutrophication: A phenomenon of nutrient accumulation in a lake or landlocked body of water. This occurs naturally over many years, but has recently accelerated because of fertilizer runoff from farms and sewage input. Algal blooms result and their decay removes dissolved oxygen, eliminating aerobic organisms such as fish, and may cause accumulation of sulfide in the water.

F

FACOSH: The FACOSH is a source for federal agency safety and health programs and policies. The work of the advisory council uses the expertise of its members and provides assistance to the secretary of labor and OSHA, and leadership to federal agencies, in an effort to reduce the number of worker injuries and illnesses in the federal government. Advises the secretary of labor on appropriate policies and initiatives to enhance occupational safety and health in the federal sector, and actively endorses these policies and initiatives.

FDA: US Food and Drug Administration, which is involved in the regulation of pesticides in the United States, particularly enforcement of tolerances in food and feed products.

Fecundity: The term includes: (i) ability to produce offspring frequently and in large numbers; (ii) in demography, the physiological ability to reproduce; (iii) the ability to produce offspring within a given period of time.

Federal Advisory Council on Occupational Safety and Health refer (FACOSH):

Federal Hazardous Substances Act (FHSA): The Federal Hazardous Substances Act (15 U.S.C. 1261–1278), administered by the CPSC, requires that certain household products that are "hazardous substances" bear cautionary labeling to alert consumers to potential hazards that those products present and inform them of the measures they need to protect themselves from those hazards. Any product that is toxic, corrosive, flammable, or combustible, an irritant, a strong sensitizer, or that generates pressure through decomposition, heat, or other means requires labeling, if the product may cause substantial personal injury or substantial illness during or as a proximate result of any customary or reasonable foreseeable handling or use, including reasonable foreseeable ingestion by children.

Federal Food, Drug, and Cosmetic Act (FFDCA): The regulatory act (P.L. 75–717, June 25, 1938) is the basic authority intended to ensure that foods are pure and wholesome, safe to eat, and produced under sanitary conditions; that drugs and devices are safe and effective for their intended uses; that cosmetics are safe and made from appropriate ingredients; and that all labeling and packaging is truthful, informative, and not deceptive. The FDA is primarily responsible for enforcing the FFDCA, although the USDA also has some enforcement responsibility. The EPA establishes limits for concentrations of pesticide residues on food under this act.

Federal register: A federal document containing current presidential orders or directives, agency regulations, proposed agency rules, notices, and other documents that are required by statute to be published for wide public distribution. The federal register is published each federal working day. The USDA publishes its rules, notices, and other documents in the federal register. Final regulations are organized by agency and programs in the CFR.

Feromone (pheromone): Chemical substance used in olfactory communication between organisms of the same species eliciting a change in sexual or social behavior.

Fertility: Ability to conceive and to produce offspring: for litter-bearing species, the number of offspring per litter is used as a measure of fertility. Reduced fertility is sometimes referred to as subfertility.

Fetotoxicity: Toxicity to the fetus.

Fibrosis: Abnormal formation of fibrous tissue.

FID: Flame ionization detector.

FIFRA: The Federal Insecticide, Fungicide, and Rodenticide Act was enacted on June 25, 1947. The act instructs the EPA to regulate: (1) the registration of all pesticides used in the United States, (2) the licensing of pesticide applicators, (3) re-registration of all pesticide products, and (4) the storage, transportation, disposal, and recall of all pesticide products.

First-pass effect: A chemical alteration resulting from biotransformation of a xenobiotic before it reaches the systemic circulation. Such biotransformation by the liver is referred to as a hepatic first-pass effect.

Flammable: (i) Any material that can be ignited easily and that will burn rapidly; (ii) as defined in the FHSA regulations at 16 CFR § 1500.3(c)(6)(ii), a substance having a flashpoint above 20°F (–6.7°C) and below 100°F (37.8°C). An extremely flammable substance, as defined in the FHSA regulations at 16 CFR § 1500.3(c)(6)(i), is any substance with a flashpoint at or below 20°F (–6.7°C).

Flashpoint: (i) The lowest temperature at which evaporation of a substance produces enough vapor to form an ignitable mixture with air; (ii) the minimum temperature at which a liquid or a solid produces a vapor near its surface sufficient to form an ignitable mixture with the air; the lower the flash point, the easier it is to ignite the material.

Fluorosis (fluoridosis): Adverse effects of fluoride, as in dental or skeletal fluorosis.

Fly ash: (i) Fly ash is one of the residues generated in the combustion of coal. Fly ash is generally captured from the chimneys of coal-fired power plants; small solid ash particles from the non-combustible portion of coal fuel. (ii) Fly ash contains environmental toxins in significant amounts. These include arsenic, barium, beryllium, boron, cadmium, chromium, chromium VI, cobalt, copper, fluorine, lead, manganese, nickel, selenium, strontium, thallium, vanadium, and zinc.

Foci: The medical term applied to a small group of cells occurring in an organ and distinguishable, either in appearance or histochemically, from the surrounding tissue.

Fetus: (i) Fetus is the developing mammal or other viviparous vertebrate, after the embryonic stage and before birth; (ii) a term used to describe a developing human infant from approximately the third month of pregnancy until delivery; (iii) this term in medicine is applied to the young of mammals when fully developed in the womb. In human beings, this stage is reached after about 3 months of pregnancy. Prior to this, the developing mammal is at the embryo stage.

Food additive: (i) Any chemical substance or mixture of substances other than the basic foodstuff present in a food as a result of any phase of production, processing, packaging, storage, transport, or handling. The USDA allows food additives in meat, poultry, and egg products only after they have received FDA safety approval. Food additives are regulated under the authority of the FFDCA and are subject to the Delaney Clause. (ii) Any chemical substance, not normally consumed as a food by itself and not normally used as a typical ingredient of a given food, whether or not it has nutritive value, that is added intentionally to food for a technological (including organoleptic) purpose in the manufacture, processing, preparation, treatment, packing, packaging, transport, or holding of the food. Addition results, or may be reasonably expected to result (directly or indirectly), in the substance or its byproducts becoming a component of, or otherwise affecting, the characteristics of the food to which it is added. Food additive does not include "contaminants" or chemical substances added to food for maintaining or improving nutritional quality.

Food allergy: Hypersensitivity reaction to chemical substances in the diet to which an individual has previously been sensitized.

Foodborne illnesses: Illnesses caused by pathogens that enter the human body through foods.

Food chain: (i) Sequence of organisms in an ecosystem, each of which uses the next, lower member of the sequence as a food source. The food web, pyramid-shaped structure illustrating the feeding order in nature wherein each organism feeds on the next lowest creature; feeding relationships between predators and prey wherein animals and plants get food in an ecosystem. (ii) Sequence of transfer of matter and energy in the form of food from organism to organism in ascending or descending trophic levels.

Food web: The network of interconnected food chains in an ecosystem.

Foot and mouth disease (FMD): A highly contagious viral disease of cattle and swine, as well as sheep, goats, deer, and other cloven-hoofed ruminants. Although rarely transmissible to humans, FMD is devastating to livestock and has critical economic consequences with potentially severe losses in the production and marketing of meat and milk.

Frame-shift mutation: Such a mutation is a change in the structure of DNA causing the transcription of genetic information into RNA to be completely altered because the start point for reading has been changed. In other words, the reading frame for transcription has been altered.

Fugacity: Broadly speaking, this word is applied to the tendency of a substance to move from one environmental compartment to another. Originally, the term was applied to the tendency of a gas to expand or escape and related to its pressure in the system being studied.

Fugitive emissions: Air pollutants released to the air other than those from stacks or vents; typically small releases from leaks in plant equipment such as valves, pump seals, flanges, sampling connections, etc.

Fumigants: Chemical substances that are vaporized in order to kill or repel pests.

Fungicide: A pesticidal chemical substance used to control or destroy fungi on food or grain crops.

Fungus: Funguses, or fungi, are types of plants that have no leaves, flowers, or roots. Both words, funguses and fungi, are the plural of fungus.

G

gamma-GT: Gamma-glutamyltranspeptidase.

Ganglion: A collection of nerve cells outside the brain or spinal cord.

Gastroenteritis: Inflammation of the stomach and intestine.

Gastrointestinal: Pertaining or communicating with the stomach and intestine.

Gavage: Administration of materials directly into the stomach by esophageal intubation.

GC: Gas chromatography.

Gelatin: Thickener from collagen that is derived from the skin, tendons, ligaments, or bones of livestock. It may be used in canned hams or jellied meat products, as well as non-food products such as photography and medicine.

Gene: The part of the DNA molecule that carries the information defining the sequence of amino acids in a specific polypeptide chain. Gene is the fundamental unit of heredity.

Generally regarded as safe (GRAS): Phrase used to describe the US FDA philosophy that justifies approval of food additives that may not meet the usual test criteria for safety, but have been used extensively and have not demonstrated to cause any harm to consumers.

Generic: The chemical name of a drug; a term referring to the chemical makeup of a drug rather than to the advertised brand name under which the drug may be sold; a term referring to any drug marketed under its chemical name without advertising.

Genetic toxicology (genotoxicology): The study of chemical substances that can cause harmful heritable changes in the genetic information carried by living organisms in the form of DNA.

Genome: (i) The general term for all the genes carried by a cell; (ii) complete set of chromosomal and extra-chromosomal genes of an organism, a cell, an organelle, or a virus, i.e., the complete DNA component of an organism. The term genome includes both the DNA present in the chromosomes and that in subcellular organelles (e.g., mitochondria or chloroplasts). It also includes the RNA genomes of some viruses.

Genomics: The term describes (i) the science of using DNA- and RNA-based technologies to demonstrate alterations in genes expression; (ii) methods and information on the consequences for genes expression of interactions of the organism with environmental stress, xenobiotics, and other factors.

Genotoxic: Adjective applied to any chemical substance that is able to cause harmful changes to DNA. Such changes may lead to a transformed cell that can form a malignant tumor.

Genotype: Genetic constitution of an organism as revealed by genetic or molecular analysis; the complete set of genes possessed by a particular organism, cell, organelle, or virus.

Germ-free animal: Animal grown under sterile conditions in the period of postnatal development: such animals are usually obtained by Cesarean operation and kept in special sterile boxes in which there are no viable microorganisms (sterile air, food, and water are supplied).

Germinal aplasia: Complete failure of gonad development.

Glia, glial cells: Glia or glial cells are also called *neuroglia* (Greek for "glue"). Glia are non-neuronal cells and provide support and protection for neurons. Glial cells provide nutrition, maintain homeostasis, form myelin, and participate in signal transmission in the nervous system. In the human brain, glia outnumber neurons by about 10 to 1.

Glomerular filtration rate: Volume of ultra-filtrate formed in the kidney tubules from the blood passing through the glomerular capillaries divided by time of filtration.

Glomerular filtration: Formation of an ultra-filtrate of the blood occurring in the glomerulus of the kidney.

Glomerular: Pertaining to a tuft or cluster, as of a plexus of capillary blood vessels or nerve fibers, especially referring to the capillaries of the glomerular of the kidney.

Glomerulus: Tuft or a cluster, as of a plexus of capillary blood vessels or nerve fibers, e.g., capillaries of the filtration apparatus of the kidney.

Glutathione (GSH): Glutathione is a tripeptide. It contains an unusual peptide linkage between the amino group of cysteine and the carboxyl group of the glutamate side chain. Glutathione is an antioxidant, meaning it protects cells from the toxicological effects of free radicals. Glutathione exists in reduced (GSH) and oxidized (GSSG) states. Glutathione has recently been used as an inhibitor of melanin in the cosmetics industry.

Goiter: Non-cancerous enlargement of the thyroid gland, visible as a swelling at the front of the neck, which is often associated with iodine deficiency.

Goitrogen: Any chemical substance (such as thiouracil) that induces the formation of a goiter.

Gonadotropic: Pertaining to effects on sex glands and on the systems that regulate them.

Good agricultural practice in the use of pesticides (GAP): Nationally authorized safe uses of pesticides under actual conditions necessary for effective and reliable pest control. GAP includes a range of levels of pesticide applications up to the highest authorized use, applied in a manner that leaves a residue that is the smallest amount practicable. Authorized safe uses include nationally registered or recommended uses and the occupational health and environmental safety considerations. Actual conditions include any stage in the production, storage, transport, distribution, and processing of food commodities and animal feed.

Good laboratory practice principles (GLP): Fundamental rules incorporated in OECD guidelines and national regulations concerned with the process of effective organization and the conditions under which laboratory studies are properly planned, performed, monitored, recorded, and reported.

Good manufacturing practice principles (GMP): Fundamental rules incorporated in national regulations concerned with the process of effective organization of production and ensuring standards of defined quality at all stages of production, distribution, and marketing. Minimization of waste and its proper disposal are part of this process.

GPE: Greatest potential for human exposure.

Graded effect: An effect that can usually be measured on a continuous scale of intensity or severity and its magnitude related directly to dose. A graded effect increases in severity with an increase in dose or exposure concentration.

Granuloma: A granular growth or tumor, usually of lymphoid and epithelial cells.

Growth: An increase in the size of an organism or part of an organism. Growth of an organism may stop at maturity, as in the case of humans and other mammals. Also, growth may continue throughout life, as in many plants. In humans, certain body parts, like hair and nails, continue to grow throughout life. Growth is usually the result of an increase in the number of cells.

GSH (glutathione): (i) A chemical substance produced in the body, and consumed in foods (spinach and parsley). It is a tripeptide formed by the combination of three amino acids: cysteine, glutamic acid, and glycine. Glutathione is used in the body to make glutathione peroxidases that act as antioxidants, protecting RBCs from damage and destruction by mopping up toxic free radicals. It is also needed for the action of insulin. (ii) Glutathione, an antioxidant helps to protect cells from reactive oxygen species such as free radicals and peroxides. Glutathione is also nucleophilic at sulfur and attacks poisonous conjugate acceptors.

Guinea pig maximization test: A skin test for screening for possible contact allergens. It is considered to be a useful model for predicting likely moderate and strong sensitizers in humans.

H

Haemosiderin: The name of the iron-protein molecule that is a source of iron for hemoglobin synthesis and other processes requiring iron.

Hair follicle: A shaft or opening on the surface of the skin through which hair grows.

Half life (t1/2): Time required for the concentration of a reactant in a given reaction to reach a value that is the arithmetic mean of its initial and final (equilibrium) values. For a reactant that is entirely consumed, it is the time taken for the reactant concentration to fall to one half of its initial value. The half life of a reaction has meaning only in special cases, e.g., (i) for a first-order reaction, the half life of the reactant may be called the half life of the reaction; (ii) for a reaction involving more than one reactant, with the concentrations of the reactants in their stoichiometric ratios, the half life of each reactant is the same, and may be called the half life of the reaction.

Hallucination: A sight, sound, smell, taste, or touch that a person believes to be real but is not real. Hallucinations can be caused by nervous system disease, certain drugs, or mental disorders.

Hamartoma: A benign (not cancer) growth made up of an abnormal mixture of cells and tissues normally found in the area of the body where the growth occurs.

Hand-foot syndrome: A condition marked by pain, swelling, numbness, tingling, or redness of the hands or feet. It sometimes occurs as a side effect of certain anticancer drugs. Hand-foot syndrome is also called palmar-plantar erythrodysesthesia.

Haploid: The term applied to a cell containing only one set of chromosomes.

Hard palate: The front, bony part of the roof of the mouth.

Harmful: Chemical substances causing or capable of causing harmful effects.

Harmful substance: A chemical substance that, following contact with an organism/individual, can cause ill health or adverse effects either at the time of exposure or later in the life of present and future generations.

Hormone: (i) A chemical substance, usually a peptide or steroid, produced by one tissue and conveyed by the bloodstream to another to effect physiological activity, such as growth or metabolism; (ii) a chemical substance released by one or more cells that affects cells in other parts of the organism/animal/humans; (iii) one of the chemical messengers produced by the endocrine glands whose secretions are liberated directly into the bloodstream and transported to a distant part or parts of the body, where they exert a specific effect for the benefit of the body as a whole.

Hormone therapy (hormone treatment): Treatment that adds, blocks, or removes hormones. For certain conditions (such as diabetes or menopause), hormones are given to adjust low hormone levels. To slow or stop the growth of certain cancers (such as prostate and breast cancer), synthetic hormones or other drugs may be given to block the body's natural hormones. Sometimes, surgery is needed to remove the gland that makes a certain hormone.

Hazard: (i) The term hazard usually refers to anything that has the ability to cause injury or the potential to cause injury. The hazard associated with a potentially toxic chemical substance is a function of its toxicity and the potential for exposure to the chemical substance. The probability of exposure to the chemical substance is a risk factor. (ii) A hazard is a situation that poses a level of threat to life, health, property, and the environment.

Hazard communication standard: An OSHA regulation that requires chemical manufacturers, suppliers, and importers to assess the hazards of the chemicals they make, supply, or import, and to inform employers, customers, and workers of these hazards through a MSDS.

Hazard evaluation: Establishment of a qualitative or quantitative relationship between hazard and benefit, involving the complex process of determining the significance of the identified hazard and balancing this against identifiable benefit.

Hazardous substance: As defined in the FHSA at 16 CFR § 1500.3(b)(4)(i)(A), any substance or mixture of substances that is toxic, corrosive, an irritant, a strong sensitizer, flammable or combustible, or generates pressure through decomposition, heat, or other means, if it may cause substantial personal injury or illness during or as a proximate result of any customary or reasonably foreseeable handling or use, including reasonably foreseeable ingestion by children.

Hazardous waste: A subset of solid wastes that pose substantial or potential threats to public health or the environment and meet any of the following criteria:

- Is specifically listed as a hazardous waste by the US EPA
- Exhibits one or more of the characteristics of hazardous wastes (ignitability, corrosiveness, reactivity, and/or toxicity)
- Is generated by the treatment of hazardous waste; or is contained in a hazardous waste

HAZCHEM: Acronym for "HAZardous CHEMicals." The Hazchem emergency action.

HDPE (high density polyethylene): High density polyethylene is more rigid and harder than lower density materials. It also has a higher tensile strength four time that of low density polyethylene. It is three times better in compressive strength. HDPE meets FDA requirements for direct food contact applications. It is also accepted by the USDA, the NSF, and the Canadian Department of Agriculture.

Health: A condition describing (a) the state of complete physical, mental, and social well-being, and not merely the absence of disease or infirmity; (b) a state of dynamic balance in which an individual's or a group's capacity to cope with the circumstances of living is at an optimal level; (c) a state characterized by anatomical, physiological, and psychological integrity, ability to perform personally valued family, work, and community roles; (d) an ability to deal with physical, biological, psychological, and social stress; (e) a feeling of well-being and freedom from the risk of disease and untimely death; and (f) a sustainable steady state in which humans and other living organisms can coexist indefinitely (when discussed in the context of ecology).

Health advisory level (HAL): A non-regulatory agency in the United States dealing with health-based reference levels of traces of chemical substances (usually in parts per million per milligram per liter) in drinking water at which there are no adverse health risks when ingested over various periods of time.

Health assessment: An evaluation of available data on existing or potential risks posed by a Superfund site. Every site on the National Priorities List has a health assessment prepared by the Agency for Toxic Substances and Disease Registry.

Health-based exposure limit: Maximum concentration or intensity of exposure that can be tolerated without significant effect (based only on scientific and not economic evidence concerning exposure levels and associated health effects).

Health hazard: (i) Health hazard means a chemical substance for which there is statistically significant evidence based on at least one study conducted in accordance with established scientific principles that acute or chronic health effects may occur in exposed employees; (ii) the term "health hazard" includes chemical substances such as carcinogens, toxic and or highly toxic agents, irritants, corrosives, sensitizers, hepatotoxins, nephrotoxins, neurotoxins, and many other chemical agents that act on the hematopoietic system, and agents that damage the lungs, skin, eyes, and mucous membranes.

Health surveillance: Periodic medico-physiological examinations of exposed workers with the objective of protecting health and preventing occupationally related disease.

Heavy metal: A term used commonly in the toxicological literature but having no generally agreed meaning, sometimes even applied to non-metals, and therefore a source of confusion and should be avoided. The term "metal" is adequate without the qualifying adjective, but may be misleading since it implies a solid material when toxicological concern is mostly for the ionic form or another chemical species.

HECD: Hall electron capture detector.

Hematemesis: Vomiting of blood.

Hematoma: (i) A pool of clotted or partially clotted blood in an organ, tissue, or body space, usually caused by a broken blood vessel; (ii) localized accumulation of blood, usually clotted, in an organ, space, or tissue, due to a failure of the wall of a blood vessel.

Hematopoiesis: The formation of new blood cells.

Hematopoietic agents: Chemical substances that act on the blood or hematopoietic system, decrease hemoglobin function, deprive the body tissues of oxygen.

Heme: The part of certain molecules that contains iron. The heme part of hemoglobin is the substance inside RBCs that binds to oxygen in the lungs and carries it to the tissues.

Hemoglobin: The substance inside RBCs that binds to oxygen in the lungs and carries it to the tissues.

Hemophilia: Group of hereditary disorders in which affected individuals fail to make enough of certain proteins needed to form blood clots.

Hemoptysis: Coughing or spitting up blood from the respiratory tract.

Hemorrhage: Loss of blood from damaged blood vessels. A hemorrhage may be internal or external, and usually involves a lot of bleeding in a short time.

Heparin: A chemical substance that slows the formation of blood clots. Heparin is made by the liver, lungs, and other tissues in the body and can also be made in the laboratory. Heparin may be injected into muscle or blood to prevent or break up blood clots. It is a type of anticoagulant.

Hepatitis C: Inflammation of the liver due to the hepatitis C virus (HCV), which is usually spread by blood transfusion, hemodialysis, and needle sticks. HCV causes most transfusion-associated hepatitis, and the damage it does to the liver can lead to cirrhosis and cancer. Transmission of the virus by sexual contact is rare. At least half of HCV patients develop chronic hepatitis C infection. Diagnosis is by blood test. Treatment is via antiviral drugs. Chronic hepatitis C may be treated with interferon sometimes in combination with antivirals. There is no vaccine for hepatitis C. Previously known as non-A, non-B hepatitis.

Hepatitis: Disease of the liver causing inflammation. Symptoms include an enlarged liver, fever, nausea, vomiting, abdominal pain, and dark urine.

Hepatoctomy: Removal of the liver in part or all by surgery.

Hepatocyte: In histology terms, the name refers to a parenchymal liver cell.

Hepatotoxic: The adjective term applied to anything that is harmful to the liver.

Hepatotoxin: A chemical substance that is poisonous to the liver cells and causes liver damage, jaundice, and liver enlargement.

Herbicide: (i) The descriptor term for a chemical substance used to kill herbs—plants; (ii) a pesticide designed to control or kill plants, weeds, or grasses. Almost 70% of all pesticide used by farmers and ranchers are herbicides. These chemicals have wide-ranging effects on non-target species.

Hernia: The bulging of an internal organ through a weak area or tear in the muscle or other tissue that holds it in place. Most hernias occur in the abdomen.

High blood pressure (hypertension): Blood pressure of 140/90 or higher. High blood pressure usually has no symptoms. It can harm the arteries and cause an increase in the risk of stroke, heart attack, kidney failure, and blindness.

Highly toxic substance: As defined by the OSHA (Appendix A of 29 CFR 1910.1200) and in the FHSA regulations at 16 CFR § 1500.3(b)(6)(i), a substance with either (a) a median lethal dose (LD50) of 50 mg/kg or less of body weight administered orally to rats; (b) a median lethal dose (LD50) of 200 mg/kg or less of body weight when administered continuously on the bare skin of rabbits for 24 h or less; or (c) a median lethal concentration (LC50) in air of 200 ppm by volume or less of gas or vapor, or 2 mg/L by volume or less of mist or dust, when exposed to continuous inhalation for 1 h or less to rats.

HIPS (high impact styrene): A cost effective material with good impact resistance and can be easily colored. It extrudes with a matte uniform finish. HIPS is used for indoor applications and, if necessary, UV inhibitors can be added making it fairly weatherable. HIPS is translucent in its natural state and translucent to opaque when colored.

Histochemistry: The branch of science concerned with the chemistry of microscopically visible structures. Chemical processes used to make cell structure microscopically visible.

Histology: The branch of science related to the study of the microanatomy of tissues and their cellular structure.

Histopathology: (i) The branch of science concerned with the study of microscopic changes in diseased tissues; (ii) microscopic pathological study of the anatomy and cell structure of tissues in disease to reveal abnormal or adverse structural changes.

HIV positive: Infected with the human immunodeficiency virus (HIV), the cause of acquired immunodeficiency syndrome (AIDS).

Hodgkin disease: A cancer of the immune system that is marked by the presence of a type of cell called the Reed-Sternberg cell. The two major types of Hodgkin disease are classical Hodgkin lymphoma and nodular lymphocyte-predominant Hodgkin lymphoma. Symptoms include the painless enlargement of lymph nodes, spleen, or other immune tissue. Other symptoms include fever, weight loss, fatigue, or night sweats. Also called Hodgkin lymphoma.

Homeostasis: (i) A state of balance among all the body systems needed for the body to survive and function correctly. In homeostasis, body levels of acid, blood pressure, blood sugar, electrolytes, energy, hormones, oxygen, proteins, and temperature are constantly adjusted to respond to changes inside and outside the body, to keep them at a normal level. (ii) In medicine and biology, this term is applied to the inherent tendency in an organism toward maintenance of physiological and psychological stability.

Hormone: Chemical substance formed in one organ or part of the body and carried in the blood to another organ or part where it selectively alters functional activity.

Human ecology: Interrelationship between humans and the entire environment—physical, biological, socioeconomic, and cultural—including the interrelationships between individual humans or groups of humans and other human groups or groups of other species.

Hydrocarbons: Chemicals consisting entirely of hydrogen and carbon. Hydrocarbons contribute to air pollution problems like smog.

Hydrocephalus: The abnormal buildup of cerebrospinal fluid in the ventricles of the brain.

Hydrogen peroxide: Chemical substance used in bleaches, dyes, cleansers, antiseptics, and disinfectants. In a concentrated form, it is toxic and irritating to tissues.

Hydrolysis: Chemical reaction of a substance with water, usually resulting in the formation of one or more new compounds.

Hydronephrosis: Pathological chronic enlargement of the collecting channels of a kidney, leading to compression and eventual destruction of kidney tissue, and diminishing kidney function.

Hydrophilic: Describing the character of a substance, material, molecular entity, or group of atoms that has an affinity for water.

Hydrophobic: Describing the character of a substance, material, molecular entity, or group of atoms that are insoluble or confer insolubility in water, or resistance to wetting or hydration.

Hydrosphere: A broad term for the water above, on, or in the earth's crust, including oceans, seas, lakes, groundwater, and atmospheric moisture.

Hygiene: Science of health and its preservation.

Hyperbilirubinemia: Excessive concentration of bilirubin in the blood.

Hypercalcemia: Excessive concentration of calcium in the blood.

Hyperemia: Excessive amount of blood in any part of the body.

Hyperglycemia: (i) Excessive concentration of glucose in the blood; (ii) higher than normal amount of glucose (a type of sugar) in the blood. Hyperglycemia can be a sign of diabetes or other conditions.

Hyperkalemia: Excessive concentration of potassium in the blood.

Hyperkarotosis: A condition marked by thickening of the outer layer of the skin, which is made of keratin (a tough, protective protein). It can result from normal use (corns, calluses), chronic inflammation (eczema), or genetic disorders.

Hypernatremia: Excessive concentration of sodium in the blood.

Hyperparathyroidism: Abnormally increased parathyroid gland activity that affects, and is affected by, plasma calcium concentration.

Hyperplasia: Abnormal multiplication or increase in the number of normal cells in a tissue or organ.

Hypersensitivity: (i) A state of altered reactivity in which the body reacts with an exaggerated immune response to what is perceived as a foreign substance; (ii) state in which an individual reacts with allergic effects following exposure to a certain substance (allergen) after having been exposed previously to the same. (Note: Most common chemically induced allergies are Type I (IgE-mediated) and Type IV (cell-mediated) hypersensitivity.)

Hypersusceptibility: Excessive reaction following exposure to a given amount or concentration of a substance compared with the large majority of other exposed subjects.

Hypertension: Persistently high blood pressure in the arteries or in a circuit, e.g., pulmonary hypertension or hepatic portal hypertension.

Hypertrophy: Excessive growth in bulk of a tissue or organ through increase in size, but not in number of constituent cells.

Hypocalcemia: Abnormally low calcium concentration in the blood.

Hypoglycemia: Abnormally low blood sugar.

Hypothalamus: The area of the brain that controls body temperature, hunger, and thirst.

Hypothyroidism: Too little thyroid hormone. Symptoms include weight gain, constipation, dry skin, and sensitivity to cold.

Hypotriglyceridemia: In medicine, the term describes the situation of decreased blood triglyceride content.

Hypoxemia: A state of oxygen deficiency in the blood.

Hypoxia: (i) A state of being abnormally low in dioxygen content or tension; (ii) deficiency of dioxygen in the inspired air, in blood, or in tissues, short of anoxia.

I

IBS (irritable bowel syndrome): A disorder of the intestines commonly marked by abdominal pain, bloating, and changes in a person's bowel habits. This may include diarrhea or constipation, or both, with one occurring after the other. Also called irritable colon, mucus colitis, and spastic colon.

Ideopathic: To describe a disease of unknown cause.

Ignitable: Capable of bursting into flames; ignitable chemical substances pose a fire hazard.

Immediately dangerous to life or health concentration (IDLH): A regulatory value defined as the maximum exposure concentration of a chemical substance in the workplace from which one could escape within 30 min without any escape-impairing symptoms or any irreversible health effects. This value should be referred to in respirator selection.

Immediately dangerous to life or health concentration (IDLHC): According to the US NIOSH, the maximum exposure concentration from which one could escape within 30 min without any escape-impairing symptoms or any irreversible health effects.

Immune complex: Product of an antigen-antibody reaction that may also contain components of the complement system.

Immune function: Production and action of cells that fight disease or infection.

Immune response: (i) The general reaction of the body to substances that are foreign or treated as foreign. It may take various forms, e.g., antibody production, cell-mediated immunity, immunological tolerance, or hypersensitivity (allergy). (ii) The activity of the immune system against foreign substances (antigens).

Immune system: (i) An integrated network of organs, glands, and tissues that has evolved to protect the body from foreign substances, including bacteria, viruses, and other infection-causing parasites and pathogens. The immune system may produce hypersensitivity reactions that, in the extreme, can be fatal. If the immune system misidentifies normal body components as foreign, this leads to autoimmune disorders, such as lupus, in which the body destroys its own constituents. (ii) The complex group of organs and cells that defends the body against infections and other diseases.

Immunity: The condition of being protected against an infectious disease. Immunity can be produced by a vaccine, previous infection with the same agent, or by transfer of immune substances from another person or animal.

Immunization: The technique used to cause an immune response that results in resistance to a specific disease, especially an infectious disease.

Immunoassay: (i) A test that uses the binding of antibodies to antigens to identify and measure certain substances. Immunoassays may be used to diagnose disease. Also, test results can provide information about a disease, which may help in planning treatment, e.g., when estrogen receptors are measured in breast cancer. (ii) Ligand-binding assay that uses a specific antigen or antibody, capable of binding to the analyte, to identify and quantify substances. The antibody can be linked to a radioisotope (radioimmunoassay, RIA) or to an enzyme that catalyzes an easily monitored reaction (enzyme-linked immunosorbent assay, ELISA), or to a highly fluorescent compound by which the location of an antigen can be visualized.

Immunochemistry: The branch of science related to molecular aspects of immunology, the chemistry of antigens, antibodies, and their relationship to each other.

Immunodeficiency: The decreased ability of the body to fight infections and other diseases.

Immunoglobulin: (i) Family of closely related glycoproteins capable of acting as antibodies and present in plasma and tissue fluids; immunoglobulin E (IgE) is the source of antibody in Type I hypersensitivity (allergic) reactions; (ii) a protein (Ig) that acts as an antibody. Immunoglobulins are made by B cells and plasma cells. An immunoglobulin is a type of glycoprotein with two heavy chains and two light chains.

Immunomodulation: Modification of the functioning of the immune system by the action of a substance that increases or reduces the ability to produce antibodies.

Immunosuppression: (i) In medicine, this term is applied to inhibition of the normal response of the immune system to an antigen. A decrease in the functional capacity of the immune response may be due to (a) the inhibition of the normal response of the immune system to an antigen or (b) prevention, by chemical or biological means, of the production of an antibody to an antigen by inhibition of the processes of transcription, translation, or formation of tertiary structure. (ii) Suppression of the body's immune system and its ability to fight infections and other diseases. Immunosuppression may be deliberately induced with drugs, as in preparation for bone marrow or other organ transplantation, to prevent rejection of the donor tissue. It may also result from certain diseases such as AIDS or lymphoma or from anticancer drugs.

Immunotherapy: Treatment to boost or restore the ability of the immune system to fight cancer, infections, and other diseases. Also used to lessen certain side effects that may be caused by some cancer treatments. Agents used in immunotherapy include monoclonal antibodies, growth factors, and vaccines. These agents may also have a direct antitumor effect.

Immunotoxic: An adjective term to describe any chemical substance harmful to the immune system.

Implant: A substance or object that is put in the body as a prosthesis, or for treatment or diagnosis.

Implantation: Attachment of the fertilized ovum (blastocyst) to the endometrium and its subsequent embedding in the compact layer, occurring 6 or 7 days after fertilization of the ovum.

Incineration: (i) A method of treating solid, liquid, or gaseous wastes by burning; (ii) the destruction of solid, liquid, or gaseous wastes by controlled burning at high temperatures. Hazardous organic compounds are converted to ash, carbon dioxide, and water. Burning destroys organics, reduces the volume of waste, and vaporizes water and other liquids that the wastes may contain. The residue ash produced may contain some hazardous material, such as non-combustible heavy metals, concentrated from the original waste.

Incision: A cut made in the body to perform surgery.

Incompatible materials: Chemical substances that can react to cause a fire, explosion, violent reaction, or lead to the evolution of flammable gases, or otherwise lead to injury to people or danger to property.

Incompatible waste: A waste unsuitable for mixing with another waste or material because of reactivity hazards.

Incontinence: Inability to control the flow of urine from the bladder (urinary incontinence) or the escape of stool from the rectum (fecal incontinence).

Indication: In medicine, a sign, symptom, or medical condition that leads to the recommendation of a treatment, test, or procedure.

Induction: Increase in the rate of synthesis of an enzyme in response to the action of an inducer or environmental conditions.

Industrial waste: Unwanted materials produced in or eliminated from an industrial operation and categorized under a variety of headings, such as liquid wastes, sludge, solid wastes, and hazardous wastes.

Infection: (i) The growth of a parasitic organism within the body. (A parasitic organism is one that lives on or in another organism and draws its nourishment there from.) A person with an infection has another organism (a "germ") growing within his/her system, drawing its nourishment from the affected individual. (ii) Invasion and multiplication of germs in the body. Infections can occur in any part of the body and can spread throughout the body. Germs may be bacteria, viruses, yeast, or fungi. They can cause a fever and other problems, depending on where the infection occurs. When the body's natural defense system is strong, it can often fight the germs and prevent infection. Some cancer treatments can weaken the natural defense system.

Inferior vena cava: The large vein that empties into the heart. It carries blood from the legs and feet and from organs in the abdomen and pelvis.

Inflammation: Reaction of the body to injury or to infectious, allergic, or chemical irritation; characterized by redness, swelling, heat, and pain resulting from dilation of the blood vessels accompanied by loss of plasma and leukocytes (white blood cells) into the tissues. This is a protective reaction to injury, disease, or irritation of the tissues.

Infusion: (i) A method of putting fluids, including drugs, into the bloodstream (also called intravenous infusion); (ii) therapeutic administration of a fluid other than blood, usually saline as a solution into a vein.

Ingestion: (i) Process of taking food and drink into the body by mouth; (ii) taking a substance into the body by mouth and swallowing it; (iii) a process of swallowing, such as eating or drinking. Chemical substances can get into or onto food, drink, utensils, cigarettes, or hands, where they can then be ingested.

Inhalation: (i) Act of drawing in of air, vapor, or gas and any suspended particulates into the lung; (ii) a process of breathing. Once inhaled, the contaminants are deposited in the lungs, taken into the blood, or both.

Inhaler: A device for giving medicines in the form of a spray that is inhaled (breathed in) through the nose or mouth. Inhalers are used to treat medical problems such as bronchitis, angina, emphysema, and asthma. They are also used to help relieve symptoms that occur when a person is trying to quit smoking.

Initiator: An agent that starts the process of tumor formation, usually by action on the genetic material.

Injury (harm or hurt): The term is applied in medicine to damage inflicted on oneself by an external agent. The injury may be accidental or deliberate, as with a needle stick injury.

Insecticide: (i) The term refers to any chemical substance used to kill insects (refer: Classification of pesticides); (ii) pesticide compound specifically used to kill or prevent the growth of insects.

Insomnia: Difficulty in going to sleep or getting enough sleep.

Integrated pest management (IPM): The use of pest and environmental information in conjunction with available pest control technologies to prevent unacceptable levels of pest damage by the most economical means and with the least possible hazard to persons, property, and the environment.

Interferon: A naturally occurring chemical substance that interferes with the ability of viruses to reproduce. Interferon also boosts the immune system.

International Agency for Research on Cancer (IARC): An agency of WHO that publishes IARC Monographs on the Evaluation of the Carcinogenic Risk of Chemicals to Humans. This publication documents reviews of information on chemicals and determinations of the cancer risk of chemicals.

Intestine: The long, tube-shaped organ in the abdomen that completes the process of digestion. The intestine has two parts, the small intestine and the large intestine. Also called bowel.

Interstitial pneumonia: Chronic form of pneumonia involving an increase of the interstitial tissue and a decrease of the functional lung tissue.

Intoxication: (i) Poisoning: pathological process with clinical signs and symptoms caused by a substance of exogenous or endogenous origin; (ii) drunkenness following consumption of beverages containing ethanol or other compounds affecting the CNS.

In vitro: A term applied to describe a study carried out in isolation from the living organism in an experimental system (applied to studies of biological functions or processes to contrast with *in vivo* studies).

In vivo: The term used to describe any study carried out within the living organism in contrast with "*in vitro*."

IPE: Intermediate potential of exposure.

Iris: The colored tissue at the front of the eye that contains the pupil in the center. The iris helps control the size of the pupil to let more or less light into the eye.

Iron: An important mineral the body needs to make hemoglobin, a substance in the blood that carries oxygen from the lungs to tissues throughout the body. Iron is also an important part of many other proteins and enzymes needed by the body for normal growth and development. It is found in red meat, fish, poultry, lentils, beans, and foods with iron added, such as cereal.

Irradiation: The use of high-energy radiation from x-rays, gamma rays, neutrons, protons, and other sources to kill cancer cells and shrink tumors. Radiation may come from a machine outside the body (external-beam radiation therapy), or it may come from radioactive material placed in the body near cancer cells (internal radiation therapy). Systemic irradiation uses a radioactive substance, such as a radiolabeled monoclonal antibody, which travels in the blood to tissues throughout the body.

Irritant: (i) A descriptor term to describe any chemical substance that causes inflammation following immediate, prolonged, or repeated contact with skin or mucous membrane; (ii) a chemical substance that causes a reversible inflammatory effect on living tissue by chemical action at the site of contact; (iii) a chemical substance that can cause irritation of the skin, eyes, or respiratory system. An irritant can cause an acute effect from a single high-level exposure, or chronic effects from repeated, low-level exposures. Some examples of irritants are chlorine, nitric acid, and various pesticides.

Ischemia: The term indicates the deficiency of blood supply to any part of the body, relative to its local requirements. Ischemia may cause tissue damage due to lack of oxygen and nutrients.

Islet of Langerhans cell: A pancreatic cell that produces hormones (e.g., insulin and glucagon) that are secreted into the bloodstream. These hormones help control the level of glucose (sugar) in the blood. It is also called endocrine pancreas cell and islet cell.

ISO Accreditation: Accreditation by the International Organization for Standardization (ISO), which develops international agreements on standards for various industries.

Isotonic: Denoting a liquid exerting the same osmotic pressure or chemical potential of water (water potential) as another liquid with which it is being compared.

Itai-itai disease: Illness (renal osteomalacia) observed in the Toyama prefecture of Japan, resulting from the ingestion of cadmium-contaminated rice. The damage occurred to the renal and skeleto-articular systems, the latter being very painful ("itai" means "ouch" in Japanese and refers to the intense pain caused by the condition).

IUPAC: The International Union of Pure and Applied Chemistry.

i.v.: Abbreviation for intravenous (used in relation to administration of drugs or other substances).

J

Jaundice: (i) A pathological condition in which the skin and the whites of the eyes become yellow, the urine darkens, and the color of stool becomes lighter than normal. Jaundice occurs when the liver is not working properly or when a bile duct is blocked. (ii) A condition characterized by the deposition of bile pigment in the skin and mucous membranes, including the conjunctivae, resulting in a yellow appearance of the patient or animal.

Jet fuel: A refined petroleum product used in jet aircraft engines. It includes kerosene-type jet fuel and naphtha-type jet fuel.

Junction: (i) A point of joining; (ii) a region of transition between semiconductor layers, such as a p/n junction, which goes from a region with a high concentration of acceptors (p-type) to one with a high concentration of donors (n-type).

K

Kava kava: A herb that is native to islands in the South Pacific. Substances taken from the root have been used in some cultures to relieve stress, anxiety, tension, sleeplessness, and problems of the menopause. Kava kava may increase the effect of alcohol and of certain drugs used to treat anxiety and depression. The US FDA advises users that kava kava may cause severe liver damage. The scientific name is *Piper methysticum*. Kava kava is also called intoxicating pepper, rauschpfeffer, tonga, and yangona.

Keloid: A thick, irregular scar caused by excessive tissue growth at the site of an incision or wound.

Keratoacanthoma: A rapidly growing, dome-shaped skin tumor that usually occurs on sun-exposed areas of the body, especially around the head and neck. Keratoacanthoma occurs more often in males. Although in most patients it goes away on its own, in a few patients it comes back. Rarely, it may spread to other parts of the body.

Ketone: A type of chemical substance used in perfumes, paints, solvents, and found in essential oils (scented liquid taken from plants). Ketones are also made by the body when there is insufficient insulin.

Ketosis: Pathological increase in the production of ketone bodies, e.g., following blockage or failure of carbohydrate metabolism.

Kidney(s): (i) One of a pair of glandular organs located in the right and left side of the abdomen, which eliminates/clears poisonous chemical substances from the blood, regulates acid concentration, and maintains water balance in the body by excreting urine. The kidneys are part of the urinary tract. The urine then passes through connecting tubes called "ureters" into the bladder. The bladder stores the urine until it is released during urination. (ii) A pair of glandular organs in the dorsal region of the vertebrate abdominal cavity, functioning to maintain proper water and electrolyte balance, regulate acid-base concentration, filter the blood of metabolic wastes that are then excreted as urine and also play a role in blood pressure regulation. (iii) The kidneys are situated on each side of the spine, they filter blood and eliminate metabolism waste in urine through the ureters.

Kidney failure: A health condition in which the kidneys stop working and are unable to remove waste and extra water from the blood or keep body chemicals in balance. Acute or severe kidney failure happens suddenly (e.g., after an injury) and may be treated and cured. Chronic kidney failure develops over many years, may be caused by conditions like high blood pressure or diabetes, and cannot be cured. Chronic kidney failure may lead to total and long-lasting kidney failure, called end-stage renal disease (ESRD). A person in ESRD needs dialysis (the process of cleaning the blood by passing it through a membrane or filter) or a kidney transplant.

Kidney function test: A biochemical test in which blood or urine samples are checked for amounts of certain substances released by the kidneys. A higher- or lower-than-normal amount of a substance can be a sign that the kidneys are not working properly; also called the renal function test.

Klinefelter syndrome: A genetic disorder in males caused by having one or more X chromosomes. Males with this disorder may have larger than normal breasts, a lack of facial and body hair, a rounded body type, and small testicles. They may learn to speak much later than other children and may have difficulty learning to read and write. Klinefelter syndrome increases the risk of developing extragonadal germ cell tumors and breast cancer.

Known human carcinogen: A chemical substance for which there is sufficient evidence of a cause and effect relationship between exposure to the material and cancer in humans.

L

Label: A display of written, printed, or graphic matter on the immediate container (not including package liners) of any food product.

Label element: One type of information that has been harmonized for use in a label, e.g., pictogram, signal word.

Labial mucosa: The inner lining of the lips.

Lacrimal gland: A gland that secretes tears. The lacrimal glands are found in the upper, outer part of each eye socket.

Lacrimation: (i) Secretion and discharge of tears; (ii) excessive production of tears when the eye is exposed to an irritant.

Lacrimal gland: The tear gland located under the upper eyelid at the outer corner of the eye. The fluid it secretes cleans and provides moisture for the cornea. It is responsible for tearing during emotional stimulation or following corneal irritation by a foreign body or chemical substance.

Lacrimal sac: The tear sac located on the side of the nose adjacent to the inner corner of the eye. Tears normally drain from the eye into the tear duct and then through the sac, finally leaving by a drain that enters the nose. The tear sac remains filled with tears when an infant has a blocked tear duct. An infection of the tear sac is called a dacryocystitis.

Lacrimator: A chemical substance that irritates the eyes and causes the production of tears.

Lactation: The secretion of milk from the breasts; the period of suckling the young until weaning.

Lactic acidosis: The buildup of lactic acid in the body. Cells make lactic acid when they use glucose (sugar) for energy. If too much lactic acid stays in the body, the balance tips and the person begins to feel ill. The signs of lactic acidosis are deep and rapid breathing, vomiting, and abdominal pain. Lactic acidosis may be caused by diabetic ketoacidosis or liver or kidney disease. Lactic acidosis is also a rare side effect of a diabetes medication called metformin.

Lactose: The sugar found in milk. The body breaks down lactose into galactose and glucose.

Lagoon: A shallow, artificial treatment pond where sunlight, bacterial action, and oxygen work to purify wastewater; a stabilization pond. An aerated lagoon is a treatment pond that uses oxygen to speed up the natural process of biological decomposition of organic wastes. A lagoon is regulated as a point source under the Clean Water Act if there is a direct surface water discharge. Some lagoons that discharge into groundwater are also regulated if they have a direct hydrogeologic connection to surface water. In other areas, historically, lagoons were used to dump various liquid, solid, and hazardous wastes from manufacturing or industrial processes. These wastes typically flooded and polluted surrounding environs or seeped underground. Such lagoons are now regulated under the RCRA but some must be cleaned up under Superfund.

Laparoscope: A thin, tube-like instrument used to look at tissues and organs inside the abdomen. A laparoscope has a light and a lens for viewing and may have a tool to remove tissue.

Large intestine: The long, tube-like organ that is connected to the small intestine at one end and the anus at the other. The large intestine has four parts: cecum, colon, rectum, and anal canal. Partly digested food moves through the cecum into the colon, where water and some nutrients and electrolytes are removed. The remaining material, solid waste called stool, moves through the colon, is stored in the rectum, and leaves the body through the anal canal and anus.

Larvicide: A chemical substance intended to kill larval life stage of an insect. Larvicides may be contact poisons, stomach poisons, growth regulators, or biological control agents.

Laryngitis: An inflammation of the mucous membrane lining the larynx that is located in the upper part of the respiratory tract.

Larynx (voice box): The area of the respiratory tract and throat containing the vocal cords, used for breathing, swallowing, and talking.

Lassitude: A feeling of tiredness, weakness, and lack of interest in daily activities.

Latent period: The period suggesting (i) a delay between exposure to a harmful substance and the manifestations of a disease or other adverse effects; (ii) a period from disease initiation to disease detection.

Latent: Hidden, dormant, inactive.

Lavage: Irrigation or washing out of a hollow organ or cavity such as the stomach, intestine, or the lungs.

Laxative (purgative): Substance that causes evacuation of the intestinal contents.

LC_{50} (median lethal concentration 50): The concentration of a chemical substance that kills 50% of a sample population; typically expressed in mass per unit volume of air.

LCn: Abbreviation to describe the exposure concentration of a toxic chemical substance lethal to n% of a test population.

LD_{50} (median lethal dose 50): The amount of a chemical substance that kills 50% of a sample population; typically expressed as milligrams per kilogram of body weight.

LDn: Abbreviation to describe the dose of a toxicant lethal to n% of a test population.

LDPE (low density polyethylene): LDPE is the first of the polyethylenes to be developed. It's an excellent material in electrical and chemical uses in low heat applications.

Lead poisoning: When lead, most often from old, peeling paint, is ingested and damages the body. In high doses, lead causes severe brain damage in children; in smaller doses lead can slow the process of a child's physical growth, cause learning disabilities, and damage to the kidneys.

Lennox-Gastaut syndrome: A severe form of epilepsy characterized by the onset in early childhood of frequent seizures of multiple types and by developmental delay.

Lesion: States of health conditions describing (i) area of pathologically altered tissue; (ii) injury or wound; (iii) infected patch of skin.

Lethal: Deadly; fatal; causing death.

Lethal concentration (LC): Concentration of a substance in an environmental medium that causes death following a certain period of exposure.

Lethal dose (LD): Amount of a substance or physical agent (e.g., radiation) that causes death when taken into the body.

Lethargy: A health condition marked by drowsiness and an unusual lack of energy and mental alertness. It can be caused by many things, including illness, injury, or drugs.

Leukemia: (i) Cancer that starts in blood-forming tissue such as the bone marrow and causes large numbers of blood cells to be produced and enter the bloodstream;

(ii) progressive, malignant disease of the blood-forming organs, characterized by distorted proliferation and development of leukocytes and their precursors in the bone marrow and blood.

Leukocyte (white blood cell/WBC): A type of immune cell. Most leukocytes are made in the bone marrow and are found in the blood and lymph tissue. Leukocytes help the body fight infections and other diseases. Granulocytes, monocytes, and lymphocytes are leukocytes.

Leukopenia: A condition in which there is a lower-than-normal number of leukocytes (white blood cells) in the blood.

Leydig cells: In the testes, the hormone-producing cells that are packed in like bunches of grapes between the seminiferous tubules in which sperms are produced; the interstitial cells in the testes that produce testosterone; they are packed between the seminiferous tubules in which the sperms grow.

Libido: Sexual desire or sex drive and sexual urge.

Limit test: Acute toxicity test in which, if no ill-effects occur at a pre-selected maximum dose, no further testing at greater exposure levels is required.

Lipophilic: Having an affinity for fat and high lipid solubility. Note: This is a physico-chemical property that describes a partitioning equilibrium of solute molecules between water and an immiscible organic solvent, favoring the latter, and which correlates with bioaccumulation.

Lipophobic: Having a low affinity for fat and a high affinity for water.

Liposome: (i) Artificially formed lipid droplet, small enough to form a relatively stable suspension in aqueous media, useful in membrane transport studies and in drug delivery; (ii) lipid droplet in the endoplasmic reticulum of a fatty liver.

Lithium: A soft metal. Lithium salts are used to treat certain mental disorders, especially bipolar (manic depressive) disorder. Lithium salts include lithium carbonate and lithium citrate.

Liver: The liver is the largest glandular organ of the body in the upper abdomen. The liver has many functions, i.e., to produce substances that break down fats, help in digestion, convert glucose to glycogen, produce urea (the main substance of urine), make certain amino acids (the building blocks of proteins), and filter harmful substances from the blood. The liver is also responsible for producing cholesterol.

Liver disease: Liver disease refers to any disorder of the liver. The liver is a large organ in the upper right abdomen that aids in digestion and removes waste products from the blood.

Liver nodule: A medical term to describe any small node or aggregation of cells within the liver.

Liver transplant: Surgery to remove a diseased liver and replace it with a healthy liver (or part of one) from a donor.

Livestock: Cattle, sheep, swine, goat, horse, mule, or other equine.

Local anesthesia: A temporary loss of feeling in one small area of the body caused by special drugs or other substances called anesthetics. The patient stays awake but has no feeling in the area of the body treated with the anesthetic.

LOEL: Lowest observed effect level.

Lowest effective dose (LED): Lowest dose of a chemical substance inducing a specified effect in a specified fraction of exposed individuals.

Lowest observed adverse effect level (LOAEL): Lowest concentration or amount of a substance (dose) found by experiment or observation, which causes an adverse effect on morphology, functional capacity, growth, development, or life span of a target organism distinguishable from normal (control) organisms of the same species and strain under defined conditions of exposure.

Lowest observed effect level (LOEL): Lowest concentration or amount of a substance (dose) found by experiment or observation, which causes any alteration in morphology, functional capacity, growth, development, or life span of target organisms distinguishable from normal (control) organisms of the same species and strain under the same defined conditions of exposure.

LPE: Lowest potential for exposure.

Lumbar puncture: A procedure in which a thin needle, called a spinal needle, is inserted into the lower part of the spinal column to collect cerebrospinal fluid or to give drugs.

Lung: One of a pair of organs in the chest that supplies the body with oxygen and removes carbon dioxide from the body. The lungs are sac-like structures where gas exchange occurs with the blood.

Lupus: A chronic, inflammatory, connective tissue disease that can affect the joints and many organs, including the skin, heart, lungs, kidneys, and nervous system. It can cause many different symptoms.

Luteinizing hormone-releasing hormone (LHRH): From the hypothalamus, a hormone that triggers the release of the pituitary gonadotropins, LH and FSH. LH frees the ovum and changes its Graafian follicle into the corpus luteum.

Lymph gland: A rounded mass of lymphatic tissue surrounded by a capsule of connective tissue. Lymph glands filter lymph (lymphatic fluid) and store lymphocytes (white blood cells). They are located along lymphatic vessels.

Lymphocyte: (i) A type of immune cell made in the bone marrow and found in the blood and in lymph tissue. The two main types of lymphocytes are B and T lymphocytes. B lymphocytes make antibodies, and T lymphocytes help kill tumor cells and help control immune responses. A lymphocyte is a type of white blood cell. (ii) Lymphocytes have a number of roles in the immune system, including antibody production, attacking and destroying cancer cells, and producing substances that kill cancer cells.

Lymphoma: (i) General term comprising tumors and conditions allied to tumors arising from some or all of the cells of lymphoid tissue; (ii) cancer of the lymphoid tissues. Lymphomas are often described as being large cell or small cell types, cleaved or non-cleaved, diffuse or nodular. The different types often have different prognoses. Lymphomas can also be referred to by the organs where they are active, such as CNS lymphomas, which are in the CNS, and GI lymphomas, which are in the gastrointestinal tract. The types of lymphomas most commonly associated with HIV infection are called non-Hodgkin's lymphomas or B cell lymphomas.

Lysimeter: Laboratory column of selected representative soil or a protected monolith of undisturbed field soil with facilities for sampling and monitoring the movement of water and chemical substances.

Lysis: The breakdown of a cell caused by damage to its plasma (outer) membrane. It can be caused by chemical or physical means (e.g., strong detergents or high-energy sound waves) or by infection with a virus strain that can lyse cells.

Lysosome: (i) A sac-like compartment inside a cell that has enzymes that can break down cellular components that need to be destroyed; (ii) membrane-bound cytoplasmic organelle containing hydrolytic enzymes. The release of these enzymes from lysosomes damaged by xenobiotics can cause autolysis of the cell.

M

Macronutrient: Any food that contains calories and can, therefore, generate hormonal responses. Protein, carbohydrate, and fat are macronutrients.

Macrophages: (i) Macrophages are white blood cells within tissues, produced by the division of monocytes; (ii) macrophages are large phagocytic cells found in connective tissue, especially in areas of inflammation; (iii) the properties of macrophages include phagocytosis and antigen presentation to T cells.

Macroscopic (gross) pathology: Studies of diseased tissue changes that are visible to the naked eye.

Macula: A specialized part of the retina that is cone-dominated. The macula is used for all detailed visual tasks. The center of the macula is called the fovea. If a disease process harms or destroys the macula, vision is usually reduced to 20/200 (legal blindness).

Macular degeneration: (i) A health condition in which there is a slow breakdown of cells in the center of the retina (the light-sensitive layers of nerve tissue at the back of the eye). This blocks vision in the center of the eye and can cause problems with activities such as reading and driving. Macular degeneration is most often seen in people who are over the age of 50. (ii) Damage or breakdown of the macula that is an area at the back of the eye that controls central vision. It may be caused by injury or aging and while it does not progress to total blindness, patients with macular degeneration require special optical aids to enlarge distant and near objects.

Macular edema: A swelling (edema) in the macula, an area near the center of the retina of the eye that is responsible for fine vision or reading vision. Macular edema is a common complication associated with diabetic retinopathy.

Malaise: A kind of vague feeling of bodily discomfort.

Malignancy: (i) A term for diseases in which abnormal cells divide without control and can invade nearby tissues. Malignant cells can also spread to other parts of the body through the blood and lymph systems. There are several main types of malignancy. Carcinoma is a malignancy that begins in the skin or in tissues that line or cover internal organs. Sarcoma is a malignancy that begins in bone, cartilage, fat, muscle, blood vessels, or other connective or supportive tissue. Leukemia is a malignancy that starts in blood-forming tissue such as the bone marrow, and causes large numbers of abnormal blood cells to be produced and enter the blood. Lymphoma and multiple myeloma are malignancies that begin in the cells of the immune system. CNS cancers are malignancies that begin in the tissues of the brain and spinal cord. (ii) A condition of cancerous growth, a mass of cells showing uncontrolled growth, a tendency to invade and damage surrounding tissues. (iii) A malignancy is a tumor that is cancerous and growing. (iv) Population of cells showing both uncontrolled growth and a tendency to invade and destroy other tissues.

Malignant: (i) A medical term used to describe a severe and progressively worsening disease—cancer; (ii) an adjective to describe cells in a cancerous growth.

Mania: Emotional disorder (mental illness) characterized by an expansive and elated state (euphoria), rapid speech, distorted ideas, decreased need for sleep, distractibility, grandiosity, poor judgment, and increased motor activity.

Margin of safety (MOS): Ratio of the no observed adverse effect level (NOAEL) to the theoretical or estimated exposure dose (EED).

Mass mean diameter (MMD): Diameter of a particle with a mass equal to the mean mass of all the particles in a population.

Mast cell: A type of white blood cell.

Material safety data sheet (MSDS): Compilation of information required under the US OSHA Hazard Communication Standard on the identity of hazardous substances, health and physical hazards, exposure limits, and precautions.

Master record identification number (MRID): A unique cataloging number assigned to an individual pesticide study at the time of its submission to US OSHA.

Material safety data sheet (MSDS): Printed material concerning a hazardous chemical substance or extremely hazardous substance, including its physical properties, hazards to personnel, fire and explosion potential, safe handling recommendations, health effects, fire-fighting techniques, reactivity, and proper disposal. Originally established for employee safety by the OSHA.

Maximum allowable concentration (admissible, acceptable concentration) (MAC): Regulatory value defining the concentration that if inhaled daily (in the case of workpeople for 8 h with a working week of 40 h, in the case of the general population 24 h) does not, in the present state of knowledge, appear capable of causing appreciable harm, however long delayed during the working life or during subsequent life or in subsequent generations.

Maximum allowable concentration (MAC): (i) An exposure concentration not to be exceeded under any circumstances; (ii) MAC is the level of a chemical substance/ pollutant that is considered harmless to healthy adults during their working hours, assuming they breathe uncontaminated air at all other times.

Maximum average daily concentration of an atmospheric pollutant: The peak daily average concentration of an air pollutant. The highest of the average daily concentrations recorded at a definite point of measurement during a certain period of observation.

Maximum contaminant level (MCL): Under the Safe Drinking Water Act (USA), primary MCL is a regulatory concentration for drinking water that takes into account both adverse effects (including sensitive populations) and technological feasibility (including natural background levels): secondary MCL is a regulatory concentration based on "welfare," such as taste and staining, rather than health, but also takes into account technical feasibility. Note: MCL goals (MCLG) under the Safe Drinking Water Act do not consider feasibility and are zero for all human and animal carcinogens.

Maximum permissible concentration (MPC): Same as maximum allowable concentration (MAC).

Maximum permissible daily dose: Maximum daily dose of chemical substance whose penetration into a human body during a lifetime will not cause diseases or health hazards that can be detected by current investigation methods and will not adversely affect future generations.

Maximum permissible level (MPL): The level, usually a combination of time and concentration, beyond which any exposure of humans to a chemical or physical agent in their immediate environment is unsafe.

Maximum residue limit (MRL): (i) MRLs are estimated for individual pesticides or veterinary drug residues in various food commodities. They are based on good agricultural practice (GAP) (pesticides) or good practice in the use of veterinary drugs in which the product has been used in an efficacious manner and appropriate withdrawal periods have been followed. They are expressed as either the parent compound or a metabolite that is, or is representative of, the residue of toxicological concern in the food commodity. MRLs are not based on toxicological data, but crude estimates of their toxicological significance are usually made by comparing the ADI with a calculation of the total intake of the residue based on the MRLs and food intake data of these commodities for which MRLs have been established. (ii) The MRL is usually determined by measurement, following a number (in the order of 10) of field trials, where the crop has been treated according to GAP and an appropriate pre-harvest interval.

Maximum residue limit for pesticide residues (MRL): Maximum contents of a pesticide residue (expressed as milligram per kilogram fresh weight) recommended by the CAC to be legally permitted in or on food commodities and animal feeds. The MRL values are based on data obtained following GAP and foods derived from commodities that comply with the respective MRLs are intended to be toxicologically acceptable.

Maximum tolerated dose (MTD): The highest dose of a drug or treatment that does not cause unacceptable side effects. The maximum tolerated dose is determined in clinical trials by testing increasing doses on different groups of people until the highest dose with acceptable side effects is found.

Mechanisms of toxicity: The biochemical method by which a chemical reacts in a living organism/animal/human.

Median effective concentration (EC50): Statistically derived median concentration of a chemical substance in an environmental medium expected to produce a certain effect in 50% of test organisms in a given population under a defined set of conditions. (Note: ECn refers to the median concentration that is effective in n% of the test population.)

Median effective dose (ED_{50}): Statistically derived median dose of a chemical or physical agent (radiation) expected to produce a certain effect in 50% of test organisms in a given population or to produce a half-maximal effect in a biological system under a defined set of conditions (Note: EDn refers to the median dose that is effective in n% of the test population.)

Median effective dose (ED_{50}): Statistically derived single dose of a chemical substance that can be expected to cause a non-lethal effect of a defined size in 50% of a given population of organisms under a defined set of experimental conditions.

Median lethal concentration (LC_{50}): Statistically derived median concentration of a substance in an environmental medium expected to kill 50% of organisms in a given population under a defined set of conditions.

Median lethal dose (LD_{50}): Statistically derived single dose of a chemical that can be expected to cause death in 50% of a given population of organisms under a defined set of experimental conditions. This figure has often been used to classify and compare toxicity among chemical substances, but its value for this purpose is doubtful.

Medical waste: All wastes from hospitals, clinics, or other healthcare facilities ("Red Bag Waste") that contain or have come in contact with diseased tissues or infectious microorganisms. Also referred to as infectious waste, which is hazardous waste with infectious characteristics, including contaminated animal waste, human blood and blood products, pathological waste, and discarded sharps (needles, scalpels, or broken medical instruments).

Medication: (i) A drug or medicine; (ii) the administration of a drug or medicine.

Mediastenum: The area between the lungs. The organs in this area include the heart and its large blood vessels, the trachea, the esophagus, the thymus, and lymph nodes, but not the lungs.

Meiosis: (i) Process of "reductive" cell division, occurring in the production of gametes, by means of which each daughter nucleus receives half the number of chromosomes characteristic of the somatic cells of the species; (ii) a special form of cell division in which each daughter cell receives half the amount of DNA of the parent cell. Meiosis occurs during formation of egg and sperm cells in mammals.

Melanin: Pigment that gives color to skin and eyes and helps protect it from damage by ultraviolet light.

Melanocyte: A cell in the skin and eyes that produces and contains the pigment called melanin.

Mercurialism (Mad Hatter syndrome): Chronic poisoning because of exposure to mercury, often by breathing its vapor but also by skin absorption and, less commonly, by ingestion. CNS damage usually predominates in mercury poisoning.

Mesothelioma: Malignant spreading tumor of the mesothelium of the pleura, pericardium, or peritoneum, arising as a result of the presence of asbestos fibers. It is diagnostic of exposure to asbestos.

Metabolic activation: Biotransformation of a substance to a more biologically active derivative.

Metabolic enzymes: Proteins that catalyze chemical transformations of body constituents and, in more common usage of xenobiotics.

Metabolic activation: Biotransformation of relatively inert chemical substances to biologically reactive metabolites.

Metabolism: (i) The term indicates the sum of (total of) the physical and chemical changes that take place in living organisms. These changes include both synthesis (anabolism) and breakdown (catabolism) of body constituents. In a narrower sense, the physical and chemical changes that take place in a given chemical substance within an organism. It includes the uptake and distribution within the body of chemical compounds, the changes (biotransformations) undergone by such chemical substances, and the elimination of the compounds and their metabolites. (ii) Metabolism is the whole range of biochemical processes that occur within the living organism/animal. The term is commonly used to refer specifically to the breakdown of food and its transformation into energy.

Metabolite: Intermediate or product resulting from metabolism.

Metaplasia: Abnormal transformation of an adult, fully differentiated tissue of one kind into a differentiated tissue of another kind.

Metastasis: The term metastasis means (i) the movement of bacteria or body cells, especially cancer cells, from one part of the body to another, resulting in change in location of a disease or of its symptoms from one part of the body to another; (ii) growth of pathogenic microorganisms or of abnormal cells distant from the site of their origin in the body.

Methemoglobin: Derivative of hemoglobin that is formed when iron(II) in the heme porphyrin is oxidized to iron(III); this derivative cannot transport dioxygen.

Mycotoxin: Toxin produced by a fungus, e.g., aflatoxins, tricothecenes, ochratoxin, and patulin.

Micro-albuminuria: Presence of small amounts of albumin in the urine.

Micromercurialism: Early or subclinical effects of exposure to elemental mercury detected at low exposure levels.

Micronucleus test: Test for mutagenicity in which animals are treated with a test agent after which time the frequency of micronucleated cells is determined; if a test group shows significantly increased levels of micronucleated cells compared to a control group, the chemical is considered capable of inducing chromosomal damage.

Microorganisms: (i) A form of life that can be seen only with a microscope; including bacteria, viruses, yeast, and single-celled animals; (ii) bacteria, yeasts, simple fungi, algae, protozoans, and a number of other organisms that are microscopic in size. Most are beneficial but some produce disease. Others are involved in composting and sewage treatment.

Microsome: Artefactual spherical particle, not present in the living cell, derived from pieces of the endoplasmic reticulum present in homogenates of tissues or cells. (Note: Microsomes sediment from such homogenates (usually the S9 fraction) when centrifuged at $100,000 \times g$ for 60 min: the microsomal fraction obtained in this way is often used as a source of monooxygenase enzymes.)

Minamata disease: Neurological disease caused by methylmercury, first seen in subjects ingesting contaminated fish from Minamata Bay in Japan.

Mind: The important seat of the faculty of intelligence and reason that thinks, reasons, perceives, wills, and feels. (In neuroscience, there is no duality between the mind and body.)

Minimal risk level (MRL): Estimate of daily human exposure to a hazardous substance that is likely to be without appreciable risk of adverse non-cancer health effects over a specified duration of exposure: this substance-specific estimate is used by ATSDR health assessors to identify contaminants and potential health effects that may be of concern at hazardous waste sites.

Mites: Tiny eight-legged animals that live off plants, animals, or stored food.

Miticides: Kill mites that feed on plants and animals

Mitosis: Process by which a cell nucleus divides into two daughter nuclei, each having the same genetic complement as the parent cell: nuclear division is usually followed by cell division.

Mixed function oxidases: These are an important set of oxidizing enzymes involved in the metabolism of many foreign compounds, giving products of different toxicity from the parent compound.

Modified PPO (polyphenelyene oxyde and styrene): Noryl is a family of products based on alloying PPO and styrene in many different combinations. It has the unique ability to perform at a wide temperature range without sacrificing key properties. Noryl can remain stable under load from $-40°F$ to $+265°F$.

Molecule: The smallest unit of a chemical substance that can exist alone and retain the character of that chemical substance.

Molluscicide: A chemical substance used to kill mollusks, snails, and slugs.

Monoclonal: Pertaining to a specific protein from a single clone of cells, all molecules of this protein being the same.

Monoclonal antibody: Antibody produced by cloned cells derived from a single lymphocyte.

Mononuclear phagocyte system (MPS): The set of cells consisting of macrophages and their precursors (blood monocytes and their precursor cells in bone marrow). The term has been proposed to replace reticuloendothelial system, which does not include all macrophages and does include other unrelated cell types.

Monooxygenase (mixed-function oxidase): Enzyme that catalyzes reactions between an organic compound and molecular oxygen in which one atom of the oxygen molecule is incorporated into the organic compound and one atom is reduced to water; involved in the metabolism of many natural and foreign compounds giving both unreactive products and products of different or increased toxicity from that of the parent compound.

Morbidity: The relative incidence of a particular disease. In common clinical usage, any disease state, including diagnosis and complications, is referred to as morbidity.

Mordant: Chemical substance that fixes a dyestuff in or on a material by combining with the dye to form an insoluble compound; used to fix or intensify stains in a tissue or cell preparation.

Mortality: (i) The ratio of deaths in an area to the population of that area, within a particular period of time; (ii) the death rate in a population or locality.

Motor neuron disease: Diseases characterized by a selective degeneration of the motor neurons of the spinal cord, brainstem, or motor cortex. Clinical subtypes are distinguished by the major site of degeneration. In amyotrophic lateral sclerosis, there is involvement of upper, lower, and brainstem motor neurons. In progressive muscular atrophy and related syndromes, the motor neurons in the spinal cord are primarily affected. With progressive bulbar palsy, the initial degeneration occurs in the brainstem. In primary lateral sclerosis, the cortical neurons are affected in isolation.

MRID: Master Record Identification.

MS: Mass spectrometry.

Multigeneration study: (i) Toxicological studies conducted in at least three generations of the test organism/animal. The test animals are exposed (usually continuous) to a candidate chemical substance and evaluated. (ii) Toxicity test in which two to three generations of the test organism are exposed to the substance being assessed. (iii) Toxicity test in which only one generation is exposed and effects on subsequent generations are assessed.

Mutagen: (i) A chemical substance or physical agent that causes mutations in the organism/animal; (ii) a chemical substance capable of changing genetic material in a cell.

Mutagenesis: (i) A process by which the genetic information of an organism/animal is changed in a stable manner, either in nature or experimentally by the use of chemical substances or radiation; (ii) a process of production of mutations possibly leading to transformation and carcinogenesis.

Mutagenic: (i) Capable of inducing mutation (used mainly of extra-cellular factors such as x-rays or chemical pollution); (ii) causing mutations; mutagenic substances may also be carcinogenic.

Mutagenicity: (i) A change in the genetic material of a living organism, usually in a single gene, which can be passed on to future generations; (ii) the property of a physical, chemical, or biological agent to induce mutations in living tissue.

Mutation: Any heritable change in genetic material. This may be a chemical transformation of an individual gene (a gene or point mutation), which alters its function. On the other hand, this change may involve a rearrangement, or a gain or loss of part of a chromosome, which may be microscopically visible. This is designated a chromosomal mutation. Most mutations are harmful.

Mycotoxin: Toxin produced by a fungus, e.g., aflatoxins, tricothecenes, ochratoxin, and patulin.

Mydriasis: Extreme dilation of the pupil of the eye, either as a result of normal physiological response or in response to a chemical exposure.

Myelosuppression: Reduction of bone marrow activity leading to a lower concentration of platelets, red cells, and white cells in the blood.

N

Nanogram: A measure of weight. One nanogram weighs a billion times less than one gram and almost a trillion times less than a pound.

Nanoparticle: Microscopic particle whose size is measured in nanometers, often restricted to so-called nanosized particles (NSPs; <100 nm in aerodynamic diameter), also called ultrafine particles.

Nanotechnology: (i) The field of research that deals with the engineering and creation of things from materials that are less than 100 nanometers (one billionth of a meter) in size, especially single atoms or molecules. Nanotechnology is being studied in the detection, diagnosis, and treatment of cancer. (ii) Scientific discipline involving the study of the actual or potential danger presented by the harmful effects of nanoparticles on living organisms and ecosystems, of the relationship of such harmful effects to exposure, and of the mechanisms of action, diagnosis, prevention, and treatment of intoxications.

Narcotic: (i) A chemical substance used to treat moderate to severe pain. Narcotics are like opiates, such as morphine and codeine, but are not made from opium. They bind to opioid receptors in the CNS. Narcotics are now called opioids. (ii) A drug that causes insensibility or stupor. A narcotic induces narcosis, from the Greek *narke* for "numbness or torpor." (iii) A drug such as marijuana that is subject to regulatory restrictions comparable to those for addictive narcotics.

Nasogastric: Describes the passage from the nose to the stomach. For example, a nasogastric tube is inserted through the nose, down the throat and esophagus, and into the stomach.

Nasopharynx: The upper part of the throat behind the nose. An opening on each side of the nasopharynx leads into the ear.

National Academy of Sciences (NAS): An institution created by Congress in 1863 to provide science-based advice to the government. The sister organizations associated with NAS are the National Academy of Engineers, the Institute of Medicine, and the National Research Council. The Academies and the Institute are honorary societies that elect new members to their ranks each year. The bulk of the institutions' science policy and technical work is conducted by the National Research Council (NRC), created expressly for that purpose. The NRC's Board on Agriculture addresses issues confronting agriculture, food, and related environmental topics.

National Cancer Institute (NCI): National Cancer Institute is a part of the National Institutes of Health (NIH) of the United States, Department of Health and Human Services, and is the federal government's principal agency for cancer research. NCI conducts, coordinates, and funds cancer research, training, health information dissemination, and other programs with respect to the cause, diagnosis, prevention, and treatment of cancer.

National Fire Protection Association (NFPA): An organization that provides information about fire protection and prevention and developed a standard outlining a hazard-warning labeling system that rates the hazard(s) of a material during a fire (health, flammability, and reactivity hazards).

National Institute of Health (NIH): A federal agency in the United States that conducts biomedical research in its own laboratories; supports the research of non-federal scientists in universities, medical schools, hospitals, and research institutions throughout the country and abroad; helps in the training of research investigators; and fosters communication of medical information.

National Institute for Occupational Safety and Health (NIOSH): US federal agency of the CDC that investigates and evaluates potential hazards in the workplace. NIOSH is also responsible for conducting research and providing recommendations for the prevention of work-related illness and injuries.

National Toxicology Program (NTP): US federal interagency program that coordinates toxicological testing programs, develops and validates improved testing methods, and provides toxicological evaluations on substances of public health concern.

Nausea: (i) A feeling of sickness or discomfort in the stomach that may come with an urge to vomit. Nausea is a side effect of some types of cancer therapy. (ii) Nausea is the urge to vomit. It can be brought on by many causes including systemic illnesses, such as influenza, medications, pain, and inner ear disease.

Nebulizer: A device used to turn liquid into a fine spray.

Necrosis: (i) Mass death of areas of tissue surrounded by otherwise healthy tissue; (ii) the sum of morphological changes resulting from cell death by lysis and (or) enzymatic degradation, usually accompanied by inflammation and affecting groups of cells in a tissue.

Nematicide: A group of chemical substance used to kill nematodes, the parasites that infect agricultural crops and farm animals.

Nematodes: Roundworms are also known as nematodes and affect human health. Most parasitic roundworm diseases are transmitted to humans through soil. Cysticercosis, an infection caused by the larval form of the pork tapeworm, *Taenia solium*, is recognized as an increasingly important cause of severe neurologic disease in developed countries because of eating undercooked pork, through contaminated water or food, or hand to mouth. Symptoms of cysticercosis include, but are not limited to, muscle pains, lumps under skin, and blurred vision. The symptoms typically occur months to years after the infection. Humans who ingest *T. solium* eggs through contaminated food/undercooked pork become infected with cysticerocosis, which later spreads to all human organs, typically the CNS.

Neoplasia: (i) Abnormal and uncontrolled cell growth; (ii) new and abnormal formation of tissue as a tumor or growth by cell proliferation that is faster than normal and continues.

Neoplasm: An abnormal mass of tissue that results when cells divide more than they should or do not die when they should. Neoplasms may be benign or malignant; also called tumor. Any new formation of tissue associated with disease such as a tumor.

Nephritis: Inflammation of the kidney, leading to kidney failure, usually accompanied by proteinuria, hematuria, edema, and hypertension.

Nephrocalcinosis: A form of renal stone disease where the kidney tissue is characterized by foci of calcification in addition to numerous deposits of calcium phosphate and calcium oxalate.

Nephropathy: Any disease or abnormality of the kidney.

Nephrotoxic: (i) Chemical substances that are poisonous or damaging to the kidney; (ii) any chemical substance harmful to the kidney.

Nerves: A bundle of fibers that receives and sends messages between the body and the brain. The messages are sent by chemical and electrical changes in the cells that make up the nerves.

Nerve cell (neuron): Type of cell that receives and sends messages from the body to the brain and back to the body. The messages are sent by a weak electrical current.

Nervous system: The organized network of nerve tissue in the body. It includes the CNS (the brain and spinal cord), the peripheral nervous system (nerves that extend from the spinal cord to the rest of the body), and other nerve tissue.

Neural: Pertaining to a nerve or to the nerves.

Neuroglia (glial cell): Any of the cells that hold nerve cells in place and help them work the way they should. Types of neuroglia include oligodendrocytes, astrocytes, microglia, and ependymal cells.

Neuroleptic malignant syndrome (NMS): A life-threatening condition that may be caused by certain drugs used to treat mental illness, nausea, or vomiting. Symptoms include high fever, sweating, unstable blood pressure, confusion, and stiffness.

Neuron(e): Nerve cell, the morphological and functional unit of the central and peripheral nervous systems.

Neuropathy (peripheral neuropathy): (i) Any disease of the central or peripheral nervous system; (ii) nerve problem that causes pain, numbness, tingling, swelling, or muscle weakness in different parts of the body. It usually begins in the hands or feet and gets worse over time. Neuropathy may be caused by physical injury, infection, toxic substances, disease (such as cancer, diabetes, kidney failure, or malnutrition), or drugs, including anticancer drugs.

Neuropeptide: A member of a class of protein-like molecules made in the brain. Neuropeptides consist of short chains of amino acids, with some functioning as neurotransmitters and some functioning as hormones.

Neurotoxicity: (i) Adverse effects on the structure or function of the central and/or peripheral nervous system caused by exposure to a toxic chemical. Symptoms of neurotoxicity include muscle weakness, loss of sensation and motor control, tremors, cognitive alterations, and autonomic nervous system dysfunction. (ii) Able to chemically produce an adverse effect on the nervous system: such effects may be subdivided into two types: (1) CNS effects (including transient effects on mood or performance and pre-senile dementia such as Alzheimer's disease) and (2) peripheral nervous system effects (such as the inhibitory effects of organophosphorus compounds on synaptic transmission).

Neurotoxin: A substance that induces an adverse effect on the structure and/or function of the central and/or peripheral nervous system.

Neurotransmitter: A chemical that is made by nerve cells and used to communicate with other cells, including other nerve cells and muscle cells.

Neutrophil: A type of immune cell that is one of the first cell types to travel to the site of an infection. Neutrophils help fight infection by ingesting microorganisms and releasing enzymes that kill the microorganisms. A neutrophil is a type of white blood cell, a type of granulocyte, and a type of phagocyte.

Niacin (nicotinic acid and vitamin B3): A nutrient in the vitamin B complex that the body needs in small amounts to function and stay healthy. Niacin helps some enzymes work properly and helps the skin, nerves, and the digestive tract stay healthy. Niacin is found in many plant and animal products. It is water soluble (can dissolve in water) and must be taken in every day. Insufficient niacin can cause a disease called pellagra (a condition marked by skin, nerve, and digestive disorders). A form of niacin is being studied in the prevention of skin and other types of cancer. Niacin may help to lower blood cholesterol.

Nicotine: An addictive, poisonous chemical found in tobacco. It can also be made in the laboratory. When it enters the body, nicotine causes an increased heart rate and use of oxygen by the heart, and a sense of well-being and relaxation. It is also used as an insecticide.

NIH: Refer: National Institutes of Health.

NIOSH: The US National Institute for Occupational Safety and Health, a federal agency that conducts research on occupational safety and health questions and makes recommendations to federal OSHA on new standards for controlling toxic chemicals in the workplace.

Nitric acid: A toxic, corrosive, colorless liquid used to make fertilizers, dyes, explosives, and other chemicals.

No observed effect level (NOEL): The greatest concentration or amount of a chemical found by experiment or observation that causes no detectable adverse alteration of morphology, functional capacity, growth, development, or life span of the target organism. The maximum dose or ambient concentration that an organism can tolerate over a specified period of time without showing any detectable adverse effect and above which adverse effects are apparent.

No effect dose (subthreshold dose) (NED): The amount of a chemical substance that has no effect on the organism.

No effect level (NEL): Maximum dose of a chemical substance that produces no detectable changes under defined conditions of exposure. This term is the same as NOAEL (no observed adverse effect level) and NOEL (no observed effect level).

NOEC: No observed effect concentration.; NOEL: No observed effect level.

Non-malignant (benign): Not cancerous and non-malignant tumors may grow large but do not spread to other parts of the body.

Non-occupational exposure: Environmental exposure outside the workplace to substances that are otherwise associated with particular work environments and/or activities and processes that occur there.

No observed adverse effect level (NOAEL): Greatest concentration or amount of a substance, found by experiment or observation, which causes no detectable adverse alteration of morphology, functional capacity, growth, development, or life span of the target organism under defined conditions of exposure.

Non-prescription (over-the-counter, OTC): A medicine that can be bought without a prescription (doctor's order). Examples include analgesics (pain relievers) such as aspirin and acetaminophen.

Non-target organism: Any organism/animal for which a pesticide was not intended to control.

NPRI: National Pollutant Release Inventory.

Nucleus: (i) The structure in a cell that contains the chromosomes. The nucleus has a membrane around it, and is where RNA is made from the DNA in the chromosomes. (ii) Compartment in the interphase eukaryotic cell bounded by a double membrane and containing the genomic DNA, with the associated functions of transcription and processing.

Nutracetical: Food or dietary supplement that is believed to provide health benefits.

Nutrient: A chemical compound (such as protein, fat, carbohydrate, vitamin, or mineral) contained in foods. These compounds are used by the body to function and grow.

Nutrition: (i) The science or practice of taking in and utilizing foods; (ii) the taking in and use of food and other nourishing material by the body. Nutrition is a three-part process. First, food or drink is consumed. Second, the body breaks down the food or drink into nutrients. Third, the nutrients travel through the bloodstream to different parts of the body where they are used as "fuel" and for many other purposes. To give the body proper nutrition, a person has to eat and drink enough of the foods that contain key nutrients.

Nystagmus: Involuntary, rapid, rhythmic movement (horizontal, vertical, rotary, mixed) of the eyeball, usually caused by a disorder of the labyrinth of the inner ear or a malfunction of the CNS.

O

Obese: Having an abnormally high, unhealthy amount of body fat.

Obesity: A condition marked by an abnormally high, unhealthy amount of body fat.

Occupational environment: Surrounding conditions at a workplace.

Occupational exposure limit (OEL): Regulatory level of exposure to substances, intensities of radiation, etc., or other conditions, specified appropriately in relevant government legislation or related codes of practice.

Occupational exposure: Experience of substances, intensities of radiation, etc., or other conditions while at work.

Occupational hygiene: The applied science concerned with the recognition, evaluation, and control of chemical, physical, and biological factors arising in or from the workplace that may affect the health or well-being of those at work or in the community.

Occupational medicine: Specialty devoted to the prevention and management of occupational injury, illness and disability, and the promotion of the health of workers, their families, and their communities.

Occupational Safety and Health Administration (OSHA): (i) US federal agency that develops and enforces occupational safety and health standards for all general, as well as, construction and maritime industries and businesses in the United States; (ii) US Department of Labor agency responsible for administering the Occupational Safety and Health Act (P.L. 91-596). According to the OSHA, farming is the nation's most hazardous occupation. Agriculture is the largest occupational group in the United States, with some 10–20 million people depending on one's criteria of "agriculture." The intrinsically seasoned nature of many segments of agriculture not only causes the size of this workforce to vary temporally and often geographically via migrant work groups, but usually also has major effects on the nature and intensity of the work itself. The OSHA has issued safety standards relating to agricultural operations.

Ocular: The adjective applied to anything pertaining to the eye.

Odor threshold: The lowest concentration of a substance in air that can be detected. Odor thresholds are highly variable because of the differing ability of individuals to detect odors.

OECD: The Organization of Economic Cooperation and Development (OECD). A Paris-based intergovernmental organization with 29 member countries. The forum in which governments develop common solutions to various social problems, including issues related with the management of toxic chemical substances.

Office for Human Research Protections (OHRP): The office within the US Department of Health and Human Services (DHHS) that protects the rights, welfare, and well-being of people involved in clinical trials. It also makes sure that the research follows the law 45 CFR 46 (Protection of Human Subjects).

Office of Pesticide Programs (OPP): EPA office that registers and regulates pesticides.

Ointment: A substance used on the skin to soothe or heal wounds, burns, rashes, scrapes, or other skin problems.

Olfactometer: Apparatus for testing the power of the sense of smell. The parts of the body involved in sensing smell, including the nose and many parts of the brain. Smell may affect emotion, behavior, memory, and thought.

Oliguria: Excretion of a diminished amount of urine in relation to fluid intake.

Omentum: Parts of the body involved in sensing smell, including the nose and many parts of the brain. Smell may affect emotion, behavior, memory, and thought.

Oncogene: (i) A gene that is a mutated (changed) form of a gene involved in normal cell growth. Oncogenes may cause the growth of cancer cells. Mutations in genes that become oncogenes can be inherited or can be caused by exposure to substances in the environment that cause cancer. (ii) Gene that can cause neoplastic transformation of a cell. Oncogenes are slightly changed equivalents of normal genes known as proto-oncogenes.

Oncogenesis: Production or causation of tumors.

Oncogenic: Capable of producing tumors in animals, either benign (non-cancerous) or malignant (cancerous).

Oncology: The scientific discipline concerned with the study of cancer.

Onset: In medicine, the first appearance of the signs or symptoms of an illness as, for example, the onset of rheumatoid arthritis. There is always an onset to a disease but never to the return to good health. The default setting is good health.

Opioids: Opioids are chemical substances (a type of alkaloid) used to treat moderate to severe pain. Opioids are like opiates, such as morphine and codeine, but are not made from opium. Opioids bind to opioid receptors in the CNS. Opioids used to be called narcotics.

Ophthalmic: Pertaining to the eye.

OPPT (Office of Pollution Prevention and Toxics): The term OPPT refers to the US EPA under the Toxic Substances Control Act (TSCA) and the Pollution Prevention Act of 1990. Under these laws, the US EPA evaluates new and existing chemicals and their risks, and finds ways to prevent or reduce pollution before it gets into the environment. The US EPA also manages a variety of environmental stewardship programs that encourage companies to reduce and prevent pollution.

Optic chiasma: A place in the brain where some of the optic nerve fibers coming from one eye cross optic nerve fibers from the other eye.

Oral cavity: The term refers to the mouth. It includes the lips, the lining inside the cheeks and lips, the front two thirds of the tongue, the upper and lower gums, the floor of the mouth under the tongue, the bony roof of the mouth, and the small area behind the wisdom teeth.

Organelle: (i) In cell biology the term means any structure that occurs in cells and has a specialized function; (ii) microstructure or separated compartment within a cell that has a specialized function, e.g., ribosome, peroxisome, lysosome, Golgi apparatus, mitochondria (structures that make energy for the cell), lysosomes (sac-like containers filled with enzymes that digest and help recycle molecules in the cell), nucleus (a structure that contains the cell's chromosomes and is where RNA is made), and nucleolus (the most obvious and clearly differentiated nuclear subcompartment, where ribosome biogenesis takes place, but it is becoming clear that the nucleolus also has non-ribosomal functions).

Organic: Chemically, a compound or molecule containing carbon bound to hydrogen. Organic compounds make up all living matter. The term organic is frequently used to distinguish "natural" products or processes from man-made "synthetic" ones. Thus, natural fertilizers include manures or rock phosphate, as opposed to fertilizers synthesized from chemical feedstocks. Likewise, in organic farming, pests are controlled by cultivation techniques and the use of pesticides derived from natural sources (e.g., rotenone and pyrethrins, both from plants) and the use of natural fertilizers (e.g., manure and compost). Some consumers, alleging risks from synthetic chemicals, prefer organic food products. The FACT Act of 1990 required the USDA to define organic foods for marketing purposes and implement a National Organic Program.

Organoleptic and organoleptic test: The term is associated with an organ, especially a sense organ as of taste, smell, or sight and is useful in nutritional technology.

Osteomalacia: An adverse health condition marked by softening of the bones because of impaired mineralization, with excess accumulation of osteoid. The symptoms include pain, tenderness, muscular weakness, anorexia, and loss of weight, resulting from deficiency of vitamin D and calcium.

Osteoporosis: Significant decrease in bone mass with increased porosity and increased tendency to fracture.

Over-the-counter: Medications/drugs that are sold without a prescription from a doctor.

Oxidative stress: Adverse effects occurring when the generation of reactive oxygen species (ROS) in a system exceeds the system's ability to neutralize and eliminate them; excess ROS can damage a cell's lipids, protein, or DNA.

Oxidizer: A chemical substance that causes the ignition of combustible materials without an external source of ignition; oxidizers can produce oxygen, and therefore support combustion in an oxygen-free atmosphere.

P

p.o.: The abbreviation for *per os* meaning oral administration.

Palpitation: The term means (i) unduly rapid or throbbing heart beat that is noted by a patient, which may be regular or irregular; (ii) undue awareness by a patient of a heart beat that is otherwise normal.

Parakeratosis: In medicine, this term is applied to the imperfect formation of horn cells of the epidermis.

Paralysis: Loss or impairment of motor function.

Parasympathetic: In medicine, this term relates to the parasympathetic nervous system, which stimulates digestive secretions, slows the heart, constricts the pupils of the eyes, and dilates blood vessels or its disturbances.

Parenchyma(-al): In medicine, this term is applied to a specific or functional component of a gland or organ.

Parenteral dosage: Method of introducing substances into an organism avoiding the gastrointestinal tract (subcutaneously, intravenously, intramuscularly etc.).

Paresis: Slight or incomplete paralysis.

Paresthesia: Abnormal or unexplained tingling, pricking, or burning sensation on the skin.

Particulate matter: (i) A general term used in the context of pollution to describe airborne solid or liquid particles of all sizes. The term aerosol is recommended to describe airborne particulate matter. (ii) Particles in air, usually of a defined size and specified as PMn, where n is the maximum aerodynamic diameter (usually expressed in micrometers) of at least 50% of the particles.

Partition coefficient: The partition coefficient is the constant ratio that is found when a heterogeneous system of two phases is in equilibrium; the ratio of the concentrations of the same molecular species in the two phases is constant at constant temperature and pressure.

Passive safety: A passive safety feature is a system that works with no human intervention or energy application to mitigate the consequences of an accident.

Pasteurization: The process of destroying microorganisms that could cause disease. This is usually done by applying heat to food. Three processes used to pasteurize foods are flash pasteurization, steam pasteurization, and irradiation pasteurization.

Patch test: Test for allergic sensitivity in which a suspected allergen is applied to the skin on a small surgical pad. Patch tests are also used to detect exposure to pesticides.

PEEK (polyetheretherketone): PEEK is a semi-crystalline thermoplastic. It is a very high-end exotic plastic with some of the best properties available of any thermoplastic.

Percutaneous: The term means entry of a chemical substance through the skin following application on the skin.

Perfumes: Perfumes when added to detergents, clothes, and raw soap to provide an aesthetic appeal and a fragrance/scent. In some domestic products, perfumes are added to mask the odor of raw ingredients.

Permeability: Ability or power to enter or pass through a cell membrane.

Permissible exposure limit (PEL): (i) Recommendation by the US OSHA for a TWA concentration that must not be exceeded during any 8-h work shift of a 40-h working week; (ii) the legally enforceable maximum amount or concentration of a chemical that a worker may be exposed to under OSHA regulations; (iii) workplace exposure limits for contaminants established by the OSHA.

Peroxide former: A chemical substance that reacts with air or oxygen to form explosive peroxy compounds that are shock, pressure, or heat sensitive.

Peroxisome: The term in cell biology to describe the organelle present in the cytoplasm of eukaryotic cells, characterized by its content of catalase and other oxidative enzymes such as peroxidase.

Persistence: The term describes the attribution or the property of a chemical substance and the length of time that the substance remains in a particular environment before it is physically removed or chemically or biologically transformed.

Persistent inorganic pollutant (PIP): An inorganic chemical substance that is stable in the environment and liable to long-range transport, may bio-accumulate in human and animal tissue, and may have significant impacts on human health and the environment. For example, arsenides, fluorides, cadmium salts, and lead salts. It is important to remember that some inorganic chemicals, like crocidolite asbestos, are persistent in almost all circumstances, but others, like metal sulfides, are persistent only in unreactive environments; sulfides can generate hydrogen sulfide in a reducing environment or sulfates and sulfuric acid in oxidizing environments. As with organic substances, persistence is often a function of environmental properties.

Persistent organic pollutant (POP): An organic chemical substance that is stable in the environment and liable to long-range transport, may bio-accumulate in human and animal tissue, and may have significant impacts on human health and the environment. POPs include dioxin, PCBs, DDT, tributyltin oxide (TBTO), and many others. The Stockholm Convention on Persistent Organic Pollutants was adopted at the Conference of Plenipotentiaries, held from 22 to 23 May, 2001, in Stockholm, Sweden; by signing this convention, governments have agreed to take measures to eliminate or reduce the release of POPs into the environment.

Personal protective equipment (PPE): (i) The term PPE includes all clothing and other work accessories designed to create a barrier against workplace hazards. Examples include safety goggles, blast shields, hard hats, hearing protectors, gloves, respirators, aprons, and work boots. (ii) Any clothing and/or equipment used to protect the head, torso, arms, hands, and feet from exposure to chemical, physical, or thermal hazards.

Pesticide: (i) A chemical substance used to kill, control, repel, or mitigate any pest. Insecticides, fungicides, rodenticides, herbicides, and germicides are all pesticides. The EPA regulates pesticides under the authority of FIFRA. In addition, under FIFRA, a substance used as a plant regulator, defoliant, or desiccant is defined as a pesticide and is regulated accordingly. All pesticides must be registered and carry a label approved by the US EPA. (ii) Chemical substances intended to repel, kill, or control any species designated a "pest," including weeds, insects, rodents, fungi, bacteria, or other organisms. The family of pesticides includes herbicides, insecticides, rodenticides, fungicides, and bactericides.

Pesticide residue: (i) A film of pesticide left on the plant, soil, container, equipment, or handler after application of the pesticide; (ii) traces of pesticide found in man or animals or in food and water following use of a pesticide. The term includes any specified derivatives, such as degradation and conversion products, metabolites, reaction products, and impurities considered to be of toxicological significance.

Pests: Pests are organisms that may adversely affect public health or attack agricultural crops, stored food grains, and other important materials of human use.

pH: A measure of the acidity or basicity (alkalinity) of a material when dissolved in water; expressed on a scale from 0 to 14.

Phagocytosis: Phagocytosis is the term to describe (i) the ingestion of microorganisms, cells, and foreign particles by phagocytes, e.g., phagocytic macrophages; (ii) a process by which particulate material is endocytosed by a cell.

Pharmacist: A professional who fills prescriptions, and in the case of a compounding pharmacist, makes them. Pharmacists are familiar with medication ingredients, interactions, cautions, and hints.

Pharmacodynamics: Broadly, this is the science concerned with the study of the way in which xenobiotics exert their effects on living organisms. Such a study aims to define the fundamental physicochemical processes that lead to the biological effect observed.

Pharmacogenetics: The study of the influence of genetic factors on the effects of drugs on individual organisms.

Pharmacokinetics: The science that describes quantitatively the uptake of drugs by the body, the biotransformation they undergo, their distribution and metabolism in tissues, and the elimination of the drugs and their metabolites from the body. Both total amounts and tissue and organ concentrations are considered. The term "toxicokinetics" is essentially the same term applied to xenobiotics other than drugs.

Pharmacology: The science that discusses the use and effects of drugs. Pharmacology has been subdivided into pharmacokinetics and pharmacodynamics.

Pharynx: The term pharynx includes the throat, the part of the digestive tract between the esophagus below and the mouth and nasal cavities above and in front.

Phase 1 reactions: This group of reactions comprises every possible stage in the enzymic modification of a xenobiotic by oxidation, reduction, hydrolysis, hydroxylation, dehydrochlorination, and related reactions.

Phase 2 reactions: This group of reactions comprises all reactions concerned with modification of a xenobiotic by conjugation. (See conjugate.)

Phase 0 clinical trial: A Phase 0 (zero) clinical trial is designed to study the pharmacodynamic and pharmacokinetic properties of a drug. In a Phase 0 trial, a limited number of doses, and much lower doses of the drug are administered, therefore there is less risk to the participant.

Phase I clinical trial: Phase I clinical trials are done to test a new biomedical or behavioral intervention in a small group of people (20–80) for the first time, to determine the metabolism and pharmacologic actions of the drug in humans, safety, side effects associated with increasing doses, and if possible, early evidence of effectiveness. Phase I trials are closely monitored and may be conducted in patients or healthy volunteers.

Phase II clinical trial: Phase II clinical trials are done to study the biomedical or behavioral intervention in a large group of people (several hundred) to determine efficacy and to further evaluate safety. They include controlled clinical studies of the effectiveness of a drug for a particular indication or indications in patients with the disease or condition under study and determination of common, short-term side effects and risks associated with the drug. Phase II studies are typically well controlled and closely monitored.

Phase III clinical trial: Phase III studies are expanded controlled and uncontrolled studies performed after preliminary evidence of drug effectiveness has been obtained. They are intended to gather additional information about the effectiveness and safety needed to evaluate the overall benefit-risk relationship of the drug and to provide an adequate basis for physician labeling. These studies usually include from several hundred to several thousand subjects.

Phase IV clinical trial: Phase IV studies are postmarketing studies (generally randomized and controlled) carried out after licensure of a drug. These studies are designed to monitor the effectiveness of an approved intervention in the general population and to collect information about any adverse effects associated with widespread use.

Phenotype: This term is applied to the appearance or constitutional nature of an organism as contrasted with its genetic potential, the genotype. (See genotype.)

Phlegm: Mucus that is coughed up from the lungs.

Photophobia: An abnormal sensitivity to light.

Photosensitization: Sensitization or heightened reactivity of the skin to sunlight, usually due to the action of certain drugs.

Phototoxicity: Adverse effects produced by exposure to light energy, especially those produced in the skin.

PIB: Piperonyl butoxide.

Pill: In pharmacy, a medicinal substance in a small round or oval mass meant to be swallowed. A pill often contains a filler material and a plastic substance, such as lactose, which permits the pill to be rolled by hand or machine into the desired form and may then be coated with a varnish-like substance.

Pinocytosis: Type of endocytosis in which soluble chemical substances are taken up by the cell and incorporated into vesicles for digestion.

Plasma: The term in biological science to describe (i) fluid component of blood in which the blood cells and platelets are suspended; (ii) fluid component of semen produced by the accessory glands, the seminal vesicles, the prostate, and the bulbo-urethral glands; (iii) cell substance outside the nucleus, i.e., the cytoplasm.

Plasmid: Autonomous, self-replicating, extra-chromosomal, circular DNA molecule present in bacteria and yeast. The plasmids replicate autonomously each time a bacterium divides and are transmitted to the daughter cells. The DNA segments are commonly cloned using plasmid vectors.

Plasticizer: Plasticizers are materials added during the manufacturing process to increase flexibility/fluidity. Safer plasticizers provide better biodegradability and less biochemical effects. The safer types of plasticizers include, but are not limited to, acetylated monoglycerides, used as food additives; alkyl citrates, used in food packaging, medical products, cosmetics, and children's toys; triethyl citrate (TEC); acetyl triethyl citrate (ATEC), higher boiling point and lower volatility than TEC; tributyl citrate (TBC); acetyl tributyl citrate (ATBC) compatible with PVC and vinyl chloride copolymers; trioctyl citrate (TOC) used for gums and controlled release medicines; acetyl trioctyl citrate (ATOC) used for printing ink; trihexyl citrate (THC) compatible with PVC and is also used for controlled release medicines; acetyl trihexyl citrate (ATHC) compatible with PVC; butyryl trihexyl citrate (BTHC, trihexyl *o*-butyryl citrate) compatible with PVC; trimethyl citrate (TMC) compatible with PVC.

Pleura: The lining of the lung.

Ploidy: The term indicating the number of sets of chromosomes present in an organism.

Plumbism: The term associated with chronic poisoning caused by absorption of lead or lead salts.

Pneumoconiosis: The term to describe lung damage owing to fibrosis of the lungs, which develops because of (prolonged) inhalation of inorganic or organic dusts. Different causative factors are linked with different but specific types of pneumoconiosis. These include: (i) anthracosis: from coal dust; (ii) asbestosis: from asbestos dust; (iii) byssinosis: from cotton dust; (iv) siderosis: from iron dust; (v) silicosis: from silica dust; (vi) stannosis: from tin dust.

Pneumonia: An inflammatory infection that occurs in the lungs.

Pneumonitis: The term used to describe the inflammation of the lung.

Poison: The term used to indicate a chemical substance or an agent that, taken into or formed within the organism, impairs the health of the organism or animal and may kill it.

Pollutant: A pollutant is described as an undesirable chemical substance, as a solid, liquid, or gaseous condition, physical matter, or sound that disturbs the normal environmental medium. The "undesirability" of a pollutant is modulated with its concentration. Low concentrations of most chemical substances are tolerable and/or even essential. (i) A primary pollutant is one emitted into the atmosphere, water, sediments, or soil from an identifiable source; (ii) a secondary pollutant is a pollutant formed by chemical reaction in the atmosphere, water, sediments, or soil.

Pollution: (i) Introduction of pollutants into a solid, liquid, or gaseous environmental medium; (ii) any chemical substances in water, soil, or air that degrade the natural quality of the environment, offend the senses of sight, taste, or smell, or cause a health hazard. The usefulness of the natural resource is usually impaired by the presence of pollutants and contaminants.

Pollution prevention: Actively identifying equipment, processes, and activities that generate excessive wastes or use toxic chemicals and then making substitutions, alterations, or product improvements. Conserving energy and minimizing wastes are pollution prevention concepts used in manufacturing, sustainable agriculture, recycling, and clean air/clean water technologies.

Polycarbonate: Polycarbonates are characterized by an excellent combination of toughness, transparency, heat and flame resistance, and dimensional stability. Polycarbonates are the material of choice for business equipment, lighting, signage, safety and medical devices.

Polyclonal antibody: Antibody produced by a number of different cell types.

Polydipsia: A health condition indicating chronic excessive thirst.

Polysulfone: Polysulfone is a high performance amorphous resin that can withstand repeated "autoclaving" cycles (steam sterilization under high pressure) and is used extensively in medical applications.

Polyuria: A health condition indicating excessive production and discharge of urine.

POPs: Persistent organic pollutants (POPs) are chemicals, chiefly compounds of carbon, which persist in the environment, bioaccumulate through the food chain, and pose a risk of causing adverse effects to human health and the environment.

Porphyria: Disturbance of porphyrin metabolism characterized by increased formation, accumulation, and excretion of porphyrins and their precursors.

Potentiation: Potentiation is the ability of one chemical substance to enhance the activity of another chemical substance to an extent greater than the simple summation of the two expected activities.

PP (polypropylene): Polypropylene is an economical material that offers a combination of outstanding physical, chemical, mechanical, thermal, and electrical properties not found in any other thermoplastic.

Practical certainty: The concept of practical certainty involves the determination of a numerically specified low risk or socially acceptable risk that may be used in decision making where absolute certainty is not possible.

Precursor: A chemical substance from which another, usually more biologically active, chemical substance is formed.

Predicted environmental concentration: The estimated concentration of a chemical in an environmental compartment calculated from available information on its properties, its use and discharge patterns, and the quantities involved.

Preneoplastic: Before the formation of a tumor.

Prescription: A physician's order for the preparation and administration of a drug or device for a patient. A prescription has several parts. They include the superscription or heading with the symbol "R" or "Rx," which stands for the word *recipe* (meaning, in Latin, to take); the inscription, which contains the names and quantities of the ingredients; the subscription or directions for compounding the drug; and the signature, which is often preceded by the sign "s," standing for *signa* (Latin for mark), giving the directions to be marked on the container.

Primary: First or foremost in time or development. The primary teeth (the baby teeth) are those that come first. Primary may also refer to symptoms or a disease to which others are secondary.

Probability: The likelihood that something will happen. For example, a probability of less than .05 indicates that the probability of something occurring by chance alone is less than 5 in 100, or 5%. This level of probability is usually taken as the level of biologic significance, so a higher incidence may be considered meaningful. The abbreviation for probability is p.

Procarcinogen: A chemical substance that has to be metabolized before it becomes a carcinogen.

Prodrug: Precursor converted to an active form of a drug within the body.

Prognosis: The expected course of a disease; the patient's chance of recovery. The prognosis predicts the outcome of a disease and therefore the future for the patient. His prognosis is grim, for example, while hers is good.

Prokaryote: A prokaryote is an organism, e.g., a mycoplasma, a blue-green alga or a bacterium, whose cells contain no membrane-bound nucleus or other membranous organelles. (See eukaryote.)

Promoter: (i) An agent that increases tumor production by another substance when applied to susceptible organisms after their exposure to the first substance; (ii) a term used in oncology to describe an agent that induces cancer when administered to an animal or human being who has been exposed to a cancer initiator.

Protein binding: The property of having a physicochemical affinity for protein.

Proteinuria: Excretion of excessive amounts of protein (derived from blood plasma or kidney tubules) in the urine.

Psychotropic drugs: Drugs or chemical substances that are known to exert an effect on the mind and are capable of modifying mental activity.

Public health: The science and art of (i) preventing disease; (ii) prolonging life; and organized community efforts for (a) the sanitation of the environment; (b) the control of communicable infections; (c) the education of the individual in personal hygiene; (d) the organization of medical and nursing devices for the early diagnosis and preventive treatment of disease; and (e) the development of the social machinery to ensure everyone a standard of living adequate for the maintenance of health, so organizing these benefits as to enable every citizen to realize his/her birthright of health and longevity.

Public health surveillance: The ongoing systematic collection, analysis, and interpretation of data relating to public health.

Pulmonary: Pertaining to the lung(s).

Pulmonary alveoli: The pulmonary alveoli are minute air-filled sacs in the vertebrate lung, thin walled and surrounded by blood vessels.

Pulmonary function: How well the lungs are working including expanding and contracting (inhaling and exhaling) and exchanging oxygen and carbon dioxide efficiently between the air (or other gases) within the lungs and the blood.

PVC or vinyl (polyvinyl chloride): Polyvinyl chloride (PVC) is a widely used, moderately priced thermoplastic polymer material that provides good clarity. Rigid PVC can be used in both interior and exterior applications. PVC is a tough, durable, versatile material available in clear to opaque colors with a moderately lustrous finish.

PVDF (polyinylidene fluoride): PVDF is a highly chemically resistant crystalline thermoplastic. It is used extensively in the chemical processing industry because of its properties.

Pyrexia: Condition in which the temperature of a human being or mammal is above normal.

Pyrogen: Any chemical substance that produces fever.

Pyrophoric: A liquid that, even in small quantities, is liable to ignite within 5 min after coming in contact with air.

Pyrophoric solid: A solid that, even in small quantities, is liable to ignite within 5 min after coming in contact with air.

Pyrotechnic article: An article containing one or more pyrotechnic substances.

Pyrotechnic substance: A substance or mixture of substances designed to produce an effect by heat, light, sound, gas, smoke, or a combination of these as the result of non-detonative, self-sustaining exothermic (heat-related) chemical reactions.

Q

Quality assurance: All those planned and systematic actions necessary to provide adequate confidence that a product or service will satisfy given requirements for quality.

Quality control: Quality control involves important inputs such as (i) operational techniques and activities that are used to fulfill requirements for quality; (ii) procedures incorporated in experimental protocols to reduce the possibility of error, especially human error. This is a requirement of GLP.

Quantitative structure-activity relationship (QSAR): A quantitative structure-biological activity model derived using regression analysis and containing physicochemical constants, indicator variables, or theoretically calculated values as parameters.

Quantitative structure metabolism relationship (QSMR): The quantitative association between the physicochemical and/or the structural properties of a substance and its metabolic behavior.

R

Radiation toxicology: The scientific study involving research, education, prevention and effects of radiation, (i) the field of study to know the injury caused by low and high doses of ionising radiation (x-rays, gamma rays, alpha rays) on cell systems and laboratory animals.

Radioactive material: A material whose nuclei spontaneously give off nuclear radiation.

Rancid/rancidity: Oxidation/breakdown of fat that occurs naturally, causing an undesirable smell and taste.

Rate: (i) The measure of the frequency with which an event occurs in a defined population in a specified period of time; (ii) rate at which this occurs.

REACH (Registration, Evaluation, and Authorization of Chemicals): "On October 28th 2004 the European Commission adopted its legislative proposal for sweeping reform in chemicals policy, called REACH. The legislation if enacted requires that all chemicals used in commerce over 1 ton per year have basic toxicity and risk information within an 11 year period and that chemicals of very high concern be treated like drugs, with only uses approved by government authorities being permitted."

Reactivity: The capacity of a substance to combine chemically with other substances.

Readily biodegradable: Arbitrary classification of chemical substances that have passed certain specified screening tests for ultimate biodegradability; these tests are so stringent that such compounds will be rapidly and completely biodegraded in a wide variety of aerobic environments.

Readily combustible solid: Powdered, granular, or pasty substance or mixture, which is dangerous if it can be easily ignited by brief contact with an ignition source, such as a burning match, and if the flame spreads rapidly.

Recalcitrance: Ability of a substance to remain in a particular environment in an unchanged form.

Receptor: Molecular structure in or on a cell, which specifically recognizes and binds to a compound and acts as a physiological signal transducer or mediator of an effect.

Recombinant DNA technology: Methods involving the use of restriction enzymes to cleave DNA at specific sites, allowing sections of DNA molecules to be inserted into plasmid or other vectors and cloned in an appropriate host organism (e.g., a bacterial or yeast cell).

Recombinant DNA: (i) The term recombinant DNA refers to a collection of techniques for creating (and analyzing) DNA molecules that contain DNA from two unrelated organisms; (ii) DNA made by transplanting or splicing DNA into the DNA of host cells in such a way that the modified DNA can be replicated in the host cells in a normal fashion.

Recommended dietary allowance (RDA): The daily dietary intake level of a nutrient that is considered sufficient to meet the requirements of nearly all (97%–98%) healthy individuals in each life-stage and gender group. The RDAs are established by the Food and Nutrition Board of the National Academy of Sciences.

Recommended exposure level (REL): The highest allowable regulatory airborne concentration of a toxicant. This exposure concentration is not expected to injure workers. It may be expressed as a ceiling limit or as a TWA.

Recommended limit: This regulatory value is the maximum concentration of a potentially toxic chemical substance that is believed to be safe. Such limits often have no legal backing, in which case a control or statutory guide level may be set, which should not be exceeded under any circumstances.

Recovery: A process leading to partial or complete restoration of a cell, tissue, organ, organism, or animal following its damage from exposure(s) to a harmful chemical substance or agent.

RED: Re-registration eligibility decision.

Regulatory dose: The term used by the US EPA to describe the expected dose resulting from human exposure(s) to a chemical substance at the level at which it is regulated in the environment.

Relative risk (RR): This term has different meanings and depends on context. (i) Ratio of the risk of disease or death among the exposed to the risk among the unexposed: this usage is synonymous with "risk ratio"; (ii) alternately, the ratio of the cumulative incidence rate in the exposed to the cumulative incidence rate in the unexposed, i.e., the cumulative incidence ratio; (iii) sometimes used as a synonym for "odds ratio."

Remedy: Some kind of process that consistently helps treat or cure a disease. From the Latin *remedium*, meaning that heals again and again, as a medicine or therapy that relieves pain, cures disease, or corrects a disorder.

Renal plasma flow: Volume of plasma passing through the kidneys in unit time.

Renal: This term has association to anything with the kidneys.

Repellent: Any chemical substance used mainly to repel/drive away blood-sucking insects in order to protect man and animals. The term is also used for chemical substances that repel mammals, birds, rodents, mites, plant pests, and other organisms.

Reproductive toxicant: Chemical substance or preparation that produces non-heritable adverse effects on male and female reproductive function or capacity and on resultant progeny.

Reproductive toxicity: Adverse effects on sexual function and fertility in adult males and females, as well as developmental toxicity in offspring (International Programme on Chemical Safety (IPCS) Environmental Health Criteria 225, Principles for Evaluating Health Risks to Reproduction Associated with Exposure to Chemicals).

Reproductive toxicology: The branch of scientific study that studies the effects of chemical substances on the adult reproductive and neuroendocrine systems, the embryo, fetus, neonate, and pre-pubertal mammal.

Residual: Something left behind. With residual disease, the disease has not been eradicated.

Residues: The pesticide remaining in a product after natural or other processes.

Respirable dust (respirable particles): Mass fraction of dust (particles) that penetrates to the unciliated airways of the lung (the alveolar region).

Respiratory sensitizer: A substance that induces hypersensitivity of the airways following inhalation of the substance.

Respiratory toxicity: Adverse health effects on the structure or function of the respiratory system caused by exposure to a toxic chemical substance. Respiratory toxicants are known to produce a variety of acute and chronic pulmonary conditions, including local irritation, bronchitis, pulmonary edema, emphysema, and cancer.

Restricted use pesticides: A pesticide that can only be sold to or used by certified applicators.

Reticuloendothelial system (RES): (i) The system of cells with the ability to take up and retain certain dyes and particles ingested into a living animal. This term has generally been replaced by the term mononuclear phagocyte system. (ii) A group of cells having the ability to take up and sequester inert particles and vital dyes, including macrophages and macrophage precursors, specialized endothelial cells lining the sinusoids of the liver, spleen, and bone marrow, and reticular cells of lymphatic tissue (macrophages) and bone marrow (fibroblasts).

Ribonucleic acid (RNA): RNA is the generic term for polynucleotides, similar to DNA but containing ribose in place of deoxyribose and uracil in place of thymine. These molecules are involved in the transfer of information from DNA, programming protein synthesis, and maintaining ribosome structure. The four main types of RNA are heterogeneous nuclear RNA (hRNA), messenger RNA (mRNA), transfer RNA (tRNA), and ribosomal RNA (rRNA).

Risk: (i) A measure of the chance that damage to life, health, property, or the environment will occur. (ii) The probability that damage to life, health, and/or the environment will occur as a result of a given hazard (such as exposure to a toxic chemical). Some risks can be measured or estimated in numerical terms, e.g., one chance in a hundred. (iii) The term risk must not be confused with the term "hazard." It is most correctly applied to the predicted or actual frequency of occurrence of an adverse effect of a chemical or other hazard.

Risk assessment: (i) Risk assessment is the identification and quantification of the risk resulting from the specific use or occurrence of a chemical, taking into account the possible harmful effects on individuals or society of using the chemical in the amount and manner proposed and all the possible routes of exposure. Quantification ideally requires the establishment of dose-effect and dose-response relationships in likely target individuals and populations. (ii) The process of estimating the severity and likelihood of harm to human health or the environment occurring from exposure to a substance or activity that, under plausible circumstances, can cause harm to human health or the environment. (iii) Risk assessment is an organized process used to estimate the amount of risk of adverse human health effects from exposure to a toxic chemical substances and how likely or unlikely it is that the adverse effect will occur. How reliable and accurate this process is depends on the quantity and quality of the information that goes into the process. The four steps in a risk assessment of a toxic chemical substance are (a) hazard identification, (b) dose-response assessment, (c) exposure assessment, and (d) risk characterization.

Risk communication: The process of exchanging information about levels or significance of health or environmental risk.

Risk evaluation: Risk evaluation is the establishment of a qualitative or quantitative relationship between risks and benefits, involving the complex process of determining the significance of the identified hazards and estimated risks to those organisms or people concerned with or affected by them.

Risk management: (i) Risk management is the decision-making process involving considerations of political, social, economic, and engineering factors with relevant risk assessments relating to a potential hazard, in order to develop, analyze, and compare regulatory options and to select the optimal regulatory response for safety from that hazard. Essentially, risk management is the combination of three steps: risk evaluation; emission and exposure control; risk monitoring. (ii) The process of evaluating policy alternatives in view of the results of risk assessment and selecting and implementing appropriate options to protect public health. Risk management determines what action to take to reduce, eliminate, or control risks. This includes establishing risk assessment policies, regulations, procedures, and a framework for decision making based on risk. (iii) The process of actually trying to reduce risk, i.e., from a toxic chemical substance, and/or trying to keep it under control. Risk management involves not just taking action, but also analyzing and selecting among options and then evaluating their effect.

Rodenticide: (i) A descriptor term for any chemical substance used to kill rodents; (ii) a pesticide or other agent used to kill rats and other rodents or to prevent them from damaging food, crops, or forage.

S

S9 fraction: Supernatant fraction obtained from an organ (usually the liver) homogenate by centrifuging at $9000 \times g$ for 20 min in a suitable medium; this fraction contains cytosol and microsomes.

Safety assessment/safety evaluation: The process of evaluating the safety or lack of safety of a chemical substance in the environment based on its toxicity and current levels of human exposure.

Safety factor: See uncertainty factor.

Safety: The practical certainty that injury will not result from exposure to a hazard under defined conditions: in other words, the high probability that injury will not result. In the context of toxicology, the term safety denotes the high probability that injury will not result from exposure to a substance under defined conditions of quantity and manner of use, ideally controlled to minimize exposure.

Salmonella: A pathogenic, diarrhea-producing bacterium that is the leading cause of human foodborne illness among intestinal pathogens. It is commonly found in raw meats, poultry, milk, and eggs, but other foods can carry it. Under 1996 rules published by the USDA to control pathogens in meat and poultry, all plants that slaughter food animals and produce raw ground meat products must meet established pathogen reduction performance standards for *salmonella* contamination. The standards, which took effect in January 1998, vary by product. Plants where USDA testing indicates contamination rates are above the national standard will be required to take remedial actions.

Sanitizer: Chemical or physical agents that reduce microorganism contamination levels present on inanimate environmental surfaces.

SARS (severe acute repository syndrome): SARS is a respiratory disease in humans caused by the SARS coronavirus. The initial symptoms are flu like and may include fever, myalgia, lethargy, cough, sore throat, gastrointestinal disorders, and several other non-specific symptoms.

Sclerosis: Hardening of an organ or tissue, especially due to excessive growth of fibrous tissue.

Screening: The term screening means a test or tests, examination(s), or procedure(s) conducted to know the undetected abnormalities, unrecognized (incipient) diseases, or health defects. Pharmacological or toxicological screening consists of a specified set of procedures to which a series of chemical compounds are subjected to characterize pharmacological and toxicological properties and to establish dose-effect and dose-response relationships.

Secondary containment: An empty chemical-resistant container/dike placed under or around chemical storage containers for the purpose of containing a spill should the chemical container leak.

Sedative: A chemical substance that exerts a soothing or tranquillizing effect on the organism or animal.

Sediment: Topsoil, sand, and minerals washed from the land into water, usually after rain or snow melt. Sediments collecting in rivers, reservoirs, and harbors can destroy fish and wildlife habitat and cloud the water so that sunlight cannot reach aquatic plants. Loss of topsoil from farming, mining, or building activities can be prevented through a variety of erosion-control techniques.

Self-heating substance: A solid or liquid substance, other than a pyrophoric substance, which, by reaction with air and without energy supply, is liable to self-heat; this substance differs from a pyrophoric substance in that it will ignite only when in large amounts (kilograms) and after long periods of time (hours or days).

Self-reactive substance: A thermally unstable liquid or solid substance liable to undergo strongly exothermic decomposition even without participation of oxygen (air). This definition excludes substances or mixtures classified under the GHS as explosive, organic peroxides, or as oxidizing.

Sensitization: (i) This term indicates the exposure to a substance (allergen) that provokes a response in the immune system such that disease symptoms will ensue on subsequent encounters with the same substance; (ii) the immune response whereby individuals become hypersensitive to chemical substances, pollen, dandruff, or other agents that make them develop a potentially harmful allergy when they are subsequently exposed to the sensitizing material (allergen).

Serum (blood serum): Watery proteinaceous portion of the blood that remains after clotting.

SGOT: Serum glutamic-oxalic transaminase.

SGPT: Serum glutamic-pyruvic transaminase.

Shigella: A bacterium carried only by humans that causes an estimated 300,000 cases of diarrheal illnesses in the United States per year. Poor hygiene, especially poor hand washing, causes *Shigella* to be passed easily from person to person via food. Once it is in food, it multiplies rapidly at room temperature.

Short-term exposure limit (STEL): According to the American Conference of Governmental Hygienists, this is the time-weighted average (TWA) airborne concentration to which workers may be exposed for periods up to 15 min, with no more than four such excursions per day and at least 60 min between them. (Also refer: TWA (time-weighted average).)

Siderosis: (i) The pneumoconiosis resulting from the inhalation of iron dust; (ii) excess of iron in the urine, blood, or tissues, characterized by hemosiderin granules in urine and iron deposits in tissues.

SIDS (sudden infant death syndrome): SIDS is a syndrome marked by the symptoms of sudden and unexplained death of an apparently healthy infant aged one month to one year.

Silicosis: Pneumoconiosis resulting from inhalation of silica dust.

SimHaz: Simple Hazard Tool (to identify those chemical substances that pose a high or low hazard).

Sister chromatid exchange (SCE): (i) The process producing reciprocal exchange of DNA between the two DNA molecules of a replicating chromosome; (ii) the reciprocal exchange of chromatin between two replicated chromosomes that remain attached to each other until anaphase of mitosis; used as a measure of mutagenicity of substances that produce this effect.

Skeletal fluorosis: Osteosclerosis due to fluoride.

Skin corrosion: The production of irreversible damage to the skin following the application of a test substance for up to 4 h.

Skin irritation: The production of reversible damage to the skin following the application of a test substance for up to 4 h.

Skin sensitizer: A substance that induces an allergic response following skin contact. The definition for "skin sensitizer" is equivalent to "contact sensitizer."

Slurry: A pumpable mixture of solids and fluid.

Sodium: Elemental sodium first isolated by Sir Humphry Davy in 1806. Sodium is the most common electrolyte found in animal blood serum. Sodium is a very important and essential element for all animal life and for some plant species. Sodium reacts exothermally with water. When burned in air, sodium forms sodium peroxide (Na_2O_2). Sodium is naturally abundant in combined forms, especially in common salt, and is used in the production of a wide variety of industrially important compounds, i.e., glass, metal, paper, petroleum, soap, and textile. Hard soaps are generally sodium salt of certain fatty acids while potassium produces softer or liquid soaps. Sodium is an essential dietary mineral. Excessive intake of sodium is associated with health disorders—hypertension.

Solid waste: Any solid, semi-solid, liquid, or contained gaseous materials discarded from industrial, commercial, mining, or agricultural operations, and from community activities. Solid waste includes garbage, construction debris, commercial refuse, sludge from water supply or waste treatment plants, or air pollution control facilities, and other discarded materials. The Resource Conservation and Recovery Act (RCRA), enacted in 1976, is the principal federal law in the United States governing the disposal of solid waste and hazardous waste.

Soluble threshold limit concentration (STLC): STLC is used to determine the hazardous waste characterization under California State regulations as outlined in Title 26 of the California Code of Regulations (CCR).

Solvent: A substance, usually a liquid that acts as a dissolving agent or that is capable of dissolving another substance. In solutions of solids or gases in a liquid, the liquid is the solvent. In all other homogeneous mixtures like liquids, solids, or gases dissolved in liquids, solids in solids, and gases in gases, the solvent is the component of greatest amount.

Specialty biocides: Specialty biocides are those chemical substances provided for end uses, e.g., swimming pools, spas, and industrial water treatment (excludes chlorine/hypochlorites that are reported separately); disinfectants and sanitizers (including industrial/institutional applications and household cleaning products). Specialty biocides also include biocides for adhesives and sealants, leather, synthetic latex polymers, metal working fluids, paints and coatings, petroleum products, plastics, and textiles. These categories of end usage are covered by FIFRA. There are other end uses of specialty biocides that are regulated under the FFDCA, including hospital/medical antiseptics, food/feed preservatives, and cosmetics/toiletries.

SPR: Structure property relationship.

Stimulant: A chemical or drug, such as caffeine or nicotine, which temporarily accelerates physiological activity.

STLC (soluble threshold limit concentration): An analysis to determine the amount of each analyte that is soluble in the "Waste Extraction Test" (WET) leachate.

Stochastic: The adjective applied to any phenomenon obeying the laws of probability.

Stomach: The sac-shaped digestive organ that is located in the upper abdomen, under the ribs. The upper part of the stomach connects to the esophagus, and the lower part leads into the small intestine.

Structure activity relationship (SAR): The traditional practice of medicinal chemistry to modify the effect, potency, or activity of bioactive chemical compounds by modifying their chemical structure. Medical chemists use the chemical techniques of synthesis to insert new chemical groups into the biomedical compound and test the modifications in their biological effect. The study helps to identify and determine the chemical groups responsible for evoking a target biological effect in the organism/animal. The analysis of the dependence of the biological effects of a chemical on its molecular structure produces a structure-activity relationship. Molecular structure and biological activity are correlated by observing the results of systematic structural modification on defined biological endpoints.

Subacute (subchronic) toxicity: A toxicity test to determine the adverse effects occurring as a result of repeated daily dosing of a chemical, or exposure to the chemical, for part of an organism's life span (usually not exceeding 10%). With experimental animals, the period of exposure may range from a few days to 6 months.

Suggested no adverse response level (SNARL): This regulatory value defines the maximum dose or concentration that, on the basis of current knowledge, is likely to be tolerated by an organism without producing any adverse effect.

Suicidal: Pertaining to suicide; the taking of one's own life. As in a suicidal gesture, suicidal thought, or suicidal act. An "online lifeline for suicidal undergrads" may help prevent college students from committing suicide.

Suicide: The act of causing one's own death. Suicide may be positive or negative and it may be direct or indirect.

Surface water: All water is naturally open to the atmosphere (rivers, lakes, reservoirs, ponds, streams, seas, and estuaries) and all springs, wells, or other collectors are directly influenced by surface water.

Surfactant: A descriptor term for any chemical substance that lowers surface tension, i.e., a detergent chemical substance that promotes lathering.

Syndrome: A set of signs and symptoms that tend to occur together, which reflect the presence of a particular disease or an increased chance of developing a particular disease.

Synergism: The adverse effect or risk from two or more chemical substances interacting with each other is greater than what it would be if each chemical substance acts separately.

Synergistic effect: A synergistic effect is any effect of two chemical substances acting together that is greater than the simple sum of their effects when acting alone: such chemical substances are said to show synergism.

Systemic: Affecting many or all body systems or organs; not localized in one spot or area.

Systemic injection: The acute systemic injection test looks for potential toxic effects as a result of a single-dose injection in mice. Mice will be injected intraperitoneally or intravenously and observed for signs of toxicity immediately after injection up to 72 h post-injection.

Systemic toxic effects: Toxic effects produced by chemical substances are generally categorized according to the site of the toxic effect. In some cases, the effect may occur at only one site. This site is referred to as the specific target organ. In other cases, toxic effects may occur at multiple sites. This is referred to as systemic toxicity.

Systemic toxicity: Systemic toxicity (acute) evaluates the potential adverse effects of medical devices on the body's organs and tissues that are remote from the site of contact. Depending on the type of device being tested, topical, inhalation, intravenous, intraperitoneal, or oral administration of extracts or implantation of the device in the animal is observed for toxicity. There are four categories: acute (24 h), subacute (14–28 days) subchronic (90 days or 10% of an animal's life span), and chronic (anything longer). This is outlined in ISO 10993-11.

T

Tablespoon: An old-fashioned but convenient household measure of capacity. A tablespoon holds about three teaspoons, each containing about 5 cc, so a tablespoon = about 15 cc of fluid.

Tachycardia: Abnormally fast heart beat.

Tachypnoea: Abnormally fast breathing.

Taeniacide: A chemical substance intended to kill tapeworms.

Target organ dose: The amount of a potentially toxic substance reaching the organ chiefly affected by that substance.

Teratogen: (i) Teratogens are those chemical substances/agents that induce structural or functional changes in offspring when consumed by the mother during or before pregnancy. (ii) An agent that, by acting on the developing embryo or fetus, can cause a structural anomaly. Teratogens act with specificity and produce specific abnormalities at specific times during gestation. (iii) Teratogenic chemical substances lead to structural and/or functional birth defects. The effects of teratogenic compounds are time dependent as well as dose dependent. At low doses there can be no effect, at intermediate doses the characteristic pattern of malformations may result, and at high doses the embryo will be killed.

Teratogenicity: The capability of an agent/chemical substance to produce fetal malformation.

Teratogenesis: Teratogenesis is a prenatal toxicity characterized by structural or functional defects in the developing embryo or fetus.

Teratogenic chemicals/agents: Acetretin, Alcohol, 4-Aminodiphenyl, Benzene, Benzidine, Carbamazepine, Chromium trioxide, Cocain, Coumarin anticoagulants,

Diazepam, Diethylstilbestrol, Hydantoins, Isotretinoin, Lithium, Lysergic acid diethylamide (LSD), 2-Naphthylamine, 4-Nitrodiphenyl, Thalidomide, Tretinoin, Valproate, and many others.

Target: The term in biological science refers to the organism, organ, tissue, cell, or cell constituent that is subject to the action of an agent.

TDI (tolerable daily intake): TDI is an estimate of the amount of a chemical substance in air, food, or drinking water that can be taken daily over a lifetime without an appreciable health risk. TDIs are calculated on the basis of laboratory toxicity data to which uncertainty factors are applied. TDIs are used for chemical substances that do not have a reason to be found in food in contrast to chemical substances such as additives, pesticide residues, or veterinary drugs found in foods.

TEAM: Total exposure assessment methodology.

Telomere: Structure that terminates the arm of a chromosome.

Temporary acceptable daily intake: Value of the ADI proposed for guidance when data are sufficient to conclude that use of the chemical substance is safe over the relatively short period of time required to generate and evaluate further safety data, but are insufficient to conclude that use of the chemical substance is safe over a lifetime.

Temporary safe reference action level (TSRAL): The regulatory value defining the inhalational exposure level in the workplace that is safe for a short time, but should be reduced as soon as possible or appropriate respiratory protection employed.

Teratogen: (i) A descriptor term applied to any chemical substance that can cause non-heritable birth defects; (ii) an agent that, when administered prenatally to the mother, induces permanent structural malformations or defects in the offspring: (iii) a chemical substance that may cause non-heritable genetic mutations or malformations in the developing embryo or fetus when a pregnant female is exposed to the chemical substance.

Teratogenesis: The production of non-heritable birth defects.

Teratogenic: Chemical substances that produce non-heritable birth defects, is said to be teratogenic.

Teratology: Study of malformations, or serious deviations from normal development in organisms.

Testing of chemicals: (i) In terms/discipline of toxicology, testing of chemical substances means procedures and evaluation of the therapeutic and potentially toxic effects of chemical substances by their application through relevant routes of exposure with appropriate organisms, animals, or biological systems and to relate the effects to dose following application; (ii) in chemistry, qualitative or quantitative analysis by the application of one or more fixed methods and comparison of the results with established standards.

Therapeutic index: The ratio between toxic and therapeutic doses. The higher the ratio, the greater the safety of the therapeutic dose.

Therapy: Treatment of an illness, disease, or disability. Therapy may be scientifically proven/unproven to treat an illness. Unproven treatments are also called alternative therapy.

Threshold limit value (TLV): (i) Term used by the ACGIH to express the recommended exposure limits of a chemical to which nearly all workers may be repeatedly exposed to, day after day, without adverse effect. (ii) This is a guideline value defined by the ACGIH to establish the airborne concentration of a potentially toxic substance to which it is believed that healthy working adults may be exposed to safely through a 40-h working week and a full working life. This concentration is measured as a TWA concentration. They are developed only as guidelines to assist in the control of health hazards and are not developed for use as legal standards. TLV is the concentration of a potentially toxic substance that should not be exceeded during any part of the working exposure (ACGIH).

Threshold: (i) Dose or exposure concentration of a chemical substance below which a defined (adverse) effect will not occur; (ii) a level of chemical exposure below which there is no adverse effect and above which there is a significant toxicological effect.

Threshold limit value (TLV): TLV is a reserved term of the ACGIH. The TLV of a chemical substance is a level at which it is believed a worker can be exposed day after day for a working lifetime without adverse health effects. The TLV is a recommendation by the ACGIH, with only a guideline status; TLV should not be confused with exposure limits having a regulatory status.

Thrombocytopenia: Decrease in the number of blood platelets.

Time-weighted average concentration (TWA): This is a regulatory value defining the concentration of a substance to which a person is exposed in ambient air, averaged over a period, usually 8 h. For a person exposed to 0.1 mg/m^3 for 6 h and 0.2 mg/m^3 for 2 h, the 8-h TWA is $(0.1 \times 6 + 0.2 \times 2)/8$, which equals 0.125 mg/m^3.

T-lymphocyte: Cell that possesses specific cell surface receptors through which it binds to foreign substances or organisms, or those that it identifies as foreign, and which initiates immune responses.

Tolerable daily intake (TDI): TDI is applied to chemical contaminants in food and drinking water. The presence of contaminants is unwanted and they have no useful function, differing from additives and residues where there is deliberate use resulting in their presence. TDIs are calculated on the basis of laboratory toxicity data with the application of uncertainty factors. A TDI is thus an estimate of the amount of a substance (contaminant) in food or drinking water that can be ingested daily over a lifetime without appreciable health risk.

Tolerance: (i) The ability of a living organism/animal/human to withstand adverse conditions, such as pest attacks, weather extremes, or pesticides; (ii) the maximum amount of a pesticide allowable in a food or feed product before it is considered adulterated, usually specified in parts per million; (iii) tolerance is the ability to experience exposure to potentially harmful amounts of a substance without showing an adverse effect; (iv) adaptive state characterized by diminished effects of a particular dose of a substance: the process leading to tolerance is called "adaptation"; (v) in food toxicology, the dose that an individual can tolerate without showing an effect; (vi) ability of an organism to survive in the presence of a toxic chemical substance: increased tolerance may be acquired by adaptation to constant exposure.

Tonic-clonic seizure: The most common type of seizure among all humans and age groups. Tonic-clonic seizure is categorized into several phases beginning with vague symptoms hours or days before an attack. It is also sometimes termed grand mal seizures.

Topical application: Medicines applied directly to the surface of the body.

Topical effect: Consequence of application of a substance, medicine, or toxic chemical (for testing) to the surface of the body that occurs at the point of application.

Total dissolved solids (TDS): The quantity of dissolved material in a given volume of water.

Total threshold limit concentration (TTLC): An analysis to determine the total concentration of each target analyte in a sample. The samples are analyzed using published US EPA methods. When any target analyte exceeds the TTLC limits, the waste is classified as hazardous and its waste code is determined by the compound(s) that failed TTLC.

Toxemia (blood poisoning): (i) Condition in which the blood contains toxins produced by body cells at a local source of infection or derived from the growth of microorganisms; (ii) pregnancy-related condition characterized by high blood pressure, swelling and fluid retention, and proteins in the urine.

Toxic: The term to describe any chemical substance able to cause injury to living organisms/animals as a result of physicochemical interaction.

Toxic chemical substance: Chemical substances that can cause severe illness, poisoning, birth defects, disease, or death when ingested, inhaled, or absorbed by living organisms.

Toxic dose: Amount of a chemical substance that produces intoxication without lethal outcome.

Toxic effects: Toxicity is complex with many influencing factors and dosage is the most important. Xenobiotics are chemical substances foreign to the body, causing many types of toxicity by a variety of mechanisms. A chemical substance in itself is toxic and is referred to as a "parent" compound. Other chemical substances must be metabolized (chemically changed within the body) before they cause toxicity to animal or humans.

Toxic substance: In general, as defined in the FHSA regulations at 16 CFR § 1500.3(b)(5), any substance (other than a radioactive substance) that has the capacity to produce personal injury or illness to man through ingestion, inhalation, or absorption through any surface of the body. This term is further defined by the OSHA and in the FHSA regulations.

Toxicant: A descriptor applied to any chemical substance that is potentially toxic.

Toxicity assessment: The process of defining the nature of injuries that may be caused to an organism by exposure to a given chemical and the exposure concentration and time dependence of the chemically induced injuries. The aim of the assessment is to establish safe exposure concentration limits in relation to possible time of exposure.

Toxicity test: Experimental study to evaluate the adverse effects of exposure(s) of a living organism or animal to a chemical substance for a defined duration under defined conditions.

Toxicity: The term "toxicity" is used in two different senses. (i) The capacity to cause injury to a living organism; and (ii) the adverse effects of a chemical on a living organism. (See adverse effect.) The severity of toxicity produced by any chemical is directly proportional to the exposure concentration and the exposure time. This relationship varies with the developmental stage of an organism and with its physiological status. (iii) The extent, quality, or degree to which a chemical substance can be poisonous or harmful to humans or other living organisms.

Toxicodynamics: The term toxicodynamics refers and describes the physiological mechanisms by which toxins/chemical substances are absorbed, distributed, metabolized, and excreted.

Toxicokinetics: Toxicokinetics determines the relationship between the systemic exposure of a chemical substance in experimental animals and the related toxicity. It is used primarily for establishing relationships between exposures in toxicology experiments in experimental animals and the corresponding exposures in humans.

Toxicological data sheet: A document that provides important information about the toxicology of a chemical substance, its production, application, properties, methods of identification, storage, precautions, and disposal.

Toxicology: Branch of medicine dealing with the study of poisons and toxic chemical substances. The study includes understanding the chemical nature of poisons, their origin and preparation, their physiological action, tests to recognize them, the pathological changes that these chemical substances produce, the antidotes, and their recognition by postmortem evidence.

Toxification: Metabolic conversion of a potentially toxic chemical substance to a product that becomes more toxic.

Toxin: A toxin is a fairly complex and highly toxic organic substance produced by a living organism. (See toxicant.)

TPE (thermoplastic elastomer): Hytrel is an elastomer made by DuPont. It is a flexible thermoplastic that can be used to replace rubber and flexible urethanes.

Transcription: A process by which the genetic information encoded in a linear sequence of nucleotides in one strand of DNA is copied into an exactly complementary sequence of RNA.

Transdermal skin patch: A medicated adhesive patch placed on the skin to release a dose of medication through the skin into the bloodstream.

Transformation (neoplastic): Neoplastic transformation is the conversion of normal cells into tumor cells. Frequently, this is the result of a genetic change (mutagenesis); the same term is used to describe the genetic modification of bacteria for use in biotechnology.

Transformation: The process that involves (i) alteration of a cell by incorporation of foreign genetic material and its subsequent expression in a new phenotype; (ii) conversion of cells growing normally to a state of rapid division in culture resembling that of a tumor; (iii) chemical modification of substances in the environment.

Transgene: Gene from one source that has been incorporated into the genome of another organism.

Transplant: The grafting of a tissue from one place to another, just as in botany a bud from one plant might be grafted onto the stem of another. The transplanting of tissue can be from one part of the patient to another (autologous transplantation), as in the case of a skin graft using the patient's own skin; or from one patient to another (allogenic transplantation), as in the case of transplanting a donor kidney into a recipient.

TRI: The US Toxics Release Inventory. Under Section 313 of the Emergency Planning and Community Right-To-Know Act of 1986 (EPCRA).

TRI (Toxic Release Inventory): A database of annual toxic releases from certain manufacturers compiled from EPCRA Section 313 reports. Manufacturers must report annually to the EPA and the states the amounts of almost 350 toxic chemicals and 22 chemical categories that they release directly to air, water, or land, inject underground, or transfer to off-site facilities. EPA compiles these reports and makes the information available to the public under the "Community Right-to-Know" portion of the law.

TRI chemicals: A list of about 650 toxic chemical substances or chemical categories included in the Toxics Release Inventory (TRI). In general, TRI chemicals are ones that the US EPA has found can be reasonably anticipated to cause acute or chronic adverse human health effects, or adverse environmental effects.

Trohoc: The term is used to describe an epidemiological study that starts with the outcome and looks backward for the causes. (The term is disapproved by the majority of epidemiologists.)

Trophic level: The term means the amount of energy in terms of food that an organism needs. Organisms not needing organic food, such as plants, are said to be on a low trophic level, whereas predator species needing food of high energy content are said to be on a high trophic level. The trophic level indicates the level of the organism in the food chain.

TSCA: The Toxic Substances Control Act (TSCA) of 1976. In theory, this law gave the US EPA the power to test, regulate, and screen nearly all chemicals produced or imported into the United States.

TTLC (total threshold limit concentration): An analysis to determine the total concentration of each target analyte in a sample.

Tumor (neoplasm): The term to describe any growth of tissue forming an abnormal mass. Cells of a benign tumor will not spread and will not cause cancer. Cells of a malignant tumor can spread through the body and cause cancer.

Tumorigenic: The term to describe chemical substances/physical agents that cause tumor formation.

TWA (time-weighted average): (i) The concentration of a chemical contaminant averaged over a workday (usually 8 h long). TWA is measured in a workplace by sampling the breathing zone for the whole workday. (ii) The average concentration of a chemical substance in air over the total exposure period of time, usually an 8-h workday.

U

UCM: Urographic contrast medium.

UDP: Uridine diphosphate.

Ulcer: A kind of defect often associated with inflammation, occurring locally or at the surface of an organ or tissue owing to sloughing of necrotic tissue.

Ultem (polyetherimide, PEI): Ultem was introduced by General Electric in the 1970s; it is an amorphous resin. Ultem can withstand multiple "autoclaving" cycles (steam sterilization under high pressure). Autoclaving is the primary process used in hospitals to sterilize surgical tools and instruments.

Ultrafine particle: Particle in air of aerodynamic diameter less than 100 nm.

Ultraviolet rays: Radiation from the sun that can be useful or potentially harmful. UV rays from one part of the spectrum (UV-A) enhance plant life. UV rays from other parts of the spectrum (UV-B) can cause skin cancer or other tissue damage. The ozone layer in the atmosphere partly shields us from ultraviolet rays reaching the earth's surface.

Uncertainty safety factor (UF): This term may be used in either of two ways depending on the context. (i) Mathematical expression of uncertainty applied to data that are used to protect populations from hazards that cannot be assessed with high precision. (ii) With regard to food additives and contaminants, a factor applied to the NOEL to derive the ADI (the NOEL is divided by the safety factor to calculate the

ADI). The value of the safety factor depends on the nature of the toxic effect, the size and type of the population to be protected, and the quality of the toxicological information available.

Uptake: Entry of a substance into the body, an organ, a tissue, a cell, or into the body fluids by passage through a membrane or by other means.

Urate oxidase: The hepatic peroxisomal enzyme that catalyzes the oxygen-mediated conversion of uric acid into allantoin.

Urticaria: Vascular reaction of the skin, marked by the transient appearance of smooth, slightly elevated patches (wheals, hives) that are redder or paler than the surrounding skin and often attended by severe itching.

USDA (US Department of Agriculture): The department of the federal government responsible for enhancing the quality of life for the American people by supporting the production of agriculture. This mission is achieved through: (1) ensuring a safe, affordable, nutritious, and accessible food supply; (2) caring for agricultural, forests, and range lands; (3) supporting sound development of rural communities; (4) providing economic opportunities for farm and rural residents; (5) expanding global markets for agricultural and forest products and services; and (6) working to reduce hunger in the USA and throughout the world.

US EPA (United States Environmental Protection Agency): The US EPA is an agency of the federal government of the United States charged to regulate chemicals and protect human health by safeguarding the natural environment, namely, air, water, and land. The US EPA conducts environmental assessment, research, and education. It has the primary responsibility for setting and enforcing national standards under a variety of environmental laws, in consultation with state, tribal, and local governments.

V

Vacuole: A membrane-bound cavity within a cell.

Validity of a measurement: Expression of the degree to which a measurement measures what it purports to measure.

Vapor: The gaseous phase of matter normally from liquid/solid material and observed at normal temperature; most organic solvents evaporate and produce vapors.

Vasoconstriction: Decrease of the caliber of the blood vessels leading to a decreased blood flow.

Vasodilation: Increase in the caliber of the blood vessels, leading to an increased blood flow.

Vehicle: Chemical substance(s) solvents or water or suspending agents used to formulate active ingredients for administration or use to conduct toxicological tests.

Venom: Animal toxin generally used for self-defense or predation and usually delivered by a bite or sting.

Vent: The connection and piping through which gases enter and exit a piece of equipment.

Ventilation: A term normally used to describe (i) a process to supply a building or room with fresh air; (ii) a process to exchange air between the ambient atmosphere and the lungs; (iii) the amount of air inhaled per day (in terms of physiology); (iv) the oxygenation of blood.

Ventricular fibrillation: A term to describe an irregular heart beat characterized by uncoordinated contractions of the ventricle.

Vermicide: A chemical substance intended to kill intestinal worms (round worms and others).

Vertigo: Dizziness; an illusion of movement as if the external world were revolving around an individual or as if the individual were revolving in space.

Vesicant: (i) A chemical substance that produces blisters on the skin.

Vesicle: A small, bladder-like, membrane-bound sac containing aqueous solution or fat— term in cell biology. A blister-like elevation on the skin containing serous fluid— a term in pathology.

Vinyl chloride: A chemical substance used in producing some plastics, which is believed to be oncogenic.

Virtually safe dose (VSD): (i) Human exposure over a lifetime to a carcinogen that has been estimated, using mathematical modeling, to result in a very low incidence of cancer, somewhere between zero and a specified incidence, e.g., one cancer in a million exposed people; (ii) the dose of a chemical substance corresponding to the level of risk determined and accepted by regulatory agencies; the dose-to-risk relationship is based on a chemical dose-response curve.

Virucide: Chemical substance(s) used to control viruses.

Volatile organic chemicals (VOC): (i) Any organic compound having, at 293.15 K, a vapor pressure of 0.01 kPa or more, or having a corresponding volatility under the particular condition of use. VOCs contribute significantly to photochemical smog production and certain health problems. (ii) Any organic chemical compound that has a high enough vapor pressure under normal conditions to significantly vaporize and enter the atmosphere. (iii) The list of VOCs is huge and varied. Although ubiquitous in nature and in modern industrial society, they may also be harmful or toxic. (iv) Volatile organic compounds are produced (a) naturally through biological mechanisms—metabolism; (b) directly by the use of fossil fuels, namely, gasoline; or (c) indirectly by automobile exhausts. VOCs are useful as fuels, solvents, scents, precursors, propellants, refrigerants, drugs, pesticides, and markers.

Volatile: Any chemical substance that evaporates quickly.

Volatility: (i) The tendency of a liquid to evaporate into a gas or vapor form. On inhalation, organic solvents are in the form of vapors. (ii) Volatility in the context of chemistry/physics/thermodynamics is a measure of the tendency of a substance to vaporize. It has also been defined as a measure of how readily a substance vaporizes. At a given temperature, substances with higher vapor pressures will vaporize more readily than substances with a lower vapor pressure.

Vulnerable zone: An area over which the airborne concentration of a chemical substance is accidentally released and could reach the level of concern.

W

Waste (Chemical waste): Any chemical substance that is discarded deliberately or otherwise disposed of on the assumption that it is of no further use to the primary user.

Wastewater: The spent or used water from a home, community, farm area, or industry that contains dissolved or suspended matter.

Wasting syndrome: A disease marked by weight loss, atrophy of muscular and other connective tissues that is not directly related to a decrease in food and water intake.

Water pollution: Occurs when water sources are polluted/contaminated with harmful or objectionable material that damages the quality of water.

Water reactive material: A chemical substance that reacts with water that could generate enough heat for the item to spontaneously combust or explode. The reaction may also release a gas that is either flammable or presents a health hazard.

Water table: The boundary between the saturated and unsaturated zones. Generally, the level to which water will rise in a well (except artesian wells).

Weight-of-evidence for toxicity: The extent to which the available biomedical/toxicological data support the hypothesis that a chemical substance causes a defined toxic effect in humans.

Wheezing: Breathing with a rasp or whistling sound; a sign of airway constriction or obstruction.

Withdrawal: Physical and psychological symptoms that follow the discontinuance of an addicting drug. The symptoms associated with smoking cessation include cravings to smoke, irritability and anxiety.

Withdrawal effect: The effect of adverse events following withdrawal from a person or animal of a drug that they have been chronically exposed to or on which they have become dependent, e.g., drug abuse or drug addict.

World Health Organization (WHO): World Health Organization is the directing and coordinating authority for health within the United Nations system. It is responsible for providing leadership on global health matters. The role of WHO includes (i) providing leadership on matters critical to health and engaging in partnerships where joint action is needed; (ii) shaping the research agenda and stimulating the generation, translation, and dissemination of valuable knowledge; (iii) setting norms and standards and promoting and monitoring their implementation; (iv) articulating ethical and evidence-based policy options; (v) providing technical support, catalyzing change, and building sustainable institutional capacity; (vi) monitoring the health situation and assessing health trends. WHO's constitution came into force on April 7, 1948. More than 8000 people from more than 150 countries work for the organization in 147 country offices, six regional offices, and at the headquarters in Geneva, Switzerland.

X

x-Disease: Hyperkeratotic disease in cattle following exposures to chlorinated dibenzo-p-dioxins, naphthalenes, and related compounds.

Xenobiotic metabolism: Xenobiotic metabolism is the sum of the physical and chemical changes that affect foreign substances in living organisms from uptake to excretion.

Xenobiotic: (i) A xenobiotic is a chemical substance that is not a natural component of the organism exposed to it; (ii) a substance with a chemical structure foreign to a given organism. These include drugs, foreign substances, compounds, or exogenous substances.

Z

Zero air: Atmospheric air purified to contain less than 0.1 ppm total hydrocarbons.

Zero tolerance: In food safety policy, a "zero tolerance" standard generally means that if a potentially dangerous substance (whether microbiological, chemical, or other) is present in or on a product, that product will be considered adulterated and unfit for human consumption. In the meat and poultry inspection program, "zero tolerance" usually refers to the USDA's rule that permits no visible signs of fecal contamination (feces) on meat and poultry carcasses.

Zoocide: A chemical substance intended to kill animals.

Zygote: The term means (i) a fertilized egg resulting from the fusion of two gametes; (ii) a cell obtained as a result of complete or partial fusion of cells produced by meiosis.

Appendix 1   Abbreviations

A

AA:	Ascorbic acid
AAF:	2-acetamidofluorene
AAN:	Acetoacetanilide
ABS:	Acrylonitrile-butadiene-styrene
ACE:	Angiotensin-converting enzyme
Ach:	Acetylcholine
AChE:	Acetylcholinesterase
AcP:	Acid phosphatase
ACTH:	Adrenocorticotropic hormone
AD:	Alzheimer's disease
ADH:	Alcohol dehydrogenase; anti-diuretic hormone
ADI:	Acceptable daily intake
ADP:	Adenosine diphosphate
AE:	Acid equivalent
AEP:	Auditory evoked potential
AEP:	Aminoethylpiperazine
AFID:	Alkali flame ionization detection
AGDCI:	Agricultural Data Call-In
AGE:	Allyl glycidyl ether
Ah:	Aromatic hydrocarbon
AHH:	Aryl hydrocarbon hydroxylase
Ai:	Active ingredient
AIBN:	2,2'-azobis(2-methylpropionitrile)
AIDS:	Acquired immune deficiency syndrome
AIN:	Acute interstitial nephritis
AKD:	Alkyl ketene dimer
ALA:	Aminolevulinic acid
ALAD:	Aminolevulinic acid dehydratase
ALAS:	o-aminolevulinic acid synthetase
ALAT:	Alanine aminotransferase
ALC:	Approximate lethal concentration
ALD:	Approximate lethal dose
ALMS:	Atomic line molecular spectrometry
ALT:	Alanine aminotransferase
AM:	Alveolar macrophages
AMP:	Amperometric titration
AMP:	2-amino-2-methyl-1-propanol
AMPA:	Aminomethylphosphonic acid

ANTU:	Alpha-naphthylthiourea
AP:	Alkaline phosphatase
aPAD:	Acute population adjusted dose
APCD:	Ammonium 1-pyrrolidinecarbodithioate
APDM:	Aminopyrine-N-demethylase
APP:	Ammonium polyphosphate

B

BA:	2-Bromoacrolein
BAC:	N,N'-bis(acrylyl)cystamine
BAL:	2,3-dimercapto-1-propanol
BCG:	Bromocresol Green
BCS:	N-(benzyloxycarbonyloxy)succinimide
BDU:	5-bromo-2'-deoxyuridine
BHA:	Butylated hydroxyanisole
BHT:	Butylated hydroxytoluene
BKF:	Benzo (k)fluoranthene
BNP:	3-nitropropionic acid
BNS:	Beta-nitrostyrene
BOA:	Butylated hydroxyanisole
BTM:	Benzyltrimethylammonium chloride
BTZ:	Phenylbutazone
BZI:	Benzimidazole
BZT:	Benzenthonium chloride

C

CA:	Chrysanthemic acid
CACP:	Cisplatin; **CPDC:** cisplatin; **CPDD:** cisplatin
CAF:	2-chloroacetophenone
CALLA:	Common acute lymphoblastic leukemia antigen
cAMP:	Cyclic adenosine monophosphate
CAP:	2-chloroacetophenone
CAPS:	3-(cyclohexylamino)-1-propanesulfonic acid
CAS:	Chemical abstracts service
CBDCA:	Carboplatin
CBF:	Cerebral blood flow
CCC:	(2-chloroethyl) trimethylammonium chloride
CDA:	Cetyldimethylethylammonium bromide
CDDP:	Cisplatin
cDNA:	Complementary DNA
CDNB:	1-chloro-2, 4-dinitrobenzene
CE:	Capillary electrophoresis
CEC:	Commission of the European Communities
CEFIC:	European Council of Chemical Industry Federations

CFC:	Chlorofluorocarbon
CFR:	Code of federal regulations
CG:	Phosgene
ChE:	Cholinesterase
CHO:	Chinese hamster ovary
CI:	Confidence interval
Ci:	Curie(s)
CIN:	Chronic interstitial nephritis
CKSCC:	Cystic keratinizing squamous cell carcinoma
CLD:	Chemiluminescence nitrogen detector
CLM:	Chemiluminescence method
CLV:	Ceiling value
CMI:	Cell-mediated immunity
cGMP:	Cyclic guanosine monophosphate
CNC:	Copper naphthenate/2-chloroacetophenone
CPCA:	Dicofol
CPIA:	2-(4-chlorophenyl)isovaleric acid
CS:	o-chlorobenzylidene malononitrile
CTAB:	Hexadecyltrimethylammonium bromide
Cl-Vacid:	2-(4-chlorophenyl)isovaleric acid
Cl$_2$CA:	3-(2,2-dichlorovinyl)-2,2-dimethylcyclo-propanecarboxylic acid
CMI:	Cell-mediated immunity
CML:	Cell-mediated lympholysis
CNS:	Central nervous system
COD:	Chemical oxygen demand
CO-Hb:	Carboxyhaemoglobin
COPD:	Chronic obstructive pulmonary disease
COT:	Committee on Toxicity (United Kingdom)
cP:	Centipoise(s)
CP:	Coproporphyrins
CPA:	Cyclopropane carboxylic acid
cPAD:	Chronic population adjusted dose
CPK:	Creatine phosphokinase
CRMs:	Certified reference materials
CSF:	Cerebrospinal fluid
CSM:	Committee on Safety of Medicines (United Kingdom)
CV:	Coefficient of variation

D

DAB:	4-dimethylaminobenzene
DABA:	L-2,4-diaminobutyric acid
DABCO:	1,4-diazabicyclo(2.2.2)octane
DACPM:	4,4'-methylenebis(2-chloroaniline)
DADPE:	4,4'-oxydianiline

DADPM:	4,4'-methylenedianiline
DADT:	N,N'-diallyltartardiamide
DAPM:	4,4'-methylenedianiline
DASD:	4,4'-diamino-2,2'-stilbenedisulfonic acid
DBCP:	1,2-dibromo-3-chloropropane
DBT:	Dibutyltin
DCAA:	Dichloroacetic acid
DCB:	Dichlorobenzene
DCDD:	2,7-dichlorodibenzo-p-dioxin
DCDP:	Cisplatin
DCEE:	2-chloroethyl ether
DCMC:	Diuron
DCMU:	Diuron
DCNB:	3,4-dichloronitrobenzene
DCS:	2-phenylphenol sodium salt tetrahydrate
DCVC:	S-(1,2-dichlorovinyl)-L-cysteine
DDC:	Diethyldithiocarbamate
DDPt:	Cisplatin
DDT:	Dichlorodiphenyltrichloroethane
DEAE:	Diethylaminoethanol
DEDC:	Sodium diethyldithiocarbamate trihydrate
DEDK:	Sodium diethyldithiocarbamate trihydrate
DEHA:	Bis(2-ethylhexyl) adipate
DEHP:	Dioctyl phthalate
DEN:	Diethylnitrosamine
DETA:	N,N-diethyl-N-toluamide
DES:	Diethylstilbestrol
DFB:	Diflubenzuron
DFP:	Diisopropylfluorophosphate (a delayed neurotoxin)
DFR:	Dislodgeable foliar residue
DHES:	Dioctyl sebacate
DHPN:	N-bis(2-hydroxypropyl) nitrosamine
DiBP:	di-iso-butyl phthalate
DIDT:	5,6-dihydro-3H-imidazo (2,1-C)-1,2,4-dithiazole-3-thione
DIT:	Diiodotyrosine
DITP:	2'-deoxyinosine-5'-triphosphate
DL:	Detection limit
DMA:	Dimethylamine
DMAB:	Borane-dimethylamine complex
DMAC:	Dimethyl acetamide
DMBA:	Dimethylbenzathraline
DMDS:	Methyl disulfide
DMF/DMFA:	Dimethylformamide

DMHP:	Dimethyl phosphite
DMNA:	N-nitrosodimethylamine
DMMP:	Dimethyl methylphosphonate
DMP:	Dimethylphenol
DMPT:DMSA:	Succinic acid 2,2-dimethylhydrazide
DMSO:	Dimethyl sulfoxide
DNA:	Deoxyribonucleic acid
DNCB:	1-chloro-2,4-dinitrobenzene
DNFB:	2,4-dinitrofluorobenzene
DNPF:	2,4-dinitrofluorobenzene
DNPH:	2,4-dinitrophenyl hydrazine
DOC:	Dissolved organic carbon
DPPA:	Diphenylphosphoryl azide
DPPE:	1,2-bis(diphenylphosphino)ethane
DRES:	Dietary risk evaluation system
DS:	Diethyl sulfate
DTBP:	Di-t-butyl peroxide
DTMC:	Dicofol
DTNB:	5,5'-dithiobis(2-nitrobenzoic acid)
DTPA:	Diethylenetriamine pentaacetic acid/pentetic acid
DWEL:	Drinking water equivalent level

E

EB:	Ethyl benzene
EBDC:	Ethylene bisdithiocarbamate
EBK:	Ethyl n-butyl ketone
EC:	Effective concentration; electron capture; emulsifiable concentrate
EC50:	Median effective concentration
ECD:	Electron capture detector

G

GLDH:	Glutamate dehydrogenase
GOT:	Glutamic-oxaloacetic transaminase
GPMT:	Guinea-pig maximization test
GPT:	Glutamic-pyruvic transaminase
GS:	Glutamine synthetase
GSH:	Glutathione-SH
GST:	Glutathione-S-transferase
GTB:	Glomerular tubular balance
GV:	Guidance value
GWP:	Global-warming potential

H

h:	Hour(s)
ha:	Hectare
Hb:	Hemoglobin
HBCD:	Hexabromocyclododecane
HBDH:	Hydroxybutyric dehydrogenase
H & E:	Hematoxylin and eosin
HCB:	Hexachlorobenzene
HCBD:	Hexachloro-1,3-butadiene
HCFC:	Hydrochlorofluorocarbon
HCFH-22:	Chlorodifluoromethane ($CHClF_2$)
HCG:	Human chorionic gonadotropin
HCH:	Hexachlorocyclohexane
HDL:	High density lipoprotein
HDPE:	High density polyethylene
HEAL:	Human exposure assessment location
HECD:	Hall electron capture detector
HEV:	High endothelial venule
HEX:	Hexachlorocyclopentadiene
HGPRT:	Hypoxanthine-guanine phosphoribosyltransferase
HIPS:	High impact polystyrene
HIV:	Human immunodeficiency virus
HLB:	Hydrophilic-lipophilic balance
HLV:	Hygienic limit value
hnRNA:	Heterogeneous nuclear RNA
HPCA:	Human progenitor cell antigen
HPI:	Cyclohexane-1,2-dicarboximide
HPLC:	High-performance liquid chromatography
HPTLC:	High-performance thin-layer chromatography
HQ:	Hydroquinone
HAS:	Heat-stable antigen; human serum albumin
HSG:	Health and safety guide

I

IARC:	International Agency for Research on Cancer
IC:	Ion chromatography
ICAM:	Intercellular adhesion molecule
IC$_{50}$:	Median inhibitory concentration
ICD:	International Classification of Diseases
ICG:	Indocyanine Green
ICST:	Isolated cold stress testing
i.d.:	Internal diameter
IFN:	Interferon

Ig:	Immunoglobulin
IL:	Interleukin
ILO:	International Labor Organization
ILSI:	International Life Science Institute
im:	Intramuscular (injection)
ip:	Intraperitoneal (injection)
IP:	Isopropylamine
IPCS:	International Program on Chemical Safety
IQ:	Intelligence quotient
IQC:	Internal quality control
IR:	Infrared
IRPTC:	International Register of Potentially Toxic Chemicals
ISO:	International Organization for Standardization
IT:	Isomeric transition
IU:	International unit
IUPAC:	International Union of Pure and Applied Chemistry
iv:	Intravenous (injection)

J

JECFA:	Joint FAO/WHO Expert Committee on Food Additives
JMPR:	Joint FAO/WHO Meeting on Pesticide Residues

K

Kcal:	Kilocalorie(s)
keV:	Kiloelectron volt(s)
K_{ow}**:**	Octanol/water partition coefficient

L

LAP:	Leucine aminopeptidase
LAQL:	Lowest analytically quantifiable level
LC:	Liquid chromatography
LC_{50}**:**	Median lethal concentration
LD_{01}**:**	Lethal dose for 1%
LD_{50}**:**	Median lethal dose
LDH:	Lactate dehydrogenase
LDL:	Low density lipoprotein
LDPE:	Low density polyethylene
LDQ:	Lowest detectable quantity
LEI:	Lifetime exposure intensity
LEV:	Local exhaust ventilation
Lf:	Limit flocculation
LFA:	Lymphocyte function-related antigen
LFP:	Lavage fluid protein

LH:	Luteinizing hormone
LI:	Labeling index
LIF:	Laser-induced fluorescence; leukemia inhibitory factor
LLD:	Lowest lethal dose
LMS:	Linear multistage model
LMW:	Low molecular weight
LOAEL:	Lowest-observed-adverse-effect level
LOD:	Limit of determination
LOEL:	Lowest-observed-effect level
LRNI:	Lower reference nutrient intake (UK)
LSC:	Liquid scintillation counter
LT$_{50}$:	Median lethal time
LTP:	Long-term potentiation

M	
MAA:	Methoxyacetic acid
MAC:	Maximum allowable concentration
MAD:	Maximum allowable deviation
MAK:	Maximum workplace concentration
MALT:	Mucosa-associated lymphoid tissue
MAM:	Methylazoxymethanol
MAP:	Mutagenic activity profile
MAOI:	Monoamine oxidase inhibitor
MARC:	Monitoring and Assessment Research Centre (UK)
MARE:	Monoclonal anti-rat immunoglobulin E
MAS:	Molecular absorption spectrometry
MAT:	Mean absorption time
MATC:	Maximum acceptable toxicant concentration
MBC:	Minimum bactericidal concentration
MBDE:	Mass balance differential equation
MBK:	Methyl n-butyl ketone
MBP:	Monobutyl phthalate
MBT:	Monobutyltin
MC:	Methyl chloroform
3-MC:	3-methylcholanthrene
MCD:	Microcoulometric detection
MCH:	Mean cell hemoglobin
MCHC:	Mean cell hemoglobin concentration
mCi:	Millicurie
MCL:	Melanotic cell lines
MCPA:	4-chloro-o-tolyoxyacetic acid
MCV:	Mean cell volume
MDA:	Malondialdehyde
MDI:	Methylene-diphenyl diisocyanate

MDMA:	Methylenedioxymethamphetamine
2-ME:	2-methoxyethanol
2-MEA:	2-methoxyethyl acetate
MeB$_{12}$:	Methylcobalamin
MED:	Minimum effective dose; minimal erythemal dose
MEHP:	Monoethylhexyl phthalate
MEK:	Methyl ethyl ketone
mEq:	Milliequivalent
MeV:	Megaelectron volt(s)
MFO:	Mixed-function oxidase
MHC:	Major histocompatibility complex
MIBK:	Methyl isobutyl ketone
MIC:	Minimal inhibitory concentration
MIT:	Methylisothiocyanate; monoiodotyrosine
MLD:	Minimum lethal dose
MLR:	Mixed lymphocyte response assay
MLSS:	Mixed liquor suspended solids
MMA:	Methoxyacetic acid
MMAD:	Mass median aerodynamic diameter
MMH:	Monomethylhydrazine
mmHg:	Millimeter(s) of mercury
MMMF:	Man-made mineral fiber
MNNG:	N-methyl-N-nitro-N-nitrosoguanidine
MNU:	N-methyl-N-nitrosurea
mPa:	Millipascal (7.5×10^{-6} mmHg)
MPC:	Maximum permissible limit
MQL:	Minimum quantifiable limit
MRBIS:	Mean running bias index score
MRL:	Maximum residue limit
mRNA:	messenger RNA
MRVIS:	Mean Running Variance Index Score
MS:	Mass spectrometry
MSD:	Mass selective detection
MSW:	Municipal solid waste
MTBE:	Methyl tertiary butyl ether
MTC:	Maximum tolerable or acceptable concentration
MTD:	Maximum tolerated dose
MTE:	Mild toxic encephalopathy
MTI:	N-(hydroxymethyl)-3,4,5,6-tetrahydrophthalamide

N

NAA:	Neutron activation analysis
NAC:	N-acetyl cysteine

NAD:	Nicotinamide adenine dinucleotide
NADP:	Nicotinamide adenine dinucleotide phosphate
NADPH:	Reduced nicotinamide adenine dinucleotide phosphate
NBU:	N-nitrosobutylurea
NCAM:	Neural cell adhesion molecule
NCI:	National Cancer Institute (USA); Negative ion chemical ionization
NCV:	Nerve conduction velocity
ND:	Not detectable
NDDC:	Sodium diethyldithiocarbamate
NDMA:	Nitrosodimethylamine
NDMC:	Sodium dimethyldithiocarbamate
NEQUAS:	National External Quality Assessment Scheme (UK)
NFT:	Neurofibrillary tangle
Ng:	Nanogram (10^{-9} g)
NIOSH:	National Institute for Occupational Safety and Health (USA)
NK:	Natural killer
nm:	Nanometer
NMCL:	Nonmelanolic cell lines
NMN:	N-methylnicotinamide
NMOR:	N-nitrosomorpholine
NMR:	Nuclear magnetic resonance
NNM:	N-nitrosomorpholine
NO:	Nitrogen oxide
NOAEL:	No-observed-adverse-effect level
NOEC:	No-observed-effect concentration
NOEL:	No-observed-effect level
NOL:	No-observed lethal concentration
NPD:	Nitrogen-phosphorus sensitive detector
NPSH:	Non-protein sulfhydryl
NSAID:	Non-steroidal anti-inflammatory drugs
NSD:	Nitrogen selective detector
NTA:	Nitrilotriacetic acid
NTE:	Neuropathy target esterase
NTEL:	No-toxic-effect level
NTP:	National Toxicology Program (USA)

O

OCT:	Ornithine carbamoyltransferase
ODC:	Ornithine decarboxylase
ODP:	Ozone-depletion potential
OECD:	Organization for Economic Co-operation and Development
OEL:	Occupational exposure limit
OER:	Oxygen enhancement ratio
OES:	Occupational exposure standard; optical emission spectrometry

OET:	Open epicutaneous test
OP:	Organophosphate
OPIDN:	Organophosphate-induced delayed neuropathy
OR:	Odds ratio
OSC:	Oil-enhanced suspension concentrate
OSHA:	Occupational Safety and Health Administration (USA)
OVA:	Ovalbumin
OZT:	Oxazolidinethione

P

PA:	Polyamides
PAA:	Photon activation analysis
2-PAM:	Chloride pralidoxine (2-pyridine aldoxime methyl)-chloride p-aminohippurate
PAH:	Polycyclic aromatic hydrocarbon
PALS:	Periarteriolar lymphocyte sheath
PAN:	Peroxyacetyl nitrate; polyacrylonitrile
PAS:	Periodic acid Schiff stain
PB:	Phenobarbital
PBA:	Phenoxybenzoic acid
PBalc:	3-phenoxybenzyl alcohol
PBald:	3-phenoxybenzaldehyde
PBB:	Polybrominated biphenyl
PBDD:	Polybrominated dibenzodioxin
PBDE:	Polybrominated diphenyl ether
PBDF:	Polybrominated dibenzofuran
PBI:	Protein-bound iodine
PBPK:	Physiologically based pharmacokinetics
PBT:	Polybutylene terephthalate
PCBD:	S-(1,2,3,4,4-pentachloro-1,3-butadienyl)
PCA:	Para-chloroaniline (4-chloroaniline); passive cutaneous anaphylaxis
PCB:	Polychlorinated biphenyl
PCDD:	Polychlorinated dibenzodioxin
PCDF:	Polychlorinated dibenzofuran
PCDPE:	Polychlorinated diphenylether
PCE:	Polychromatic erythrocytes
PCOM:	Phase contrast optical microscopy
PCPY:	Polychlorinated pyrene
PCQ:	Polychlorinated quaterphenyl
PCT:	Porphyria cutanea tarda
PCV:	Packed-cell volume
PD:	Plasma desorption
PDG:	Phosphate-dependent glutaminase
PE:	Polyethylene

PEF:	Peak expiratory flow
PEG:	Pneumoencephalography
PEL:	Permissible exposure limit
PET:	Polyethylene terephthalate
PFC:	Plaque-forming cell
pg:	Picogram (10^{-12} g)
PG:	Prostaglandin
pH:	Negative logarithm of the hydrogen ion concentration
PHF:	Paired helical filaments
PHS:	Prostaglandin-H-synthetase
PIB:	Piperonyl butoxide
PIC:	Picrotoxin
PID:	Photo-ionization detection
PIXE:	Proton-induced x-ray emission
pKa:	Negative logarithm of the dissociation constant
PMBA:	p-methylbenzyl alcohol
PMN:	Polymorphonuclear leukocyte
PMTDI:	Provisional maximum tolerable daily intake
PMR:	Proportional mortality rate
PMSG:	Pregnant mare serum gonadotrophin
PNS:	Peripheral nervous system
pO2:	Plasma partial pressure (concentration) of oxygen
POCP:	Photochemical ozone-creation potential
PoG:	Proteoglycan
POS:	Psycho-organic syndrome
PP:	Polypropylene
ppb:	Parts per billion
ppm:	Parts per million
ppt:	Parts per trillion
PSD:	Passive sampling device
PSPS:	Pesticides Safety Precautions Scheme (UK)
PT:	Prothrombin time
PTH:	Parathyroid hormone
PTT:	Partial thromboplastin time
PTU:	Propylenethiourea
PTWI:	Provisional tolerable weekly intake
PTZ:	Pentylenetetrazole
PVC:	Polyvinyl chloride
PYR:	Pyrene

Q

QA:	Quality assurance
QAP:	Quality assurance program

QC: Quality control
QSAR: Quantitative structure-activity relationship

R
RACB: Reproductive assessment by continuous breeding
RAST: Radioallergosorbent test
RBC: Red blood cell
RBP: Retinal binding protein
RDA: Recommended dietary allowance (USA)
REL: Recommended exposure limit
RER: Rough endoplasmic reticulum
RIA: Radio-immuno assay
RIPT: Repeat insult patch test
RMA: Reflex modification audiometry
RNA: Ribonucleic acid
RNI: Reference nutrient intake (UK)
ROC: Reactive organic carbon
RPN: Renal papillary necrosis
RR: Relative risk
RTECS: Registry of Toxic Effects of Chemical Substances
RUBISCO: Ribulose 1,5-biphosphate carboxylase

S
S9: $9000 \times g$ supernatant
SAM: S-adenosylmethionine
SAP: Serum alkaline phosphatase
SAR: Structure-activity relationship
sc: Subcutaneous
SC: Suspension concentrate
SCE: Sister chromatid exchange
SCF: Stem-cell factor
SCID: Severe combined immuno-deficiency
SCOPE: Scientific Committee on Problems of the Environment of the International Council of Scientific Unions
SD: Standard deviation
SDAT: Senile dementia of Alzheimer type
SE: Standard error
SEM: Standard error of the mean; scanning electron microscopy
SER: Smooth endoplasmic reticulum
SFC: Supercritical fluid chromatography
SFS: Subjective facial sensation
SGOMSEC: Scientific Group on Methodologies for the Safety Evaluation of Chemicals
SGOT: Serum glutamic-oxaloacetic transaminase

SGPT:	Serum glutamic-pyruvic transaminase
SHE:	Syrian hamster embryo
SIM:	Selected ion monitoring
SIMS:	Secondary ion mass spectrometry
SLE:	Systemic lupus erythematosus
SMA:	Sequential multiple analyzer
SMR:	Standardized mortality ratio
SOP:	Standard operating procedure
SPF:	Specific pathogen free
SPM:	Suspended particulate matter
S-PMA:	S-phenyl-mercapturic acid
SRT:	Simple reaction time
SSB:	Single strand breaks
STEL:	Short-term exposure limit

T	
2,4,5-T:	2,4,5-trichlorophenoxyacetic acid
TADI:	Temporary acceptable daily intake
TAN:	Tropical ataxic neuropathy
TAP:	Trialkyl/aryl phosphate
TBBPA:	Tetrabromobisphenol A
TBG:	Thyroxine-binding globulin
TBP:	Tributyl phosphate
TBPP:	Tris (2,3-dibromopropyl) phosphate
TBT:	Tributyltin
TBTO:	Tributyltin oxide
TCA:	Tricarboxylic acid cycle
TCD:	Thermal conductivity detection
TCDD:	2,3,7,8-tetrachlorinated dibenzo-p-dioxin
TCDF:	2,3,7,8-tetrachlorinated dibenzofuran
TCE:	1,1,1-trichloroethylene
TCP:	Trichlorophenol; tricresyl phosphate
TCPP:	Tris(1-chloro-2-propyl)phosphate
TCR:	T-cell receptor
TDC:	Thermal conductivity detection
TDI:	Tolerable daily intake; toluene diisocyanate
TDLAS:	Tuneable diode laser absorption spectrometry
TEA:	Tetraethyl ammonium; thermal energy analyzer
TEAC:	Tetraethylammonium chloride
TEAM:	Total exposure assessment methodology
TEF:	Toxicity equivalency factor
TEPP:	Tetraethyl-pyrophosphate
TGA:	Thermogravimetric analysis
TH:	Thyroid hormone

THF:	Tetrahydrofolate
TI:	Tolerable intake
TLC:	Thin-layer chromatography
TLV:	Threshold limit value
TMCP:	Tri-meta-cresyl phosphate
TMDI:	Theoretical maximum daily intake
TML:	Tetramethyl lead
TMRL:	Temporary maximum residue limit
TMT:	Trimethyltin
TNF:	Tumor necrosis factor
TOCP:	Tri-ortho-cresyl phosphate
TOD:	Total oxygen demand
TPA:	12-O-tetradecanoylphorbol-13-acetate
TPCP:	Tri-para-cresyl phosphate
TPI:	3,4,5,6-tetrahydrophthalimide
TPIA:	3,4,5,6-tetrahydrophthalic acid
TPN:	Total parenteral nutrition
TPO:	Thyroid peroxidase
TPP:	Triphenyl phosphate
TPTA:	Triphenyltin acetate
TPTH:	Triphenyltin hydroxide
tRNA:	Transfer ribonucleic acid
TRP:	Tubular reabsorption of phosphate
TSH:	Thyroid-stimulating hormone (thyrotropin)
TSP:	Total suspended particulate
TST:	Temperature sensitivity
TT:	Toxicity threshold
TWA:	Time-weighted average

U	
UCM:	Urographic contrast medium
UCL:	Upper confidence limit
UDP:	Uridine diphosphate
UDPGA:	UDP-glucuronic acid
UDPGT:	UDP-glucuronosyltransferase
UDS:	Unscheduled DNA synthesis
UF:	Uncertainty factor
ULV:	Ultra-low volume
UNEP:	United Nations Environment Program
UNICEF:	United Nations Children's Fund
US ATSDR:	US Agency for Toxic Substance and Disease Registry
UV:	Ultraviolet
UVB:	Ultraviolet B

V

VC:	Vinyl chloride
VCAM:	Vascular cell adhesion molecule
VER:	Visual evoked response
VHH:	Volatile halogenated hydrocarbon
VLA:	Very late antigen
VLDL:	Very low-density lipoproteins
VOC:	Volatile organic carbon compound
VSD:	Virtually safe dose
VTR:	Vibration threshold
v/v:	Volume per volume

W

WAIS:	Weschler Adult Intelligence Scale
WBC:	White blood cell
WG:	Water-dispersible granule
WHO:	World Health Organization
WISN:	Warfarin-induced skin necrosis
WP:	Wettable powder
w/v:	Weight per volume

X

XRF:	X-ray-generated atomic fluorescence
XRFS:	X-ray fluorescence spectroscopy

Z

ZPP:	Zinc protoporphyrin

Appendix 2   Toxicity Rating: Units of Measurements

Concept	Description of Unit	Abbreviation
Dosage to test animals or human volunteers/people	Milligrams of compound per kilogram of body weight	mg/kg
Concentration of storage in tissue or of residue in food or water	Parts of compound per million parts of tissue, food, or water by weight	ppm
Concentration in air	Milligrams of compound per cubic meter of air	mg/m^3
Concentration in formulation	Parts of compound per hundred parts of formulation (weight/volume)	Percentage (%)
Rate of application to a surface	Milligrams of compound per square meter of area	mg/m^2

Source: Dikshith, T.S.S. and Diwan, P.V. 2003. *Industrial Guide to Chemicals and Drug Safety.* Wiley, J. Hoboken, NJ.

Appendix 3 Conversion Tables for Units of Measures

Concentration in Tissue, Food, or Water[a]	ppm
1 ng/g	0.001
1 µg/100 g (µg %)	0.01
1 µg/kg	1
1 µg/g	1
1 µg/100 mg	10
1 mg/100 g (mg %)	10
1 grain/pound	142.9
1 mg/g	1,000
1% (concentration)	10,000
1 g/pound	2,204.6
1 oz/100 lb	624
1 oz/60 lb bushel	1,041.6

Concentration in Air[b]	mg/m³
1 µg/L	1
1 g/1,000 cu. ft.	35,315
1 ppm (of compound in air by volume)[c]	

[a] The concentration [C] of a molar solution expressed as ppm is C = [molecular weight (g/L)] × 1000.

[b] The concentration [C] of a saturated vapor expressed as mg/m³ at any given temperature is C = 53.8 × vapor pressure × molecular weight when the vapor pressure at the same temperature is expressed as mm Hg.

[c] This is an expression for concentration frequently used in industrial hygiene. It is based on the assumption that the material in question exists as a gas or vapor and expresses the number of volumes of compound per million volumes of air. The value may be calculated by the formula:

$$\text{ppm} = \frac{\text{Observed concentration (mg/L)} \times 24{,}450}{\text{Molecular weight of compound}} = \frac{\text{mg/m}^3 \times 24.45}{\text{Molecular weight}}$$

The figure, 24,450 mL, is the gram molecular volume of a gas at a pressure of 760 mm Hg and a temperature of 25°C.

Concentration in Formulation		
1 pound/gallon	119.8	G/L
1 pound/gallon	12.0	% w/v

Rate of Application		
1 pound/acre	10.4	mg/sq. ft.
1 pound/acre	112.1	mg/m²
1 pound/acre	1.121	kg/ha
1 mg/square foot	10.8	mg/m²

Appendix 4 *Toxicity Rating of Chemical Substances*

Toxicity Rate	Dose	Adult Average Measure
Practically non-toxic	>15 g/kg	More than one quart
Slightly toxic	5–15 g/kg	Between a pint and quart
Moderately toxic	0.5–5 g/kg	Between an ounce and a pint
Very toxic	50–5000 mg/kg	Between a teaspoonful and ounce
Extremely toxic	5–50 mg/kg	Between seven drops and a teaspoonful
Super toxic	<5 mg/kg	A taste (less than seven drops)

Source: Gleason, M.N., Gosslin, R.E., Hodge, H.C. and Smith, R.O. *Clinical Toxicology of Commercial Products: Acute Poisoning*, 3rd edition. Williams and Williams, Baltimore, MD, 1969; Dikshith, T.S.S. and Diwan, P.V. 2003. *Industrial Guide to Chemicals and Drug Safety.* Wiley, J. Hoboken, NJ.

Note: Probable oral lethal dose for a human adult.

Appendix 5   WHO Classification of Pesticide Hazards

LD50 for the rat (mg/kg body weight)

Class[a]	Oral Route		Dermal Route	
	Solid	Liquid	Solid	Liquid
I Extremely hazardous	4 or less	20 or less	10 or less	40 or less
I Highly hazardous	5–50	20–200	10–100	40–400
II Moderately hazardous	50–500	200–2000	100–1000	400–4000
III Slightly hazardous	Over 500	Over 2000	Over 1000	Over 4000

[a] The terms solid and liquid denote the physical state of the product/formulation under class.

Appendix 6 Material Damages Caused by Air Pollution

Materials	Air Pollutants	Effects
Metals	SO_2, acid gases	Corrosion, spoilage of surface, loss of metals, tarnishing
Building materials	SO_2, acid gases, particulates	Discoloration leaching
Paint	SO_2, H_2S	Discoloration
Textiles	SO_2, acid gases	Deterioration reduced tensile strength and fading
Textile dyes	NO_2, ozone	Deterioration reduced tensile strength and fading
Rubber	Ozone, oxidants	Cracking weakening
Leather	SO_2, acid gases	Disintegration powdered surface
Paper	SO_2, acid gases	Embitterment
Ceramics	Acid gases	Change in surface appearance

Source: Dikshith, T.S.S. and Diwan, P.V. 2003. *Industrial Guide to Chemicals and Drug Safety*. Wiley, J. Hoboken, NJ.

Appendix 7 Major Global Chemical Disasters

With the enlargement and ambitious growth of chemical industries around the world, in order to meet national and international needs and requirements many kinds of chemical substances are manufactured, stored, and transported, often with less care and few precautions. Global productivity of different chemical substances is now several hundred millions tons annually?. A large number of chemical substances so produced are known to be toxic and potentially hazardous to humans and the environment. Instances indicate that chemical accidents have occurred all over the world because of negligence and improper management of chemical substances. This has led to disasters where human health and environmental safety have been in jeopardy. The ecological and environmental catastrophe of 1986 in Schweizerhalle, Switzerland, is an example. A huge volume of industrial and toxic chemical waste was dumped into the Rhine River. Many more such incidents could be listed as global chemical disasters. The following are some of the well-documented environmental disasters involving chemical substances. More in literature.

Year	Location	Chemical Substance	Deaths	Injuries
1944	Cleveland, USA	–	–	–
1974	Decatur, USA	Propane explosion	7	152
1974	Flixborough, UK	Explosion in caprolactum plant	28	89
1975	Beek, the Netherlands	Propylene explosion	14	107
1976	Seveso, Italy	Dioxin/TCDD	–	193
1977	Chicago, USA	Hydrogen sulfide release	8	29
1978	Santa Cruz, Mexico	Methane fire	52	–
1978	Xilatopec, Mexico	Gas explosion in transit	100	150
1978	Los Alfaques	Propylene transfer	216	200
1979	Three Mile Island, USA	Nuclear reactor accident	–	–
1979	Novosibirsk, USSR	Chemical plant accident	300	–
1980	Sommerville, USA	PCl_3 accident	–	300
1981	Tacoa, Venezuela	Oil explosion	145	–
1982	Taff, USA	Acrolein explosion	–	–
1984	Sao Paulo, Brazil	Petro pipeline explosion	508	–
1984	Ixhuatepec, Mexico	LPG tank explosion	452	4248
1984	Bhopal, India	Leakage in pesticide plant	2500	–
1986	Chernobyl, USSR	Nuclear reactor accident	325	300
1986	Devnya, Bulgaria	Fire in a chemical complex	17	19
1988	North Sea, UK	Piper alpha oil rig explosion	166	–
1990	–	–	–	–

Source: Dikshith, T.S.S., (Ed.). 2009. *Safe Use of Chemicals: A Practical Guide*. CRC Press, Boca Raton, FL. pp. 227–228. (Appendix. 3).

Appendix 8   *Chemical Substances, Global Movement, Transport, and Disasters*

Toxic Chemical	Country	Year
Crude oil, phenol	Markus Hook, USA	1975
Ammonia	Deer Park and Houston, USA	1976
Hydrogen bromide	Rockwood, USA	1977
Chlorine leakage	Youngstown, USA	1977
Propylene	San Carlos, Spain	1977
Oil, gas	Bantry Bay, Ireland	1979
Chlorine explosion	Mississauga, Canada	1979
Nitric acid leakage	Blythe, USA	1981
Chlorine	Montanas, Mexico	1981
Silicon tetra chloride	San Francisco, USA	1981
Butadine	Melbourne, Australia	1982
LPG explosion	Egypt	1983
Ammonia	Matamoros, Mexico	1984
PCB	Kenora, Canada	1985
Gasoline	Tamil Nadu, India	1985
Phosphoric acid	Maimisburg, USA	1985
Lead oxide	Hemel Hampstead, UK	1985
Phosphorous oxychloride	Pittsburg, USA	1987
Chlorine	Annan, USSR	1987
Gasoline	Herborn, Germany	1987
Explosives explosion	Arzamas, USSR	1988
Pesticides	Chakhnounia, USSR	1988
Explosives	Sverdiorsk, USSR	1988
Toxic chemicals storage	Shenzhen, China	1993

Sources: UNEP compilation, *Manual on Emergency Preparedness*, Ministry of E&F, Government of India. Workshop on Transportation of Dangerous Goods, New Delhi, India October 18, 2001.

Appendix 9 Effects of Alcoholism (Excessive Drinking) and Vehicle Driving

- Measurable amount of alcohol in blood
- Inability to deal with crisis
- Tendency for own accidents
- Excessive caution
- Increased distractibility
- Irregular tendency in road driving
- Wrong judgment while driving
- Color blindness at traffic signal points
- Prone to accidents→human fatalities

Appendix 10   Chemical Substances, Industries, and Pollution

As members of the global society, we are living in a time of globalization and experiencing the effects of global pollution. Today, no country in the world can enjoy a pristine environment because irreparable pollution problems are spreading far beyond the consumer market.

Many sites of the massive industrialization that took place in the twentieth century in North America, Europe, and elsewhere are now trickling residues of toxic chemical substances into the environment. Owing to outdated technologies, lack of pollution control measures, and improper disposal of industrial waste materials, organic chemical substances and mercury from petrochemical and industrial complexes are contaminating our cities. In the mining industry, waste rock and untreated water from chromite mines have had significant long-term effects on local water sources and the air and soil systems. Large numbers of industrial estates, including chemical weapons and industrial manufacturing sites, discharge heavy metals, pesticides, and toxic chemical substances of lead and mercury. Synthetic organic pesticides are compounds that include insecticides, fungicides, and other pests that inhibit human conduct. Mining and smelting operations have devastated areas with particulates and heavy metal pollution. Of these pollutants, chlorophenoxy acids, organophosphates, carbamates, and chlorinated hydrocarbons are the most dominant.

- Aluminium smelters produce a quantity of fluoride waste that, unless carefully controlled, can be very toxic to vegetation around the smelter plants. A major environmental hazard associated with integrated steel mills is the pollution produced in the manufacture of coke, which is an essential intermediate product in the reduction of iron ore in a blast furnace.

- Caustic soda is a caustic metallic base mostly used as a strong chemical base in the manufacture of pulp and paper, textiles, drinking water, soaps and detergents, and as a drain cleaner.

- India is the second largest cement producing country. The main source of pollution in the cement industry is dust emissions. During the process of cement manufacture, a considerable amount of dust is emitted at almost every stage. Other pollutants include gases, noise, and vibration when operating machinery and during blasting in quarries, and damage to countryside from quarrying. It is believed that the cement industry produces about 5% of global man-made CO_2 emissions, of which 50% is from the chemical process, and 40% from burning fuel.

- Copper smelters emit staggering amounts of toxic pollution, including lead, arsenic, and selenium.

- The main source of pollution from distilleries is the spentwash, which is a byproduct of the fermentation and distillation of molasses. The disposal of these effluents from distilleries into rivers and streams is a matter of major environmental concern.

- The need for wastewater management and pollution control can hardly be over-emphasized in the pesticides industry, which has a diverse raw material-product mix and complex nature of pollutants.

- Tanning is the process of making leather, which does not easily decompose, from the skins of animals, which do. Often this process uses tannin, an acidic chemical compound. Other hazardous chemicals discharged by the leather industries include arsenic, cadmium, mercury, nickel, and chrome VI. These pollutants contaminate underground water streams and have a devastating impact on the environment and public health.

- The pesticides industry includes a large group of diverse products. It has been identified as a highly-polluting industries needing pollution control as a priority.

- Waste from petrochemical manufacturing plants contains suspended solids, oils and grease, phenols, and benzene. Solid waste generated by petrochemical processes contains spent caustic and other hazardous chemicals implicated in cancer.

- The pulp and paper industry is one of the largest and most polluting industries in the world, causing water, air, and solid waste pollution. Other concerns include the use of chlorine-based bleaches and the resultant toxic emissions to air, water, and soil.

- The combustion of coal contributes the most to acid rain and air pollution, and has been connected with global warming. Owing to the chemical composition of coal, there are difficulties in removing impurities from the solid fuel prior to its combustion. Electricity generation using carbon-based fuels is responsible for a large fraction of carbon dioxide (CO_2) emissions worldwide. Another problem related to coal combustion is the emission of particulates that have a serious impact on public health. Also, coal contains low levels of uranium, thorium, and other naturally occurring radioactive isotopes whose release into the environment leads to radioactive contamination.

- The production of sulfide zinc ores produces large amounts of sulfur dioxide and cadmium vapor. Smelter slag and other residues of the process also contain significant amounts of heavy metals.

- Metallurgical manufacturing industries, namely, zinc, lead, copper, aluminum, and steel

- Paper, pulp, and newsprint

- Pesticides

- Refineries

- Fertilizers

- Paints

- Dyes, and color pigments

- Leather tanning and leather processing

- Rayon and synthetic fiber manufacturing

- Sodium/potassium cyanide associated industries

- Basic drugs manufacturing

- Foundry industries

- Storage batteries (lead-acid type)

- Acids/alkalies
- Plastics fabrication
- Rubber-synthetic manufacturing
- Cement processing and manufacturing
- Asbestos and associated operations
- Fermentation industry
- Electroplating industry

Appendix 11   Classification of Industries and Pollution Control

Industries with high pollution indices are classified under different categories, i.e. category RED, category ORANGE, and category GREEN. These industries have different capacities based on their production turnover, which classify them as large, medium, and small. These industries need periodical inspections to comply with the laws of the pollution control authorities of the concerned country. The inspections are covered under the following laws:

- The Water (Prevention & Control of Pollution) Act 1974
- The Air (Prevention & Control of Pollution) Act 1981
- The Water (Prevention & Control of Pollution) Cess Act 1977
- The Environment (Protection) Act 1986

All large-scale industries have been placed under category RED and the frequency of their inspection is once every six months. All medium-scale industries that have been placed under category RED have a frequency of inspection once every three months. All medium-scale industries placed under category ORANGE will have a yearly inspection. All medium-scale industries placed under category GREEN will be inspected every second year. All small-scale industries placed under the RED category will be inspected every six months. All small-scale industries under the ORANGE category will be inspected once a year; and all small-scale industries that have been placed under category GREEN will be inspected once every two years.

CATEGORY: RED

1. Lime manufacture (pending decision on proven pollution control device and Supreme Court's decision on quarrying)
2. Ceramics
3. Sanitary wares
4. Tyres and tubes
5. Refuse incineration
6. Large flour mills
7. Vegetable oils including solvent extracted oils
8. Soap without steam boiling process
9. Synthetic detergent formulations
10. Steam-generating plants

11. Manufacture of machineries, machine tools, and equipment
12. Manufacture of office and household equipment and appliances
13. Involving use of fossil fuel combustion
14. Industrial gases (only nitrogen, oxygen, and CO_2)
15. Miscellaneous without involving use of fossil fuel combustion
16. Optical glass
17. Petroleum storage and transfer facilities
18. Surgical and medical products including prophylactics and latex products
19. Malted food
20. Manufacture of power-driven pumps, compressors, refrigeration units, fire-fighting equipment, etc.
21. Acetylene (synthetic)
22. Glue and gelatine manufacture
23. Metallic sodium
24. Photographic films, papers, and photographic chemicals
25. Plant nutrients (manure)
26. Ferrous and non-ferrous metals extraction, refining, casting, forging, alloy making, processing, etc.
27. Dry coal processing/mineral processing/mineral processing industries like ore sintering, beneficiation, pelletization, etc.
28. Phosphate rock processing plants
29. Cement plants with horizontal rotary kilns
30. Cement plant with vertical shaft kiln technology—pending certification of proven technology on pollution control
31. Glass and glass products involving use of coal
32. Petroleum refinery
33. Petrochemical industries
34. Manufacture of lubricating oils and greases
35. Synthetic rubber manufacture
36. Coal, oil, nuclear, and wood-based thermal power plants
37. Hydrogenated vegetable oils for industrial purposes
38. Sugar mills industries
39. Kraft paper mills and related industries
40. Coke-oven by-products and coal tar distillation products
41. Alkalies, caustic soda, and potash manufacturing industries
42. Electrothermal products (artificial abrasives and calcium carbide) industries
43. Phosphorus and associated industries
44. Organic, inorganic acids, and their salts manufacturing
45. Nitrogen compounds, cyanides, cyanamides, and related compounds
46. Manufacturing of explosives, detonators, and fuses related industries

47. Phthalic anhydride manufacturing industries
48. Processes associated with the manufacturing of chlorinated hydrocarbons, chlorines, fluorine, bromine, iodine, and their compounds
49. Fertilizer industry
50. Paper board and straw board related industry
51. Manufacture of pesticides (insecticides, herbicides and the formulations)
52. Manufacture of basic drugs and formulations
53. Manufacture of alcohols industrial and potable
54. Manufacture of leather, tanning, and processing
55. Processes and operations associated with coke, coal liquification, and fuel gas making industries
56. Industries associated with the manufacturing of fiber glass and processing
57. Refractories
58. Manufacture of pulp, wood pulp, mechanical or chemical processing
59. Manufacture of pigments, dyes, and their intermediates
60. Manufacture of industrial carbons, graphite electrodes, anodes, midget electrodes, graphite blocks, graphite crucibles, gas carbons, activated carbon, synthetic diamonds, carbon black, channel black, and lamp black
61. Manufacture of electrochemicals (other than those covered under alkali group)
62. Manufacture of paints, enamels, and varnishes
63. Manufacture of polypropylene
64. Manufacture of polyvinyl chloride
65. Manufacture of chlorates, perchlorates, and peroxides
66. Manufacture of polishes
67. Manufacture of synthetic resin and plastic products

CATEGORY: ORANGE

1. Electroplating
2. Galvanizing
3. Manufacture of mirror from sheet glass and photoframing
4. Surgical gauze and bandages
5. Cotton spinning and weaving
6. Wires, pipes, extruded shapes from metals
7. Automobile servicing and repair stations
8. Restaurants
9. Ice cream
10. Mineralized water and soft drinks bottling plants
11. Formulations of pharmaceuticals
12. Dyeing and printing (small units)
13. Laboratory ware

14. Wire drawing (cold process)
15. Bailing straps
16. Steel furniture
17. Fasteners etc.
18. Potassium permanganate
19. Surface coating industries
20. Fragrance, flavors, and food additives
21. Aerated water/soft drink
22. Light engineering industry excluding fabrication, electroplating
23. Small textile industry
24. Plastic industry
25. Chemical industry
26. Readymade garment industry
27. Flour mills
28. Bleaching
29. Degreasing
30. Phosphating
31. Dyeing
32. Pickling
33. Tanning
34. Polishing
35. Cooking of fibers, digesting, desizing of fabric
36. Unhairing, soaking, deliming, and bating of hides
37. Washing of fabric
38. Trimming, cutting, juicing, and blanching of fruits and vegetables
39. Washing of equipment and regular floor washing, using large quantities of cooling water
40. Separating milk and whey
41. Steeping and processing of grain
42. Distillation of alcohol
43. Stilage evaporation
44. Slaughtering of animals, rendering of bones, washing of meat
45. Juicing of sugar cane, extraction of sugar
46. Filtration, centrifugation, distillation, pulping, and fermenting of coffee beans
47. Processing of fish
48. DM plant exceeding 20 kL per day capacity
49. Pulp making, pulp processing, and paper making
50. Coking of coal
51. Washing of blast furnace flue gases

CATEGORY: GREEN

1. Washing used sand by hydraulic discharge
2. Atta-chakkies
3. Rice millers
4. Ice boxes
5. Dal (pulses) mills
6. Groundnut (peanut) decorticating (dry)
7. Chilling
8. Tailoring and garment making
9. Cotton and woolen hosiery
10. Apparel making
11. Handloom weaving
12. Shoe lace manufacture
13. Gold and silver thread and zari (textile) work
14. Gold and silver smithy
15. Leather footwear and leather products excluding tanning and hide processing
16. Musical instruments manufacture
17. Sports goods
18. Bamboo and cane products (only dry operations)
19. Cardboard box and paper products (paper and pulp manufacture excluded)
20. Insulation and other coated papers (paper and pulp manufacture excluded)
21. Scientific and mathematical instruments
22. Furniture (wooden and steel)
23. Assembly of domestic electrical appliances
24. Radio assembling
25. Fountain pens
26. Polyethylene, plastic, and PVC goods through extrusion/molding
27. Rope (cotton and plastic)
28. Carpet weaving
29. Assembly of air coolers, conditioners
30. Assembly of bicycles, baby carriages, and other small non-motorized vehicles
31. Electronics equipment (assembly)
32. Toys
33. Candles
34. Carpentry—excluding saw mill
35. Cold storages (small scale)
36. Oil ginning expelling (no hydrogeneration and no refining)
37. Jobbing and machining
38. Manufacture of steel trunks and suit cases

39. Paper pins and U-clips
40. Block making for printing
41. Optical frames
42. Tyre retreading
43. Power looms and handlooms (without dying and bleaching)
44. Printing press
45. Garments stitching, tailoring
46. Thermometer—making
47. Footwear (rubber)
48. Plastic processed goods
49. Medical and surgical instruments
50. Electronic and electrical goods
51. Rubber goods industry

Source: Dikshith, T.S.S. and Diwan, P.V. 2003. *Industrial Guide to Chemicals and Drug Safety.* Wiley, J. Hoboken, NJ.

Appendix 12 Pesticides and Mammalian Toxicity

Group and Name Pesticide	Nature of Toxicity
Organochlorine Pesticides	
Aldrin	General malaise, anxiety, irritability, vomiting, convulsions
Benzene hexachloride	Hyperexcitability, neurological disorders, myoclonic jerks, aplastic anemia, hepatotoxicity, neurotoxicity, cerebral seizures
Chlordane	Generalized convulsions, reproductive toxicity, birth defects, loss of consciousness, change in EEG pattern, hepatic disorders, neurological disturbances, mutagenicity, carcinogenicity
DDT	Loss of weight, anorexia, tremors parasthesia, hepatotoxicity, reproductive toxicity, cancer
Dicofol	Nausea, vomiting, muscular weakness
DDD	Ataxia, confusion, abnormal walk, mild anemia
Dieldrin	Violent headache, muscular pain, reproductive toxicity, birth defects, cancer
Dimethoate	Cancer, mutagenicity, reproductive toxicity, birth defects
Endrin	Nausea, dizziness, headache, hyperexcitability, abdominal discomfort
Endosulfan	Agitation, diarrhea, foaming, vomiting, hyperplexia, muscle twitching, cyanosis, chronic toxicity
Heptachlor	Myoclonic jerking, psychological disorders, irritability, anxiety, carcinogenicity
Isodrin	Motor hyperexcitability, intermittent muscle twitching
Lindane (Gamma HCH)	Neurotoxicity, hepatotoxicity
Methoxychlor	Neurotoxicity, hepatotoxicity
Telodrin	Nausea, vomiting, hyperexcitability
Toxaphene	Loss of consciousness, epileptiform convulsions, carcinogenicity
Organophosphate Pesticides	
Azinphos-methyl	Neurotoxicity, depression, slurred speech
Bromophos ethyl	Neurotoxicity, depression, slurred speech
Chlorpyrifos	Neurotoxicity, depression, slurred speech
Crotoxyphos	Neurotoxicity, depression, slurred speech
Demeton	Mutagenicity, birth defects
Diazinon	Neurotoxicity
Dichlorvos	Neurotoxicity, depression, slurred speech
Dimethoate	Muscle weakness, respiratory distress
Ediphenphos	Dizziness, vomiting, nausea
Ethion	Discomfort, vomiting, muscular twitching, nausea, nervousness, convulsions
Fenitrithion	Tremors, fatigue, memory loss, lethargy

(continued)

(continued)

Group and Name Pesticide	Nature of Toxicity
Fensulfothin	Vomiting, diarrhea, muscular twitching, pulmonary edema, convulsions, coma
Fenthion	Muscle weakness, respiratory distress, neurotoxicity
Methamidophos	Muscle weakness, respiratory distress
Mevinphos	Mutagenicity
Monocrotophos	Muscle weakness, respiratory distress
Parathion (ethyl)	Headache, miosis, nervousness, salivation, diarrhea, respiratory distress, convulsions, coma, cancer, mutagenicity
Parathion methyl	Diarrhea, salivation, nervousness, respiratory distress, convulsions, chronic toxicity, mutagenicity
Phosmet	Cancer, mutagenicity
Phosphomidon	Respiratory distress, nervousness, diarrhea, convulsions, salivation, paralysis, coma
Quinolphos	Respiratory distress, nervousness, diarrhea, convulsions, salivation, paralysis, coma
Carbamate Pesticides	
Aldicarb	Extremely toxic even in very small concentrations
Carbaryl	Mutagenicity, nephronotoxicity
Chlorpropham	Mutagenicity
Fenvalerate	Cancer
Methomyl	Chronic toxicity, mutagenicity
Chlorophenoxy compounds	
2,4-D and 2,4,5-T	Nausea, dizziness, vomiting
Synthetic Pyrethroids	
Cypermethrin	Burrowing, sinous writhing
Deltamethrin	Clonic seizures
Fenpropanthrin	Dermal tingling
Fenvalerate	Profuse salivation, enhanced startle response
Permethrin	Prostration
Phenorthrin	Whole body tremor
Resmethrin	Enhanced startle response
Tetramethrin	Aggressiveness

Sources: Dikshith, T.S.S. and Diwan, P.V. 2003. *Industrial Guide to Chemicals and Drug Safety*. Wiley, J. Hoboken, NJ.

Appendix 13 Pesticides and Hormone Disturbances in Mammals

Organochlorines Compounds

Alachlor, Benzene hexachloride, Cycloodines: Aldrin, Chlordane, Dieldrin, Endrin, Endosulfan, Heptachlor, Isodrin, Telodrin and Toxaphene, DDT and metabolites, Dicofol, Dimethoate, Lindane, Methoxychlor, Mirex, Pentachlorophenol, Perthane

Organophosphorous Compounds

Azinphosmethyl, Bromophos ethyl, Chlorpyrifos, Crotoxyphos, Demeton, Diazinon, Dichlorvos, Ethion, Fenitrothion, Fensulfothin, Fenthion, Flusulfothin, Methamidophos, Mevinphos, Monocrotophos, and Dichrotophos, Oxamyl, Phorate, Parathion ethyl, Parathion methyl, Phosphomidon, Quinolphos Temephos

Carbamate Compounds

Aldicarb, Benomyl, Carbaryl, Chlorpropham, Fenvalerate, Methomyl

Chlorophenoxy Compounds

2,4-D, 2,4,5-T

Source: Dikshith, T.S.S. 2009. *Safe Use of Chemicals: A Practical Guide.* CRC Press, Boca Raton, FL.

Appendix 14 *Global Regulatory Agencies and Chemical Substances*

Source	Regulatory Act
Air	Clean Air Act
	National Ambient Air Quality Standards
	Air Toxics (HAPS)
	Motor Vehicle Emission Standards
Water	Safe Drinking Water Act
	Clean Water Act
Food	Food Quality Protection Act
	Federal Food, Drug and Cosmetic Act
Agrochemicals/pesticides	Federal Insecticide Fungicide and Rodenticide Act (FIFRA), United States
	Insecticide Act [Central Insecticide Board (CIB)] Ministry of Agriculture, Government of India, India Control of Pesticides Act, No. 33 of 1980
	Parliament of the Democratic Socialist Republic of Sri Lanka
Hazardous waste	Resources Conservation and Recovery Act
	Comprehensive Environmental Response
	Compensation and Liability Act
	Superfund Amendments and Reauthorization Act

Sources: US Environmental Protection Agency, United States. Food and Drug Administration, United States, 1999; Dikshith, T.S.S. and Diwan, P.V. 2003. *Industrial Guide to Chemicals and Drug Safety*. Wiley, J. Hoboken, NJ.

Appendix 15   Global Development of Pesticides

Period/Era	Chemicals and Location
Era of Natural Products	
900	Arsenites (China)
1690	Tobacco (Europe)
1800	Pyronthroids (Caucasus)
1848	Derris Root (Malaysia)
Era of Fumigants, Inorganics, Petroleum Products	
1854	Carbon disulfide (France)
1867	Paris Green (USA)
1892	Lead arsenate (USA)
1918	Chloropicrin (France)
1932	Methyl bromide (France)
Era of Modern Synthetic Insecticides	
1939	DDT (Germany)
1941	BHC (France)
1944	Parathion (Germany)
1945	Aldrin (USA)
1947	Dimetan (Switzerland)
1958	Sevin (USA)
Era of Hormone Mimics and Pheromones, Re-Birth of Botanical Insecticide	
1967	First Juvenile Hormone Analogue (USA)
Era of Microbial Insecticides	
1980	Avermectin Bia

Sources: Dikshith, T.S.S. and Diwan, P.V. 2003. *Industrial Guide to Chemicals and Drug Safety*. Wiley, J. Hoboken, NJ.
Dikshith, T.S.S. 2009. *Safe Use of Chemicals: A Practical Guide*. CRC Press, Boca Raton, FL.

Appendix 16   Signs and Symptoms of Insecticide Poisoning

Organ System	Signs[a]/Symptoms[b]
CNS, somatomotor	Twitch, tremor, ataxia, convulsion, rigidity, flaccidity, restlessness, general motor activity, reaction to stimuli, headache, dreams, poor sleep, nervousness, dizziness
Autonomic	Miosis, mydriasis, salivation, lacrimation
Respiratory	Discharge, rhinorrhea, bradypnea, dyspnea, yawning, constriction of chest, cough, wheezing
Ocular	Ptosis, exophthalmos, dimness, lacrimation, conjunctival redness
Gastrointestinal	Diarrhea, vomiting
General side effects	Temperature, skin texture and color, cyanosis

Source: T.S.S. Dikshith 2009. *Safe use of chemicals. A practical guide.* CRC press Bocoration., FL.

[a]   Signs in animals.

[b]   Symptoms in humans.

Appendix 17   Pesticides, Signs of Toxicity, and Parts of the Body Affected

Component	Site Affected	Toxic Signs
Parasympathetic (muscarinic)	Exocrine glands	Lacrimation, increased salivation
	Eyes	Miosis, blurred vision, "bloody tears"
	GI tract	Nausea, vomiting, diarrhea
	Respiratory trace	Bronchial secretions, rhinorrhea, dyspnea
	Cardiovascular	Tachycardia, decreased blood pressure
	Bladder	Urinary incontinence
Parasympathetic and sympathetic (nicotinic)	Cardiovascular	Tachycardia, increased blood pressure
Somatic motor (nicotinic)	Skeletal muscles	Fasciculations, ataxia, paralysis
Brain (ache receptors)		CNS lethargy, tremors, convulsions, dyspnea, depression of respiratory center, cyanosis

Appendix 18   Pesticide Exposures, Behavioral, and Non-Behavioral Changes

Behavioral
Anxiety and irritability
Depression
Memory deficit
Reduced concentration
Insomnia
Linguistic disturbance

Non-Behavioral
Tremor
Ataxia
Paralysis
Paraesthesia
Polyneuritis

Source: Dikshith, T.S.S. 2009. *Safe Use of Chemicals: A Practical Guide*. CRC Press, Boca Raton, FL.

Appendix 19   Classification of Carcinogens by Different Agencies

ACGIH[a]	EU[b]	GHS[c]	IARC[d]	NTP[e]
A1	Category 1	Category 1A	Group 1	KC[f]
A2	Category 2	Category 1B	Group 2A	RS[g]
A3	Category 3	Category 2	Group 2B	
A4			Group 3	
A5			Group 4	

[a] ACGIH: American Conference of Governmental Industrial Hygiene.
[b] EU: European Union.
[c] GHS: Global Harmonization System.
[d] IARC: International Agency for Research on Cancer.
[e] NTP: National Toxicology Program.
[f] KC: Known carcinogen.
[g] RC: Reasonably suspected carcinogen.

Appendix 20 Classification of Benign and Malignant Tumors in Mammals

Tissue	Benign Tumor	Malignant Tumor
Connective tissue		
Adult fibrous	Fibroma	Fibrosarcoma
Bone	Osteoma	Osteosarcoma
Cartilage	Chondroma	Chondrosarcoma
Embryonic fibrous	Myxoma	Myxosarcoma
Fat	Lipoma	Liposarcoma
Endothelium		
Blood vessels	Hemangioma	Hemangiosarcoma
Lymph vessels	Lymphangioma	Lymphangiosarcoma
Epithelium		
Glandular	Adenoma	Adenocarcinoma carcinoma
Squamous	Squamous cell papilloma	Squamous cell carcinoma
Transitional	Transitional cell papilloma	Transitional cell carcinoma
Hematopoietic		
Bone marrow	Not recognized	Leukemia
Lymphoreticular		
Lymph nodes	Not recognized	Lymphosarcoma
Muscle		
Skeletal muscle	Rhabdomyoma	Rhabdomyosarcoma
Smooth muscle	Leiomyoma	Leiomyosarcoma
Nervous system		
Glial cells	Glioma	Malignant glioma (glioblastoma)
Nerve sheath	Neurilemmoma	Neurogenic sarcoma

Source: United States Environmental Protection Agency, *Guidelines for Carcinogen Risk Assessment*, Federal Register, 51, 33992–34003, 1986.

Appendix 21 *Organochlorine Pesticides and Carcinogenicity*

Pesticides	IARC	NTP	USEPA	Type of Changes
Aldrin	3	–	B2	Mouse liver tumors
Chlordane heptachlor	3	–	B2	Mouse liver tumors
Kepone (chlordecone)	2B	E	–	Rat, mouse liver tumors
DDT	2B	E	B2	Mouse liver, lung tumors, lymphomas; rat liver tumors; no tumors in three hamster studies
Dieldrin	3	E	B2	Mouse liver tumors
Endrin	3	–	–	No evidence of tumor
Lindane	–	E	B2/C	Mouse liver tumors
Mirex	2B	E	B2	Mouse, rat liver tumors and thyroid tumors
Toxaphene	2B	E	B2	Mouse, rat liver tumors

Source: Dikshith, T.S.S. and Diwan, P.V. 2003. *Industrial Guide to Chemicals and Drug Safety*. Wiley, J. Hoboken, NJ.

Note: B2: probable human carcinogen (no human evidence); 2B: possibly carcinogenic to humans; C: possible human carcinogen; 3: not classifiable as to carcinogenicity in humans; E: reasonably anticipated to be carcinogenic to humans.

Appendix 22 *Regulatory Standings of Organochlorine Pesticides*

Pesticides	United States Regulations
Aldrin and dieldrin	USEPA cancelled use on all products
Chlordane, heptachlor	Voluntarily cancelled; used for the control of white ants on power transformers; existing stocks for termite control
DDT	Cancelled use on all products
Endrin	Voluntary cancellation on all products
Kepone	Cancelled use on all products
Lindane (gamma BHC)	Restricted use in plant nurseries, pet shampoos, livestock sprays, seed treatment, household sprays, flea collars, hardwood logs, avocados
Mirex	Cancelled use on all products; existing stocks used for ants on pineapples in Hawaii
Toxaphene	Cancelled use on all products; existing stocks used for cattle dip, pineapples in Puerto Rico, bananas in Virgin Islands; emergency use on corn, cotton, and small grains

Source: USEPA, Suspended, Cancelled, and Restricted Pesticides, NTIS, PB92-231240, 1990. R. T. Meister, *Farm Chemicals Handbook*. Meister, Willoughby, OH, 1992.

Appendix 23　Federal Laws: Chemical Substances

The United States has a complex set of chemical safety statutes and regulations that are administered by a number of federal agencies. The principal statutes are briefly described below.

Toxic Substances and Control Act (TSCA): Regulates industrial chemicals, including heavy metals. Identifies and controls industrial chemical hazards that are toxic to human health and the environment. Administered by the US Environmental Protection Agency (EPA).

Pesticides: In the United States, pesticides are regulated by a myriad of laws and agency rules. No less than 14 different federal acts control some aspect of the manufacture, registration, distribution, use, consumption, and disposal of pesticides. The bulk of pesticide regulation falls under the Federal Insecticide, Fungicide and Rodenticide Act (FIFRA). This legislation governs the registration, distribution, sale, and use of pesticides. The US EPA is responsible for the administration of FIFRA and for establishing rules and regulations consistent with the acts intent.

Federal Insecticide, Fungicide and Rodenticide Act (FIFRA): FIFRA makes it a violation to use a pesticide in a manner inconsistent with its label, including the specified uses. FIFRA was revised and strengthened substantially by the Food Quality Protection Act in August 1996. Administered by the EPA, requires the agency to register all pesticides sold in the United States. FIFRA requires the US EPA to balance the risks of pesticide exposure to human health and the environment against the benefits of pesticide use to society and the economy. A pesticide registration will be granted if, after careful consideration of the health, economic, social, and environmental costs and benefits, the benefits of the pesticides use outweigh the costs of its use.

Federal Food, Drug and Cosmetic Act (FFDCA): Regulates the establishment of pesticide tolerances (maximum residue levels). FFDCA was revised and strengthened substantially by the Food Quality Protection Act in August 1996. Administered by the US EPA and the Food and Drug Administration.

Food Quality Protection Act (FQPA): Amends both FIFRA and FFDCA to make a more consistent, protective regulatory system that is supported by sound science. It mandates a single, health-based standard for all pesticides in all foods and provides special protection for infants and small children. Several attempts have been made to address the problem in the United States through "Circle of Poison" legislation. These bills, none of which have been passed, were designed to stop companies from exporting banned and unregistered pesticides, as well as to introduce tougher testing standards to keep these pesticide residues from showing up in consumers' food. The Planning and Community Right-to-Know Act (EPCRA) requires local emergency planning for responses to industrial chemical or pesticide accidents; requires industries to notify their communities and states of releases; provides information from companies about possible industrial chemical or pesticide hazards in the facility's community; and mandates a national inventory of toxic chemical releases (Toxics Release Inventory, TRI). Administered by the US EPA.

Source: USDA/Extension Service/National Agricultural Pesticide Impact Assessment Program, Oregon State University, Oregon, USA.

Appendix 24   Insecticide Act 1968
Government of India, India

In 1968, the Republic of India enacted an act called the Insecticides Act, 1968, to regulate the import, manufacture, sale, transport, distribution, and use of insecticides with a view to preventing risk to human beings or animals, and for matters connected therewith. The Insecticides Act, 1968, extends to the whole of India. Subsequently, the Insecticides Act, 1968, was further amended in 2000 as the Insecticides (amendment) Act, 2000.

List of Insecticides

1. Acrylonitrile
2. Aldrine (1: 2: 3: 4: 10: 10=hexachloro-1: 4a, 5, 8, 8A; -hexhydro-1′: 4: 5: 8-dimethanonaphthalene)
3. [Allethrin and its stereo isomers] ally homologue of cinerin
4. Aluminium phosphide
5. Amiton
6. Antu (alpha-naphthyl thiourea)
7. Armaite [2p-tert-butyphenoxy) isopropyl 1-2 chloroethyl sulfite]
8. Barium carbonate
9. Barium fluro silicate
10. BHC (benzene hexachloride (1,2,3,4,4,6-hexachlorohexane))
11. Bis-dimethylamino flourophosphine oxide
12. Calcum arsenate
13. Calcium cyanide
14. Captan (N-trichloromethy mercapto-4-cyclohexane), 1,2-discarboximide carbaryl (1-naphthy-N-methyl carbamate)
15. Carbon disulfide
16. Carbon tetrachloride
17. Chlorbenside (p-chlorobenzyl-p-chlorophyenyl sulfide)
18. Chlorobis ethyl amino triznine
19. Chlorodane (1,2,3,4,5,6,7,8, 8-octachlore-2-3,3a-4-7,7a-hexahydro-4-7-methanoina-dance)
20. Chlorobenzilate (ethyl 4,4-dischaorbenzilate)
21. Choloro-IPC
22. Chloropeicrin
23. Chlorofenson (p-chlorophenyl-p-chlorobenzene sulfonate)
24. S-(p-chlorophenythio) methy-o-odiethyl phosphorodithioate (rithion)

25. CIPC [isopropyl-N 93-chlorophenyl) carbanamate]

26. CMU (Manuron)

27. Copper arsenate

28. Copper cynanide

29. Copper napthanate

30. Copper sulfate

31. Coumachlor [3-a-acetony]-4-chlorobenzyl-4-hydroxy coumarin)]

32. Copper oxychloride

33. Cuprous oxide

34. Dalapon (sodium 2,2, dichloropropionate)

35. D-D mixture

36. DDD (dichloro diphenyl dichloroethane)

37. DDT [a mixture of 1 1-trichloro-2-bis (p-chlorophenyl) ethane and 1-1, 1-trichloro-2 (o-chlorophenyl1-2 (p-chlorophenyl) ethane]

38. DDVP (2-dichlorovinyl dimethyl phosphate)

39. Demeton-O (O, O-diethyl-S[(2-ethylthiol-ethyl/phosphorothioate)

40. Demeton-O (S, O-diethyl-S-[2-ethylshiol-ethyl) phosphorothioate)

41. Diazinon (O, O-diethyl-O [e-isopropyl-methyl-4-pyrimidinyl] phosphorothioate)

42. Dibrom (1, 2-dibromo, 2,2,2-dichloroethyl phosphate)

43. Dichlorophenoxy acetic acid (2,4-D)

44. Dieldrin (1,2,3,4,10; 10-hexachloro-6, 7-epoxy-1, 4a: 5: 6: 7: 8: 8a octahydro-1: 4: 5: 8-dimethanonaphthalene)

45. Dimethoate (O, O-dimethyl-S (N-methylcarbamoyl methyl) phosphorodithioate)

46. Dipterex (O, O-dimethyl-2, 2, 2-trichloro hydroxy ethyl phosphonate)

47. DNOC (dinitro-ortho-compound) (3: 5-dinitro-o-cresol)

48. EDCT mixture (ethylene dichloride carbon tetrachloride mixture)

49. Ekatin

50. Endrin (1,2,3,4,10,10-hexachloro-6, 7-epoxy-1, 4, 4a, 5, 6, 7, 8, 8a octahydro-1, 4-endo-endo 5-8dimethanonphthalene)

51. EPN (O-ethyl-O-p nitriphenyl benzence thiopnosophonate)

52. Ethyoxy ethyl mercury chloride

53. Ethyl di-n-prophylthiolcarbamate (Eptam)

54. Ethyl mercury phosphate

55. Ethyl mercury chloride

56. Ethylene dibromide

57. Ethylene dichloride

58. Fenson (parchlorophenyl benzence sulfonate)

59. Fenthion (3-methyl-4-methyl thiophenyl phosphorothionate)

60. Ferbam (ferric dimethyl dithion carbamate)

61. Gusathion(O,O-dimethylS(4-oxo-1,23-benzotrizinyl-3-methylphosphorothionate)

62. Heptchlor (1,4,5,6,8,8-heptachloro-4-7-methano-3a, 4, 7, 7a-tetrahydoriindene)
63. HETP (hexaethyl tetraphosphate)
64. Hexachlorobenzene
65. Hydrogen cyanide
66. Hydrogen phosphide
67. Lead arsenate
68. Lime sulfur (calcium polysulfide, water-free sulfur, calcium thiosulfate mixture)
69. Lindane (gamma, BHC)
70. Malathion S-(1-2Bis (ethorxycarboyl ethyl), O-dimethyl-phosphoro-dithioate)
71. Maleichydrazide (1, 2 dihydropyropyridazine 3, 6-dione)
72. Menenb maganese ethylene bisdithiocarbamate)
73. MCPA (4-chloro-2 methyl phenoxy acetic acid)
74. Mercuric chloride
75. Metaldehyde
76. Metsystox
77. Methoxychlor (1,1,1-trichloro, 2,2-di-p-methoxyphenylethane)
78. Methoxy ethyl mercury chloride
79. Methyl bromide
80. Methyl dementon (Dmeton-methyl and Dimeton-methyl)
81. Methyl mercury chloride
82. Methyl parathion (O, O-dimethyl-O-p-nitrophenolthiophosphate)
83. Metox (chlorulficide)
84. Nabam (disodium ethylene-1, 2 bisdithiocarbamate)
85. Nicotine sulfate
86. Octa methyl pyrophosbhoramide
87. Para-dichloro benzene
88. Parathion (O, O-diethyl O-p-nitrophenylthiohosphate)
89. Paris Green (copper aceto arsenite)
90. Pentachloronitrobenzene (PCNB)
91. Pentachlorophenol
92. Phenyl mercury acetate
93. Phenyl mercury chloride
94. Phenyl mercury urea
95. Phosdrine
96. Phthalimidomethyl-O-O-dimethyl pshosphorodithioate (Imidan)
97. Piperony butoxide (butyl carbityl) (6-propyl piperonyl) ether O
98. Pival (2-pivalyl-indine 1-3-dione)
99. Potassium cyanide
100. n-Propyl ethyl-n-butyl thiocarbamate (Tillam)

101. Phyrethrines (active principles of *Chrysanthemum cinerariae folium*)
102. Rotenone
103. Ryania
104. Sodium fluoroacetate
105. Sodium cyanide
106. Sodium fluro silicate
107. Sulfur (wettable or colloidal sulfur)
108. Strychnine
109. Sulfoxide (1,2-methylene-dioxy-4(2-octylsulfinyl) propyl benzene)
110. TCA (tichlor acetic acid sodium and ammonium salts)
111. Tedion (tetrachlor diphenyl sulfone)
112. TEPP (tetraethyl pyrophysophate)
113. Tetrachloro-p-benzoquinone
114. Thanite
115. Thiram (bis(dimethyl thiocarbamyl) disulfide)
116. Tolyl mercury acetate
117. Trichlorphon
118. Thriothoresyl phosphate
119. Thallium sulfate
120. Thiometon
121. Toxaphene (chlorinated camphene containing 67%–69% chlorine)
122. Trichlorophenoxy acetic acid (e, 4,5-T)
123. Warfarin (3-a-acetonyl benzyl-4-hydroxy-coumarin)
124. Zinc phosphide
125. Zimet
126. Zineb (zinc ethylene bis-dithiocarbamate)
127. Ziam (zinc dimethyl dithiocarbamate)
128. Zulate
129. Acrokin 2 propenal or acryladehyde
130. Actellic (primphosemethyl) 2-diethylamino-6-methylpyrimidin-4-yl dimethyl phosphorothionate
131. Afugondiethylmethylexthoxycarbonylpyrazolopyrimidine-yl-phosphorothionate
132. Alachlor 2-chloro-2, 6-diethyle-n-(methoxymethyl)-acetanilide
133. Aldicarb 2-methyl-2-(methylthio) propional-dehyde-o-(methylcarbamoyl) oxime
134. Amidithion S-(N-2-methoxythyl-carlomoyl-methyl) dimethyl phosphorothiolothionate
135. Amitrole 3-amino-1,2,4-triazole
136. Ammonium sulfamate
137. Asulam methyl-N-(4-aminobenzenesulfonyl) carbamate
138. Atrazine 2-chloro-4-ethylamino-6-isopropylamino-1,3,5-triazine

139. Aureofungin

140. Azinhos-ethyl S-(3,4-dihydro-4oxobenzo-(d)-(1,2,3)-trinnine-3-yl-methyl) diethyl phosphorothio-lothionate

141. Barban 4-chloro-2-ynyl-3-chloriphenyl carbamate

142. Barium polysulfide barium polysulfide

143. Bassa O-secondary-butyllphenylmethyl carbamate

144. BCPE (chlorphenithel) 1,1-bis(4-chlorophenyl)—ethanol

145. Bemomyl methyl-N-benzimidazol-2-yl-N-(butylearbemoyal carbamate)

146. Bensulide -S (O,O-di.isoproplyl phosphorodithionate) cester with N-(2-mercaptoethyl) benzene sulfenamide

147. Binapacryl 2-(1-methyl-n-propyl)-4, 6-dinitr phenyl-2-2-methylcrobonate

148. Bromacil 5-bromo-6-methyl-3-(1-methylpropyl) uracil

149. Bromopyrazen 5-amino-4-bromo-2-phenylpyridazin-3-one

150. Bromoxynil 3,5 diboromo-4-hydroxybenzonitrile

151. Brozone methyl bromide and chloropicrin in petroleum solvent

152. Buturon 3-(4-chlorophenyl) 1-methyl-1-(10methyl prop-2-ynyl) urea

153. Butylate S-ethyl-N, N-disobutylthiocrabamate

154. Cadmium-based compounds (cadmium chloride, cadmium sulfate, cadmium succinate)

155. Captafel N-O(1, 1, 2, 2-tetrachlorotheylthio) cyelohex-4-one-1, 2-diocarboxymide

156. Carbefuran 2,3-dihydro-2, 2-dimethyl-7-benezofurmyayl methylcarbamate

157. Carbophenothion S[(p-chlorophenylithio)]-methyl, o-diethyl phosphordithiotate

158. Carboxin (DCMO) 5,6-(dihydro-2-methyl-1-1,4-oxathin-3-car boxanilide

159. Chinomethionate 6-methyl-2-oxo-1,3-dithio (4, 5-b) quinoxaline

160. Chloramben 3-amine-2,5-dichlorobenzonic acid

161. Chlorobufam (BIPC) 1-methyl-2-propynyl-m-chlorocarbonilate

162. Chlorfenyinphos 2-chloro-1 (2,4-dichlorophenyl)-vinyl diethylphosphate

163. Chloromequat chloride (2-chloroethyl) trimethylammonium chloride

164. Chloroneb 1,4-dichloro-2,5-dimethexybenzine

165. Chloropropane

166. Chloroxur-n N'-4-(-4 chlorophenoxy) phenyl-N N-dimethyl-ureat

167. Citicide chlorinated turpene

168. Citowett alkyl aryl polyglycol ether

169. Clonitralid 5, dichloro-4-'nitro-salicylic-anilide-ethanolamine

170. Copper hydroxide

171. Coumafuryl 3-(a-acetonylfurfuryl)-4-hydrocyceumarin

172. Coumaphos 3-chloro-4-methyl-7-coumarinyl diethyl phosphorothionate

173. Coumatetralyl 4-hydroxy-3(1,2,3,4-tetraphydrol-naphthyl) coumarin

174. Coyden 3,5-dichloro, 6-dimethyl-4-pyridinol

175. CPAS 4-chlorophenyl, 1,2,4,5-trichlorophyny-lazosulfide

176. Cyclomoph N-cyclodedecyl-2, 6-dimethyl-morpholinacetate
177. Cyclurin (OMU) N'-cycle-octyl-N-dimethyl-urea
178. Cytrolane 2-(diethoxy phosphraylimino) 4-methyl-1,3-dithiolane
179. Decarbofuran 2,3-dihydro-2-methylobenzofuren-7-yl-methyl carbamate
180. Decazolin 1-(alpha, alpha-dimethyl-beta-acetoxy-propionly-3-isopropyl-2,4-dioxodeoa-hydroquinzaoline)
181. DEET N, N-diethyl-m-toluamide
182. Dibromochloropropane 1, 2-dibromo-3-chloropropane
183. Dicamba 3, 6-dichloro-2-methyoxybenzoic acid
184. Dichlorbenil 2, 6-diclorobanza nitrile
185. Dichlofenthion 0-(2, 4-dichlorophenyl) O, O-diethyl phosphorothioate
186. Dichlone 2, 3-dichloro-1, 4-naphthoquinanone
187. Dicoloropropane 1, 3-dichloropropane
188. Dicloran 2, 6-dichloro-4-nitroaniline
189. Dicofol 2, 2, 2-trichloro-1, 1-di-(4-chlorophenyl) ethanol
190. Dicrotophos dimethyl phosphate ester with (E)-3-hydroxy-N, N-dimethyl-cis-crotonamide dimethyl phosphate
191. 2, 4-DB 4-(2, 4-dichlorophenoxy) butyric acid
192. Difenphos (Abate) O, O, O', O'-tetramethyl O, O'-thiodi-p-phenylene phosphorothiaote
193. Dikar A blend of Ditghane M-45 and Tech. Karathane
194. Dimas (Alar) N-dimethylamino succinic acid
195. Dinocap mixture of 4 and 5 parts of 2,4-dinitro-6-octophyany crotonates to 2 parts of the isomer 2,6 dinitro-4-octyl phenyl crotonates
196. Dinoseb 4-dinitro-6-s-butylphenol
197. Dinoseb acetate 2, 4-dinitro-6-5-butyl phenol acetate
198. Dioxathion S-S-1, 4-dioxane-2, 3-ylidene (bis, (o, o-diethyl) phos phosphorothiolo-thionate
199. Diphacinene 2-diphenylacetyl 1, 3 indanediene
200. Diphenamid N,N-dimethyl-2, 2-diphenylacetamide
201. Disulfoton-diethyl s-1(2-2thylthio) ethyl) phosphorothiolothionate
202. Diuron N'-3, 4-dicholorophenyl) NN-dimethyl urea
203. DMPA O-(2, 4-dicholorphenyl) O' methyl N-isopropyl)-phosphoroamidithicate
204. Dodine didecylguanidine-monoacetate
205. Dodomorph 4-cyclododeeyl-2, 6-dimethyl-morpholine
206. Drat (chlorophacinone) 2-(a-chlophenyl-a-pheylacetyl) indane-1, 3-dione
207. DSMA (sisodiun methanearsonate)
208. Dursban- O, O-diethyl)-(3,5, 6-trichloro-2 pyridyl) phosphorothioate
209. Dusting sulfur
210. Ediphenphos O-ethyl-s, s-diphenyl-dithiophosphate

211. Endosulfan 6,7,8,9,10, 10-hexachloro-1, 5, 5a, 6,9, 6a hexahydro-6, 9-methano-2, 4 3-benzo (e)- ,dioxatheiepin-3-oxide

1. Endothall 7-oxabicylo (2, 2, 1)-heptane-2, 3-dicarboxylate
2. EPTC S-ethyl-dipropylthiocarbamate
3. Erobn 2-(2, 4, 5-trichlorophenoxy) methyl 2, 2-dicholoropionate
4. Ethion Tetraethyl SS; methylene his phosphorothioate
5. Ethrel 2-chloroethane phosphonic acid
6. Fenac sodium 2, 3, 6-tricholorophenylacetate
7. Fenazeflor phenyl, 5, 6-dichloro-2-triflouromethyl benzimidazole-1-carboxylate
8. Fenitrothion dimethyl 3-methyl-4-nitrophenyl phosphorothioate
9. Fensulfothion diethyl 4-(methyl sulfinyl) phenyl phosphorothionate
10. Fentinacetate triphenyltin acetate
11. Fentin chloride triphenyltin chloride
12. Fentin hydroxide triphenyltin hydroxide
13. Folex SSS-tributyl phosphorotrithioate
14. Formethion S-(N-formethylcarbamoyl methyl) OO-dimethyl phosphorodithioate
15. Fonofos (Dyfonate) O-ethol-s-phenyl ethyl phosphorodithioate
16. Fujithion O, O-dimethyl-s-parachlorophenyl phosphorothis
17. Gibberellins Gibberellic acid
18. Indole acetic and butyric acids
19. Ioxynil (Pantrol) 3, 5-di-iodo-4-hydroxy benzethitrile
20. Isobenzan 1, 3, 4, 5, 6, 7, 7, 7-octachloro 1, 3, 3a, 4, 7, 7a-hexahydro-4, 7, methanois olbenzofuran
21. Isononuron N; (hexahydro-4, 7, -methanoindan-lyl-) NN dimethyl urea
22. Kitazin O-O-di-isopropyl-s-benzyle thiophosphate
23. Lenacil 3-cyclothexyl-5, 6-trimethylenuracil
24. Linuron N-(3, 4-dichlorophenyl)-N-methoxy-N-methyl urea
25. Lucel 5, 6, 7, 8-tetrachloroquinoxaline
26. Machete (Butachlor) (2-chloro-2, 6, -diethyl-N-(butoxmethyl)-acetanilide)
27. MCPB 4 (4-chloro-2-methylphenoxy) butyric acid
28. Menazon S-(4-6-diamino-1, 3, 5-triazin-2-yl methyl) dimethyl=phosphorothiolotionate
29. Methamidophos O-S-dimethyleaster amide of thiophosphoric acid
30. Metam sodium N-methyldithiocarbamic acid
31. Methomyl S-methyl N-(methylolcarbamyl) oxy thioacetimidate
32. Methylmetiram ammonium complex with Zn-(N' N 1, 2-prophylenebis-(dithiocarbamate and N', N'-poly-1, 2-propylene-bis (thiocarbamoyl)-di-sulfide
33. Metiram ammonium complex with Zn-(N', N-1 2-ethylenebis, (dithiocarbamate) and N-N-poly-1, 2-propylene-bi (thiocarbamoyl)-di-sulfide
34. Metoxuron N (3-chloro-4-methoxyphenyl)-N, N-demethyl urea

35. MIPCIN 2-incorpylphenyl-N-methyl carbamate
36. Menap O-ethyl S, S-dipropyl phosphordithioate
37. Molinate S-ethyl-N-hexaphydro-I hazepinuthiol-carbamate
38. Monocrotophos 3-hydroxy-N-methyl-crotonamide dimethyl phosphate
39. Monolinuron N-(4-chlorophenyl N-Methoxy-N-methyl urea
40. MSMA Monosodium methaearsonate
41. Neled 1, 2, dibromo-2, 2-dichloroethyl dimethyl phosphate
42. Naphythylacetic acid and its derivations
43. Naburon 1-butyl-3-(3, 4-dichlorophenyl)-1-methyl urea
44. Nemafos Thinozim O, O-diethyl C-2 pyrazinyl phosphorothioate
45. Neopyanmin 3, 4, 5, 6-tetrahydro-phthalimidomethyl chrysauthamate
46. Nickel Chloride
47. Nitrofen 2, 4-dichlorophenyl 4- nitrophenyl ether
48. Omethotae dimethyl 8-(n-methyl-carbamoyle methyl) phosphorothioate
49. Orthane O, S-dimethyl N-acetyl phospho-ramidothioate
50. Oxapyrazon (5-bromo-1, 6-dihydro-6-oxo-1-phenyl-4-pyridazinyl oxamicacid-compound with 2 dimethyl aminoethanol (1 & 1)
51. Oxycarboxin (DCMCD) 5, 6, dihydro-2-methyl-1, 4-oxathin-3-carboxanilide 4, 4-dioxide
52. Paraquat 1, 1-pimethyal-4, 4-bipyridyliumion
53. Pebulate 1, propyl-butyl-ethlocarbamate
54. Phenthoate S-aethoxycarbonylbanzyl-O, O-dimethyl phosphore rodithioate
55. Phorate diethyl S-(ethylthizmethyl) phospho-mthiolithicate
56. Phosalone S (6-echtoro-2-oxabenzoxolin-3yl) methyl-O, O-dietyl-phosphorothioate
57. Phosphamidon 2-chloro-2-diethylcar bamoyl-I-methyl vinyl dimethyl phosphate
58. Phosphorus paste
59. Phosmet (Imidan) O, O-dimethyl-S-phtballimide-methyphos-phorodithioate
60. Phosvel (Leptophos) pheynlglyoxylonitrite oxime O-O diethyl phosphorothioate
61. Picloram 4-amino-3, 5, 6, tricholoroplocolinic acid
62. Plictran tricyclohexyl tinhydroxide derivatives
63. Pronamide (Kerb) 3, 5-dichloro-N (1, -dimethyl-2 propynyl) benexamide
64. Propanil 3, 4, -dichloropropionanilide
65. Propargite (Omite) Prop-2-ynyl phenoxy)-eycohexyl sulfite
66. Propineb O-isoproposyphenylmethyl carbamate
67. Prynachlor O-N-butyn-(1) y; choloroacetanilide
68. Pyracarbolid 2-methyl-5, 6-dihydro-4-H-pyran-3-carboxylic anilde
69. Pyrazan (PCA) 5-amino-4-chloro-2-phenyl-3-pyridazone
70. Quinalphos O, O-diethyl quinoxalin-2-yl phosphorothioate
71. Rabicide 4, 5, 6, 7, -tetrachlorophalide
72. Ro-Neet S-ethyl N-ethyl-N-cyclothexyl-thiocarbamate

73. Ronnel O, O-dimethyl (O-2, 4, 5-trichlorophenyl) phosphorothioate 4-421 octa-chlorodipropyl ether

74. Sclex 3, -(3, 5-dichlorophenyl)-5, 5-deimethyl oxazo-lidinedione 2, 4

75. Simazine 2-chloro-4, 6-bis (ethylamine) s-trazine

76. Sindone A 1,1,4-dimethyl-4,6-di-isopropyl-indanyl ethyl ketone

77. Sindone B 1,1,4-dimethyl-4, 6-di-isopropyl-5-indanyl ethyl ketone

78. Sirmate 3, 4-and 2, 3-dicholobenzyl N-methyl carbamate

79. Swep methyl 3, 4 dichlorocarbanilate

80. Tar acid complex phenolic compounds or tar oil or creosotes

81. Travron G 2,2,2-trichlorethyl stryrene

82. Tecnazene 1,2,4,5-tetrachloro-3-nitrobenzenes

83. Terbacil 3-t-butyl-5 chloro-methyluracil

84. Tetrachlorvinphos (Gardona) 2-chloro 1-(2-4,4-trichlorophenyl) vinyl dimethyl phosphate

85. Tetram O, O-diethyl S-2 diethylamino ethyl phosphorothioate hydrogen oxalate 2,4,5-TB 4-(2, 4, 5-trichlorocophenoxyl) –butyric acid

86. Thiadizinthion (Terracur) 5-carboxymathyl-methyl-2 H-1, 3, 5 thiadiazin 2 thione

87. Thiophanate-M 1, 2-di-(3-methoxy-carbonyl-2 thiouried) benzene

88. Tranid exo-3-chloro-endo-6-cyano-2 norborman O-Ene-l-methyl-carbamoyl) oxime

89. Triallate S-2, 2, 3-trichloroallyl dis-lopropylthic-carbamate

90. Tridemorph 2,6-dimethyl-4-tridecyl morpholone

91. Tunic 2,(3, S-dichlorophenyl)-4-methyl lm 2, 4-oxadiazo-linine-3, 5, dione

92. Udonkor N-(beta-cynanothyl) monochloroacetamide

93. Vamidothion O, O-dimethyl-s-2-1, methyl-8-carbamoyl ethylthio ethgyl) phosphorothioate

94. Vegetta ethylene thiuram monosulfide

95. Vomzlate S-propyl NN-diprophyl thiolcarbamate

96. Zetran 4-dimethylamino-3, 5-xyle-N-methylcarbamate

97. Basagram 3-isopropyl-1H-2, 3-benxathiadixin 4-(3H)-one, 2-2dioxide

98. Basalin N-2-chloressthyl) N-propyl-trifhcoro 2, 6, dinitro-p-toultdine

99. Bavistin 2-(methoxy-carbamoyl) –benzimidazole

100. Campogram M 2,5-dimethyl furan 3-carbanic acid anillide and 320 gms og (Zinc)

101. Trifluralin 2,6-dinitro-N N-dipropyl-4-triflouromethyl aniline

102. Flumeturon N-(3-trifluro-methylphenyl)-N′N′-dimethyl urea

103. Metabromuron N-(P-bromophenyl)-N′-methyl-N-methexy urea

104. Mancozeb zinc manganese ethylene-bis-dithiocarbamate

105. Methabenzthiazuron 1,3-dimethyl-3 (2, benzothiozoiyl)-urea

106. Streptomycin

107. Tetracyclines

108. Terbuteryne 2-tert, butylamino 4-ethylamino-6-methylio, S-triazine)

109. Glusophosate [N-1(phosphonomethyl) glycine] present as isoprophylamine salt]
110. Bacillus thuringlenisis
111. Benthiocarb S-(4-chlorobenzyl)-N, N-diethylthiol carbamate
112. Cypermethrin x-cyano-3-phenoxybenzyl-2, 2-dimethyl-3-(2-dichlorovinyl) cyclopropane carboxylate
113. Decamethrin (S)-x-cyano-m-phenoxybenzyl (IR-3R) 3-(2, 2-dibromovinyl) dimethyl cycloprompane carboxylate
114. Fenvalerate x-cyano-m-phynoxybenzyl-x-isspropyl-pchlorphynel-acetate
115. Permethrin 3-phenoxy benzyl (cis-trans-3-(2, 2-dischlorvinyl)2, 2, dimethyl-cyclopropane carboxylate
116. Tetrachloroisophthalonitrile 1,2,4 trichloro-3, 5-dinitrobenzene]
117. Bromophos 0-(4 bromo 2, 5 dichlorcophenyl) OO-dimethyl phosphorothioate
118. Bromobhes ethyl 0-(4 bromo 2, 5dichlorophenyl) OO-di-ethyl phosphorothioate
119. Cartap 1, 3-di (carmolythio) 2-dimethyl-aminopropane
120. Dichloprophen (Antiphen) 5, 5-dichloro 2, 2, dihydroxy diphenyl methane
121. Dinobuton isopropyl-2 (1-methyl-n-propyl) 4, 6-dintrophenyl carbonate
122. Dithianon 2, 3-dicyano-1, 4-dithia-=antharquin-one
123. Ehyl formate thyl formate
124. Fenfuram 2-methyl-furan 3 carboxanilide
125. Glyphosine N, N-bis (phosphonomethyl) glycine
126. Guazatine 1, 17-diguanidino-9-azahepta decane
127. Isofanphos O-ethyl-o-(2-isproproposy-carbonyl) phenyl isopropyl phosphoramidothioate
128. Isoprofuron N, n-dimethyl-N-4-isopropyl phenyl urea
129. Magnesium phosphide
130. Matribuzin 4-amino-6-tort, butyl-3 (methykl thio, 1,2,4 triazine-5-(4H) one
131. Triforine 1, 4-tridzon-1-yl)-2-butanone
132. Triforine 1, 4,-di (2, 2, 2-trichloro-1-formami doethyl) piperazine
133. Vacor N-3 pyridyl methyl-N-p-nitrophenyl urea]
134. Amitraz 2-methyl-1,3-di (2-4xyly-limino) 2-azapropane
135. Bendiocarb -2,2-dimethylbenzo-2,3-dioxon 4-ylmentyl carbamate
136. Benzoylpropethyl - ethyl N-benzoul-N-(3, 4-dichlorophenyl) DL alaninate
137. Cyanazine -2-(4-chloro-6-ethylamino-1, 3, 5-trianzin- 2-ylamino)-2-methly propionitrile
138. Etrimofos -O-(6ethhoxy-2-ethyl-4pyrimidinyl) O,O-dimethyl phosphrothioate
139. Glyodin -2-heptadecyl-3, 4-dihydro-1, H-imidazolyl acetate
140. Oxadiazon -3-(2,4-dichloro-5-(1-methylethoxy) phenyl)-5 (1,1-demethylethyl)-1, 3, 4-oxadiazol-2 (3H)-one
141. Oxyfluorfen -2-choloro-1-(3-ethoxy-4-nitrophinoxyl 4-trifluormethyl) benzene
142. Propetamphos -(E)-O-2, isopropoxycarbonyl-1 methyl vinyl-O-methyl ethlyphosphoramidothiote

143. Scilliroside -3-B, 6B-6-acetyloxy-3 (B-D-glucophyranosylozy)8, 14-dihyroxy bufa-1-20,22, trinolide

144. Thiocyclam (hydrogenoxalate)- N, N-dimethyl-1, 23-trithian-5-amine hydrogenoxalate

145. Viclozolin -3-(3, 5-dichlofopheny)-5-ethenly-5-methyl-2, 4-oxazolodinedione

146. Benodanil- 2-iodo-N-phenylbenzamide

147. Diclofopmethyl- 2-(4-(2, 4-dichiorophenoxy) phenoxy)- propanole acid

148. Metalaxyl -methyl-(2-methoxyacetyl)-N-(2,6-xylyl)-DL-alaninate

149. Pendimethalin - N-(1-ethyl propyl)-3, 4-dimethyl 2, 6-dinitronebzeneamin

150. Fosethyl aluminium - aluminium tris (ethyl-phosphonate)

151. Thiabendazole - 2(4-thiazoloyl)-benzimidazole

152. Butrizol- 4-n-butyl-4H-1, 24-triazole

153. TCMTB - 2-(thiocynomethyl-thio) benzothiazole

154. Bromadiolone -3,3,4-bromol 1-biphenyl-4, Y-3 hydroxy-1-phenyl propyl-4, hydroxy-2 H-1 benzopyran-2 one)1

1. Flucythrinate RS-cyano-3-3phenoxy benzyle (s) 2-(4-difluomethoxy phenyl)-methylbutyrate

2. Trazophos O, O-diethyl 0-1-phenyl-1, 2, 4 triazol-3- zylphosphorothioate

3. Diflbenzuron 1, (4-chlorophgenyl)-3 (2), 6-difluorobenzoval) urea

4. Bitertanol B-(1-, 1-biphenyl) 4 yloxy)-1, (1, 1 dimethyl ethyl-l, 1, 2-triazole-l-ethrnol

5. Sethoxydim 2, 1(ethoxyimine) butyl-5 (2-ethylthoropy)-3-hydroxy 2-cyclohexen-1-one

6. Brodifaccum 2-(2(4-bromobiphenyl-4-yl) 1,2,3,4, tetrahydro 1-naphtyl) 4-hydroxyeoumarin

7. Methoprene Isopropyl (2E, 4E 11-methoxy-3, 7-11 trimethyl-2, 4-didecadienotate]

8. Isoprothiolane diesopropyl-1, 3-ditholan-2yl-idenemalenate

9. Carbosulfan (2, 3-dihydry-2, 2-dimetgyl-7-benofuranyl (dibutylamine, thio) methyl carbamate

10. Prochloraz N-propyl-N-2-(2, 4, 6, trichlorophenoxy) ethyl, imidaxole-1-carbozamide

11. Methacrifos 0-2 methoxycarbonylprop 1-enyl-0, 0, dimethyl-phosphorothioate

12. Chloroluron 3(3-chloro-p-tolyl-1, 1-dimethylurea

13. Probenazole 3-allyloxy 1, 2-benzolsothiazol-1, 1-dioxide

14. Tau Fluvalinate (RS)-d-cyano-3 phynoxybenzyl(R) -2(2)-chloro-4-trifluoromethyl-aniline) -3 methyl butanoate

15. DEPA N-diethyl-phenyl acetamide

16. Fenpropathrin -d cyano-3 phenoxybenzyl2, 2, 3, 3, 3-tetramethyl-1-cyclopropane-carboxylate

17. Phenothrin 3 phenoxybenyl (IRS)-cis-trans-chrysanthemate

18. Kasugamycin 5-amino-2-methyl-6-(2, 3, 4, 5 ,6-pentahydroxy-cyclohexyl) tetrahy-dropyran-3-yl) amino-a-imonoace-tic-acid

19. Amidine hydrazone -tetraphydro-5, 5-dimethyl 2 (1-H) pyrmidinone-3 4(tri-fluoromethyl) phenyl (1-(2-4 trifluoromethyl) phenyl (ethenyl-2-propenylidene) hydrazone

20. Anilofos S N(4-chloro-phenyl) (N-isopropyl-carbomoyl-methyl] O, O-dimethyl-dithiophosphate

A herbal extract containing diallyl disulfide, allyl propyl disulfide, and allyl isothocyanate,

1. Phenamiphos ethyl 4-methylthio-m-tolyl

2. Fenamiphos isoprpylphosphoramidate

3. Ametryne N-ethyl -N 1 1-(methylethyl) 6-(methylthio)-1, 3, 5-triazine-2, 4-diamine

4. Prometyne N, N-1bis (1-methylethyl(6-(methylthio)-1, 3-5-triazine 2, 4-diamine

5. Fluazifop RS: 2-(495 trifluoromethyl-2-pyridyloxy phenoxy), proppionic acid

6. Boric acid- ortho boric acid

7. Bromopropylate - isopropyl 4, 4' dibromobenzilate

8. Resmethrin- 5-benzyl-3-furylmentyl (IRS, 3RS, 1RS, 3SR)-2-dimethyl-3-(2-methlyprop-1-enyl), cyclopropanecarboxylate

9. Azamethiphos - S-6 chloro-2, 3-dihydro-2oxo-oxazole -b) pyridine-3- ylmethyl C, O-dimethyl ph-phorohioate

10. Metolochlor - 2-chloro-6'-ethyl-N-(2-methoxyl-1-methylethlty) acet-o-toluidide

11. Vadilamycina - IL-1,3, 4/2, 6)-2,3-dihydroxy-6-hydroxymethyl-4-(IS, 4R,5S, 63)-4,5,6 trihydroxy 3-hydroxy-methylcy-clohex-2-enylamino) cyclohexy 1-BD-gluco-pyrianoside

12. Haloxyfop methyl - methyl 2-(4)(3-chloro-5- (trifluromethyl-2-pyridinyl) oxy)phe-noxy) propanoate

13. Pimaricin - (8E,14E,16E,18E,20E) 1s,3R,5S,7S,12R,24R,25S,26R,-22(3amino-3-6-di-deoxy-B-D-mannopyrano-syloxy-1,3,26-trithydroxy-12-mwthylo-10-0x0-6,11,28-trioxatrcyclo (22,3-1-05.7)oc-tac0sa-8,14,16,18,20-pentaene-25-carboxylic acid

14. Cyhalothrin- (RS) 1-cyano-3-phenoxybenzyl (z)-(1RS,3RS)-(2-choloros-3,3-trifiuoropropenyl)-2,2-domethylcyclopropane-carboxylate

15. Cholecalcirerol-9,10-seocholesta-5,7,10(10)-trein-3beteol;activate-7-dehydrochosterol

16. Tolclofos,methyl-O-(2,6-dichloro-4 mehtyl-phenyl) O,O-dimethyl phosphorothioate

17. Piperophos-S-2-methylpiperidinocerbonyl-methyl-O,O-dipropylphos-phorodithicate

18. Chlorpyiphos-methyl O,O-demethyl O-3,5,6-trichloro-pyrideylphosphoro-thioate

GSR 858 (E) dated August 12, 1988 – in exercise of the powers conferred by sub-clause (ii) of Cl © of Sec 3 of the Insecticides Act, 1968 (46 of 1968), the Central Government, after consultation with the Central Insecticides Board hereby includes the following substances in the Schedule to the said Act, namely:

1. Cypermethrin: (SIR-cis-and R-IS (IUPAC) (S) –alphacyano-3-phenoxybenzyl -cis-isomer) (Alphamethrin) (IR,3R)-(2, 2-dichlorivinyl-2, 2-dimethyl 2-dimethyl cyclopropane carboxylate & (R) -alpha-cyano-3-phenoxybenzyl (IS, 3S)-3-cyclopropane carboxylate

2. Benfuracarb: (IUPAC) ethyl N-2, 3-dihydro-2, 2-dimethyl benzofuron-7-yl oxycarbonyl (methyl)

3. Cyphenothrin : (IUPAC) (RS)-0-cyano-3-phenoxybenzyl (IR)-cis, trans-chrysanthemate

4. Cyfluthrin: (IUPAC) cyano-(4-fluro-3-phenoxyphenyl)-methyl-3-2 (2, 2-dichloro-ethenyl) 2, 2-dimethyl-cyclopropane carboxylate

5. Dimethazone: (IUPAC) 2-(2-chlorophenyl) 4-4-dimethyl-3-isoxazolidinone)

6. Dienochlor: (IUPAC) perchloro-1,1-bicyclopenta-2,4-dience

7. Ephofenprox: 2-(4-ethroxyphenyl)2-methylopropyl-3-phenoxybenzyl ether

8. Flocoumafen : 4-hydroxy-3-(1, 2, 3, 4,-tetrahydro-3-(4-4-trifluromethyl benzyloxy)= (phenyl) –1-napthyl) coumarin)

9. Fenarimol: (IUPAC) (+/-) -2, 40dichloro-alpha-(pyrimidin-5-yl) benzhydryl alcohol

10. Fluroxypyr (IUPAC): 4-0amino-3, 5-dichloro-6-fluro-2-pyridyloxy acetic acid

11. Flufenoxuron: 1-(4-2-chloro-4-(trifluoro-methyl) phenoxy-2-flourophenyl) 3(2-6-difluorobenzoyl) urea

12. Filept (IUPAC): N-(trichloromethylthio)phthalimide

13. Clopyralid: (IUPAC) 3, 6-dichloropyridine-2-carboxylic acid

14. Myglobutanil: (IUPAC) 2-P-chlorophenyl 2-(1H-1, 2,4- triazol-1-yl-methyl) hoxaneritriale

15. Methedathion: S-2, 3-dihydro-5-methoxy-2-oxo-1, 3, 4-thiadizol-3-yl-methyil 0-0 dimethyl phosphorodithioate

16. Oxadixyl: (IUPAC) 2-methoxy-N-(2-oxo-1, 3-oxazolidin-3 yl) acet-2, 6-xylidide

17. Penconazoie: (IUPAC) 1(2, 4-dichloro-b -propylphenethyl)-1 H-1, 2, 4, -triazole

18. Pretilachlor: (IUPAC) 2-chloro-2, 6-diethyl-N(2-propoxyethyl) acetanilide

19. Prallethrin: RS-2-methyl-3-(2-propynyl) 4-oxocyclopent-2-enyl (IR)-cis, trans-chly-santhemat

20. Puridate: (IUPAC) 6-chloro-3-phenyl pyridazine-4-yl S-octvl thiocarbonate

21. Pyrooluilon: (IUPAC) 1, 2, 5, 6-tetrahydropyrrolo (3, 2, 1-I-j) quinolon-4-one

22. Pyridaphenthion: O, O-diethyl 2,3-dihydro-3-Oxm-2-phenyl-6- pyridazinyl-phosphorothioate

23. Sulprofos: (IUPAC) O-ethyl O-4 (methylthio) phenyl S-propyl phosphorodithioate

24. Sebuphos: S, S-di-sec-butyl O-ethyl phosphorodithioate

25. Thriadimenol: (IUPAC) I-(4-chloro-phenoxy)-3, 3-dimethyl-1-yl) butan-2-ol

26. Tricyclazole: (IUPAC) 5-methyl-1, 2, 4-triazole (3,4-b) (1, 3) benzothiazole

27. Tridiphane: (IUPAC) (RS)-2-(3, (IUPAC) (RS) -, (3, 5-dizhlor phenyl), 2-2, 2, -, trichloroethyl oxirane

GSR 577 (E), dated August 26, 1993: In exercise of the powers conferrred by sub-clause (ii) of Cl(e) of the Insecticide Act, 1968 (46 og 1968) the Central Government after consultation with the Central Insecticide Board, hereby includes the following sub stances in the Schedule to the said Act, namely:

1. Blasticidin 'S': 1-(4-amino-1, 2-dihydro-2-ox-opyrimidin -1-yll-4 (S)-3- amino-5 (1-methylguanidino) valermido) 1,2,3,4-tetradeoxy-D-erythro-hex-2-enopyranuronci acid

2. Difethialone: ((bromo-4-(bipheny-1-1)-yl-4)3 tetrahydro-1,2,3,4-naphthyl-1)3 hydroxy-4,2H-1benzophenyl)-2

3. Imazalil magnate: allyl 1-(2,4-d-chlorophenyl)-2-imidazol-1-ylethyl ether

4. Naproanilide: 2-(2.napthyloxy)propionanilide

5. Phenothiol: S-ethyl(4-chloro-o-tolyloxy)thicacetate

6. Terbufos: S-(1,1-dimethylethyl)thio)methyl)o,)-di-ethyl phosphorodithicate. or S-tert-butyl thiomethyl-O,)-diethyl phosphorodithicate

7. Butocarboxim: 3-(methyl thio_-O-((methyl amino) carbonyl)oxime-2 butanone

8. Imidacloprid: 1-(6-chloro-3-pyridinyl)-methyl)-N-nitroimidazol-idin-2-ylideneamine

9. Piroxofop propinyl: 2-propynyl(R)-2 (4-(5-chloro-3-fluroro-2- pyridinyloxy -phenoxy -propionate

Appendix 25   *Workplace Chemical Substances, Health Effects, and Symptoms of Poisoning*

Hazardous substances are used in many workplaces today. Working people are discovering that they need to know more about the health effects of chemical substances that they are exposed to in one way or another and more so because of job requirements and work areas. Students, laboratory workers, semi-skilled and skilled occupational workers, and supervisors may find this a useful and ready desk reference to understand important technical information about the safety of different chemical substances and to contain chemical disasters at the workplace.

Acetic acid: Corrosive, irritation to eyes, skin, and respiratory tract.

Amino ethyl carbazole: Harmful, nausea, headache, vomiting, and coughing, causes eye and skin irritation, irritating to mucous membranes and upper respiratory tract, central nervous system (CNS) depression, narcotic effect, chest pains, difficulty breathing, damage to the heart. Possible mutagen and possibly carcinogenic.

Ammonium acetate: Caution! Causes irritation to skin, eyes, and respiratory tract, harmful if swallowed, gastrointestinal (GI) tract irritation, abdominal pain, nausea, vomiting, flaccidity of facial muscles, tremor, generalized discomfort, anxiety, and impairment of motor performance.

Ammonium bicarbonate (ammonium hydrogen carbonate): Warning! Harmful if inhaled. Causes irritation to skin, eyes, nose, throat, and respiratory tract. Vapors released upon decomposition cause irritation of the upper respiratory tract, coughing, vomiting, and redness of the mucous membranes, restlessness, tightness in the chest, pulmonary edema, weak pulse, and cyanosis.

Ammonium chloride: Harmful. Causes irritation to skin, redness, itching, pain, irritation to the respiratory tract, coughing, shortness of breath, irritation to the GI tract, nausea, vomiting, and diarrhea.

Ammonium hydroxide: Causes severe skin irritation, inflammation, skin burns, deep penetrating ulcers of the skin, skin may become stained, thickening of the skin. Harmful if swallowed, may cause severe and permanent damage to the digestive tract, GI tract burns, throat constriction, vomiting, convulsions, and shock. Inhalation causes severe irritation of the upper respiratory tract, coughing, burns, breathing difficulty, and possible coma.

Beta-mercaptoethanol and 2-mercaptoethanol: Flammable. Toxic to upper respiratory tract, eyes, and CNS. Repeated or prolonged exposure produces target organs damage. Repeated exposures cause deterioration of health. Severe irritation to eyes/skin/respiratory tract.

Bleach (10% diluted solution): Irritation to eyes and skin.

Calcium chloride: Irritation to eyes, skin, and respiratory tract.

Carbenicillin: Irritation to eyes, skin, and respiratory tract. Sensitization.

Cesium chloride: Irritation to eyes, skin, and respiratory tract.

Chloroform: Carcinogen. Irritation to eyes, skin, and respiratory tract.

Diaminobenzidine: Carcinogen. Irritation to eyes, skin, and respiratory tract.

Dimethylformamide: Irritation to eyes, skin, and respiratory tract.

Dimethyl sulfoxide: Irritation to eyes and skin.

Ethanol: Flammable. Irritation to eyes, skin, and respiratory tract.

Ethidium bromide: Irritation to eyes, skin, and respiratory tract.

Ethylenediaminetetraacetic acid: Mild irritation to eyes, skin, and respiratory tract.

Fetal bovine serum: Irritation to eyes, skin, and respiratory tract.

Glucose-Tris-EDTA: Irritation to eyes, skin, and respiratory tract.

Hydrochloric acid: Corrosive. Burns to eyes, skin, and respiratory tract.

Hydrofluoric acid: Corrosive. Burns to eyes, skin, and respiratory tract.

Hydrogen peroxide: Oxidizer. Burns to eyes and skin.

Magnesium acetate: Mild irritation to eyes, skin, and respiratory tract.

Magnesium chloride: Irritation to eyes and skin.

Magnesium sulfate: Irritation to respiratory tract.

Methanol: Flammable. Poison. Irritation to eyes, skin, and respiratory tract. Headache, vertigo, lethargy, confusion, coma, and seizures. Nausea, vomiting, and abdominal pain.

N,N-Dimethyl formamide: Flammable. Vapor and/or mist a strong irritant to eyes, skin, mucous membranes, upper respiratory tract, stomach pains, vomiting, diarrhea, nausea, dizziness, headache, and muscle weakness. Damage to the kidneys, a potent liver toxin, changes in urine composition, weight loss and weight gain, cancer hazard!

Phenylmethylsofonyl fluoride: Severe irritation to eyes, skin, and respiratory tract.

Phosphate buffered saline-Tween: Irritation to eyes, skin, and respiratory tract.

Piperazine-N, N′-bis(2-ethanesulfonic acid): Irritation to eyes, skin, and respiratory tract.

Polyacrylamide gel electrophoresis running buffer: Irritation to eyes, skin, and respiratory tract.

Polyvinyl alcohol: Irritation to skin.

Potassium acetate: Mild irritation to eyes, skin, and respiratory tract.

Potassium chloride: Mild irritation to eyes and respiratory tract. Exposures and ingestion at high amounts may cause effects on the cardiovascular system, resulting in cardiac dysrhythmia.

Potassium hydroxide: Corrosive. **Poison! Danger!** Causes severe burns to skin, eyes, respiratory tract, and GI tract. Chemical substance is extremely destructive to all body tissues, may cause fatal injury by accidental ingestion and is harmful if inhaled.

Potassium iodide: Headache, irritation of mucous membranes, irritation to eyes, skin, coughing, shortness of breath, respiratory tract irritation, weakness, and anemia.

Potassium phosphate, dibasic: Mild irritation to eyes, skin, and respiratory tract.

Potassium phosphate, monobasic: Mild irritation to eyes, skin, and respiratory tract.

Sodium acetate: Mild irritation to eyes, skin, and respiratory tract.

Sodium azide: Poison. Irritation to eyes, convulsions, low blood pressure, loss of consciousness, lung injury, respiratory failure, and fatal injury.

Sodium bicarbonate: Irritation to eyes, skin, and respiratory tract.

Sodium carbonate: Corrosive. Burns to eyes, skin, and respiratory tract.

Sodium chloride: Irritation to respiratory tract.

Sodium citrate: Irritation to eyes, skin, and respiratory tract.

Sodium dodecyl sulfate: Irritation to eyes, skin, and respiratory tract.

Sodium fluoride: Toxic. Irritation to eyes, skin, and respiratory tract.

Sodium hydroxide: Corrosive. Burns to eyes, skin, and respiratory tract.

Sodium lauryl sulfate: Irritation to eyes, skin, and respiratory tract.

Sodium phosphate, monobasic: Irritation to eyes, skin, and respiratory tract.

Sodium phosphate, dibasic: Mild irritation to eyes, skin, and respiratory tract.

Sodium citrate, saline: Irritation to eyes, skin, and respiratory tract.

Sulfuric acid: Corrosive. Burns to eyes, skin, and respiratory tract.

2,4,6-trichlorophenol: Caution. Severe skin and eye irritant. Harmful if ingested. Experimental carcinogen and a possible human carcinogen.

Tris(hydroxymethyl)aminomethane: Irritation to eyes, skin, and respiratory tract.

Tris buffered saline: Irritation to eyes, skin, and respiratory tract.

Tris buffered saline-Tween: Irritation to eyes, skin, and respiratory tract.

Tris ethylenediamine-tetraacetic acid: Irritation to eyes, skin, and respiratory tract.

Tris HCL/acetic acid/EDTA buffer: Irritation to eyes, skin, and respiratory tract.

Tris HCL/boric acid/EDTA buffer: Irritation to eyes, skin, and respiratory tract.

Tris hydrochloride: Irritation to eyes, skin, and respiratory tract.

Tris-acetate-EDTA buffer: Irritation to eyes, skin, and respiratory tract.

Tris-borate-EDTA buffer: Irritation to eyes, skin, and respiratory tract.

Vesphene: Irritation to eyes and skin.

Zeocin: Irritation to eyes, skin, and respiratory tract.

Appendix 26   Signs and Symptoms of Toxicity in Humans

(predictable from species of laboratory animal studies)

Clinical side effect		Clinical side effect	
Drowsiness	Yes	Hypertension	Yes
Anorexia	Yes	Nausea	No
Insomnia	Yes	Depression	Yes
Dizziness	No	Fatigue	No
Increased appetite	Yes	Sedation	Yes
Constipation	Yes	Tremor	Yes
Dry mouth	Yes	Tinnitus	No
Perspiration	Yes	Nervousness	Yes
Weight gain	Yes	Dermatitis	Yes
Epigastric distress	No	Hypotension	Yes
Headache	No	Vertigo	No
Vomiting	Yes	Heartburn	No
Palpitation	Yes	Weakness	Yes
Diarrhea	Yes	Blurred vision	Yes
Skin rash	Yes	Lethargy	Yes

Source: Dikshith, T.S.S. (Ed.) 2009. *Safe use of chemicals: A practical guide.* CRC Press, Boca Raton, FL. (With Copyright permission)

Appendix 27   Industrial Gas Cylinders and Safety

Gases	Distinctive Color	Body Band
Oxygen (O_2)	Black	None
Nitrogen (N_2)	Gray	Black
Carbon dioxide (CO_2)	Black	White
Ammonia (NH_3)	Black	Red and yellow
Freon-12 (CCl_2F_2)	Bottom end gray; neck end violet	None
Argon (Ar)	Blue	None
Chlorine (Cl_2)	Yellow	None
Hydrogen (H_2)	Red	None
Acetylene (C_2H_2)	Maroon	None
Air	Gray	None

Source: Dikshith, T.S.S. and Diwan, P.V. 2003. *Industrial Guide to Chemicals and Drug Safety.* Wiley, J. Hoboken, NJ. Other published literature.

Appendix 28 *Industrial Chemicals, Inhalation Toxicity, and Lung Diseases*

Chemical Substances	Sources of Exposures	Lung Damage
Asbestos	Mining, construction, ship building, manufacture of asbestos-containing materials	Asbestos, lung cancer
Aluminum dust	Manufacture of aluminum products, fireworks, ceramics, paints, electrical goods, abrasives	Fibrosis
Aluminum	Manufacture of abrasives, smelting	Fibrosis initiated from short exposures
Ammonia	Ammonia production, manufacture of fertilizers, chemical production, explosives	Irritation
Arsenic Pb$_3$ (AsO$_4$)$_2$	Manufacture of pesticides, pigments, glass, alloys	Lung cancer, bronchitis, laryngitis
Beryllium	Ore extraction, manufacture of alloys, ceramics	Dyspnea, interstitial granuloma, fibrosis, corpulmonale, chronic disease
Boron	Chemical process	Acute CNS
Cadmium oxide (fume dust)	Welding, manufacture of electrical equipment, alloys, pigments, smelting	Emphysema
Carbides of tungsten, titanium, tantalium	Manufacture of cutting edges on tools	Pulmonary fibrosis
Chlorine	Manufacture of pulp and paper, plastics, chlorinated chemicals	Irritation
Chromium (IV)	Production of chromium compounds, paint pigments, reduction of chromite ore	Lung cancer
Coal dust	Coal mining	Pulmonary fibrosis
Coke oven emissions	Coke production	Lung cancer (nine times greater than other steel workers)
Hydrogen fluoride	Manufacture of chemicals, photographic film, solvents, plastics	Irritation, edema
Iron oxides	Welding, foundry work, steel manufacture, hematite mining, jewelry making	Diffuse fibrosis
Kaolin	Pottery making	Fibrosis
Manganese	Chemical and metal industries	
Nickel	Nickel ore extraction, nickel smelting, electronic electroplating, fossil fuel	Nasal cancer, lung cancer, acute pulmonary edema (NiCO)
Osmium tetraoxide	Chemical and metal industry	
Oxides of nitrogen	Welding, silo filling, explosive manufacture	Emphysema
Ozone	Welding, bleaching flour, deodorizing	Emphysema
Phosgene	Production of plastics, pesticides, chemicals	Edema
Perchloroethylene	Dry cleaning, metal degreasing, grain fumigating	Edema

Silica	Mining, stone-cutting construction, farming, quarrying	Silicosis (fibrosis)
Sulfur dioxide	Manufacture of chemicals, refrigeration, bleaching, fumigation	
Talc	Rubber industry, cosmetics	Fibrosis, pleural sclerosis
Tin	Mining, processing of tin	
Toluene 2,4-diisocyanate	Manufacture of plastics	Decrement of pulmonary function (FEV_1)
Vanadium	Steel manufacture	Irritation
Xylene	Manufacture of resins, paints, varnishes, other chemicals, general solvent for adhesives	Edema

Source: Dikshith, T.S.S. and Diwan, P.V. 2003. *Industrial Guide to Chemicals and Drug Safety.* Wiley, J. Hoboken, NJ.

Appendix 29 Occupational Exposure to Industrial Chemical Substances and Pulmonary Diseases

Chemical Substances	Common Name of Disease	Acute Effect
Asbestos	Asbestosis	
Aluminum	Aluminosis	Cough, shortness of breath
Aluminum abrasives	Shaver's disease, corundum smelter's lung, bauxite lung	Alveolar edema
Ammonia		Immediate upper and lower respiratory tract irritation, edema
Arsenic		Bronchitis
Beryllium	Berylliosis	Severe pulmonary edema, pneumonia
Boron		Edema and hemorrhage
Cadmium oxide		Cough, pneumonia
Carbides of tungsten, titanium, tantalium	Hard metal disease	Hyperplasia and metaplasia of bronchial epithelium
Chlorine		Cough, hemoptysis, dyspnea, tracheo-bronchitis, bronchopneumonia
Chromium (IV)		Nasal irritation, bronchitis
Coal dust	Pneumoconiosis	
Coke oven emissions		Lung cancer (nine times greater than other steel workers)
Cotton dust	Byssinosis	Tightness in chest, wheezing, dyspnea
Hydrogen fluoride		Respiratory irritation, hemorrhagic pulmonary edema
Iron oxides	Siderotic lung diseases: Silver finisher's lung, hematite miner's lung, arc welder's lung	Cough
Kaolin	Kaolinosis	
Manganese	Manganese pneumonia	Acute pneumonia, often fatal
Nickel		Pulmonary edema, delayed by 2 days (NiCO)
Osmium tetraoxide		Bronchitis, bronchopneumonia
Oxides of nitrogen		Pulmonary congestion and edema
Ozone		Pulmonary edema
Phosgene		Edema
Perchloroethylene		Pulmonary edema
Silica	Silicosis, pneumoconiosis	Silicosis (fibrosis)
Sulfur dioxide		Bronchoconstriction, cough, tightness in chest
Talc	Talcosis	
Tin	Stanosis	

(continued)

471

(continued)

Chemical Substances	Common Name of Disease	Acute Effect
Toluene		Acute bronchitis, bronchospasm, pulmonary edema
Vanadium		Upper airway irritation and mucous production
Xylene		Pulmonary edema

Source: Dikshith, T.S.S. and Diwan, P.V. 2003. *Industrial Guide to Chemicals and Drug Safety.* Wiley, J. Hoboken, NJ. Other published literature.

Appendix 30   Use and Handling of Chemical Substances

Students and the general public should be aware of the elements of proper use, handling, disposal, and management of chemical substances. In fact, public education is essential to raising awareness of the need for responsible and judicious use of chemical substances. Also, improved understanding about proper use, storage, transportation, waste disposal, and overall management of chemical substances and individual ethical responsibilities would go a long way to significantly control chemical disasters both at the workplace and in the living environment at national and international levels.

During use and handling of different chemical substances, and more so of potentially toxic chemicals, students and occupational workers should remember the guidelines listed below:

- Students and workers SHOULD KNOW the physical and health hazards associated with the candidate chemical substance that he/she is using/handling.
- Students and workers SHOULD carefully READ the LABEL and the material safety data sheet (MSDS) of the candidate chemical substance before using it for the first time. Also, the appropriate standard operating procedure (SOP) should be reviewed before proceeding further.
- Students and workers SHOULD NOT work alone in the laboratory. If working alone, a supervisor must be informed.
- Students and occupational workers should use required personal protective equipment.
- Students should ALWAYS wear eye and face protection in the laboratory. Adequate eye protection is fundamental to the use of chemical substances such as housekeeping materials, wax strippers, detergent and toilet bowl cleaners, and laboratory operations involving grinding, drilling, and sawing with power tools.
- Students and occupational workers should use splash goggles with splash-proof sides or a face shield to avoid chemical splashes.
- Students and occupational workers should ensure that all chemical containers are LEGIBLY labeled.
- After working with chemical substances, students and occupational workers SHOULD WASH thoroughly the exposed body surfaces with plenty of water before leaving the workplace.
- Students and occupational workers SHOULD AVOID direct contact with any chemical substance that he/she uses/handles.
- Students and occupational workers SHOULD ALWAYS WEAR a laboratory coat and protective equipment.
- Students and occupational workers SHOULD NEVER smell, inhale, or taste a candidate chemical substance at the workplace.

- Students and occupational workers SHOULD AVOID smoking, drinking, eating, and the application of cosmetics at workplaces.
- Students and occupational workers SHOULD use/handle chemical substances with adequate ventilation or in a chemical fume hood.
- Students and occupational workers SHOULD NEVER use mouth suction to fill a pipette, but use a pipette bulb or other pipette-filling device.
- Students and occupational workers exposed to strong acids and acid gases, organic solvents and strong oxidizing agents, carcinogens, and mutagens SHOULD USE proper protective equipment to avoid skin contamination. Impervious protective equipment such as (i) rubber or nitrile gloves, (ii) rubber boots, (iii) rubberized suits, and (iv) special protective equipment must be used.

Index